FLORA OF THE GUIANAS

Edited by

M.J. JANSEN-JACOBS

Series A: Phanerogams
Fascicle 28

LEGUMINOSAE

subfamily

87. MIMOSOIDEAE
(R.C. Barneby, J.W. Grimes & O. Poncy)

taxonomic update
(G.P. Lewis)

including
Wood and Timber
(P. Détienne, J. Koek-Noorman, L.Y.Th. Westra, I. Poole & B.J.H. ter Welle)

2011
Royal Botanic Gardens, Kew

The Flora of the Guianas

is a modern, critical and illustrated Flora of Guyana, Suriname, and French Guiana designed to treat Phanerogams as well as Cryptogams of the area.

Contents: Publication takes place in fascicles, each treating a single family, or a group of related families, in the following series: A: Phanerogams; B: Ferns and Fern allies; C: Bryophytes; D: Algae; and E: Fungi and Lichens. A list of numbered families in taxonomic order has been established for the Phanerogams. Publication of fascicles will take place when available.
In the Supplementary series other relevant information concerning the plant collections from the Guianas appears, like indexes of plant collectors.

The Flora in general, follows the format of other, modern Floras such as the *Flora of Ecuador* and *Flora Neotropica*. The treatment provides fundamental and applied information; it covers, when possible, wood anatomy, chemical analysis, economic uses, vernacular names, and data on endangered species.

ORGANISATION: The Flora is a co-operative project of: Botanischer Garten und Botanisches Museum Berlin-Dahlem, *Berlin*; Institut de Recherche pour le Développement, IRD, Centre de Cayenne, *Cayenne*; Department of Biology, University of Guyana, *Georgetown*; Herbarium, Royal Botanic Gardens, *Kew*; Nationaal Herbarium Nederland, Leiden University branch, *Leiden;* New York Botanical Garden, *New York*; Nationaal Herbarium Suriname, *Paramaribo*; Muséum National d'Histoire Naturelle, *Paris*, and Department of Botany, Smithsonian Institution, *Washington, D.C.*

PUBLICATION: The *Flora of the Guianas* is a publication of The Royal Botanic Gardens, Kew. The prices of the fascicles are determined by their size. Authors are requested to submit a hard copy of their manuscript as well as an electronic version; Word and other standard word processing packages are acceptable.

INFORMATION: http://www.nationaalherbarium.nl/FoGWebsite/index.htm

Editorial office for correspondence on contributions, etc.:
M.J. Jansen-Jacobs
Nationaal Herbarium Nederland
Leiden University branch
P.O. Box 9514
2300 RA Leiden
The Netherlands
Email: M.J.Jansen-Jacobs@uu.nl

Publisher
Royal Botanic Gardens
Kew
Richmond
Surrey, TW9 3AB
U.K.

ISBN 978 1 84246 437 3

Printed in the USA by The University of Chicago Press

Contents

LEGUMINOSAE

subfamily

87. MIMOSOIDEAE

by

RUPERT C. BARNEBY (†)[1], JAMES W. GRIMES[2] & ODILE PONCY[3]

Taxonomically updated
by
GWILYM P. LEWIS[4]

Introduction: The bulk of this fascicle was submitted as clean manuscripts by R.C. Barneby (NY), J.W. Grimes (NY and US) and O. Poncy (P) over fifteen years ago (except the genus *Parkia*). For the present edition, Poncy provided a treatment of *Parkia* as well as an updated version of her former manuscript of *Inga* (in the meantime a monograph of the genus was published by T. Pennington (1997)). Rupert Barneby died in 2000 and James Grimes no longer works in botany, raising the question of what would be the most pragmatic approach to the editing of the taxonomic content of the genera written by each of them. They both published accounts of many mimosoid genera in the interim. Their contributions, along with a number of revisions by other legume specialists that together either alter the taxonomic concepts and/or update the nomenclature in the original manuscript presented to the Flora of the Guianas editorial committee. The editorial committee decided that the original text should remain largely intact and that taxonomic concepts should not be altered unless based on data subsequently published by Barneby & Grimes (e.g., the monograph of *Calliandra* by Barneby (1998) significantly changed the taxonomic concepts as submitted in manuscript for the Flora by Grimes). However, rather than rewrite taxon descriptions and keys to species it was decided to update the reader by the use of taxonomic notes (written by G.P. Lewis) and several of these are scattered throughout the fascicle. It is intended that these notes bring the reader up-to-date with the latest literature on each genus and highlight significant taxonomic realignments and nomenclatural synonymy.

[1] New York Botanical Garden, Bronx, New York 10458-5126, U.S.A.

[2] New York Botanical Garden, Bronx, New York 10458-5126, U.S.A.; Department of Botany, National Museum of Natural History, Smithsonian Institution, Washington, D.C. 20013-70162, U.S.A.

[3] Museum National d'Histoire Naturelle, Dep. Systematique et Evolution, Herbier Plantes Vasculaires, CP 39, 75231 Paris cedex 05, France. Email: poncy@mnhn.fr.

[4] Royal Botanic Gardens, Kew, Richmond, Surrey, TW9 3AB, U.K.

Selected specimens cited by Barneby and Grimes are largely housed in CAY, NY, U and US. The editors have not added additional material housed in other herbaria or collected since the original manuscript was submitted, so that, inevitably, some species records for the Guianas and some new country records will have been missed. For collections housed at P, all mimosoid specimens from the Guianas have been databased, the reader is invited to check the database (http://www.mnhn.fr/base/sonnerat.html).

Notes on descriptions: Barneby and Grimes provided for each species a leaf-formula. This formula, calculated from the largest, completely formed mature leaves from any specimen, records the maximum number of pinna- and leaflet-pairs. Thus, a leaf-formula ii-iii/5-9 indicates that the largest, matures leaves of the specimens examined had 2 or 3 pairs of pinnae, each pinna with 5 to 9 pairs of leaflets.

Used terminology and abbreviations:

Annotinous – a year old, belonging to last year.
Craspedium – a fruit in which the valves break and fall away from the sutural ribs.
Pedestal – the terminal part of an inflorescence receptacle that, when present, usually supports one or more modified flowers.
Pleurogram – a small line or fissure seen on the sides of the seeds of many species of Mimosoideae.
Stemonozone – a tube formed by the cohesion of the base of the stamen-filaments and corolla.
fl – flower
fls – flowers
infl – inflorescence
lf – leaf
lf-stk – leaf-stalk
lft – leaflet
lfts – leaflets
lvs – leaves
spms – specimens

Subfamily description:
Shrubs, trees, vines or herbs mostly pubescent at least when young, most unarmed but some thorny, spiny, or prickly; stipules present or lacking, when present quickly deciduous to persistent, sometimes forming buds covering dormant apices. Leaves paripinnate or bipinnate, pinnae 1-many pairs, opposite, or subopposite at base of lf-stk; lfts opposite or alternate, pulvinulate, uni- to multijugate, commonly inequilateral; foliar nectaries present on the petiole and/or between or just below the pinnae or leaflets (lacking in few genera); venation pinnate, palmate-pinnate, palmate,

or reduced to mid-vein. Inflorescences of capitula or spikes commonly 1) axillary to contemporaneous leaves; or 2) leaves suppressed early in the growing season (hysteranthous); or 3) leaves suppressed at the end of the growing season and inflorescence a pseudoraceme or panicle of capitula, spikes, or spiciform racemes, or umbels; or 4) inflorescence cauliflorous; receptacle bracteate, bracts deciduous or persistent; bracteoles absent. Flowers sessile or pedicellate, actinomorphic, those within any unit of inflorescence homomorphic or heteromorphic either because andromonoecious, or because the terminal 1-few enlarged, or (some *Calliandra*) both; hypanthium present or lacking; perianth 3-5-merous (random anomalies), gamosepalous, petals either united, or becoming free to the base; intrastaminal nectary or nectarial disc on receptacle present or lacking; stemonozone present or lacking; stamens either as many, or twice as many as perianth segments, or numerous, free or united into a tube beyond the stemonozone, anthers dorsifixed, opening by lateral slits, connective sometimes provided with a gland; gynoecium 1- or rarely 2(-3)-pistillate (up to 9-pistillate in *Inga* species outside the Guianas), ovary sessile or on a gynophore. Fruits strongly laterally compressed to turgid and ± cylindric, inertly dehiscent through one or both sutures, or dehiscent and contorted, or indehiscent and variously modified, or lomentiform, or a craspedium; seeds ± plump to strongly laterally compressed, arillate in one genus, though often persistent on the funicle in others, pleurogram present or lacking, when present either complete or incomplete; endosperm present or lacking; embryo sometimes anthocyanic or chlorophyllous.

Distribution: A subfamily of about 3270 species, most numerous in monsoon, savanna and desert climates, uncommon in either tropical or temperate mountains; 163 species (178 taxa) in the Guianas, with 4 additional species to be expected in the Guianas and included in this treatment.

LITERATURE

Barneby, R.C. & J.W. Grimes. 1996. Silk Tree, Guanacaste, Monkey's Earring. A generic system for the synandrous Mimosaceae of the Americas. Part I. Abarema, Albizia, and allies. Mem. New York Bot. Gard. 74(1): 1-292.

Barneby, R.C. & J.W. Grimes. 1997. Silk Tree, Guanacaste, Monkey's Earring. A generic system for the synandrous Mimosaceae of the Americas. Part II. Pithecellobium, Cojoba, and Zygia. Mem. New York Bot. Gard. 74(2): 1-149.

Barneby, R.C. 1998. Silk Tree, Guanacaste, Monkey's Earring. A generic system for the synandrous Mimosaceae of the Americas. Part III. Calliandra. Mem. New York Bot. Gard. 74(3): 1-223.

Barneby, R.C. & J.W. Grimes *et al.* 2001. Mimosaceae. In J.A. Steyermark *et al.*, Flora of the Venezuelan Guayana 6: 580-686.

Bentham, G. 1875. Revision of the suborder Mimoseae. Trans. Linn. Soc. London 30: 335-664.

Britton, N.L. & J.N. Rose. 1928. (Rosales) Mimosaceae. N. Amer. Fl. 23: 1-194.

Grimes, J.W. 2002. Mimosaceae. In S.A. Mori *et al.*, Guide to the Vascular Plants of Central French Guiana. Mem. New York Bot. Gard. 76(2): 484-510.

Irwin, H.S. 1966. Contributions to the botany of Guiana. III. Leguminosae-Mimosoideae. Mem. New York Bot. Gard. 15: 96-111.

Jansen-Jacobs, M.J. 1976. Mimosaceae. In J. Lanjouw & A.L. Stoffers, Flora of Suriname 2(2), Additions and Corrections: 611-653.

Kleinhoonte, A. 1940. Mimosaceae. In A. Pulle, Flora of Suriname, 2(2): 259-331.

Lewis, G.P. *et al.* (eds.) 2005. Legumes of the world. Royal Botanic Gardens, Kew. pp. 1-577.

Lewis, G. & P.E. Owen. 1989. Legumes of the Ilha de Maracá. Mimosoideae pp. 29-47.

Lewis, G.P. & L. Rico Arce. 2005. Tribe Ingeae. In G. Lewis *et al.*, Legumes of the world pp. 193-213.

Rico Arce, M. de L. 1999. New combinations in Mimosaceae. Novon 9: 554-556.

KEY TO THE GENERA

1 Stamens as many or twice as many as divisions of perianth, free or obscurely united below ovary 2
  Stamens more than twice as many as perianth lobes, or if rarely exactly twice as many (some *Calliandra*), this not on every flower 17

2 Inflorescences composed of large globose, biglobose or clavate capitula, often heteromorphic with fertile, bisexual and male flowers; calyx-lobes imbricate in vernation. . . . . *19. Parkia*
  Inflorescences and flowers not at once as above; if flowers fertile bisexual and male (*Neptunia*, some *Calliandra*), calyx-lobes valvate in vernation; if calyx-lobes valvate in vernation (*Pentaclethra*), flowers all fertile bisexual and inflorescence a long spike 3

3 Flowers spicate, floral receptacle 1.5(+) cm. 4
  Flowers capitulate, floral receptacle less than 1 cm 11

4 Leaflets, perianth and pod stellate-pubescent . . . . *17-17. Mimosa schomburgkii*
Leaflets, perianth and pod pubescent with simple trichomes or glabrous . . . 5

5 Flowering branchlets or stems unarmed 6
Flowering branchlets armed either on internodes or at nodes with epidermal prickles 11

6 Petiole lacking a nectary 7
Petiolar nectary present, either between first pair of pinnae or well below them 10

7 Leaflets of longer pinnae 26-56 pairs 8
Leaflets of longer pinnae 3-9 pairs 9

8 Leaflets opposite; spikes terminally paniculate; calyx-teeth imbricate; stamens dimorphic, 5 fertile, 5-9.5 mm long, 5 sterile, 17-31 mm long; pod stiffly erect, ± 25-35 x 3-5 cm, woody valves vertically striate; seeds 4-4.5 cm, not enclosed in endocarp . . . . *20. Pentaclethra*
Leaflets alternate (subopposite); spikes at anthesis solitary in leaf-axils, not in terminal panicles; calyx-teeth valvate; stamens 10, monomorphic; pod variously disposed but only rarely stiffly erect, 9.5-18 x 1.4-2.2 cm, valves smooth; seeds much smaller, endocarp becoming free from valve and enclosing seed . . . . *23. Plathymenia*

9 Free-standing trees; tendrils 0; inflorescence of axillary spikes; fruit a bean-like pod 1.2-1.6 cm wide; seeds red . . . . *3. Adenanthera*
High-climbing bush-ropes; distal pinnae of some leaves modified into tendrils; inflorescence a dense efoliate, one-sided pseudoraceme of spikes; fruit a craspedium 4-8 cm wide; seeds brown . . . . *11. Entada*

10 Leaflets at once opposite along rachis and less than 25 mm long; pod an elongate linear bean dehiscent along one suture only; seeds plano-compressed, the lustrous membranous testa lacking pleurogram; endosperm lacking . . . . *24. Pseudopiptadenia*
Leaflets either opposite or alternate, but evidently alternate when less than 25 mm long, much longer when opposite; pod indehiscent except by weathering; seeds plump, hard-coated, with pleurogram and endosperm . . . . *26. Stryphnodendron*

11 Anther-sacs incurved around a transversely dilated connective, this lacking a terminal gland; fruit a craspedium . . . . *17. Mimosa*
Anther-sacs straight, parallel, contiguous, narrow connective tipped with a caducous gland; fruit a planocompressed pod with continuous valves . . . . *21. Piptadenia*

12 Flowers of each capitulum homomorphic; stems armed or not 13
Flowers of each capitulum heteromorphic, lower ones sterile, with ± dilated staminodes, upper ones perfect; stems unarmed 16

13 Plants monocarpic, stems setose but not aculeate; fruit a craspedium . . . . *17-3. Mimosa camporum*
Plant perennial, shrubby, or arborescent 14

14 Trees, stems and leaf-stalks unarmed; fruit a dehiscent pod with continuous valves 15
Herbs and shrubs, usually armed with prickles; fruit a craspedium . . . . *17. Mimosa*

15 Bract subtending individual flowers ovate or subulate, neither stalked nor peltate at tip; anthers much less than 1mm, glabrous; pod follicular, dehiscent along one suture only; seeds thinly discoid, the submembranous, lustrous black testa lacking pleurogram; endosperm lacking. . . . . *5. Anadenanthera*
Bract subtending individual flowers stalked, peltate at tip; anthers relatively large (± 1 mm), thinly pilose; pod dehiscent through both sutures; seeds plump, hard, areolate; endosperm present . . . . *15. Leucaena*

16 Plants terrestrial; peduncles ebracteolate; receptacle straight; staminodes white; pod sessile, narrowly linear, 2.5-4 mm wide . . . . . . *10. Desmanthus*
Plants aquatic; peduncles bibracteolate near and below middle; receptacle of capitulum recurved at anthesis; staminodes yellow; pod stipitate, 8-11 mm wide . . . . *18. Neptunia*

17 Stems and/or axes of inflorescence armed with spinescent stipules, axillary thorns or epidermal prickles 18
Stems unarmed 20

18 Leaves all with one pair of pinna, each pinna with one pair of leaflets; seeds arillate . . . . *22. Pithecellobium*
Leaves all with more than one pair of pinnae, each pinnae with more than one pair of leaflets; seeds exarillate 19

19 Plants armed with stipular spines or epidermal prickles; flowers of each capitulum homomorphic; fruit straight or slightly falcate . . . . . . *2. Acacia*
Nodes of stem with axillary thorns derived from sterile lignescent peduncles; each capitulum with a heteromorphic terminal flower; fruit coiled through 1.5-4 circles . . . . *9. Chloroleucon*

20 Leaves simply paripinnate 21
Leaves bipinnate (*Zygia* commonly has one pair of long pinnae borne on a short petiole that are easily mistaken for simply – thus, opposite - pinnate leaves) 23

21 Leaves with nectaries 22
Leaves lacking nectary . . . . *7-4. Calliandra hymenaeodes*

22 Fruit indehiscent; seeds covered with white,edible pulp . . . . *14. Inga*
Fruit dehiscent; no edible pulp . . . . *28-6. Zygia inundata*

23 Petiolar nectaries wanting 24
Petiolar nectary present between or below basal-most pair of pinnae 25

24 Erect trees, subshrubs, and shrubs; axillary brachyblasts commonly present; pollen-polyads 8-celled . . . . *7. Calliandra*
Scandent shrubs; axillary brachyblasts lacking; pollen-polyads 16-celled . . . . *27. Zapoteca*

25 Plant cauliflorous, inflorescences arising from defoliate nodes of main trunk or of older branches 26
Plant not cauliflorous, inflorescences arising on current stems either with or slightly ahead of expanding subtending leaf 27

26 Inflorescence capitulate; calyx 14-20 mm. . . . *16-1. Macrosamanea kegelii*
Inflorescence capitulate or spicate; calyx much less than 14 mm . . . . *28. Zygia*

27 Specimen in flower 28
Specimen in fruit 38

28 Individual units of inflorescence of racemes or spikes, flowers on an axis at least 3 cm . . . . *1. Abarema*, in part
Individual units of inflorescence of capitula or cymose-umbels, if rarely somewhat spiciform axis less than 2.5 cm 29

29 Flowers of each partial inflorescence homomorphic 30
Flowers of each partial inflorescence heteromorphic with an enlarged terminal flower 33

30 Bracts subtending individual flowers charged on ventral side with a nectary resembling that of lf-stk . . . . *16. Macrosamanea*
Bracts subtending individual flowers without a nectary 31

31 Inflorescences precocious, peduncles arising below and earlier than leaves; pod sessile, broad-linear decurved through 0.5-2 circles . . . . *12. Enterolobium*
Inflorescences axillary to coevally expanding or fully expanded leaves. . . . 32

32 Petals united above stemonozone into a tube; fruit dehiscent along one or both sutures, valves internally colored orange or orange-red either in seed cavities or overall, seeds persistent on funicles, blue . . . . *1. Abarema*, in part
Petals free to stemonozone; fruit indehiscent, articles twisting 90° between seeds . . . . *8. Cedrelinga*

33 Lf-formula i/1, all leaves 4-foliolate. . . . . *1. Abarema*, in part
Lf-formula higher in some respect, all or most leaves at least 8-foliolate 34

34 Venation of leaflets palmate, or reduced to midrib 35
Venation of leaflets pinnate 36

35 Capitula either all axillary to coevally expanding leaves, or together paniculate above or at least beyond foliage of current year . . . . . *4. Albizia*
Capitula (either solitary or fasciculate) or efoliate pseudoracemes of capitula arising in axil of leaves of current year or from efoliate nodes below them . . . . *12. Enterolobium*

36 Ovary truncate; three genera, easily differentiated by fruits but not technically separable at anthesis . . . . *1. Abarema*, in part; *6. Balizia*, *13. Hydrochorea*
Ovary tapering at apex 37

37 Corolla of peripheral flowers at most 7 mm; pod dehiscent . . . . *1. Abarema*, in part
Corolla of peripheral flowers 9-18 mm; pod indehiscent . . . . . *25. Samanea*

38 Seeds exareolate 39
Seeds areolate 40

39 Pod diverse in curvature, texture, and compression, but endocarp internally either red, maroon, or orange-brown, contrasting in color with ± translucent, either blue, gray-blue or pale tan (sometimes partly opaque and bright white) seed-coat; seed persistent on funicle . . . . *1. Abarema*
Pod ± laterally compressed, valves either flat or low-convex, never red, maroon or orange-brown internally, seed-coat brownish, seed not persistent on funicle . . . . *16. Macrosamanea*

40 Venation of leaflets pinnate 41
Venation of leaflets palmate or reduced to midrib 44

41 Pod indehiscent, strongly biconvex or subcylindroid, continuous valves fibro-ligneous or pulpy, endocarp at least narrowly septiferous, sutures externally differentiated into prominent broad keels . . . . *25. Samanea*
Pod not as above, always strongly compressed laterally, either leathery and elastically dehiscent, tardily follicular, or lomentiform 42

42 Valves of pod tardily segmented between seeds, incipiently lomentiform . . . . *13. Hydrochorea*
Valves of pod continuous 43

43 Pod falcately recurved or coiled, endocarp internally either red, maroon, or orange-brown, contrasting in color with ± translucent, either blue, gray-blue or pale tan (sometimes partly opaque and bright white) seed-testa; seed perisistent on funicle . . . . *1. Abarema*
Pod straight or almost so, not red inside; sutures dilated, framing transversely fibrous valves; seeds not persistent on funicle . . . . *6. Balizia*

44 Pod straight, not lignescent, either lomentiform or inertly dehiscent . . . . *4. Albizia*
Pod lignescent and incurved through at least 3/4 circle, often further into a compressed spiral, indehiscent; seeds separated by complete septa . . . . *12. Enterolobium*

1. **ABAREMA**[5] Pittier, Arb. Arbust. Orden Legum. 56. 1927. – *Pithecolobium* sect. *Abaremotemo* Benth., London J. Bot. 3: 203. 1844.

Lectotype (designated by Britton & Killip, Ann. New York Acad. Sci. 35: 126. 1936): A. cochliacarpos (Gomes) Barneby & J.W. Grimes (Mimosa cochliacarpos Gomes) (= Pithecolobium auaremotemo Mart.)

Jupunba Britton & Rose, N. Amer. Fl. 23: 25. 1928.
Type: J. jupunba (Willd.) Britton & Rose (Acacia jupunba Willd.) [Abarema jupunba (Willd.) Britton & Killip]
Klugiodendron Britton & Killip, Ann. New York Acad. Sci. 35: 125, 1936.
Type: K. laetum (Benth.) Britton & Killip (Pithecolobium laetum Benth.) [Abarema laeta (Poepp. & Endl.) Barneby & J.W. Grimes]

Shrubs or trees, young branches and inflorescences densely sordid-yellow, brown or dark brown pubescent. Leaves alternate, bipinnate; leaf-formula i-xii/1-26; stipules herbaceous, lanceolate, linear-lanceolate or elliptic, caducous, or sometimes lacking; petiole and rachis terete or grooved ventrally, with nectaries between first pinna-pair and commonly between subsequent pairs; leaflets opposite, faintly to strongly bicolored, glabrous overall, or minutely ciliolate on margin and midrib beneath, or densely villous beneath, venation pinnate or in 1 species palmately 3-veined from pulvinule. Inflorescences capitulate to shortly spicate, solitary, geminate or ternate in leaf-axils, or pseudo-paniculate through suppression of subtending leaves, 9-28-flowered, flowers homomorphic or the terminal 1(-few) heteromorphic, i.e. usually wider, coarser and/or longer than peripheral ones; bracts subtending each flower (rarely lacking) linear-oblanceolate, elliptic-oblanceolate, elliptic or trullate, caducous or lowermost peristent. Calyx and corolla 5-lobed (random irregularities); androecium longer than perianth, ± = style. Pods presumably ornithochorous, dehiscence follicular in pods with greatly thickened valves, or more commonly through both sutures and valves contorting; valves internally scarcely and irregularly mottled to completely orange-red or red; seeds persistent on funicle, when mature and fresh seed-coat white and/or translucent, embryo usually aniline-blue and visible; pleurogram complete, incomplete or lacking.

[5] by James W. Grimes

Distribution: A genus of ± 45 species dispersed from S Mexico (Oaxaca, Chiapas) sparingly scattered through C America, widespread in the West Indies, through tropical S America, where most diverse in the Amazon Basin to subtropical Brazil (Rio Grande do Sul); 9 species occur in the Guianas, 1 more may be expected to occur.

Taxonomic note: Since submission of the Mimosoideae manuscript for the Flora of the Guianas, a revision of *Abarema* has been published by Barneby & Grimes (1996: 41-111). The reader is referred to this publication for additional data.

KEY TO THE SPECIES

1 Flowers in spiciform racemes, axis 6-12 cm long; pod moniliform. . . . . *3. A. curvicarpa*
Flowers capitate, shortly racemose, or pseudo-umbellate, axis of inflorescence less than 3 cm long; pod undulately broad-linear, not moniliform 2

2 Leaflets exactly 1 pair per pinna. . . . . *8. A. laeta*
Leaflets 2-22 pairs on largest pinna 3

3 Leaflets convex or bullate, dorsally pilosulous overall or at least in basal angle of midrib 4
Leaflets plane (at most margin minutely revolute), dorsal surface brown-villous, strigulose, puberulent or glabrate 5

4 Lf-formula vi-xii/12-19; larger lfts (excluding terminal pair) of any specimen 4.5-11.5 x 2-5 mm; secondary veins on lower leaf-surface 2-6 on each side of midrib. . . . . *1a. A. barbouriana* var. *barbouriana*
Lf-formula ii-v(vi)/(7)8-11(13); larger lfts (excluding terminal pair) of any specimen (9)10-18 x (3)5-10 mm; secondary veins on lower leaf-surface (5)6-9 on each side of midrib . . . . *1b. A. barbouriana* var. *arenaria*

5 Dorsal face of leaflets brown-villous . . . . *2. A. commutata*
Dorsal face of leaflets strigulose, puberulent or glabrate 6

6 Leaflets 2-6(7) pairs per pinnae, if as many as 6(7) then pinnae only 1-3 pairs per leaf 7
Leaflets (6)7-22 pairs per pinnae, if as few as (6)7 then pinnae 3-7 pairs per leaf 10

7 Calyx of peripheral flowers 5-7.3 mm long (expected in GU) . . . . *9. A. longipedunculata*
Calyx of peripheral flowers 1.7-3.6 mm long. 8

8 Pulvinules of leaflets 5-9 mm long; leaflets 2-3 pairs per pinna. . . . .
. . . . *10. A. mataybifolia*
Pulvinules of leaflets less than 3 mm long; leaflets 3-7 pairs per pinna 9

9 Larger leaflets 5-8 cm long; filaments red; pod falcate, 12.5-16 x 2.2-2.4 cm, endocarp dark red in seed-locules, tan between . . . . *6. A. gallorum*
Larger leaflets less than 3.5 cm long; filaments white; pod annulate, 6-9.5 x 1-1.6 cm, endocarp uniformly orange overall . . . .
. . . . *7b. A. jupunba* var. *trapezifolia*

10 Calyx of peripheral flowers 1.7-3 mm; corolla 4-6.2(7) mm; filaments whitish . . . . *7a. A. jupunba* var. *jupunba*
Calyx of peripheral flowers 4-9 mm; corolla 8-13 mm; filaments rose or red 11

11 Venation of 3 veins from pulvinule, thence pinnate; rachis of longer pinnae 4-7 cm . . . . *4. A. ferruginea*
Venation of leaflets essenially pinnate; rachis of longer pinnae (7)8-14(20) cm . . . . *5. A. floribunda*

1. **Abarema barbouriana** (Standl.) Barneby & J.W. Grimes, Mem. New York Bot. Gard. 74(1): 70. 1996. – *Pithecolobium barbourianum* Standl., Contr. Arnold Arb. 5: 74. 1933. Type: Panama, Zetek Trail, Barro Colorado Island, Canal Zone, Shattuck 237 (holotype F!, isotype MO!).

Trees to 30 m but often smaller, young branches and all axes of lvs and inflorescences densely sordid-pilosulous with erect or forwardly incurved, yellowish or reddish-brown hairs. Lf-formula (ii)iii-xii/(7)8-22; stipules subulate, linear-elliptic, or linear 1-3(4) mm, caducous; lf-stks of larger lvs 2-14.5 cm, petiole including pulvinus (6)7-15(17) mm, longer interpinnal segments 7-16(18) mm; pinnae strongly decrescent proximally, sometimes a little so distally, rachis of those near and above mid-lf 3-7 cm, longer interfoliolar segments 2.5-10 mm; lfts decrescent at base of pinna, thence subequilong except for broader and sometimes longer apical pair, blades bicolored, obtusangulately rhombic, rhombic-oblong or -elliptic, inequilaterally flabellate-cuneate at base, very obtuse, largest (4.5)5-21 x (2)2.6-10 mm, 1.9-2.6 times as long as wide. Peduncle 1.5-8 cm; capitula ± 12-20-flowered, clavate receptacle including short terminal pedestal 2-4 mm; bracts dimorphic, all sessile or lowest peripheral ones borne on obscure pedicel not over 0.5(0.7) mm, perianth of all 5-merous (random irregularities), densely appressed-pilosulous externally overall; peripheral fls: calyx deeply campanulate or turbinate-campanulate, 3-6 x 1.7-2.4 mm; corolla 5-8 mm; filament tube 1.4-6.5 mm; terminal fl: calyx 3.5-6 x 1.6-2.4 mm; corolla 7-12 mm; filament tube

6-16 mm. Pod usually solitary, sessile, in profile undulately broad-linear and (when well fertilized) evenly incurved through 1-2.5 circles 3.7-4.4 cm diam., (5)6-12 x (1)1.1-1.6 cm, 6-12(14)-seeded, sutures 1.4-2 mm wide, valves low-convex over each seed, either strigulose or pilosulous overall (but especially along sutures) with sordid or yellowish hairs, endocarp orange-red either overall or only in seed-cavities, dehiscence downward through both sutures, valves coiling; seeds (few seen) plump, ± 5-7 mm diam., translucent testa revealing bluish embryo, pleurogram either 0 or complete, ± 1.5 mm diam., faintly engraved.

Distribution: From C Panama to NW Colombia and northern Venezuela, SE to Amazonian Venezuela and Brazil, disjunct in Guyana and French Guiana.

1a. **Abarema barbouriana** (Standl.) Barneby & J.W. Grimes var. **barbouriana**

Pithecellobium fanshawei Sandwith, Kew Bull. 1948: 314. 1948. Type: Guyana, Bartica-Potaro Road, 107 miles, Fanshawe in Forest Dept. British Guiana 4181 (holotype K!, isotype NY!).

Lf-formula vi-xii/12-19; lfts almost always finely and often densely silky-villosulous beneath, sometimes only barbellate in anterior basal angle of midrib and along midrib, largest ones at most 11.5 x 5 mm, venation (as given in key) relatively simple. Pod either strigulose or pilosulous.

Distribution: From C Panama to NW Colombia and mountainous northern Venezuela and far disjunct in lowland Guyana and French Guiana; in non-inundated forest from near sea level to 950 m elev.; flowering August to February (GU: 4; FG: 1).

Selected specimens: Guyana: Bartica-Potaro Road, Fanshawe FD 4181 (K, NY); Mahdia Cr., Potaro R., Fanshawe FD 3490 (NY). French Guiana: Station des Nouragues, Bassin de l'Arataye R., Sabatier & Prévost 2878 (CAY, NY); Mts. des Nouragues, Bassin de l'Approuague R., Larpin 847 (CAY, NY).

1b. **Abarema barbouriana** (Standl.) Barneby & J.W. Grimes var. **arenaria** (Ducke) Barneby & J.W. Grimes, Mem. New York Bot. Gard. 74(1): 72. 1996. – *Pithecolobium arenarium* Ducke, Arq. Inst. Biol. Veg. 2: 37. 1935. Type: Brazil, near Manáos, circa Ponte do Mindú, Ducke HJBR 23233 (holotype RB!, isotypes P!, US!).

Lf-formula (ii)iii-v/(7)8-11; lfts either pilosulous overall beneath, or glabrate except for pilosulous midrib, or (locally) glabrous, largest ones 9-21 x 5-10 (locally only 3-7) mm, venation usually better developed than in var. *barbouriana*. Pod loosely strigulose with hairs up to 0.5 mm.

Distribution: Venezuela (Amazonas and Bolívar), Brazil (Amazonas), and on N-slope of Mt. Roraima, Guyana; in savanna, forest-margin habitats, and along streams, from 120-2000 m elev.; flowering October to February (GU: 1).

Representative specimen examined: Guyana: Roraima, northern foothills, Hahn 5396 (US).

2. **Abarema commutata** Barneby & J.W. Grimes, Mem. New York Bot. Gard. 74(1): 65. 1996. Type: Guyana, southern Pakaraima Mts., Waipa trail from north Kopinang Savanna, Maguire *et al.* 46100 (holotype NY!, isotype US!).

Trees 8-20 m, young stems and all axes of lvs and inflorescences densely villous-tomentulose with brown-golden or bronze hairs. Lf-formula (i) ii-iv/3-6; stipules early caducous (few seen), elliptic-oblanceolate 3.5-6.5 x 0.8-1.3 mm; lf-stks (1)2.5-15 cm, petiole (0.5)0.8-2.4 cm, one (or longest) interpinnal segment 0.5-4(5) cm; pinnae much accrescent distally, rachis of apical pair (3)4-10(12) cm, longer interfoliolar segments 10-24 mm; lfts coriaceous, dark-green, glabrous and lustrous above, paler and densely golden-brown-villous beneath, strongly accrescent distally, in outline obovate-elliptic, rhombic-elliptic or obtusely rhombic from inequilaterally cuneate base, broadly obtuse, distal pair (3)3.5-6.5 x 2.1-4.5 cm, 1.4-2 times as long as wide. Peduncle 3.5-8 cm; capitula ± 12-22-flowered, axis including terminal pedestal 4-9 mm; bracts elliptic-oblanceolate, 2.5-4 x 1-1.5 mm, caducous; fls white, heteromorphic, lower peripheral ones raised on a pedicel 0.5-1.2 mm, terminal fl sessile, 5-merous perianth of all densely silky externally; peripheral fls: calyx turbinate-campanulate, 4.5-6 x 3.3-4.5 mm; corolla 6.5-10 mm; terminal fl: calyx broadly campanulate, 5-7 mm; corolla 8-10 mm. Pod (one seen, this perhaps stunted and atypical) sessile, narrowly oblong in profile, recurved through less than 1/4-circle, plano-compressed but low-convex over each of ± 8 seeds, sutures ± 2 mm wide, valves crustaceous, glabrescent, dull brown outside, reddish overall within; seeds not seen.

Distribution: Localized on E edge of the Guayana Highlands: Venezuela (Gran Sabana, Bolívar), Guyana (Pakaraima Mts.), and Suriname (Tafelberg); in savanna-forest ecotones and in openings in scrub forest, at 825-1200 m elev.; flowering August to September (GU: 2; SU: 1).

Selected specimens: Guyana: Pakaraima Mts.: Kopinang Savanna, Maguire *et al.* 46100 (NY, US), Chimapu Savanna, Maguire *et al.* 46145 (NY, U). Suriname: Tafelberg, Maguire 24629 (NY, U, US).

Note: The specimen from Suriname herein referred to *A. commutata* was referred to *Pithecellobium villiferum* Ducke by Jansen-Jacobs (1976: 647).

3. **Abarema curvicarpa** (H.S. Irwin) Barneby & J.W. Grimes, Acta Amaz. 14, Supl.: 95. 1986. – *Pithecellobium curvicarpum* H.S. Irwin, Mem. New York Bot. Gard. 15: 107. 1966. Type: Guyana, Essequibo R., Kamuni Cr., Groete Cr., Maguire & Fanshawe 22950 (holotype NY!, isotypes G!, GH!, MO!, U not seen, US!).

   In the Guianas only: var. **curvicarpa** – Fig. 1

Trees attaining ± 30-45 m, with sometimes buttressed trunk to 4-15 dm DBH, young branchlets, all lf-axes and young inflorescences densely minutely puberulent with forwardly incurved sordid hairs. Lf-formula (v) vi-xii/14-26 (lfts 120 or more per lf); stipules not seen; lf-stks (6)9-15 cm, ventrally flattened petiole 2-3.5 cm, longer interpinnal segments ± 12-17 mm; pinnae decrescent proximally, otherwise subequilong, rachis of those at and beyond mid-lf 4-9 cm, longer interfoliolar segments 2-6.5 mm; lfts (disregarding slightly broader furthest pair) subequiform, in outline obliquely oblong, elliptic-oblong or linear-oblong, 7-18 x 1.8-6 mm. Peduncle and raceme-axis together ± 6-12 cm, raceme ± 30-50-flowered, floral axis becoming 3.5-4.5 cm; fls subhomomorphic, distal one(s) slightly broader but their androecium not modified; perianth 5-merous, greenish, externally yellow-silky-puberulent, corolla lobes tipped with white polypoid trichomes, filaments white; calyx submembranous, campanulate, that of most fls 2-2.2 x 1.6 mm, depressed-deltate teeth 0.2 mm, that of apical fl ± 2.6-3 x 2 mm; corolla 4.5-5.2 mm, ovate lobes ± 1.3 x 0.9 mm. Pod turgid, moniliform and evenly incurved through 1-2 full spirals, 10-16-seeded, in profile ± 10-16 x 1.7-2.3 cm, prominent sutures 4-8 mm wide, woody valves up to 4-8 mm thick where grossly dilated and domed over each seed, externally coarsely rugose, red, or bright yellow and red, glabrous, dehiscence tardy, primarily along ventral suture; seeds compressed-obovoid, in broad view ± 9 x 7 mm, pleurogram 0.

Distribution: Guyana (Groete Cr.) and French Guiana (Saül and Piste-de-St-Elie); in moist riverine and upland forest, known from 200-400 m elev.; flowering April to May (GU: 1; FG: 7).

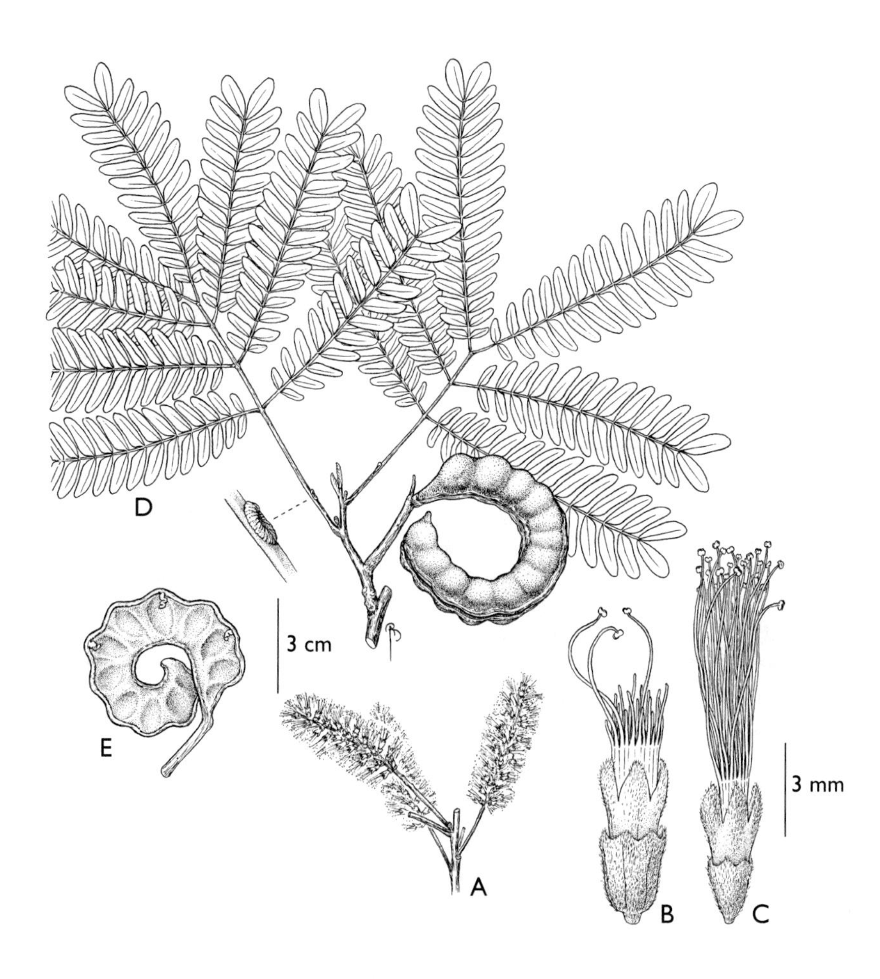

Fig. 1. *Abarema curvicarpa* (H.S. Irwin) Barneby & J.W. Grimes var. *curvicarpa*: A, inflorescence; B, terminal flower; C, peripheral flower; D, habit with pod, and leaf-nectary; E, opened pod (A-E, Sabatier 3051). Drawing by Bobby Angell; reproduced with permission from Mori *et al.*, 2002: 486.

Selected specimens: French Guiana: La Fumée Mt., near Saül, Mori 14943, 19172 (CAY, NY); Piste-de-St-Elie, Grimes 3312 (CAY, NY).

Note: The var. *rodriguesii* Barneby & J.W. Grimes is known only from the vicinity of Manaus, Amazonas, Brazil.

4. **Abarema ferruginea** (Benth.) Pittier, 3rd Conf. Interam. Agric. Caracas 360. 1945. – *Pithecolobium ferrugineum* Benth., London J. Bot. 3: 216. 1844. Type: Guyana [but very likely on the Venezuelan slope of Mt. Roraima], Ro. Schomburgk ser. II, 663 = Ri. 994 (holotype K!, isotype U not seen).

Arborescent shrubs attaining 8 m, bark gray lenticellate, branches and all lf- and inflorescence-axes densely pilosulous with golden-brown hairs. Lf-formula iii-iv/8-11; stipules linear-lanceolate to -elliptic 4-7 x 0.75-1.5 mm, quickly deciduous from expanding lf; lf-stks 5-7 cm, petiole 1.2-1.8 cm, longer interpinnal segments 1-2 cm; pinnae little graduated, 4-7 cm, longer interfoliolar segments 5-8 mm; lfts when dry brown above, pale brown beneath, puberulent along midrib on both faces, a little accrescent distally, inequilaterally rhombic-oblong-elliptic from rounded base, broadly obtuse, longer ones 18-26 x 9-14 mm, 1.8-2.2 times as long as wide; venation of 3 veins from pulvinule and thence pinnate. Peduncle 5-13 cm; capitula ± 45-60-flowered, 5-(randomly 6)-merous fls seemingly homomorphic (few whole capitula available for study) and sessile or almost so, stout pedicel less than 1.5 mm; bracts decrescent upward, lowest oblanceolate or trullate, 5-7 x 2-2.5 mm, upper ones spatulate shorter, all caducous; perianth densely appressed-silky overall, indument of corolla golden, of calyx brown; calyx deeply campanulate, 6.5-7.5 x 3.5-4 mm; corolla 12-13 mm. Pod (one incomplete sample seen) apparently like that of *A. floribunda*, not seen ripe.

Distribution: Venezuela (Gran Sabana of Bolívar), Brazil (Roraima) and Mt. Roraima in Guyana or the Venezuelan side; on wooded slopes and ridges at 1100-1525 m elev.; known to flower in December, January, and July (GU: 1).

Specimen examined: the type, see above.

5. **Abarema floribunda** (Spruce ex Benth.) Barneby & J.W. Grimes, Mem. New York Bot. Gard. 74(1): 48. 1996. – *Pithecolobium floribundum* Spruce ex Benth., Trans. Linn. Soc. London 30: 584. 1875. Type: Brazil, Rio Vaupés near Panuré, Spruce 2471 (holotype K (hb. Benth.)!).

Tree 10-20(29) m, but sometimes flowering precociously as a shrubby treelet, young parts, except lfts, densely pubescent with sordid or golden-brown hairs, but glabrate with age. Lf-formula (iii)iv-vi/8-17(21); stipules linear-lanceolate, lanceolate or narrowly ovate (3)4-13 x 1.5-4 mm, quickly deciduous; lf-stks (6)8-15(23) cm, petiole (2)3-7 cm, longer interpinnal segments 1.3-4 cm; pinnae either scarcely or strongly graduated (7)8-14(20) cm, longer interfoliolar segments 5-15 mm; young lfts subvertically imbricate along rachis but spreading in age, lfts bicolored, when dry brown above, pale brown beneath, commonly puberulent on both faces but sometimes glabrate on both faces, a little distally accrescent, oblong from inequilateral, broadly cuneate base, broadly obtuse, longer ones (10)17-28(37) x (3.5)5.5-12.5(17) mm, 2.3-3.3 times as long as wide. Peduncle 3.5-8 cm; capitula ± 40-60-flowered, clavate receptacle 5-14 x 3-6 mm, 5-(randomly 6)-merous fls heteromorphic, peripheral ones pedicellate, 1-3 terminal ones sessile and larger, perianth of all densely appressed-silky, indument of corolla golden, of calyx brown, corolla of peripheral ones purplish, of terminal ones (fide Spruce & Ducke) white, filaments rose-purple; peripheral fls: pedicel 1-5 x 0.4-0.55(0.7) mm; calyx campanulate 4-9 x 2.1-2.4 mm; corolla 8-13 mm; terminal fl (few seen): calyx 6-1.5 x 2.5-3.5 mm; corolla 15-17 mm. Pod 1-2 per capitulum, sessile or basally attenuate into a short pseudostipe, in profile broad-linear, evenly retrofalcate through 1/3-2/3-circle, 12-26(33) x (1.4)1.7-2.5 cm, coriaceous, dark brown, glabrous valves framed by stout, shallowly undulate sutures 2-3 mm wide.

Distribution: Known in typical form from lowland Amazonian Brazil and Peru (Loreto, San Martín); in forest on terra firme; in Brazil flowering in August and September (GU: 1).

Note: The foregoing description applies specifically to *A. floribunda* sensu stricto. The one collection from Guyana (Tillett & Tillett 45554) may represent a distinct taxon. It is congruent with *A. floribunda* in form of leaves and in relatively short peduncles, but approaching the almost sympatric *A. ferruginea* in smaller flowers and thin-textured leaflets.

6. **Abarema gallorum** Barneby & J.W. Grimes, Mem. New York Bot. Gard. 74(1): 51. 1996. Type: French Guiana, SW de Sinnamary, Piste-de-St-Elie (arbre E38), Lescure 877 (holotype P!, isotype CAY!). – Fig. 2

Trees to 25(30) m, bark smooth, gray-brown, branchlets, lf-axes, and especially axillary lf-buds densely tomentulose with brown hairs. Lf-formula iii/4-5; stipules (few seen) lanceolate ± 4-4.5 mm, caducous; lf-stks 8-18 cm, petiole 3-5 cm, longer interpinnal segments 3.5-6.5; pinnae distally accrescent, rachis of apical pair 9.5-13.5 cm, longest interfoliolar segment 2.5-4 cm; foliage bicolored, lfts when dry stiffly papery, lustrous dark brown-olivaceous and glabrous above, beneath paler and dull, lateral lfts oblong around a diagonal, slightly porrect midrib, at

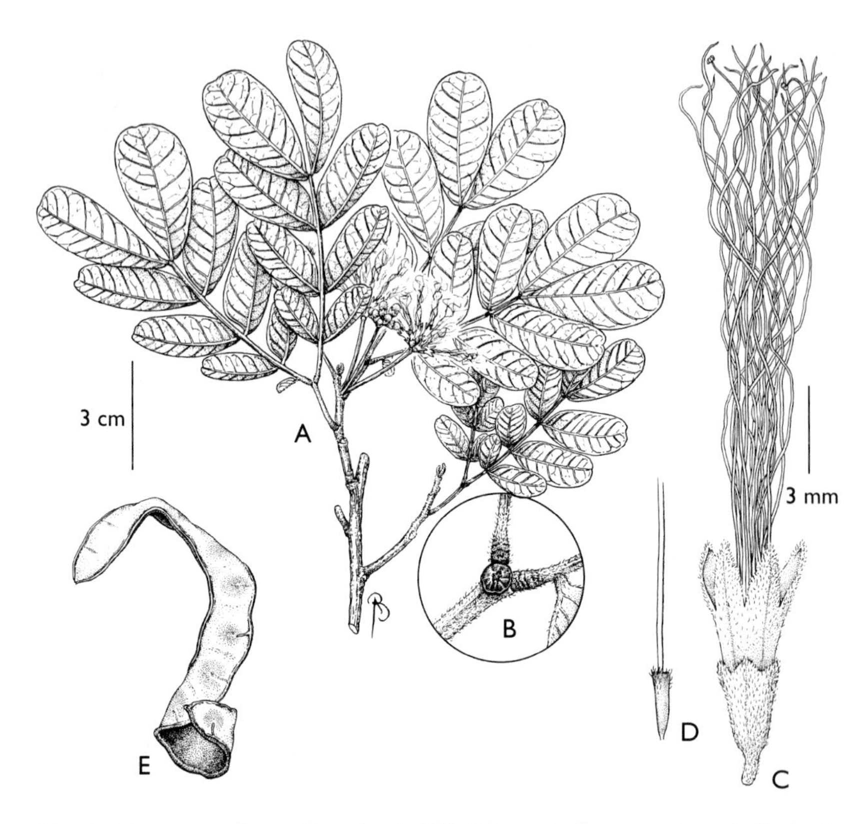

Fig. 2. *Abarema gallorum* Barneby & J.W. Grimes: A, flowering branch; B, detail of extrafloral nectary between terminal leaflet pair; C, flower; D, gynoecium; E, single valve of a dehisced pod (A-E, Sabatier 354). Drawing by Bobby Angell; reproduced with permission from NY.

base very inequilaterally rounded, at apex broadly obtuse and incipiently emarginate, blades of penultimate pair ± 5-8 x 2-3.5 cm, 2-2.5 times as long as wide, terminal pair broadly elliptic obtuse ± 6-8.5 cm. Peduncle (seen in fruit only) 4.5-7.5 cm; receptacle obovoid, including short terminal pedestal (that probably indicates a modified terminal fl) ± 5 x 3.5 mm; fls unknown, but fragmentary persistent calyx apparently ± 3 mm and glabrous; bracts caducous (not seen); fruiting pedicel of peripheral fls drum-shaped, 1.5-2.5 x 2.5-3 mm. Pod solitary, undulately broad-linear and falcately retroarcuate, 12.5-16 x 2.1-2.4 cm, contracted at base into a short pseudostipe and at apex into a short cusp, 10-14-seeded, leathery, almost plane, fuscous glabrous, evenulose valves framed by prominent sutures 1.5-2 mm wide.

Distribution: In lowland forest, known only from French Guiana (FG: 3).

Representative specimen examined: French Guiana: Station des Nouragues, Sabatier 3511 (CAY).

7. **Abarema jupunba** (Willd.) Britton & Killip, Ann. New York Acad. Sci. 35: 126. 1936. – *Acacia jupunba* Willd., Sp. Pl. 4: 1067. 1806. Type: Brazil, Pará, Sieber 44 (holotype B-WILLD 19142).

Trees (3)5-35 m (in upland Guyana sometimes small bushy tree 1.5-3 m) tall, young stems, lf-axes and inflorescences densely golden-brown- or sordid-tomentulose or -puberulent with spreading-incurved or subappressed hairs. Lf-formula i-vii/3-12; stipules linear or linear-lanceolate or elliptic, 1.5-6 x 0.4-0.7(1) mm, early deciduous from a small pallid scar; lf-stk of major lvs associated with fls 2.5-11(18) cm, of var. *trapezifolia* (0.7)1-8(14) cm, petiole (1)1.3-3.5(4) cm, longer interpinnal segments (often lacking in var. *trapezifolia*) 0.8-2.8(3.5) cm; pinnae (when more than 1 pair) distally accrescent, rachis of apical or penultimate pair longest, 2.5-9 cm, longer interfoliolar segments 6-13(14) or in var. *trapezifolia* 12-26(31) mm; lfts bicolored, on upper face glossy dark green, often purplish-speckled or mottled and (except for sometimes ciliolate costa) glabrous, beneath pallid-papillate dull and either densely or remotely but always minutely strigulose overall, decrescent toward base of pinnae, toward apex either scarcely or greatly accrescent, in outline varying from obtusely rhombic to rhombic-elliptic or -obovate from inequilaterally cuneate or subrectangular base, obtuse or emarginate, terminal pair (1.3)1.6-6.5(7) x 1-3.6(4.4) cm, (1.3)1.5-2.1(2.3) times as long as wide. Peduncle mostly solitary or geminate, seldom ternate, (1)2.5-7.5(10) cm; capitula or capituliform racemes ± (10)15-45-flowered, either compact or with one or more fls downwardly displaced on peduncle, receptacle including very short terminal pedestal 2-11 mm; bracts linear-oblanceolate, oblanceolate or elliptic, 0.7-1.6 mm, deciduous, or lowest longer and more persistent; fls ordinarily heteromorphic, one (2-3) terminal ones larger and coarser, with modified androecium, but these sometimes modified only as to androecium or abortive and early deciduous, or exceptionally wanting, perianth of all fls 5-merous, densely brownish-silky overall with appressed hairs; peripheral fls: pedicel of proximal ones mostly 0.4-1.2 mm, rarely subobsolete, that of distal ones often a little shorter; calyx narrowly or widely turbinate-campanulate, 1.7-3 x 1.4-2.1(2.4) mm; corolla 4-6.2(7) mm; terminal fl(s): sessile, calyx broadly campanulate, (1.8)2.2-3.4 x 2-2.6 mm, corolla (5.5)6.5-8.5 mm. Pod 1-2 per capitulum, sessile or almost so, in profile broad-linear, decurved through 2/3-nearly

2 circles, when well fertilized 6-9.5 x 1-1.6 cm, 8-12-seeded (by random abortion of ovules often shorter); seeds plumply lentiform, in broad view as long as or a little longer than wide, 5-8 mm diam., pleurogram 0.

Distribution: In upland and non-inundated lowland, primary and secondary forest and forest-savanna ecotone, widespread over tropical S America from Colombia to Bolivia, through Venezuela to Trinidad and Tobago and the Guianas, in Amazonian Brazil to Rondônia and Mato Grosso, disjunct in coastal forest of Bahia; Lesser Antilles south from Guadeloupe.

7a. **Abarema jupunba** (Willd.) Britton & Killip var. **jupunba**

Pithecolobium brongniartii Duchass. & Walp., Flora 36: 232. 1853. Type: Guadeloupe, Duchassaing s.n. (lectotype P!).
Mimosa vaga sensu Aubl., Hist. Pl. Guiane 2: 945. 1775; non Linnaeus, 1753. Spm. authent., BM = NY Neg. 141!

Lf-formula iii-vii/(6)7-12, and terminal lfts of larger lvs (13)16-35(36) x 10-19(21) mm.

Distribution: Interruptedly dispersed around and within the Amazon Basin from Ecuador, NE Bolivia and E Peru east to the delta, through inter-Andean valleys of Colombia, and to the Maracaibo Basin in Venezuela; reappearing in lowland forest near the coast of French Guiana and in Bahia, Brazil, and disjunct on Lesser Antilles; in the Guianas below 400 m, in virgin and disturbed, lowland and upland forest, in forest-savanna ecotone, and on rocky river banks, but not entering seasonally flooded forest; flowering all months of the year (FG: 4).

Selected specimens: French Guiana: SE of Cayenne, Cacao, Oldeman 1583 (CAY, NY); Petit Saut, Sabatier 2213 (CAY, NY).

Vernacular name: French Guiana: tamalin (fide Oldeman).

7b. **Abarema jupunba** (Willd.) Britton & Killip var. **trapezifolia** (Vahl) Barneby & J.W. Grimes, Mem. New York Bot. Gard. 74(1): 69. 1996. – *Mimosa trapezifolia* Vahl, Eclog. Amer. 3: 36, t. 28. 1807. Type: Trinidad, Ryan (holotype C!).

Mimosa atakta Steud., Flora 26: 758. 1843. Type: Suriname, Hostmann & Kappler 479 (holotype P!, isotype NY!, U!)
Pithecolobium benthamianum Miq., Linnaea 18: 592. 1844. Type: Suriname, near Paramaribo, Focke 812 (holotype U!).

Lf-formula i-iii(iv)/3-6(7), and terminal lfts (34)35-65(70) x 19-36 mm.

Distribution: Replacing var. *jupunba* in parts of north-eastern S America: in the Atlantic lowlands and upland interior of Guyana, (on Tafelberg attaining 500 m), and NW French Guiana; in lowland and upland rainforest, in gallery forest and shrub islands within savanna climax, and occasional along streams (but not seasonally inundated) from 500-1400 m elev.; flowering all months of the year (GU: 31; SU: 18; FG: 2).

Selected specimens: Guyana: Pakaraima Mts., Chimapu Cr., Maguire 46158; Pomeroon Distr., de la Cruz 1239 (NY). Suriname: near Moengo tapoe at Grote Zwiebelzwamp, Lanjouw & Lindeman 417 (NY, U); Tafelberg, North Ridge Cascade, Maguire 24660 (NY). French Guiana: Bassin du Haut-Marouini, de Granville 9543 (CAY, NY).

Vernacular names: Guyana: huruassa, horoassa, culane. Suriname: swampoetamarin. French Guiana: huruassa, horoassa.

8. **Abarema laeta** (Benth.) Barneby & J.W. Grimes, Mem. New York Bot. Gard. 74(1): 79. 1996. – *Pithecolobium laetum* Benth., London J. Bot. 3: 203, 1844. Type: Peru, Maynas near Yurimaguas, Poeppig D.2367 (holotype W!).

Arborescent shrubs, fertile at (1.5)2 m but becoming 6(-?) m tall with trunks up to 2(-?) cm diam. and yellow wood, hornotinous branchlets, lf-stks and peduncles pilosulous or subappressed- or incurved-puberulent with rufous or sordid-yellowish trichomes. Lf-formula i-ii/1, lfts either exactly 4 or exactly 8 per lf; stipules herbaceous, narrowly lanceolate, 1.5-5(7) mm, 1-3-veined dorsally, early deciduous; lf-stks (1.5)2-14 cm, narrowly shallowly grooved ventrally, petiole proper (1.5)2-10(12) cm, one interpinnal segment, when present, (2.5)3-7 cm, ordinarily a little longer than petiole, sometimes as long or a trifle shorter; pinna-rachises (4)6-20(31) mm, widely shallowly grooved ventrally; pulvinule (on dorsal side) (1)1.5-3 mm, papery lfts lustrous above, dull beneath, either glabrous overall, or minutely ciliolate on margin and on midrib beneath, blades of distal (or only) pair of pinnae subsymmetrically elliptic or (ob)ovate-elliptic from cuneate base, caudately acuminate, larger ones (including acumen) 10-18(20) x 3-7(7.7) cm, (2)2.1-2.5(3.9) times as long as wide, anterior one of each pair a little longer than its fellow, all cuneate and usually revolute at very base, acumen 1-2.5(3) cm. Peduncle 3-7 cm, naked or bearing under capitulum one or more empty bracts, occasionally with a flower distant from rest; capitula 12-28-flowered, subglobose or clavate receptacle 2-7 mm long; floral bracts thinly herbaceous, linear-lanceolate, 2.5-8 x 0.4-1.5 mm (rarely abbreviate, ovate and only 1.5 mm), tardily deciduous from a small spur; fls sessile or almost so, heteromorphic, 1 or rarely 2 terminal ones a little larger and with long-exserted stamen-tube, all greenish, externally either rufescent-pilosulous overall or glabrate except for minutely barbellate or puberulent calyx- and corolla-lobes; peripheral fls: perianth (4)5-merous; calyx turbinate, 2-4 mm; corolla (4.2)4.7-7 mm; androecium white, fading orange; terminal fl: slightly wider and coarser than peripheral ones, corolla scarcely longer (up to 6-8 mm), but staminal sheath 7-16 mm long, at least 1 mm longer than corolla-lobes. Pod usually 1 per capitulum, subsessile but abruptly narrowed into a short neck at base, in broad profile oblong or broad-linear, (5)5.5-13(16) x 1.4-2.2 cm, 2.5-7(10) times as long as wide, when short nearly straight but when longer evenly recurved through up to 3/4 circle, at first plano-compressed and framed by scarcely undulate sutures, becoming umbonate over developing seeds, stiffly papery valves green turning tan-brown, densely cross-venulose, venules subcontiguous; seed plumply ellipsoid, 10-11 x 6-8 x 5 mm, pleurogram lacking.

Distribution: In W and C Amazonia, SE Colombia, E Peru, extreme SW Venezuela, and Brazil, and outside this range known from remote stations in W Venezuela, in SE Suriname and adj. French Guiana, and in SW Pará, Brazil; in the understory of wet, virgin and disturbed equatorial forest, on terra firme, mostly between 100-600 m, but to 700 m in French Guiana; flowering all year (SU: 1; FG: 1).

Selected specimens: Suriname: near airstrip at Oelemari R., Wessels Boer 1089 (NY, U). French Guiana: Mont Atachi Bacca, région de l'Inini, de Granville 10930 (CAY, NY).

9. **Abarema longipedunculata** (H.S. Irwin) Barneby & J.W. Grimes, Mem. New York Bot. Gard. 74(1): 52. 1996. – *Pithecellobium longipedunculatum* H.S. Irwin, Acta Bot. Venez. 2: 223. 1967. Type: Venezuela, Bolívar, Auyan-tepui, Steyermark 93316 (holotype NY!).

Shrubs and trees 2-20 m, with densely rusty-tomentellous growing tips but early glabrate. Lf-formula i-iii/3-5; stipules linear or linear-lanceolate 1.5-2.5 mm, early deciduous, absent from fruiting spms; lf-stks stout 0.7-5.5 cm, petiole 0.7-2.2 cm, longer interpinnal segment(s) 0-1.8 cm; pinnae when more than one pair accrescent distally, rachis of distal (or only) pair 4-7 cm, longer interfoliolar segments 9-23 mm; lfts dark above, paler beneath, distally accrescent, asymmetrically elliptic, oblong-, or obovate-elliptic from antically cuneate, postically rounded base, obtuse or obscurely emarginate, larger ones 3.8-6.5 x 2-3.5 cm, 1.4-2 times as long as wide; venation pinnate and reticulate, gently porrect midrib subcentric, major secondary veines 9-12 pairs. Peduncle (3)5-12.5 cm; capitula ± 20-25-flowered, clavate receptacle 4-8 mm; bracts (caducous) not seen; fls (known only fallen) subsessile and obscurely pedicellate, probably heteromorphic but terminal one not observed, perianth of peripheral ones 5-merous, densely brown-puberulent externally, corolla more thinly so below lobes; peripheral fls: calyx narrowly campanulate 5-7.3 x 2-2.3 mm; corolla 8-9.5 mm; androecium not seen complete, color (probably white) uncertain. Pods up to 6 per capitulum, sessile or cuneately narrowed at base into an obscure pseudostipe, in profile linear, when well fertilized 7-10.5 x 1.3-1.7 cm, 10-13-seeded, gently evenly retrofalcate through less than 1/4-circle, plano-compressed leathery, glabrous reddish valves becoming low-convex over each developing seed, when dry fuscous, coarsely cross-veined and reticulate, framed by evenly arched or (peripherally) shallowly undulate sutures 1.6-2 mm wide, endocarp internally orange-red at seed-cavities and pallid between them; seeds plumply oblong-ellipsoid, in broad profile ± 8.5-10.5 x 5.5-7 mm, pleurogram 0.

Distribution: Endemic to Venezuelan Guayana (Bolívar), known from Meseta Jáua, C. Auyántepui and C. Venamo; on rocky summits, on cliff ledges, and sandstone table mountains, at (1100)1400-2100 m elev.; flowering January and February; expected in Guyana.

10. **Abarema mataybifolia** (Sandwith) Barneby & J.W. Grimes, Mem. New York Bot. Gard. 74(1): 78. 1996. – *Pithecellobium mataybifolium* Sandwith, Kew Bull. 1948: 313. 1948. Type: Guyana, Bartica-Potaro road, 107 miles, Fanshawe 1496 (= FD 4232) (holotype K!).

Trees 8-20 m, with trunk 8-30 cm DBH, glabrous except for densely brown-sordid-puberulent axillary buds and inflorescences. Lf-formula i-ii/2-3; stipules not seen, perhaps very small caducous, but possibly lacking (scar obscure); lfts 8-16 per lf; lf-stk 1-8 cm, true petiole including wrinkled pulvinus 1-5 cm, interpinnal segment, when present, up to 4.5 cm; rachis of pinnae 4.5-12 cm, one or the longer of two interfoliolar segments (2)2.5-6 cm; chartaceous lfts olivaceous sublustrous above, paler beneath, slightly accrescent distally, subequilaterally ovate or elliptic-ovate from broad-cuneate or, especially on anterior side, rounded base, shortly obtusely acuminate, blades of distal pair 8-14 x 3-7.5 cm, 1.8-2.3(2.5) times as long as wide. Peduncle 2-5 cm; capitula ± 15-20-flowered, subglobose or shortly clavate receptacle ± 3 mm; perianth 5-merous, rusty-pilosulous externally; fls dimorphic, peripheral ones pedicellate, terminal one (few seen) sessile and a trifle coarser; peripheral fls: calyx campanulate, 2-3.6 mm, cuneately narrowed at base into a pedicel 2-3.6 mm; corolla narrowly trumpet-shaped, 6.3-7.4 mm; terminal fl: corolla hardly longer than rest, but staminal column exserted up to 2 mm; corolla of all fls greenish, filaments white, brownish in fading. Pod 1-2 per capitulum, in profile broad-linear, evenly recurved through half or nearly a complete circle, ± 10-20 x (1.3)1.8-2.6 cm, 9-14-seeded, at first plano-compressed and framed by thickened, very shallowly undulate sutures, brown or fuscous, glabrous valves becoming low-convex over each seed, coarsely transverse- and reticulate-venose externally, within red-crimson overall; seeds plumply lentiform-globose, in broader view 8-9 x 7-8 mm, pleurogram 0.

Distribution: Scattered in E equatorial S America in lat. 8°N-5°S, from the Barima R. in Guyana to NE Pará in Brazil; in virgin and second-growth, non-inundated forest below 170 m elev.; flowering August to November (GU: 3; FG: 2).

Selected specimens: Guyana: Sibaruni Cr., left bank Demerara R., Fanshawe FD 6333 (NY); Matthews Ridge, Barima R., Maguire 39332 (NY). French Guiana: Station Nouragues, Sabatier 2869 (CAY, NY); Piste-de-St-Elie, interfluve Sinnamary/Counamama, Sabatier & Prévost 3628 (CAY, NY).

Vernacular name: Guyana: huruassa.

2. **ACACIA**[6] Mill., Gard. Dict. Abr. ed. 4. 1754.

Lectotype: A. nilotica (L.) Willd. ex Delile (Mimosa nilotica L.)

Trees, shrubs or lianes, one species with brachyblasts (*A. farnesiana*), others with homomorphic shoots, these armed either with epidermal prickles or stipular spines, inflorescence a compound raceme or panicle of spikes or capitula. Leaves alternate, bipinnate; lf-formula i-xxiv/11-35; stipules (when known) herbaceous and quickly deciduous, or modified into spines; petiole and sometimes lf-rachis between ultimate few to many pinna-pairs charged with a nectary; lfts opposite, equi- or inequilateral in shape, pulvinulate; lft-venation of a simple midrib, or rarely palmate from petiolule. Spikes or capitula fasciculate, at any node peduncles ± equal or very inequal in length; involucre present or lacking, bracts deciduous or subpersistent; fls 5-merous. Calyx tubular or tubular-campanulate; corolla united into a tube or rarely free to stemonozone; stemonozone present

[6] by James W. Grimes

or lacking; stamens numerous, free or ± united at base, intrastaminal nectary present or lacking; ovary sessile or elevated on a gynophore. Pods flattened, elliptic in outline, dehiscence inert, through both sutures, or pod a loment and breaking into 1-seeded indehiscent segments; seeds elliptic, round or obovate, exotesta ± hard, tan, light brown or dark brown, pleurogram U-shaped or almost complete.

Distribution: About 1450 species worldwide, in tropical and subtropical zones, in very xeric or very moist habitats; 5 species are known to occur in the Guianas, 1 of which is introduced, a 6th species may be expected to occur.

Taxonomic note: The above description applies only to the species in the Guianas.
Receipt of the Mimosoideae manuscript for the Flora of the Guianas long predates the segregation of the genus *Acacia* into a number of reinstated genera. The nomenclature and synonymy for species presented below largely follows that given by Rico Arce (2007) in her checklist and synopsis of American species of *Acacia*, and synonymy is cross-referenced to Seigler, Ebinger & Miller (2006).

LITERATURE

Grimes, J.W. 1992. Description of Acacia tenuifolia var. producta (Leguminosae, Mimosoideae), a new variety from the Guianas, and discussion of the typification of the species. Brittonia 44: 266-269.

Rico Arce, M. de L. 2007. A checklist and synopsis of American species of Acacia (Leguminosae: Mimosoideae). Tlalpan, Mexico. 207 pp.

Seigler, D.A. *et al.* 2006. The genus Senegalia (Fabaceae: Mimosoideae) from the New World. Phytologia 88(1): 38-93.

KEY TO THE SPECIES

1 Inflorescences spicate 2
  Inflorescences capitulate 4

2 Rachis of inflorescence 15 cm or longer; corolla 6-7 mm (expected in the Guianas) . . . . *1. A. alemquerensis*
  Rachis of inflorescence 4 cm or shorter, corolla 3.5 mm or shorter 3

3 Rachis of inflorescence less than 1 cm; pinnae xxi-xxxv pairs . . . . 6b. *A. tenuifolia* var. *producta*
Rachis of inflorescence 3-3.9 cm; pinnae v-xi pairs . . . . 2. *A. articulata*

4 Plants armed with stipular spines; inflorescence involucrate 5
Plants armed with epidermal prickles, but no stipular spines; involucre 0 6

5 Shoot-system dimorphic, with long-shoots and short-shoots; lf-formula of long-shoots v-vi(xvii)/16-20(35), that of short-shoots i-iii(vi) 11-16(21) . . . . 3. *A. farnesiana*
Shoot system monomorphic, lf-formula viii-xxiv/25-36 (introduced in FG) . . . . 4. *A. macracantha*

6 Corolla 3-5 mm, pubescent; lf-formula v-xii(xxii))/14-40(68) . . . . 5. *A. polyphylla*
Corolla 1.85-2.5 mm, glabrous or very sparingly pubescent on lobes; lf-formula xxi-xxxv/43-107 7

7 Inflorescence 13-23-flowered; gynophore 0.8-1 mm . . . . 6a. *A. tenuifolia* var. *tenuifolia*
Inflorescence 31-43-flowered; gynophore 0.25-0.51 mm. . . . . 6b. *A. tenuifolia* var. *producta*

1. **Acacia alemquerensis** Huber, Bol. Mus. Paraense Hist. Nat. 5: 380. 1909. Type: Brazil, Pará, in forest near Alemquer, Ducke (holotype, MG (not seen)).

Lianas with bicolored leaflets, lfts and axes of lvs and infl moderately, fls densely pubescent with short white hairs, younger stems charged with short, recurved prickles, these becoming scarce on old wood and on terminal paniculate inflorescence of long white spikes. Lf-formula x-xi/14-22; stipules not seen; lf-stks 16.7-25 cm long, petiole 3.0-5.5 cm, charged just above pulvinus with a ± round to shortly elliptic mound shaped or mammiform nectary up to 5 mm on longest axis; interpinnal segments 11-21 mm; pinnae 8-11.5 cm, elliptic lfts 12-21 x 5-7 mm, inequilaterally truncate at base, broadly round or round-acuminate to slightly emarginate at apex, midvein forwardly displaced, both surfaces covered with short, fine, white hairs, lower surface more densely so. Axis of infl to 60 cm, spikes 1-3-fasciculate at nodes, peduncle 0.9-1.7 cm, rachis 15-22.7 cm, ± 170-178-flowered; bracts not seen; fls sessile, irregularly disposed along rachis, calyx tubular 3.0-3.75 mm, corolla tubular or tubular-funnelform 6-7 mm; stemonozone 0; intrastaminal nectary 0.2-0.45 mm, stamens free above nectary; ovary glabrous or very sparingly pubescent at base, elevated on a gynophore 1.85-2.0 mm. Fruit not seen.

Distribution: Brazil (Amapá, Rondônia and Amazonas); on terra firme in forest and forest margins below 200 m; flowering June to July; to be expected in the Guianas.

Taxonomic note: Rico Arce (2007) places *Senegalia alemquerensis* (Huber) Seigler & Ebinger as a synonym.

2. **Acacia articulata** Ducke, Arch. Jard. Bot. Rio de Janeiro 3: 73. 1922. Syntypes: Brazil, in flooded forest along the Gurupatuba R. near Montealegre, Ducke 16038, 16494 (RB (not seen)).

Small trees to 6 m, young stems, lf-, infl-axes, and calyx glabrate to uniformly pubescent with short, white, gently curved hairs, stems with paired or solitary, recurved prickles 6-9 mm long, these lacking from terminal racemose to shortly paniculate inflorescence of white-flowered spikes. Lf-formula v-xi/18-38; stipules not seen; lf-stk 7.0-12.2 cm, petiole 1.1-2.7 cm, charged just above pulvinus or above mid-petiole with a patelliform, sometimes wrinkled, round to elliptic nectary 1.15-2.75 x 0.75-1.15 mm; interpinnal segments 7-12 mm; pinnae 4.9-8.2 cm; lfts inequilaterally rhombic-elliptic, 6-9.25 x 0.88-2.75 mm, apex inequilaterally acute or round acuminate, midrib diagonal, forwardly displaced at base of blade, ± centric at apex. Main axis of inflorescence up to 30 cm, peduncles 2-5-fasciculate, maturing sequentially, the first formed apparently always of greater size, 0.9-1.3 cm, rachis 3.0-3.9 cm, 30-73-flowered; bracts spatulate or ovate-spatulate, sometimes somewhat carinate at apex, deciduous; fls tubular- or funnelform-campanulate, pubescent at least apically, calyx 1.65-2.05 mm, corolla 2.5-3.25 mm, intrastaminal nectary 0.3-1.55 mm, stemonozone 0.25-0.8 mm, ± equal to or only half as long as intrastaminal nectary, stamens free above nectary; ovary villose, on a gynophore 1.49-1.65 mm. Pod a loment, elliptic, 9.5-14.2 x 1.4-1.7 mm, flattened, articulate between seeds, breaking into individual segments between seeds, but valves indehiscent; seeds (very few seen) round-elliptic to elliptic in profile 7-10.7 x 4-7 mm, seed-coat shiny brown, almost complete pleurogram at center of seed-face.

Distribution: Venezuela (Zulia, Barinas, Guarico) and Brazil (Pará, Amazonas, Roraima); in riverine or disturbed, humid forests in heavy soils, from 0-100 m; flowering June to August; the only collection from Suriname (Corantijn R.) is immature (SU: 1).

Representative specimen examined: Suriname: Along Corantijn R., Rombouts 66 (NY).

Taxonomic note: Seigler *et al.* (2006) placed *Acacia articulata* as a synonym of *Senegalia rostrata* (Humb. & Bonpl. ex Willd.) Seigler & Ebinger (which is based on *Acacia rostrata* Humb. & Bonpl. ex Willd.), but Rico Arce (2007) retained *A. articulata* and *A. rostrata* as separate taxa.

3. **Acacia farnesiana** (L.) Willd., Spec. Pl. 4: 1083. 1805. – *Mimosa farnesiana* L., Spec. Pl. 1: 521, 1753. Type: "Habitat in Domingo".

Shrubs or densely branched trees ± 3-4 m tall, armed at nodes with straight, divergent, stipular spines, axes of young lvs, peduncles and fl-buds densely to moderately pubescent, becoming glabrous with age; shoots and lf-formula dimorphic, long-shoots leafy when young, otherwise lvs and infls crowded on axillary brachyblasts, lvs on long-shoots of higher formula than those of short-shoots; infls only rarely solitary or geminate in axils of long-shoots. Stipular spines 0.4-4.6 cm, with age gray, black toward apex, long-persistent; long-shoots: lf-formula (iv)v-vi(xiv, xvii)/16-20(35), lf-stk 4.8-7.7 cm, with aculeus to 3 mm surpassing pinna-pair, petiole 6-10 mm, interpinnal segments 7-9 mm; pinnae slightly decrescent, subopposite, 1.8-2.4 cm; lfts subsessile on rachis, slightly paler on lower surface, in outline narrowly oblong 4-5 x 0.75-1.25 mm, midrib ± centric to slightly forwardly displaced, basally ciliolate; short-shoots: lf-formula i-iii(vi)/11-16(21), lf-stk 13.5-20 mm, petiole 7-11 mm, interpinnal segments 5-9 mm; pinnae 15-17 mm; lfts as on long-shoots. Peduncle 0.8-2.5 cm, elongating slightly and becoming stouter in fruit; capitula 25-103-flowered, fls sessile; bracts dimorphic, lowermost fused to form a quickly deciduous 5-lobed involucre to 1.12 mm, remainder abruptly spatulate, ± 0.5 mm, ciliolate, subpersistent; calyx cylindro-campanulate, membranous, 5-lobed, 1.25-2.0 mm, lobes darker and somewhat cucullate; corolla 2.25-3.25 mm; intrastaminal nectary 0.0-0.2 mm, stamens weakly to strongly united into a tube slightly shorter than corolla; ovary attenuate at base into a short stipe to 0.2 mm. Pod 1-2 per capitulum, arcuate-reflexed or ± erect, substipitate, ± falcate-terete, 3.5-7 cm, when fully ripe to 1.5 cm diam., valves coarsely cross-venulose and dark-shiny at maturity, internally ± mealy, septate; seeds elliptic to ± round, in profile to 8 mm diam., to 5 mm broad, though often distorted by mutual pressure, testa tan to dark brown with an almost complete pleurogram.

Distribution: Probably of American origin, though now widely naturalized in tropical regions of the world; in a wide variety of habitats, but where disturbed, from 0 to over 2400 m elev.; flowering February to December, but most prolifically from June to November; collected in all 3 Guianas (GU: 1; SU: 2; FG: 1).

Selected specimens: Guyana: coastlands, Berbice, Jenman 5250 (NY). Suriname: Corantijnpolder near Nieuw Nickerie, Lanjouw 638 (NY). French Guiana: Cayenne, near ORSTOM, Grimes 3266 (CAY).

4. **Acacia macracantha** Willd., Sp. Pl. 4: 1080. 1806. Type: "in America meridionale" (holotype IDC fiche 7449, 19168).

   Acacia flexuosa Humb. & Bonpl. ex Willd., Sp. Pl. 4: 1082. 1806. Type: "Habitat in Cumaná" (holotype IDC fiche 7449, 19172).

Trees 2-9 m tall, brachyblasts only poorly developed and most reproductive and vegetative growth occurs on long shoots, young growth glabrous or pubescent with short white hairs. Leaves monomorphic, lf-formula viii-xxiv/25-36; stipules spinulose, straight, usually somewhat reddish, on younger branches 2-9 mm, but older ones becoming flattened and to 6.0 cm long; lf-stk 5.0-21.0 cm, petiole 0.6-1.3 cm, interpinnal segments 3-8 mm; pinnae subopposite, rachis of longer ones 19-30 mm; lfts slightly decrescent, elliptic, 2.0-4.25 x 0.5-1.0 mm, midrib ± centric, blade somewhat paler on lower surface. Capitula geminate to 10-fasciculate in nodes on long-shoots, rarely solitary or geminate on poorly developed brachyblasts; peduncles 0.8-6.5 cm, becoming stouter with age; capitula 65-81-flowered, fls sessile; bracts dimorphic, lowest usually forming a scarcely lobed involucre to 0.5 mm; calyx 1.0-1.5 mm, 5 lobes somewhat darker and more pubescent; corolla 2.0-2.3 mm, lobes like those of calyx; intrastaminal nectary 0; stamens free or ± weakly and unevenly united above nectary into a tube about equal to corolla; ovary sessile, but attenuate at base. Pod indehiscent, when fully ripe 1-3 per capitulum, 7.1-11.8 cm, ± straight, terete or slightly compressed, subsessile on an attenuate base, but not truly stipitate, valves wrinkled or somewhat venulose, pubescent when young, commonly glabrous with age, mealy mesocarp forming septa; seeds elliptic to round, in outline to 8 mm in broadest dimension, though often distorted by mutual pressure, seed-coat shiny brown with nearly complete pleurogram.

Distribution: From southern USA and Mexico, through Venezuela and Colombia to Bolivia and Peru; evidently introduced near Cayenne in French Guiana.

Representative specimen examined: French Guiana: vicinity of Cayenne, Broadway 282 (NY).

Taxonomic note: Rico-Arce (2007) includes *Vachellia macracantha* (Humb. & Bonpl. ex Willd.) Seigler & Ebinger as a synonym of *Acacia macracantha*.

5. **Acacia polyphylla** DC., Cat. Hort. Monsp. 74. 1813. Type: [hortus monspessulani] (holotype G-DC, IDC fiche 2562. 425: II, 7).

Acacia glomerosa Benth., London J. Bot. 1: 521. 1842. Syntypes: Brazil, Rio de Janeiro, Guillemin 809; Brazil, Piauí, Gardner 1940; Brazil, Rio S. Francisco, Claussen (K!).

Scandent shrubs to 5 m, or more commonly tall trees to 30 m, old wood charged with (rarely lacking) recurved, erect or acroscopic prickles, these commonly also on branchlets and lf-stks, plants glabrate or most commonly pubescent on infl-axes, ventral surface of lf-stks, lower surface of lfts, and always on calyx and corolla. Lf-formula v-xii(xxii)/14-40(68), lft-size is not reciprocally correlated with lf-formula, lf-size or lft-number; stipules triangular to linear-triangular, to 3 x 1 mm, very quickly deciduous and lacking from most specimens; lf-stk 5.2-26.5 cm, petiole 2.0-4.2 cm; interpinnal segments 0.6-2.9 cm; pinnae 5.5-12.5 cm; lfts lanceolate-elliptic from an inequilateral truncate base, (3.5)7-29 x (0.95)1.25-11 mm, midrib forwardly displaced to divide blade 1:2.3-6.6. Infl a pseudopanicle of capitula 2-10 fasciculate at nodes; peduncle 3.5-20 mm; bracts obovate to spatulate-obovate, 0.3-0.75 mm, tardily deciduous; flowers densely (moderately) appressed pubescent; calyx 1.5-2.25 mm; corolla 3.0-5.0 mm, 1.6-2.3 times greater than calyx; intrastaminal nectary 0.55-0.75 mm; ovary villose, on a gynophore 1.5-1.75 mm. Pod reddish brown when dried, 1-several per capitulum, in profile elliptic from a stipitate base; 9-25 x 1.5-5 cm, valves coarsely venulose, somewhat lustrous, sutures moderately to strongly ribbed, dehiscence inert, through both sutures; seeds (very few seen) round to depressed obovate, to 10 mm in largest diam., seed-coat shiny brown, pleurogram almost complete.

Distribution: From the SW coast of Mexico (as *A. glomerosa*) south, widespread in S America to Argentina and Paraguay; in gallery-forest, secondary forest, or surviving in cultivated fields, sometimes planted, from near sea-level to over 2000 m, but mostly below 500 m; flowering September to May; in the Guianas known only from the Kanuku Mts. and the Rupununi R. in Guyana (GU: 6).

Selected specimens: Guyana: Kanuku Mts., in drainage of Takutu R., A.C. Smith 3086; Sand Cr., Rupununi R., FD 5641 (= WB 85) (NY).

Vernacular name: Guyana: parika (Creole).

Note: This taxon has traditionally been divided into the 2 species cited in the protologue. *A. glomerosa* was considered to be a species with fewer pinnae and larger leaflets. However, the dimensions of the leaf overlap, with no discontinuities in numbers of parts. Furthermore, the size of the leaflets is totally independent of the number of pinnae, the size of the leaf-stalk, or the number of leaflets. *A. polyphylla* is apparently an extremely variable species as concerns the leaf-architecture, though homogeneous in characteristics of the flower. The fruit, though varying considerably in size, is consistent in texture, and in having a ribbed suture.

Taxonomic note: Rico Arce (2007) included *Senegalia polyphylla* (DC.) Britton & Rose and *Senegalia glomerosa* (Benth.) Britton & Rose as synonyms of *Acacia polyphylla*.

6. **Acacia tenuifolia** (L.) Willd., Sp. Pl. 4(2): 1091. 1806. Type: "Habitat in India Occidentale" (see Grimes (1992) for discussion of typification).

Prostrate to weakly erect shrubs or most commonly lianes with sensitive leaves and panicles of white or cream colored capitula or short-spikes, stems, lf- and infl-axes armed with recurved, yellow to dark brown prickles. Lf-formula xxi-xxxv/43-107, lft- and pinna-number not inversely correlated; stipules not seen; lf-stk 11.2-23 cm, petiole 1.0-2.4 cm, charged at base with a most commonly sessile patelliform, rarely stipitate-pateliform, or stipitate-ureceolate nectary 2.25-4 mm on long axis, commonly with smaller, stipitate nectaries between ultimate 13 to terminal pair(s) of pinnae; interpinnal segments 3.5-7 mm; pinnae 3.2-5.0 cm, lfts elliptic, 2.3-3.8 x 0.3-0.65 mm, either equi- or inequilateral at base and commonly ciliolate, apex inequilaterally acute or round-acuminate, midrib forwardly displaced almost to margin, or displaced basally and apically centric (diagonal), or centric, sometimes larger lfts faintly palmately veined from petiolule, lfts sensitive, folding ventri-apically along pinna-rachis. Peduncles 2-5-fasciculate, maturing sequentially, the first formed always attaining greater size, 4.5-14 mm at anthesis, commonly with an extra-floral bract on distal 1/2; capitula or spikes 12-43-flowered; bracts spatulate or ovate-spatulate 0.33-0.55 mm, subpersistent; fls tubular-campanulate, sessile or rarely on a stipe to 0.25 mm, calyx 1.07-2.0 mm, pubescent, corolla 1.85-2.5 mm, lobes darker in color; intrastaminal disc to 0.5 mm, stamens weakly united above nectary into a tube ± equal to corolla; ovary pubescent, on a gynophore 0.25-1.0 mm. Fruit elliptic 10.2-12.5 x 1.9-3.1 cm, flat, coarsly wrinkled, dark brown, sutures ribbed, dehiscence inert, through both sutures; seeds (few seen mature) elliptic ca. 12.5 x 7.5 mm, seed-coat shiny brown, areola lighter incolor, pleurogram centric, U-shaped.

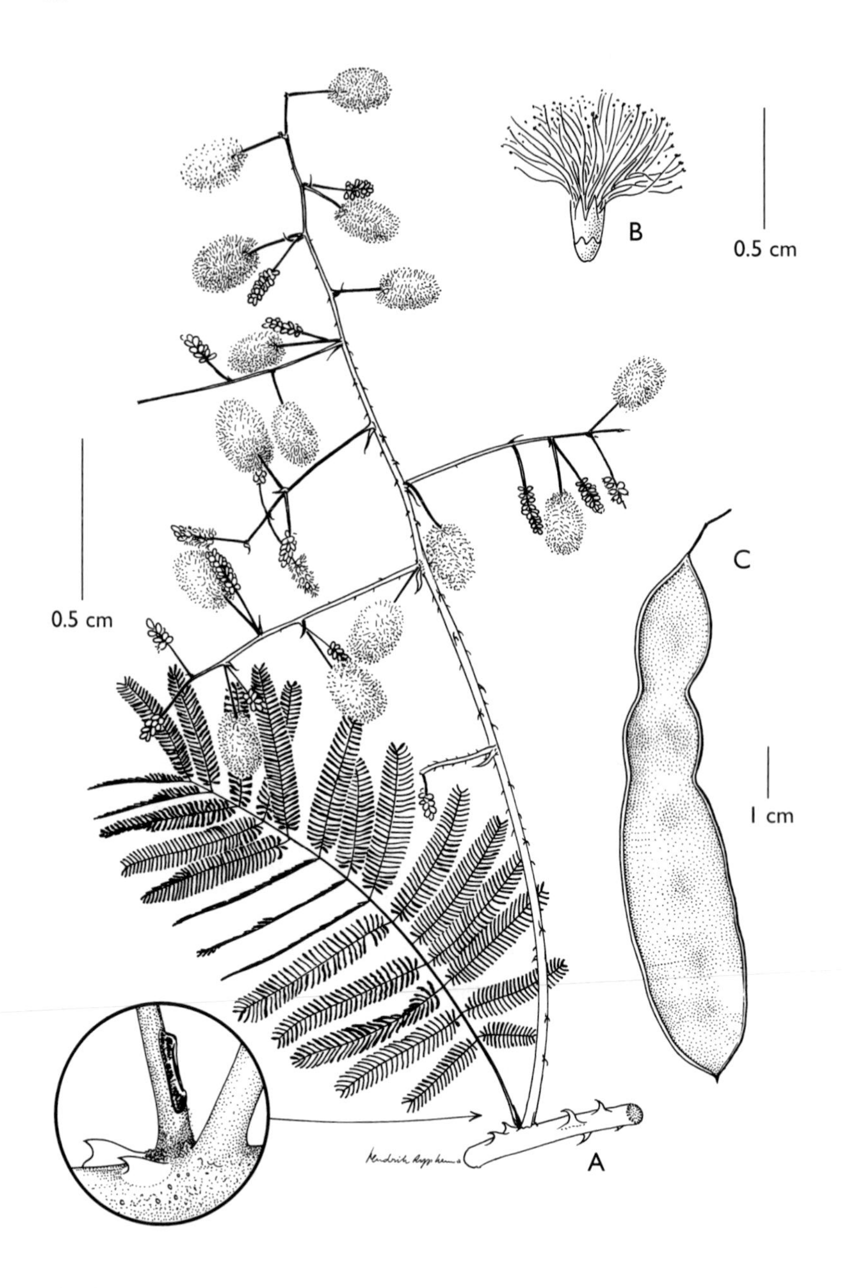

Fig. 3. *Acacia tenuifolia* (L.) Willd. var. *tenuifolia*: A, flowering branch, detail of extrafloral nectary (in circle); B, flower; C, pod (A-B, van Andel 4504; C, Jansen-Jacobs *et al*. 5825). Drawing by Hendrik Rypkema; reproduced with permission from U.

Distribution: The taxa passing under the name *Acacia tenuifolia* (L.) Willd., and it is certain there are more than one outside the Guianas, are dispersed from C America through S America to Argentina; *Acacia tenuifolia* sensu stricto is widespread in the Guianas, Irwin (1966) recognized 2 groups in the Guianas, one of which is formally recognized below as var. *producta*; most commonly on river banks, but occasionally above water-level on terra firme, from 0-1300 m; flowering May to November.

6a. **Acacia tenuifolia** (L.) Willd. var. **tenuifolia** – Fig. 3

Acacia paniculata Willd., Sp. Pl. 4(2): 1074. 1806. Lectotype: Brazil, Pará, Hoffmannsegg (B).

Receptacle capitate, to 3.5 mm diam., 13-23-flowered; gynophore 0.8-1 mm.

Distribution: Widespread in tropical and subtropical S America; collected in all 3 Guianas (GU: 5; SU: 1; FG: 4).

Selected specimens: Guyana: basin of Rupununi R., A.C. Smith 2388 (NY); Marudi Mts., Mazoa Hill, Stoffers 217 (NY). Suriname: Suriname R., Sauvain 315 (NY). French Guiana: Piste de Saül à Belizon, de Granville B4636 (NY); Bassin de Haut-Marouini, de Granville 9866 (NY).

Vernacular names: Guyana: angico. Suriname: aka maka tataj, wacht-een-beetje. French Guiana: queue lezard.

Taxonomic note: Seigler *et al.* (2006) placed *Acacia tenuifolia* (L.) Willd. as a synonym of *Senegalia tenuifolia* (L.) Britton & Rose, but Rico Arce (2007) included *Senegalia tenuifolia* (L.) Britton & Rose as a synonym of *Acacia tenuifolia* var. *tenuifolia*.

6b. **Acacia tenuifolia** Willd. var. **producta** J.W. Grimes, Brittonia 44: 267. 1992. Type: Suriname, Lucie R., Irwin *et al.*, 55373 (holotype NY).

Receptacle shortly stipitate, 3.5-9 mm, 31-43-flowered, gynophore 0.25-0.51 mm.

Distribution: Endemic to the Guianas; on riverbanks and in riverine forest, 50-200 m elev. (SU: 3; FG: 2).

Selected specimens: Suriname: Nickerie Distr., area of Kabalebo Dam project, Lindeman & de Roon 703 (NY); Bakhuis Mts., Kabalebo airstrip, Florschütz & Maas 2498 (NY). French Guiana: Cr. Grand Tamouri à 200 m de Camopi R., Oldeman & Sastre 228 (NY); Région de Paul Isnard, piste au tour de Citron, Feuillet 322 (CAY).

Taxonomic note: Rico Arce (2007) included *Senegalia tenuifolia* var. *producta* (J.W. Grimes) Seigler & Ebinger as a synonym of *Acacia tenuifolia* var. *producta*.

3. **ADENANTHERA**[7] L., Sp. Pl. 384. 1753 & Gen. Pl., ed. 5: 181. 1754. Type: A. pavonina L.

Distribution: A genus of 13 species, native in Asia, East Indies, N Australia and Pacifica; 1 species (as described below) introduced to the Americas in colonial times, now widely planted in Latin America for shade, ornament, and the decorative seeds, subspontaneous in waste places and naturalized.

LITERATURE

Nielsen, I. & P. Guinet. 1992. Synopsis of Adenanthera L. Nordic J. Bot. 12: 85-114.

1. **Adenanthera pavonina** L., Sp. Pl. 384. 1753. Type: Ceylon, Hermann (holotype BM).

Unarmed trees 4-12(20) m, with trunk exceptionally to 1 m DBH. Indument of small bronze or sordid hairs, sometimes dense on juvenile leaves, inconspicuous or almost 0 at maturity. Stipules 0. Leaves bipinnate, leaf-stalk ± 1.5-3.5 dm, pinnae 3-5 pairs, subopposite or alternate, leaflets alternate; petiolar nectaries 0; rachis of longer pinnae 8-15 cm; leaflets of longer pinnae 5-7(9) on each side of rachis and one obliquely terminal, blades broadly ovate or oblong-elliptic from inequilateral base, obtuse or minutely apiculate, larger ones ± 2-4 x 1.2-2 cm, all dull dark green above, pallid beneath, faintly penniveined from centric midrib. Inflorescence of erect, laxly many-flowered racemes axillary to coeval leaves or terminally paniculate, axis of each individual raceme, including

[7] by Rupert C. Barneby

short peduncle, ± 1-2.5 dm; bracts subulate, less than 1 mm, caducous; pedicels subhorizontal 1.3-3.5 mm, articulate close to base, ebracteolate; flowers mostly bisexual, 5-merous, 10-androus, glabrous; calyx shallowly campanulate 0.5-0.8 mm, open in bud, rim crenately denticulate; petals valvate in bud, early separating to base, elliptic acute 3.5-4 mm, yellow or greenish-ochroleucous; filaments ± as long as corolla, free to base; anthers dorsifixed, connective tipped with a caducous gland; gynoecium glabrous, ovary sessile, style scarcely exserted, stigma porate. Pod ascending or pendulous, in profile undulately linear and either evenly falcate-recurved or randomly twisted, (10)16-30 x 1.2-1.6 cm, attenuate at base, abruptly apiculate, compressed but plump, 9-13-seeded, valves firmly papery or subcoriaceous, fuscous glabrous, bullately elevated over each seed; dehiscence through length of both sutures, valves randomly twisting and coiling, exocarp separating from smooth yellowish endocarp; funicles basally dilated; seeds biconvex 8-10 mm diam, testa smooth, lustrous, scarlet with fine engraved pleurogram and thin endosperm.

Distribution: Native in SE Asia; in the Guianas known positively from Suriname (Zanderij), to be expected elsewhere in streets, gardens, and plantations, and naturalized (SU: 1).

Specimen examined: Suriname: Zanderij, ground of Land's Bosbeheer, Lems 5088 (NY).

Vernacular names: red sandalwood (former British colonies); coralitas (West Indies).

Uses: The hard strong wood is employed in carpentry and has been used in dyeing. The seeds are used as beads in necklaces and jewelry, and are reported edible after cooking.

4. **ALBIZIA**[8] Durazz., Mag. Tosc. 3(4): 10. 1772.
Type: A. julibrissin Durazz.

Unarmed, mostly deciduous trees of sympodial growth in the Guianas. Leaves alternate, bipinnate; lf-formula i-xiv(xix)/3-63; stipules usually less than 5 mm, always caducous; lft-venation palmate or palmate-pinnate. Fls heteromorphic or homomorphic, peripheral ones either sessile or pedicellate, always small, calyx less that 3.5 mm, corolla less than 6.5 mm; androecium 10-40(46)-merous, less than 2 cm long; intrastaminal disc known only in terminal flower of *A. lebbeck*. Pods

[8] by James W. Grimes

sessile or shortly stipitate, in profile broad-linear, straight or nearly so, compressed or plano-compressed, variable in texture of valves and in mode of dehiscence: a) papery and inertly dehiscent through both sutures or ventral suture only; or b) leathery, valves segmented between seeds but retained by wiry sutures (resembling pods of *Hydrochorea*); seed-funicle filiform or scarcely dilated; seed attached either basally or laterally above base, (3.5)4-11 mm, exotesta mostly light brown or ivory-white or yellowish, occasionally brown, areolate, pleurogram at middle of seed face, either U-shaped or less often complete.

Distribution: A circumtropical genus, most highly diversified in tropical America, Africa, SE Asia and Malesia, variously defined and its species consequently of indefinite number; represented in the Guianas by 4 native species in the lowland and low-montane tropics, and 1 widely naturalized species.

Taxonomic notes: Since submission of the Mimosoideae manuscript for the Flora of the Guianas, a revision of *Albizia* has been published by Barneby & Grimes (1996: 203-245). The reader is referred to this publication for additional data.
Lewis & Rico Arce (2005), give the number of species of *Albizia* as ca. 120-140 with 22 species native to tropical America.

KEY TO THE SPECIES

1 Inflorescence reduced to a simple terminal pseudoraceme; peduncles all or mostly axillary to a coeval leaf; lft-venation palmate-pinnate (cultivated in GU) . . . . *3. A. lebbeck*
Inflorescence a compound panicle or composed largely of efoliolate pseudoracemose capitula; lft-venation palmate 2

2 Leaflets of longer pinnae 8-14 pairs, longest 11-28 mm; valves of fruit tardily separating by transverse cracks into 1-seeded segments, but neither seed nor segment released from wiry persistent sutures 3
Leaflets of longer pinnae 16-63 pairs, longest 4.5-14 mm; pods inertly dehiscent through one or both sutures, or if pods as above the leaflets 20-35 pairs per pinna 4

3 Longer leaf-stalks 6-16(20)cm; rachis of longer pinnae (5)6-9(11)cm; larger leaflets 16-25(29) x 6-9(12.5) mm . . . . *5a. A. subdimidiata* var. *subdimidiata*
Longer leaf-stalks 1.5-5 cm; rachis of longer pinnae 3-5.5 cm; larger leaflets 11-14 x 3-5 mm . . . . *5b. A. subdimidiata* var. *minor*

4 Corolla externally sericeous or pilosulous, either overall, or at least beyond the tube; pod planocompressed, dehiscent through the sutures; androecium 26-30-merous . . . . *1. A. barinensis*
Corolla glabrous externally or only very thinly minutely puberulent toward tip of lobes; pod various 5

5 Androecium 20-40-merous; pod planocompressed, ripe valves papery continuous, inertly dehiscent along both sutures, these narrowly winged . . . . *4. A. niopoides*
Androecium ± 16-merous; pod lomentiform indehiscent, ripe valves transversely segemented between seeds, sutures not winged. . . . . *2. A. glabripetala*

1. **Albizia barinensis** Cárdenas, Ernstia 21: 5. 1983. Type: Venezuela, near to Punta de Piedra, Edo. Barinas, Cardenas de Guevara 2273 (isotype NY!).

Trees 6-20 m tall, with trunk attaining 3-5 dm DBH, young stems, leaves (except for glabrescent upper face of lfts) and inflorescences softly pilosulous throughout with ascending or forwardly incurved hairs. Lf-formula (vi)viii-xii(xiv)/16-24("28"); stipules linear-subulate, 1-2.5 mm, caducous, leaving a minute or barely perceptible scar; lf-stks 10-16(21) cm, petiole ± 3-4.5 cm, at middle 1-1.8 mm diam., longer interpinnal segments 12-16 mm; basalmost pinnae decrescent, others subequilong, rachis of longer ones 7-9(10) cm, longer interfoliolar segments 3.5-5 mm; lfts subsessile against rachis, dull on both faces, a little paler beneath, subequilong except at very ends of rachis, first pair reduced to linear-subulate paraphyllidia, rest oblong, larger ones 8.5-11.5(12) x 3.2-4.8 mm, 2.3-3 times as long as wide; venation of 3(4) veins from pulvinule. Primary axis of inflorescence 6-16 cm, that of branches, when present to ± 2.5 cm; peduncles (1)2-6 per node of inflorescence, longest (distal) peduncle 9-16 mm; capitula (12)16-26-flowered, globose receptacle 0.6-1 mm diam. (terminal pedestal lacking); bracts linear-spatulate or obovate, 0.5-0.8 mm, deciduous; fls homomorphic, subsessile or lowest contracted into a pedicel 0.2-0.5 mm long; perianth 5(6)-merous, finely pilosulous overall; calyx narrowly campanulate, faintly 5-veined, tube 1.4-1.8 x 1-1.2 mm; corolla 4-4.5 mm. Pod subsessile, broad-linear or linear-elliptic in profile, plano-compressed, contracted at base into a very short stipe-like neck, body (7)10-14 x 1.8-2.5 cm; dehiscence tardy, inert, through both sutures; seeds appearing horizontal but in reality pitched backward and broader than long, in broad profile elliptic, 4.5-5 x 7-9 mm, strongly compressed.

D i s t r i b u t i o n : Venezuela, middle and lower Orinoco valley, extending N to the Caribbean coast in Sucre; in seasonally dry woodland and surviving in pastures, at 200-300 m; flowering March to August (GU: 1).

S p e c i m e n e x a m i n e d : Guyana: Kanuku Mts., Iramaipang, FD 5914 (= Wilson-Browne 513) (NY).

2. **Albizia glabripetala** (H.S. Irwin) G.P. Lewis & P.E. Owen, in Legumes Ilha de Maracá 42. 1989. – *Pithecellobium glabripetalum* H.S. Irwin, Mem. New York Bot. Gard. 15: 109, fig. 5. 1966. Type: Guyana, Corantyne R., Orealla, Jenman 364 (holotype NY!).

Pithecolobium polycephalum sensu Benth., Lond. J. Bot. 5: 108. 1846, p.p. including Ro. Schomburgk ser. II, 525 (= Ri. Schomburgk 824) from Guyana, not including the type.

Trees 5-15(18) m tall, almost smooth trunk attaining 2-3 dm DBH, except for sometimes glabrate (never densely pubescent) upper face of lfts finely minutely puberulent throughout with loose pallid hairs. Lf-formula (iv)v-x/20-35; stipules (few seen) triangular-subulate, 0.4-1 mm, very early caducous; lf-stks 6-16.5 cm, petiole 2.5-4.5 cm, at middle 0.9-1.6 mm diam.; longer interpinnal segments 6-12 mm; pinnae little graduated in length, rachis of longer ones 5-8.5 cm, longer interfoliolar segments 1.2-2.2 mm; lfts dark olivaceous (brunnescent) and sublustrous above, paler dull beneath, equilong except at extreme ends of rachis and first pair commonly represented by minute subulate paraphyllidia, blades inequilaterally lance-oblong, incipiently porrect, longer ones 5-8 x 1.4-2.3 mm, 3-3.9 times as long as wide; venation of 3-4 veins from pulvinule. Primary axis and secondary divisions of inflorescence ± 8-12 cm; peduncles (1)2-3 per node, longest of each fascicle mostly 15-25 (at distal nodes somewhat shorter), capitula 9-22-flowered; fls dimorphic, peripheral ones at least shortly pedicellate, much larger central one sessile on a short pedestal, whole receptacle 1-2 mm; bracts minute, ciliolate; perianth of all fls 5-merous, or terminal fl 6-merous, glabrous except for microscopically ciliolate calyx-teeth, or for a few minute scattered hairs on corolla-lobes; peripheral fls: pedicel 0.2-1.2 mm; calyx campanulate, 0.7-1 x 0.8-0.9 mm; corolla greenish-white or faintly anthocyanic, 2.8-3.2 mm; central fl: sessile; calyx 1.4-1.6 x 1.4 mm; corolla 3.4-4.7 mm. Pod 1 or 2 per peduncle, sessile but often contracted basally, due to abortion of first seed, into a short sterile neck, in profile broad-linear, at apex broadly obtuse-apiculate, when well fertilized 11-18 x 1.5-2.2 cm, 10-14-seeded, plano-compressed, when mature transversely cracking between seeds; dehiscence not seen, much delayed or seeds perhaps released only by weathering, apparent segments not liberated from wiry sutures; seeds (few seen) transverse, in broad profile 7 x 5 mm.

Distribution: Apparently uncommon about the periphery of the Guayana Highland in Venezuela (Amazonas, Bolívar), and on upper Rio Branco in Brazil (Roraima), in Guyana on the upper Rupununi and Middle Courantyne Rs.; in low, seasonally flooded woodland, sometimes with *Mauritia* palms, on creek and river banks, and at edge of white sand savanna, below 200 m; flowering November to January (GU: 4).

Selected specimens: Guyana: Sand Creek, Rupununi R., FD 5642 (= Wilson-Browne 86) (NY); Orealla Savanna, Courantyne R., FD 5397 (= Fanshawe 2609) (NY).

3. **Albizia lebbeck** (L.) Benth., London J. Bot. 3: 87. 1844. – *Mimosa lebbeck* L., Sp. Pl. 516. 1753. Type: "Habitat in AEgypto superiore" (holotype (Brenan, Fl. Trop. E. Afr., Leguminosae subfam. Mimosoideae 147. 1959): LINN 1228.16.

Pithecolobium splitgerberianum Miq., Stirp. Surinam. Select. 5. 1851. Type: Suriname, Hostmann 459 (holotype U!)

Trees of rapid growth, mostly 3-15(18) m tall, trunk attaining 2-6 dm DBH, young branchlets and all axes of lvs and inflorescence usually finely pilosulous. Lf-formula ii-iii(iv)/4-9(10); stipules (few seen) submembranous linear-oblong, 2-4 mm, obtuse, early caducous; lf-stks (6)7-17(22) cm, petiole (3.5)4-11(12.5) cm; first (only) interpinnal segment 2-5 cm; pinnae subdecrescent proximally or subequilong, rachis of longest pair (4)6-14.5 cm, longer interfoliolar segments (1)1.2-2.2 cm; papery-membranous lfts above bright green, sublustrous, beneath paler, often glaucescent, decrescent proximally, thence subequilong, blades oblong or ovate-oblong, largest 30-48 x (8.5)11-17 mm, 1.8-3.1 times as long as wide; venation palmate and then pinnate, of 4-6 primary veins from pulvinule. Peduncles (1)2-4 per node, 3-12(13) cm; capitula 18-45-flowered, fls heteromorphic, peripheral ones pedicellate, terminal one larger and sessile, scarcely dilated receptacle 2.5-6 mm; bracts oblanceolate, linear-spatulate, or subrhombic, 1.2-4 mm, fugacious from below young fl-buds; perianth 5(6)-merous, calyx finely pilosulous, greenish corolla strigulose on lobes only; peripheral fls: pedicels (1)1.5-5.5 mm; fl-buds slenderly or plumply pyriform; calyx narrowly campanulate and often a little expanded beyond middle, 3.7-5.8 x 1.9-2.5 mm, toward base weakly 5-angulate; corolla slenderly vase-shaped,

7.7-12.3 mm; terminal fl: calyx 4.8-6 x 2.2-3.5 mm; corolla 9-14.5 mm. Pod randomly ascending, 1-5 per capitulum, sessile or almost so, in profile broad-linear or narrowly oblong-elliptic, 13.0-32 x 2.5-5.2 cm, broad-cuneate at base, rostellate or apiculate at apex, framed by essentially straight, dorsally plane or shallowly bisulcate sutures 1.2-2.2 mm wide, 8-12-seeded, valves stiffly papery, stramineous and glabrous, bullately dilated over each seed; dehiscence tardy, often after fall of pod, inert, through ventral suture only; seeds compressed but plump, in broad profile oblong-elliptic or round, ± 9-11 in greatest diam. and ± 3 mm thick, testa smooth but hardly lustrous, khaki or light reddish-brown, inversely U-shaped pleurogram occupying about one half of each face.

Distribution: Native of tropical E Asia, but now randomly circumtropical, becoming weedy in disturbed woodland and in pastureland; in the New World known from Mexico through C America to Panama, in Bermuda and Bahamas, the Greater and Lesser Antilles, and in S America from northern Colombia and Venezuela to coastal Guyana and reported (Lindeman, pers. comm.) from Suriname (GU: 1).

Representative specimen examined: Guyana: [cultivated], Jenman 2112 (NY).

Vernacular name: Guyana: woman's tongue.

4. **Albizia niopoides** (Spruce ex Benth.) Burkart, Legum. Argent. ed. 2: 542. 1952. – *Pithecolobium niopoides* Spruce ex Benth., Trans. Linn. Soc. London 30: 591. 1875. Type: Brazil, Pará, near Santarem, Spruce 1088 (holotype K! (hb. Benth.)). – Fig. 4

   Pithecolobium caribaeum Urb., Symb. Antill. 2: 260. 1900. – *Albizia caribaea* (Urb.) Britton & Rose, N. Amer. Fl. 23: 44. 1928. Type: Tobago near Aukenskeoch, Eggers 5920 (neotype K!)

Trees of 25-40 m and with a trunk 3.5-12(15) dm DBH, bark pallid smooth, peeling in flakes, young stems, ventral face of all lf-axes, and all axes of inflorescence either thinly or densely villosulous with fine, erect and incurved, pallid or yellowish hairs. Lf-formula ix-xix (seldom over xii)/28-57(63); stipules linear-subulate, 0.7-2.5 mm, very early dry fugacious; lf-stk of major lvs (4.5)6-16(21) cm, petiole (1.5)2-4.5 cm, longer interpinnal segments 8-24 cm; pinnae decrescent proximally, either furthest or next to last pair longest, rachis of longer ones 5-11.5 cm, longer interfoliolar segments 0.6-2.5 mm; narrow crowded lfts olivaceous and ± lustrous above, paler dull beneath, decrescent at base and sometimes

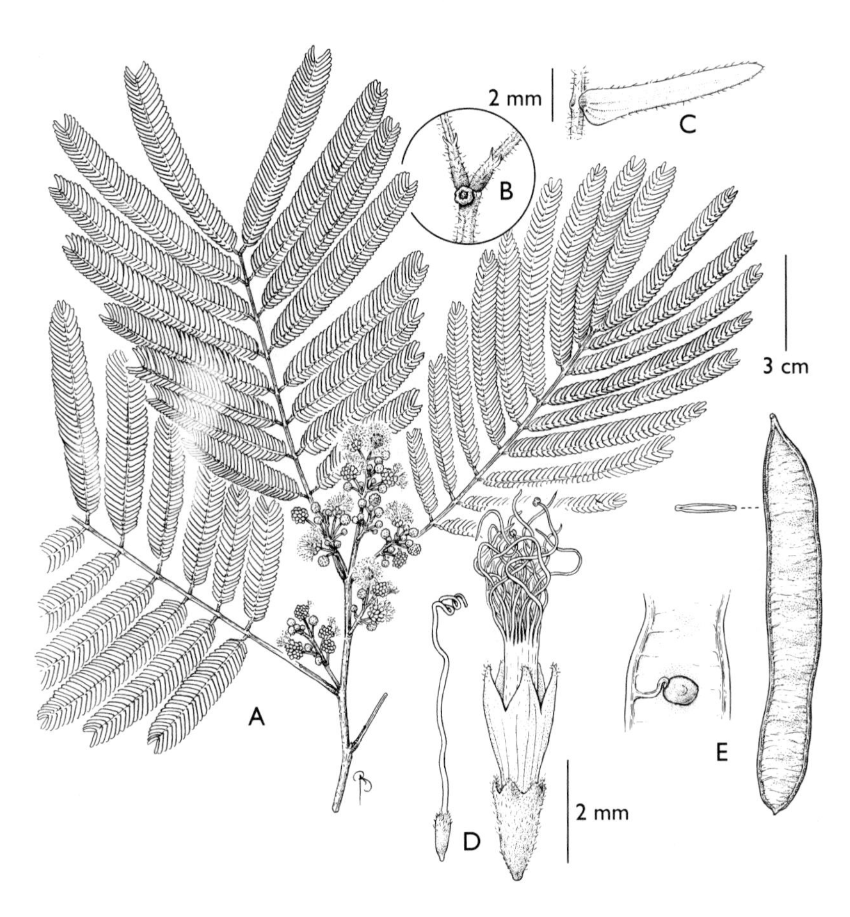

Fig. 4. *Albizia niopoides* (Spruce ex Benth.) Burkart: A, flowering branch; B, extrafloral nectary between terminal leaflet pair; C, leaflet; D, flower and gynoecium; E, pod, showing detail of immature seed (A-E, Aristeguieta 7036, from Venezuela). Drawing by Bobby Angell; reproduced with permission from NY.

also at extreme apex of pinna-rachis, otherwise subequilong, first pair commonly reduced to subulate paraphyllidia, blades linear or linear-lanceolate, larger ones 5-13 x 0.9-2.5 mm; primary venation palmate, asymmetric midrib submarginal at base of lft but becoming subcentric at its apex. Primary axis of inflorescence 1.5-6(10) cm; peduncles 2-3 per node, 4-14 (or random axillary capitula up to 25) mm; bracts spatulate, less than 1 mm, persistent; fls sessile or cuneately contracted into a very short pedicel, homomorphic or nearly so, terminal fl sometimes a trifle longer than peripheral ones but its androecium not modified except for slightly longer filament-tube, perianth greenish-white; calyx tube 0.9-

1.5(1.7) mm; corolla (2.7)3-4.5 mm, glabrous except for often thinly puberulent lobes or wholly glabrous. Pod solitary or rarely 2-3 per capitulum, sessile but abruptly contracted at base into a pseudostipe 2-8 mm long, in profile broad-linear, 8.5-16.5 x 1.4-2.6 cm, straight or nearly so, when well fertilized 9-13-seeded, body planocompressed, framed by slightly elevated sutures produced as a sharp-edged wing to (0.8)1-2.3 mm wide, dehiscence inert, through both sutures; seeds in broad profile 5.5-7.6 x 4-5 mm, pleurogram inversely U-shaped.

Distribution: Widely dispersed around, and very locally within, the Amazon Basin in Brazil, misiones in Argentina, Paraguay and E Bolivia to N Colombia, N Venezuela, Trinidad and Tobago, sparingly into the Lesser Antilles and discontinuously into S Mexico; cultivated in the Guyana Bot. Garden, Georgetown, and French Guiana (GU: 1; FG: 1).

Selected specimens: Guyana: Georgetown, Bot. Garden, Pipoly 7348 (US). French Guiana: Cr. Plomb, Loubry 1579 (CAY, NY).

Note: The var. *colombiana* (Britton & Killip) Barneby & J.W. Grimes (1996) is found in N Colombia and adjacent NW Venezuela.

5. **Albizia subdimidiata** (Splitg.) Barneby & J.W. Grimes, Mem. New York Bot. Gard. 74: 232. 1996. – *Acacia subdimidiata* Splitg., Tijdschr. Natuurl. Gesch. Physiol. 9: 112. 1842. Type: Suriname, banks of the upper Suriname R., Splitgerber 917 (holotype L!). Equated with *Pithecolobium multiflorum* by Benth., London J. Bot. 3: 220. 1844.

   Pithecolobium multiflorum sensu Benth., Trans. Linn. Soc. London 30: 591. 1875, excl. type.
   Arthrosamanea multiflora sensu Kleinhoonte in Pulle, Fl. Suriname 2(2): 326. 1940, type and some synon. excl.

Trees 5-20(25) m tall with single or multiple trunk attaining 1-3(4) dm DBH, lf-stks (at least ventrally) and axes of inflorescence always, branchlets sometimes, pilosulous with pallid hairs. Lf-formula ii-iv(v)/(7)8-13(14), in saplings pinnae sometimes up to 5-jugate; stipules deltate or triangular-ovate, 0.8-1.8 mm, very early dry caducous; lf-stks of larger lvs mostly (5)6-16(20) cm, petiole (3)3.5-7, longest interpinnal segments (1.5)2-4 cm; pinnae a little decrescent proximally, rachis of longer ones (5)6-9(11) cm, longer interfoliolar segments (5)6-11 mm; lfts glabrous facially except midrib barbellate toward base of lft on its anterior side, bicolored, olivaceous (brunnescent) and dull or sublustrous above, pallid and usually microscopically pallid-papillate beneath, lfts a little decrescent proximally, lance-oblong, oblong, semi-ovate, or asymmetrically ovate-elliptic, straight or gently incurved near apex, larger ones 16-25(29) x 6-9(12.5) mm, (2)2.2-3.3(3.6) times as long as wide; venation of (4)5-6 primary veins from pulvinule, midrib forwardly displaced to divide blade 1: 2-3. Primary axis of inflorescence 7-21 cm; peduncles fasciculate by 2-6, longer of each fascicle 3-9(11) mm; capitula (8)11-22-flowered; globose or claviform receptacle 0.6-2 mm; bracts rhombic-ovate or spatulate, 0.4-0.7 mm, persistent; fls dimorphic, central one of each head hardly longer than rest but wider and with modified androecium, perianth 5 (or that of terminal fl 6-7)-merous, all greenish-white; peripheral fls: sessile or raised on an obscure pedicel less than 0.6 mm, calyx campanulate or turbinate-campanulate, (0.65)0.7-1.2 x 0.7-1 mm; corolla (2)2.2-3.3 mm; central fl: always sessile, shallowly campanulate calyx (0.75)0.8-1.3 x 1.3-2.1 mm, corolla 2-3.7 mm. Pod solitary, subsessile or commonly contracted at base into a laterally sulcate pseudostipe 1.5-5 mm, body linear in outline but contracted between seeds, either straight or decurved, (8)9-14(15) mm wide, when well fertilized (6)8-17 cm long and (8)10-15-seeded; fruit lomentiform but not truly dehiscent, valves often cracking along sulcus between seeds but sutures continuous and attached to 1-seeded segments, seeds not individually dispersed; seeds nearly round in broad view, or obtusely quadrate or broad-ellipsoid, compressed but plump, in broad profile 5.3-7(8) mm diam.

Distribution: On river banks, in várzea forest, or at forest margins in seasonally inundated campo; NE Colombia, Venezuela (Orinoco and Maracaibo basins), lowland Guyana and Suriname, scattered through the Amazon Basin from Peru and Bolivia to Brazil (Pará and Maranhão).

5a. **Albizia subdimidiata** (Splitg.) Barneby & J.W. Grimes var. **subdimidiata**

Lvs relatively ample, longer lf-stks 6-16(20) cm, petiole (3)3.5-7 cm x 0.9-1.9 mm, rachis of longer pinnae (5)6-9(11) cm, and its interfoliolar segments up to (5)6-11 mm; larger lfts 16-25(29) x 6-9(12.5) mm.

Distribution: Range of the species (GU: 5; SU: 1).

Selected specimens: Guyana: Puruni R., FD 7729 (= JB 45) (NY); Roraima, Ro. Schomburgk ser. II, 490 (NY). Suriname: Kanariekreek, Corantijn R., Stahel & Gonggrijp BW 3029 (NY).

5b. **Albizia subdimidiata** (Splitg.) Barneby & J.W. Grimes var. **minor** Barneby & J.W. Grimes, Mem. New York Bot. Gard. 74: 234. 1996. Type: Guyana, basin of Essequibo R., Kuyaliwak Falls, A.C. Smith 2156 (holotype NY!).

Lvs smaller in all parts, longer lf-stks 1.5-5 cm, petiole 1.5-3.2 cm x 0.7-0.9 mm, rachis of longer pinnae 3-5.5 cm, longer interfoliolar segments 3-6 mm, and larger lfts 11-14 x 3-5 mm.

Distribution: In seasonally flooded savanna and at forest margins or on stream banks, apparently localized on the upper Rio Branco in Brazil (Roraima) and in the middle and upper Essequibo R. basin in Guyana (GU: 4).

Selected specimens: Guyana: S Rupununi, behind Aishalton Hospital, Stoffers *et al.* 426 (NY); N Rupununi, Davis 810 (NY).

5. **ANADENANTHERA**[9] Speg., Physis 6: 313. 1923.
Type: A. peregrina (L.) Speg. (Mimosa peregrina L.)

Piptadenia sect. *Niopa* Benth., J. Bot. (Hooker) 4: 340. 1841. – *Niopa* (Benth.) Britton & Rose, Addisonia 12: 37, pl. 403. 1927.
Type: P. peregrina (L.) Benth. [Anadenanthera peregrina (L.) Speg.]

Distribution: An American genus of 2 species, each comprising 2 varieties, widespread in continental S America and some West Indies, planted and naturalized elsewhere; 1 species in the Guianas.

Notes: Pulverized seeds of *Anadenanthera* furnish the Indians of Arawak stock their ceremonial hallucinogenic snuff and beverages.
Lack of anther-glands, to which the generic name alludes, is characteristic of *A. peregrina*, but not of the whole genus.

LITERATURE

Altschul, S. von R. 1964. A taxonomic study of the genus Anadenanthera. Contr. Gray Herb. 193: 1-65.

[9] by Rupert C. Barneby

1. **Anadenanthera peregrina** (L.) Speg., Physis 6: 313. 1923. – *Mimosa peregrina* L., Sp. Pl. 520. 1753. Type: A plant grown at Hartekamp, The Netherlands, by George Clifford (BM (Hort. Cliff.)).

In the Guianas only: var. **peregrina**

Mimosa acacioides Benth., J. Bot. (Hooker) 2: 132. 1840. Syntypes: Guyana, on upper Essequibo R., and on Rio Branco in adjoining Brazil, Ro. Schomburgk ser. I, 852 (syntype K!), ser. I, 866 (syntypes K!, NY!).

Unarmed, semideciduous, precociously flowering trees attaining ± 25 m, with roughened trunk up to 4 dm DBH and elaborately multifoliolate bipinnate leaves, new growth and all leaf- and inflorescence-axes puberulent with short fine hairs, small crowded leaflets facially glabrous, microscopically ciliolate. Stipules small, setiform, fugitive (absent from most specimens), but resting buds perulate, consisting of firm ovate scales deciduous from base of new branchlets. Lf-stks ± 1-2.5 dm, petiole ± 0.5-3 cm; a sessile elliptic, shallowly concave nectary (0.5) 1-5 mm diam. near or below mid-petiole and smaller ones on lf-stk immediately below 2-5 furthest pinna-pairs; pinnae of larger lvs 12-30-jug, (sub)opposite, rachis of longer ones ± 2.5-6.5 cm; lfts contiguous or imbricate, sessile, on longer pinnae ± 35-60-jug., blades linear or linear-lanceolate, acute, straight or subfalcate, larger ones ± 2-4 x 0.3-1.4 mm, all faintly 1-veined. Inflorescence composed of small globose capitula arising singly or fasciculate by 2-5 from lower, efoliate or hysteranthously foliate nodes of newly expanding branchlets; peduncle (1)1.5-6 cm, charged above middle (but below first flower) with a membranous, campanulate 3-4-toothed involucre, this abscissile in age but persisting as a collar threaded on peduncle; flowers homomorphic (except some functionally staminate), sessile or obscurely pedicellate, 5-merous 10-androus, perianth greenish-white or ochroleucous, minutely puberulent; calyx narrowly turbinate-campanulate 0.9-2.2 mm, bluntly 5-angulate, incurved teeth 0.25-0.45 mm; corolla 2.4-3.4 mm, lobes 0.7-1 mm, valvate in bud, remaining erect; stamens ± 7-8 mm, white filaments free to base, anthers eglandular; ovary subsessile 1-1.4 mm, glabrous or apically puberulent, style linear, 1-2 mm longer than filaments. Pod in profile broad-linear, tapering at base into a stipe, more abruptly apiculate at apex, body when well fertilized 10-22 x 1.2-1.9 cm, 10-16-seeded, either straight or gently decurved, framed by scarcely dilated sutures either shallowly constricted between seeds or (randomly) nearly straight, valves stiffly leathery, brown, glabrous, scurfy when ripe, endocarp rufous, smooth; dehiscence follicular, through ventral suture, valves narrowly gaping; seeds disciform, nearly round, nearly as wide as the pod-cavity, testa membranous, atrocastaneous, lustrous, acutely marginate but not truly winged, lacking pleurogram; endosperm 0.

Distribution: Thought by Altschul (1964) to be native only in S America, from Colombia, Venezuela and Trinidad into Brazil to Goiás, and naturalized in pre-Columbian times in Puerto Rico, Hispaniola, and Grenada, known with a few collections from the Guianas; in semi-arid or seasonally dry habitats, in savanna, at edge of gallery-forest, and in disturbed woodland, surviving deforestation in hedges and pastures, near sea level up to 1200 m; flowering with flush of new growth, dependent on seasonal rains (GU: 2; SU: 1; FG: 1).

Selected specimens: Guyana: Basin of Rupununi R., near mouth of Charwair Ck., A.C. Smith 2396 (NY); Savanna between Takutu R. and Kanuku Mts., A.C. Smith 3217 (NY). Suriname: Sipaliwini Savanna area near Brazilian frontier, Oldenburger *et al.*, 543 (NY). French Guiana: Ilet dans le saut de l'Itany, Service Forestier 7941 (CAY).

Vernacular names: Colombia and Venezuela: nopo (niopo, yopo). Brazil: parica.

Note: In the Guianas *A. peregrina* may be recognized among Mimosoideae with decamerous androecium by lack of prickles, very numerous minute leaflets, a cupular involucre above mid-peduncle, capitulate flowers, and free stamens lacking anther-glands. Its follicular pod (except for mostly undulately constricted, not straight sutures, and its thin anemochorous seeds) are most similar to those of *Pseudopiptadenia*, which differs in spicate flowers.

6. **BALIZIA**[10] Barneby & J.W. Grimes, Mem. New York Bot. Gard. 74(1): 34. 1996.
   Type: B. pedicellaris (DC.) Barneby & J.W. Grimes (Inga pedicellaris DC.)

Trees attaining 30-45 m but flowering at 10 m upward, fertile branches indeterminate, indument of young stems and lf-axes composed of short simple, usually sordid or brown hairs. Leaves alternate, bipinnate; lf-formula iii- xv/5-33; stipules linear-lanceolate or ligulate, caducous; lft-pulvinules less than 2 mm; largest lfts 6-45 mm; venation of lfts pinnate, midrib not or scarcely excentric. Receptacle of capitula, including terminal pedestal, 1.5-4.5 mm; bracts small caducous. Fls heteromorphic, peripheral ones slenderly pedicellate, mostly bisexual, terminal fl larger than rest and functionally sterile; perianth of peripheral fls 5-merous (random anomalies); filaments white, greenish-white, or roseate. Pods

[10] by James W. Grimes

(Guianan species) woody, plano-compressed, transversely fibrous, tardily follicular; seeds narrowly oblong, testa either putty-colored and ± discolored within narrowly U-shaped pleurogram or partly white and partly translucent, embryo green or anthocyanic.

Distribution: American genus with 3 species from C America (S Mexico, Belize) through Venezuela, the Guianas and the Amazon basin into the Atlantic forest of Brazil to Paraná; 1 species in the Guianas.

Taxonomic note: Rico Arce (1999), commented that her new combinations in the genus *Albizia* for *Balizia pedicellaris* and *B. elegans* (Ducke) Barneby & J.W. Grimes effectively render the whole genus *Balizia* as a synonym of *Albizia*. This generic synonymy was followed by Lewis & Rico Arce (2005, p. 211).

1. **Balizia pedicellaris** (DC.) Barneby & J.W. Grimes, Mem. New York Bot. Gard. 74(1): 37. 1996. – *Inga pedicellaris* DC., Prod. 2: 441. 1825. – *Macrosamanea(?) pedicellaris* (DC.) Kleinhoonte in Pulle, Fl. Suriname 2(2): 329. 1940. Type: French Guiana, "in Cayenna", the collector not given (holotype G-DC!). – Fig. 5

Trees attaining 40 m in height with trunk to 1 m diam DBH, bark brown-red with characteristic pockmarks; hornotinous branchlets, lf- and inflorescence-axes sordid-pilosulous or -strigulose with ascending-incurved or appressed hairs. Lf-formula vi-x (xiv)/(16)18-29(33); stipules linear-lanceolate or ligulate (2)2.5-8 x (0.3)0.5-1.3 mm, very early caducous; lf-stk of major lvs 7-18(25) cm, petiole 2-4.2(5) cm, longer interpinnal segments (7)9-17(21) mm; pinnae proximally decrescent, beyond mid-rachis subequilong, the rachis of distal ones (4)5-9(10) cm, longer interfoliolar segments 2-4 mm; lfts bicolored, marginally revolute, glabrous and often lustrous above, beneath pale brown dull, thinly pilosulous and densely pallid-papillate, strongly decrescent proximally, less so distally, first pair of ten represented by minute paraphyllidia, blades narrowly oblong, narrowly lance-oblong, or oblong-elliptic, those near mid-rachis 6-13.5(16) x 1.8-3.3(4) mm, (3)3.2-4.2(5) times as long as wide. Peduncle (2.5)3-6.5(8) cm, becoming stout and persistent into a second year; capitula ± (15)20-40-flowered, receptacle 2-4.5 mm, fls dimorphic, peripheral ones slenderly pedicellate, one or commonly 2-3 distal ones (sub)sessile, longer, and with exserted staminal tube; bracts minute, caducous long before anthesis; perianth pubescent externally, calyx sordid-puberulent overall, corolla silky dorsally, lobes white-ciliolate; peripheral fls: lower pedicels (4)4.5-7.5 mm; calyx campanulate or turbinate-campanulate (2)2.3-3 x 1.3-1.8(2) mm; corolla 5.3-7.4 mm;

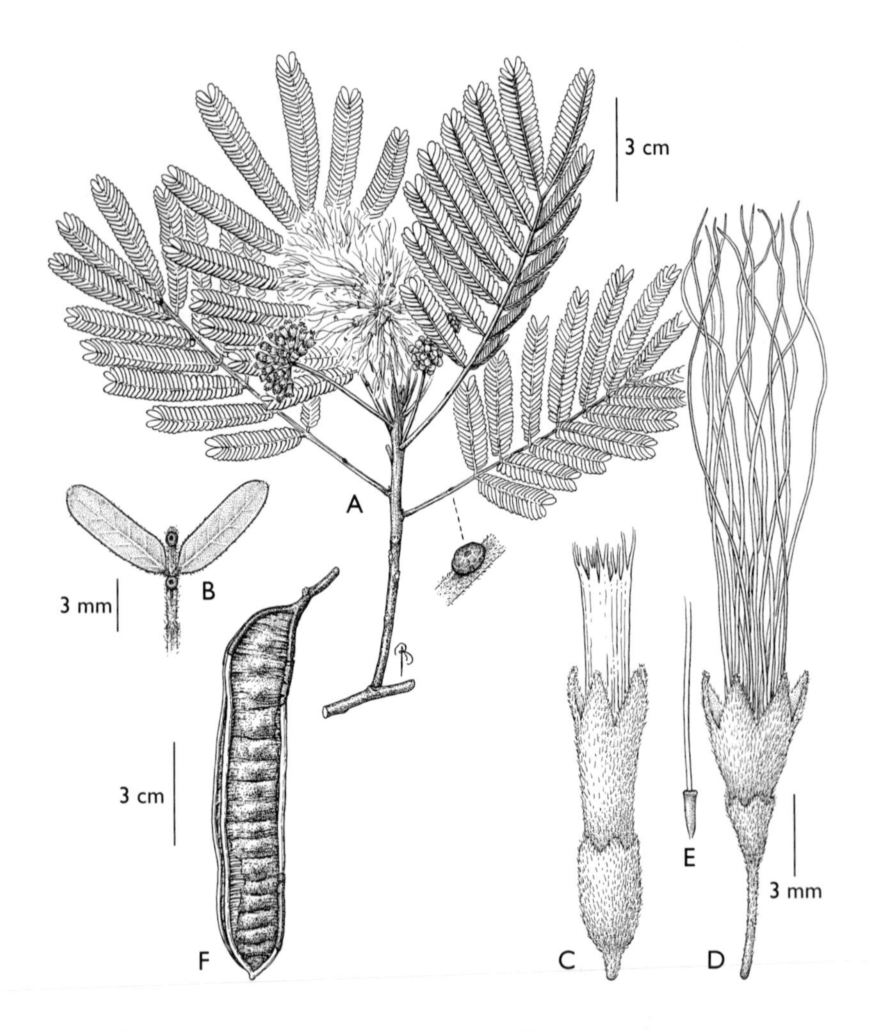

Fig. 5. *Balizia pedicellaris* (DC.) Barneby & J.W. Grimes: A, apex of flowering stem showing bipinnately compound leaves and inflorescences; note detail of extrafloral nectary near midpoint of petiole (right); B, apex of petiolule and lower part of leaflet showing extrafloral nectaries; C, distal flower showing exserted staminal tube; D, proximal flower; E, base of gynoecium; F, pod (A-E, Jansen-Jacobs *et al.* 1858; F, Stahel (Wood Herbarium Surinam) 49). Drawing by Bobby Angell; reproduced with permission from Mori *et al.*, 2002: 488.

terminal fl(s): calyx plumply campanulate from turbinate base, (3)3.4-4.6 x 2-2.8 mm; corolla (6.5)7-11 mm. Pods 1-3 per capitulum, stiffly spreading-ascending from receptacle, sessile at oblique base, in profile oblong or broad-linear, straight or almost so, abruptly contracted at apex into a linear-subulate (deciduous) beak up to 1 cm, whole 7-12.5(14) x (1.7)1.8-3.2 cm, valves lignescent, never pulpy (but sometimes resinous), at first plane and dark reddish-brown, becoming blackish-brown, framed by almost straight, plane woody sutures 2.5-4 mm wide, ripe valves consisting of: a) thin blackish exocarp breaking up into small tetragonal tesserae; b) a mid-layer of coarse transverse, parallel and subcontiguous woody fibers; and c) a crustaceous endocarp coherent between seeds but not forming septa; seeds in broad outline narrowly oblong or oblong-elliptic, (6)7.5-9 x 3-4 mm, compressed but plump.

Distribution: Discontinuously widespread in S America: relatively frequent from SE Venezuela, through the Guianas and lower Amazonian Brazil to Maranhão, through the Amazon basin to SE Ecuador, SE Colombia, Peru, and Bolivia; disjunct along the SE Brazilian coast in Bahia, Rio de Janeiro, and Sao Paulo; non-inundated primary rain-forest (GU: 5; SU: 9; FG: 19).

Selected specimens: Guyana: Takutu Cr., to Paruni R., Mozarumi R., FD 4805 (NY); Moraballii Cr., near Bartica, Sandwith 469 (NY). Suriname: Zanderij, grounds of land's bosbeheer, Lems 5089 (NY); Zanderij, Stahel 237 (NY). French Guiana: Piste-de-St-Elie, Interfluve Sinnamary/Counamama, Sabatier 1774 (CAY, NY); St. Madeleine Rd., vicinity of Cayenne, Broadway 743 (NY).

Vernacular names: Guyana: red manariballi. Suriname: manariballi herodikere. French Guiana: assao.

Taxonomic note: *Balizia pedicellaris* = *Albizia pedicellaris* (DC.) L. Rico, Novon 9(4): 555. 1999.

7. **CALLIANDRA**[11] Benth., J. Bot. (Hooker) 2: 138. 1840, nom. cons. Type: C. inermis (L.) Druce (Gleditsia inermis L.), typ. cons.

Anneslia Salisb., Parad. Lond. 64. 1807, nom. rejic.

Trees, shrubs and subshrubs, ± pubescent with simple, white or sordid hairs, or glabrous, always eglandular, in the Guianas unarmed; petiolar nectaries lacking. Lf-formula diverse and commonly intraspecifically variable; stipules in the Guianan species striately veined dorsally and persistent, covering resting buds; pinnae 1-many pairs; leaflets 1.5-many pairs, usually subsessile, venation palmate from pulvinule, or in

[11] by James W. Grimes

small leaflets reduced to midrib. Inflorescences of capitula or umbels arising axillary to coeval lvs on long-shoots and sometimes terminally pseudoracemose from suppression of distal leaves, or more commonly on axillary brachyblasts; peduncle sometimes bracteolate; bracts subtending each flower persistent at least through anthesis. Fls of each partial inflorescence homomorphic, or heteromorphic in shape and terminal ones larger in some respects, and/or peripheral ones often staminate and distal one or several bisexual; perianth 5-merous (random irregularities), calyx and corolla often striately veined or only calyx striately veined; androecium 10-polymerous, filaments united about 0.5 to twice or more length of corolla, stemonozone either obscure or forming a well differentiated hypanthium; nectarial disc lacking; anthers dorsifixed, transversely oblong to elliptic; pollen shed in 8-grained polyads unlike any other members of tribe Ingeae; ovary 1, tapering into style, stigma dilated, sometimes scarcely so, stigmatic surface either low-convex ("fungiform") or shallowly cupular, or sometimes apparently penicillate. Pods ascending on stiffened peduncle, in profile narrowly oblanceolate, straight or slightly falcate, valves plane or slightly convex, leathery or lignescent, recessed into a frame formed by massive sutural ribs; dehiscence elastic, from apex downward or from both ends, valves recurved but not laterally twisted; seeds descending on a short, dilated funicle, plumply discoid or compressed-rhomboidal, testa hard, pleurogram U-shaped or lacking.

Distribution: A tropical and marginally warm-temperate genus provisionally estimated to contain ca. 135 species, the generic limits unresolved; either all or nearly all American, most plentiful and diverse in monsoon climates and in open brush-woodland or savanna communities at low elevations; 8 species in the Guianas.

Taxonomic notes: Since submission of the Mimosoideae manuscript for the Flora of the Guianas, a revision of *Calliandra* has been published by Barneby (1998: 1-223). The reader is referred to this publication for additional data.
The number of species of *Calliandra* given above (as ca. 135) is based on Lewis & Rico Arce (2005).

Note: In the Guianas *Calliandra* may be recognized at anthesis by striate persistent stipules, basally united filaments, and lack of petiolar nectaries, and in fruit by the stipules, lack of petiolar nectaries, and the characteristic fruit.

KEY TO THE SPECIES

1 Leaves bipinnate, pinnae 1-many pairs. 2
Leaves pinnate, leaflets 1 or 2 pairs . . . . *4. C. hymenaeodes*

2 Pinnae at least 2 pairs in most or in all leaves 3
Pinnae exactly 1 pair in all leaves. 6

3 Peduncles arising from the axil of contemporary leaves or from efoliate axillary brachyblasts. . . . . *6. C. laxa* var. *stipulacea*
Peduncles disposed in terminal pseudoracemes, these efoliate except sometimes at lowest nodes; pinnae either 1-2 or more than 6 pairs; larger leaflets 3 mm wide or less 4

4 Pinnae (1)2 pairs; leaflets with ± 4 parallel veins running from base to apex; peduncles solitary or geminate; flowers subsessile, perianth densely pilosulous; hypanthium not spongy-thickened internally . . . . *5. C. rigida*
Pinnae 4-13 pairs; leaflets 1-veined; peduncles 2-6 fasciculate; flowers umbellate, pedicels 2-4.5 mm, perianth glabrous; hypanthium spongy internally. 5

5 Flowers homomorphic, 3-8 per capitulum; androecium 40-52-merous . . . . *2. C. glomerulata* var. *gomerulata*
Flowers heteromorphic, 8-18 per capitulum; androecium 11-15-merous . . . . *3. C. houstoniana* var. *calothyrsus*

6 Leaflets of each pinna exactly 3, a terminal pair and one smaller one on exterior side of rachis . . . . *1. C. coriacea*
Leaflets of each pinna 3-17 pairs . . . . *7. C. surinamensis (*including *8. C. tenuiflora)*

1. **Calliandra coriacea** (Willd.) Benth., London J. Bot. 3: 95 (with query, the plant itself unknown). 1844. – *Inga coriacea* Humb. & Bonpl. ex Willd., Sp. Pl. 4(2): 1010. 1806. Type: “Habitat in America meridionali”, Humboldt s.n. (holotype B-WILLD 19017).

Calliandra anthoniae J.W. Grimes, Brittonia 45: 25. 1993 Type: Suriname, Below Hendrik Cr., Coppename R., headquarters, Maguire 25068 (holotype NY, isotype U).
Anneslia tergemina sensu Kleinhoonte in Pulle, Flora of Suriname 2(2): 322. 1940; non *Mimosa tergemina* L. 1753; – *Calliandra tergemina* (L.) Benth., of the Lesser Antilles and Caribbean coast of South America.

Arborescent shrubs 2-20 m tall, glabrous throughout or almost so. Lf-formula i/1½; stipules all persistent, deltate, 1-2 mm long, those on short-shoots somewhat smaller, all weakly veined externally; petiole of primary lvs 1-3.5 cm; pinna-rachises (0.7)1-2 cm, odd lft of each inserted either next to or up to 4 mm distant from pinna-pulvinus; terminal pair of lfts inequilaterally elliptic from shallowly semicordate base, either obtuse-mucronulate or shallowly emarginate, ± 2-7 x 1-3.5 cm, proximal lft scarcely half as long; 3-5-veined from pulvinule, gently incurved midrib somewhat ventrally displaced. Inflorescence arising singly from condensed axillary brachyblasts, or more rarely axillary to and coeval with lvs on long-shoots, brachyblasts occasionally 2 per leaf-axil; peduncle 2-4.5 cm; capitula 11-18-flowered, receptacle 1.5-2 mm, sessile fls of subequal length but slightly broader distally, all glabrous or almost so; bracts ovate or lanceolate, 0.3-0.8 mm, persistent; fls slightly heteromorphic, terminal 1-5 with longer, somewhat more tubular corolla and longer and stouter staminal tube, female-sterile flowers apparently lacking; calyx campanulate, 1.2-2.2 mm; corolla slenderly trumpet-shaped, 8.5-11.5 (those of terminal fl to 13.5) mm, white to greenish-white(?); androecium 24-38-merous (including barren ± dilated stubs), 42-47 mm, stemonozone 0.6-0.8 mm, whitish tube 12-19 (of terminal fl(s) 24-29) mm, 2-4 mm wide at orifice, stamens deep pink or red; ovary glabrous at anthesis. Pod elastically dehiscent and with raised sutural margins (like all *Calliandra*), narrowly oblanceolate, 7-13 x 1-1.3 cm, glabrous, valves finely cross-venulose; seeds (not fully mature) oblong, 10-11.5 x 6-8 mm in outline, sublustrous testa greenish, pleurogram lacking.

Distribution: Panama, Colombia, Venezuela, Brazil (NW Pará), in the Guianas locally abundant in interior Suriname, on the Mapuera R. in Guyana, to be expected in French Guiana; on rocky river banks and islands in rapids, below 200 m; flowering May to November (GU:1, SU: 5).

Selected specimens: Guyana: Rapids below camp on Mapuera R., FD 7512 (= G 497) (NY). Suriname: Lucie R., 2 km below affluence of Oost R., Maguire *et al.* 54139 (NY); Falls in Upper Tanjimama R., Mennega 397 (NY).

2. **Calliandra glomerulata** H. Karst., Fl. Columb. 2: 5. 1862. Type: Venezuela, in mountains between Quibor and Tucujo, Karsten (holotype LE?, not seen, but illustration decisive).

In the Guianas only: var. **glomerulata**

Shrubs or trees to 15 m, inflorescences, young stems and leaf-axes red setose-strigose, fruit long red-setose strigose basally and on ribs, valves white-tomentose. Lf-formula iv-x/(20)23-29(32); stipules narrowly to broadly lanceolate-triangular or elliptic, 3-5.5 x 1.5-3.5 mm, at inception densely red-setose-strigose, becoming glabrate, indurate, persistent several years; lf-stks 4.8-8.7 cm, petiole, including enlarged wrinkled pulvinus 0.8-2.2 cm, interpinnal segments 4-8(16) mm, rachis slightly grooved ventrally, dorsally appendaged beyond terminal pair of pinnae; rachis of longer pinnae 3.1-6.4 cm, appendaged beyond terminal pair of lfts; lfts dark above, paler beneath, in outline narrowly to broadly rhombic-elliptic 3.25-9 x 1-3.5 mm, inequilaterally cordate at base, apiculate, midrib slightly ventrally displaced. Inflorescence of capitula axillary to coeval lvs or in pseudoterminal panicles, when present primary axis of pseudopanicle to 9 cm; peduncles 1-3 per node, maturing sequentially, 5.5-24 mm; receptacle globose to depressed ovate, ± 1 mm; bracts narrowly to broadly elliptic-ovate, 1-1.25 mm, those subtending lowest flowers separate but forming an involucre-like structure, these persistent, becoming reflexed; flowers 8-18 per capitulum, morphologically heteromorphic, though terminal one commonly abortive, and commonly andromonoecious, many capitula apparently lacking female fls; peripheral fls: calyx campanulate, 1.5-2.5 mm, sessile or base attenuate into a psuedostipe to 0.75 mm; corolla tubular-campanulate 5-7.5 mm, teeth ± 2 mm; androecium 11-15 merous, stemonozone 0.75-2.5 mm, staminal tube 3-6 mm, filaments white below, rose-purple above; ovary truncate oblanceolate, pubescent; terminal fl: scarcely enlarged but with longer staminal tube (to 12 mm) or much coarser and larger, and with large intrastaminal disc, female-sterile or not. Pod oblanceolate, 7.2-10 x 1.6-2.4 mm, sutural ribs straight, valves undulate, surface with transverse cracks; seed (few seen mature) broadly elliptic to ovate, 9-11 x 7-8 mm, greenish, brown-mottled.

Distribution: Throughout most of Venezuela, adj. Brazil, and in Guyana on the Kanuku Mts.; at edges of pastures, on slopes, and granitic outcrops, sometimes along streams; flowering October through March (GU: 1).

Representative specimen examined: Guyana: Western extremity of Kanuku Mts., in drainage of Takutu R., A.C. Smith 3156 (NY).

Taxonomic note: Barneby (1998: 23-26) recognised 2 varieties in *C. glomerulata*. It is var. *glomerulata*, comprising the larger leaved forms of the species, that occurs in the Guianas. The description presented here is unaltered from the original manuscript and thus includes the circumscription of both infraspecific taxa. Refer to Barneby (op. cit.) for clarification.

3. **Calliandra houstoniana** (Mill.) Standl., Contr. U.S. Natl. Herb. 23: 386. 1922. – *Mimosa houstoniana* Mill., Gard, Dict. ed. 8: Mimosa # 16. 1768. Lectotype: A fruiting specimen so annotated by H. Hernández (BM!).

   In the Guianas only: var. **calothyrsus** (Meisn.) Barneby, Mem. New York Bot. Gard. 74(3): 180. 1998. – *Calliandra calothyrsus* Meisn., Linnaea 21: 251. 1848. – *Anneslia calothyrsus* (Meisn.) Kleinhoonte in Pulle, Fl. Suriname 2(2): 323. 1940. Type: Suriname, Saramacca R., on mountains near Mariepaston Cr., Kegel 1465 (holotype NY, isotype GOET acc. to Breteler, Acta Bot. Neerl. 38(1): 79. 1989).

Shrubs and slender treelets 2-6(8) m tall, young stems either glabrous or appressed-pilosulous. Lf-formula (vi)ix-xviii/(30)34-62; stipules narrowly lanceolate 3-6 mm, ± 5-veined, deciduous; lf-stks (7)8-15 cm, petiole 1.5-3.5(4) cm, longer interpinnal segments 6-10(12) mm; pinnae subequilong, rachis of longer ones (4)5-8.5 cm, longer interfoliolar segments 0.7-2 mm; lfts darker green above, either glabrous on both faces or remotely strigulose dorsally, ciliolate, linear from shallowly obtusely auriculate base, straight, acute or acuminulate, 4-9(10) x 0.7-1.9 mm, 1-veined, midrib simple, subcentric at mid-blade. Inflorescence a terminal efoliate pseudoraceme of fasciculate, shortly pedunculate, umbelliform capitula produced well above foliage; axis of pseudoraceme 3-17 cm; peduncles 2-6 per node, (3)4-12(15) mm, each peduncle subtended by a pair of caducous stipular bracts and a rudimentary leaf; capitula umbellately 3-8-flowered, receptacle at anthesis flat or low-convex; bracts subtending each flower lanceolate, 0.5-1 mm, persistent through anthesis; flowers homomorphic; pedicels at anthesis 2-4.5 mm; perianth glabrous, greenish, ± maroon-tinged, prior to anthesis pyriform; calyx cupular, 1.8-2.3 x 2.5-3 mm, corolla 8-9 mm, ovate-oblong lobes 4-5 mm at anthesis, often further separating with age; androecium 40-52-merous, ± 4-5 cm, stemonozone 2-2.3 mm, lined with spongy tissue, tube ± 2-2.5 mm, filaments deep reddish-purple; ovary sessile, narrow-oblong, glabrous; stigma dilated to 0.5 mm diam. Pod often several per capitulum, in profile narrowly oblanceolate, 8.5-11 x 1.2-1.6 cm, glabrous or remotely inconspicuously strigulose, sutural rim smooth, valves openly and weakly cross-venulose; funicle broadly dilated; seeds (very few seen mature) obovate, ± 10 x 5 mm, pleurogram narrowly U-shaped.

Distribution: C America, cultivated and weedy in the Paleotropics; first recognized as a distinct taxon from specimens collected in 1846 by H.A.H. Kegel near the Saramacca R. in Suriname, where not seen again and hence surmised to have been introduced from C America; in brush-woodland, becoming colonial in disturbed places, along stony river banks, on roadsides, and in wasteland, native at 50-1300 m (SU: 1).

Taxonomic note: The manuscript submitted by Grimes recognised this taxon as *Calliandra calothyrsus* Meisn.

4. **Calliandra hymenaeodes** (Pers.) Benth., Trans. Linn. Soc. London 30: 537. 1875, as 'hymenaeoides'. – *Mimosa hymenaeodes* Pers., Syn. pl. 2: 262. 1806. Type: French Guiana, Cayenne, L.C. Richard (holotype P).

   Calliandra patrisii Sagot, Ann. Sci. Nat. VI, 13: 324. 1882. Type: French Guiana, Patris s.n. (holotype G-DC).

Trees of unknown stature, unique among neotropical *Calliandra* for pinnate leaves, plants nearly glabrous except for ventrally puberulent lf-stks and minutely ciliolate lfts. Lvs 2- or 4-foliolate; stipules lance-ovate, 2-5 x 1-1.5 mm, several-veined dorsally, persistent; lf-stks 6-27 mm, petiole 6-10 mm, one interfoliolar segment, when present, less than 2 cm, rachis wide, shallow, ventral groove appendaged beyond terminal pair of lfts; lfts obliquely ovate from inequilateral, broadly flabellate base, obtuse apiculate, larger ones 3.5-5.5 x 2.0-3.3 cm; 4-5-veined from pulvinule, forwardly incurved midrib ventrally displaced. Inflorescence 1-2 per lf-axil, sometimes on incipient brachyblasts; peduncle ± 2-5 cm; receptacle subglobose, less than 2 mm diam; bracts 0.5-0.75 mm, persistent through anthesis; fls sessile or almost so, heteromorphic, terminal 1(2?) somewhat enlarged, perianth subglabrous, corolla charged with a few scattered appressed hairs; calyx tubular-campanulate, 1.4-2.5 mm (that of terminal fl to 2+ mm), tube striately ± 15-veined; corolla narrowly trumpet-shaped, 6.5-10 mm, that of terminal fl(s) broadly trumpet-shaped and 10-11 mm long; androecium 12-16-merous, ± 33 mm, stemonozone 1.8 mm, staminal tube 12.5-15 mm (that of terminal fl(s) 17-19 mm) mm, tassel pink distally; ovary elliptic, glabrous, stigma not seen. Fruit not seen.

Distribution: Endemic for the Guianas, known from the Bartica-Potaro Road in Guyana, the Nassau Mts. in Suriname, and from near Cayenne, French Guiana; at forest edge; flowering May (GU: 1; SU: 2; FG: 3).

Selected specimens: Suriname: Nassau Mts., Lanjouw & Lindeman, 2568 (U). French Guiana: près de la rivière des Cascades, Grenand 2075 (CAY); Route du Tour de l'Ile, Région de Cayenne, Feuillet 3007 (CAY, U).

5. **Calliandra rigida** Benth., London J. Bot. 5: 103. 1846. – *Calliandra hookeriana* Ro. Schomburgk, Linnaea 20: 754. 1847, nom. substit. illegit. Type: Guyana, on the Carimani or Kamarang R., tributary to the Mazaruni R., Ro. or Ri. Schomburgk s.n. (holotype K, hb. Benth.).

   Calliandra pakaraimensis R.S. Cowan, Mem. New York Bot. Gard. 10: 142. 1958. Type: Guyana, Mt. Ayanganna, Pakaraima Mts., Maguire *et al.* 40561 (holotype NY).

Shrubs 1.5-2 m tall, except for facially glabrous but ciliolate lfts pilosulous throughout with gray hairs to 0.5-0.8 mm, coarse brown young twigs clothed in retro-arcuate multifoliolate lvs. Lf-formula (i)ii/27-35; stipules firmly papery, lance-ovate or ovate-acuminate 2-11 mm, densely pubescent externally, glabrous striate within, deciduous from a lunate scar; lf-stks ±15-21 mm, petiole 8-11 mm, one interpinnal segment nearly as long, ventral groove bridged at insertion of pinnae; rachis of distal pinnae 7-9.5 cm, interfoliolar segments 2-2.5 mm, lfts contiguous or narrowly imbricate; lft-pulvinules transversely lunate, 0.2-0.4 mm, blades sessile against rachis; lfts decrescent at each end of rachis, subequilong otherwise, blades linear from obtusely auriculate base, obtuse and slightly porrect at apex, those near mid-rachis ± 14-18 x 2-3 mm, all brown sublustrous ventrally, paler dull-olivaceous dorsally, 6-7-veined from pulvinule but 2-3 posterior veins very short, rest parallel, simple, produced to blade's apex, all immersed on upper face, prominulous beneath. Inflorescence a terminal, short and dense, efoliate pseudoraceme of hermispherical capitula; primary axis of pseudoraceme 4-11 cm, solitary or geminate peduncle stout, 12-23 mm, lowest sometimes subtended by a depauperate lf but rest by a stipuliform bract, pyriform receptacle 2-3 mm; bracts subtending 1 or more lowermost fls stipuliform, rest much smaller; calyx subsessile, turbinate-campanulate, ± 3 mm, densely pilosulous externally, lance-triangular teeth nearly as long as tube; corolla turbinate-campanulate, 8-8.5 mm, lobes ovate, ± 3 x 2 mm, weakly 3-veined; androecium 20-merous, ± 3.5 cm, stemonozone ± 1.7 mm, tube ± 3.5 mm, free filaments red; ovary slenderly elliptic in profile, pilosulous laterally; style to 6 cm, slightly dilated stigma scarcely 0.1 mm diam. Pod 1-3 per capitulum, ascending, in profile narrowly oblanceolate, 8-9.5 x 0.8-1 cm, straight, valves dilated frame and almost plane, alike gray-pilosulous overall, red-brown valves in addition densely minutely granular-papillate.

Distribution: Known only from the Pakaraima Mts. in Guyana; in scrub woodland, sometimes on cliffs, at 500-550 m; flowering February (GU: 3).

Specimen examined: Guyana: Cuyuni-Mazaruni Region, Chi-Chi Mt. Range, Pipoly 10251 (NY, US).

T a x o n o m i c   n o t e s : The manuscript submitted by Grimes considered *Calliandra rigida* and *C. pakaraimensis* to be conspecific. Barneby (1998: 193-196) retains the two as distinct species and both occur in Guyana. *C. rigida* is readily distinguished from *C. pakaraimensis* by its glabrous perianth, smaller leaves and leaflets, much smaller flowers and 6-7-merous androecium. The description given above encompasses both species.
The collection Pipoly 10251 is *Calliandra pakaraimensis* according to Barneby (1998).

6. **Calliandra laxa** (Willd.) Benth., Trans. Linn. Soc. London 30: 551. 1875. – *Acacia laxa* Willd., Sp. Pl. 4: 1069. 1805. Type: Venezuela, near Caracas, Bredemeyer 15 (holotype B-Willd. 19148).

   In the Guianas only: var. **stipulacea** (Benth.) Barneby. – *Calliandra stipulacea* Benth., J. Bot. (Hooker) 2: 137. 1840. Type: Guyana, on the Quitaro R. [Kwitaro], Ro. Schomburgk ser. I., 582 (holotype K! (hb. Benth.), isotype NY!).

Shrubs and small trees, 3-4 m tall, when young all parts save leaflets, bracts and calyx-tubes covered with sordid white to white spreading-incurved or straight hairs, stems tardily glabrate with age. Leaves long-persistent, commonly subtending brachyblasts of apparently many seasons' growth; lf-formula iii-v/(8)9-15; stipules firm, erect, lanceolate to ovate, 3-8.5 x 1-3.25 mm, striately veined, glabrous (with age?), pubescent at base, or pubescent overall, long-persistent; lf-stks of mature lvs (3)3.5-8(9) cm, petiole 1.2-2.5 cm, interpinnal segments 0.8-2 cm; pinnae subaccrescent distally, rachis of longer ones 4-8 cm, longer interfoliolar segments 5-7.5 mm; lfts slightly distally accrescent or not, broadly or narrowly oblong or ovate-oblong from a broadly semicordate base, obtuse of obtuse-apiculate, larger ones 13-22 x 4-8 mm, ciliate; 3-4-veined from pulvinule, midrib at least somewhat ventrally displaced, weakly branched, tertiary veins, when obvious, pinnately branched. Inflorescence coeval with lvs on long shoots, or more commonly on short, mostly efoliate brachyblasts with persistent stipules; peduncle 1.8-6 cm; bracts ± 1 mm, ciliate, deciduous; fls greenish-white, heteromorphic, distal one sessile, scarcely longer than peripheral ones, but wider and more tubular, and androecial tube ± 2.7-3.0 cm; peripheral fls shortly pedicellate (± 1 mm) or sessile toward apex of receptacle; calyx 2-3 x 1.7-2 mm, teeth 0.6-1 mm; corolla 5-7.3 mm; androecium (9)12-merous, 32-35 mm, stemonozone 1-1.4 mm, tube 4-5.5 mm, scarcely if at all exserted, filaments white at base, purplish distally; ovary glabrous at anthesis. Pod stiffly ascending,

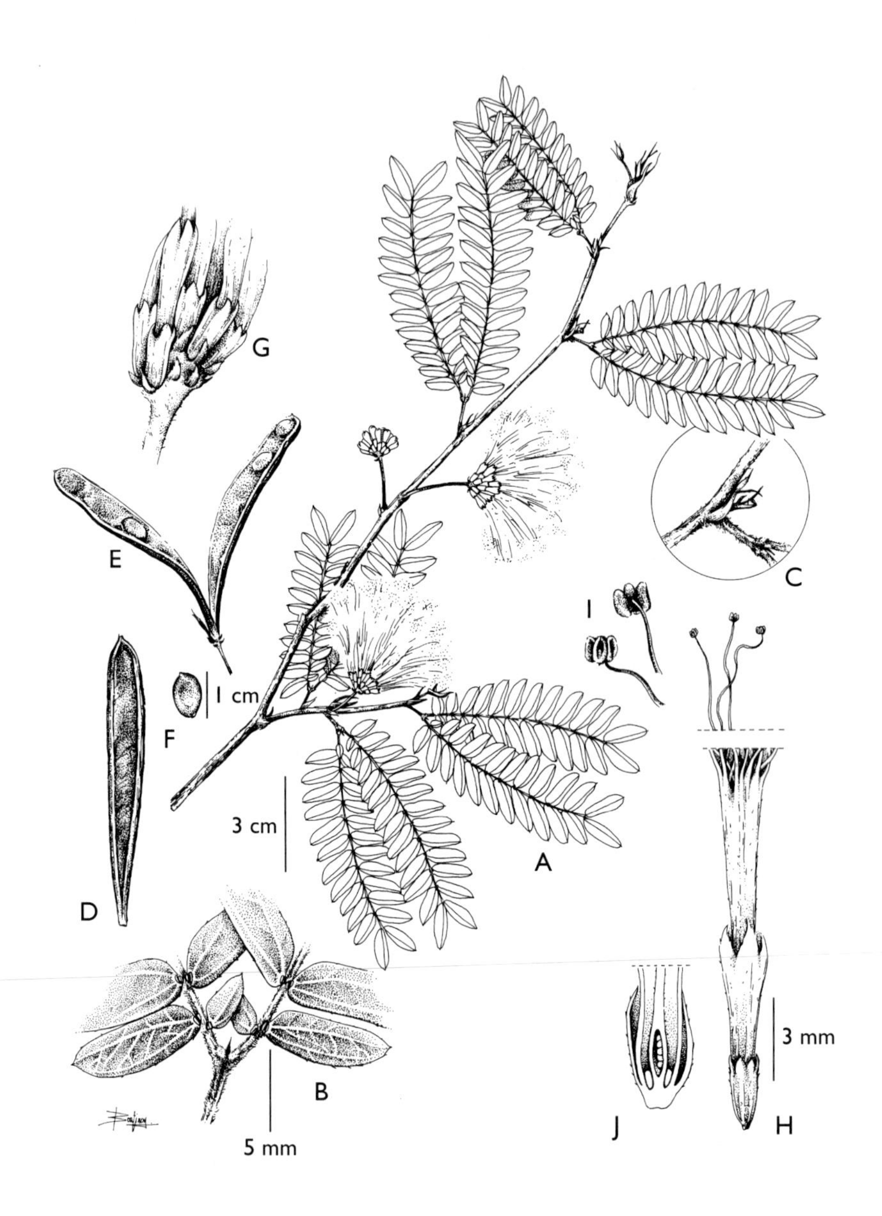

Fig. 6. *Calliandra surinamensis* Benth.: A, habit: leaves and inflorescences; B, leaf: detail of leaf-stalk, petiolules, leaflets; C, detail of stipules; D, pod; E, pod, view of the mature pod after dehiscence; F, seed; G, inflorescence, detail; H, flower; I, anthers; J, longitudinal section of basal part of flower. Drawing by Nadège Boutinon; reproduced with permission from P.

in profile narrowly oblanceolate, 9-13 x 1-1.3 cm, thinly puberulent overall, sutures low-carinate dorsally, smooth, valves recessed; seeds (immature) ± 11 x 7 mm in profile, pleurogram lacking.

Distribution: Amazonian Colombia and Brazil, Venezuela, and in Guyana known from the Potaro-Siparuni and upper Rupununi R. basins; in open forest and savanna margins at 100-750 m; flowering December through February (GU: 4).

Selected specimens: Guyana: Dadanawa, Rupununi R., Jansen-Jacobs *et al.* 2033 (NY); Potaro-Siparuni Region, Hahn 5608 (NY).

Note: Ducke (Bol. Técn. Inst. Agron. N. 36: 49. 1958) reported *Calliandra glomerulata* H. Karst. from the Kanuku Mts., but did not cite any specimens. This is probably a misidentification of *C. laxa* var. *stipulacea*, which is similar in leaf-formula and which has been collected close to the Kanuku Mts. by Jansen-Jacobs.

7. **Calliandra surinamensis** Benth., London J. Bot. 3: 105. 1844. Type: Suriname, Hostmann 171 (holotype K! = photo NY!, isotypes BM!, NY!, OXF!). – Fig. 6

Inga fasciculata Willd., Sp. Pl. 4: 1022. 1806. – *Anneslia fasciculata* (Willd.) Kleinhoonte in Pulle, Fl. Suriname 2(2): 322. 1940. – non *Calliandra fasciculata* Benth., J. Bot. (Hooker) 2: 140. 1840. Type: Brazil, Pará, [Sieber], Hoffmannsegg (holotype B-WILLD 19048, seen in microfiche!).

Arborescent shrubs flowering when (0.7)1-5 m tall, at maturity broad-crowned with widely spreading or horizontal branches, annotinous and older stems decorticate, pallid terete, long-shoots mostly barren but giving rise to fertile, either few-lvd or efoliate, axillary brachyblasts commonly clothed in imbricate stipules; annotinous long-shoots and lf-axes pilosulous-hirsutulous with sordid, forwardly appressed or spreading hairs up to 0.15-0.6 mm, thin-textured, facially glabrous, sometimes ciliolate lfts moderately bicolored, ordinarily lustrous on both faces but more so on upper. Lf-formula i/8-15(17); stipules narrowly ovate or lance-acuminate, firm, 3-several-veined dorsally, persistent; petioles (1)2-8(11) mm; rachis of longer pinnae (3.5)4-9.5 cm, longest interfoliolar segments 4-8 mm; lfts either a little decrescent only proximally and thence subequilong, or decrescent near each end, but furthest pair nearly always longer and proportionately narrower than penultimate pair, blades (disregarding terminal pair) varying from asymmetrically lance-elliptic to narrowly rhombic or rhombic-oblong from semicordate base, obtuse-apiculate, penultimate pair in well-developed leaves 9-21(23) x 3.5-5.5 mm; venation of 3-4 primary veined from pulvinule, shallowly sigmoid midrib subcentric at mid-blade or only slightly displaced, inner posterior primary vein produced at least to mid-blade and often further, outer primary ones short and weak, secondary and reticular venules prominulous on upper face, often less so beneath. Inflorescence of either hetero- or subhomomorphic flowers solitary, fls all ascending; peduncle 5-16 mm, mintuely bracteolate near or below middle; capitula 12-24-flowered, receptacle 0.5-2 mm, fls subsessile or raised on a thickened fuscous pedicel to 0.6 mm; bracts ovate or lanceolate 0.4-1.6 mm, persistent; fls subhomomorphic but peripheral ones mostly staminate, one (or several) furthest not or scarcely longer but mostly bisexual and with androecial tube more dilated; perianth glabrous or thinly brownish-strigose-pilosulous, calyx striate but corolla almost evenulose; peripheral fls: calyx deeply campanulate or a little dilated at mouth 2.1-3.2 x 0.6-1.2 mm, teeth obtuse, often unequal, 0.15-0.5 mm; corolla narrowly trumpet-shaped 6.3-8.5(-12, in Brazil) mm, ovate lobes 1-1.6 mm; androecium 10-16-merous, 34-50 mm, pallid tube 9-15 mm, exserted 1-7 mm from corolla, tassel purple-pink or red-purple; ovary of bisexual fls sessile, linear-ellipsoid, glabrous. Pod 5-11 x 0.6-0.9 cm, glabrous or thinly remotely brown-pubescent, valves transversely venulose; seeds brown or speckled, pleurogram U-shaped.

Distribution: Widespread in northern and equatorial S America, cultivated and occasionally naturalized; riverine forest, savanna and secondary scrub woodland, between 10 and 750 m; flowering October through April (GU: 3; SU: 1; FG: 2).

Selected specimens: Guyana: Barima-Waini Region, Matthews Ridge, Pipoly 8382 (NY, US); area of Ayanganna, FD 7938 (= RB 114) (NY). Suriname: Hostmann 171 (NY-fiche). French Guiana: Cr. Gabaret, Bassin de l'Oyapock, Cremers 9966 (CAY, NY); Mts. des Nouragues, Bassin de l'Approuague-Arataye, Larpin 704 (CAY, NY).

Taxonomic notes: Barneby (1998: 76) places *Calliandra tenuiflora* Benth. as a synonym of *C. surinamensis*. Other binomials placed as synonyms of *C. surinamensis* by Grimes in the manuscript submitted to the Flora have subsequently been recognised as distinct taxa by Barneby (1998). Those that represent additional *Calliandra* species records for the Guianas are:

1) *Calliandra riparia* Pittier, Arb. Arbust. Venez. 6-8: 80. 1927. Type: Venezuela, Rio San Juan, near San Juan de los Morros, Pittier 12309 (isotype NY!).
2) *Calliandra purpurea* (L.) Benth., London J. Bot. 3: 104. 1844, based directly on *Inga purpurea* (L.) Willd., Sp. Pl. 4(2): 1021. 1806. Type:

"Habitat in Martinica," which ≡ *Mimosa purpurea* L., Sp. Pl. 517. 1753. – "habitat in America meridionali." (holotype: Plumier`s figure of *Acacia frutescens non aculeata, flore purpurascente* in Plumier ed. Burmann, Pl. Amer. t. X, fig. 2). Refer to Barneby (1998: 78) for additional synonymy.

8. **Calliandra tenuiflora** Benth., Trans. Linn. Soc. London 30: 547. 1875. Type: Brazil, Rio Tapajoz near Santarem, Spruce 389 (holotype K!, isotype NY!).

Arborescent shrubs, but potentially attaining 8-12 m, lfts either glabrous or glabrous ciliolate, or exceptionally (in Suriname) papillate on upper face and pilosulous, especially along veins, beneath. Lf-formula i/5-9(11); stipules of *C. surinamensis*; petioles 2-15(18) mm; rachis of longer pinnae 4-9(11) cm, longer interfoliolar segments (6)7-12(18) mm; lfts distally accrescent, blades obtusely rhombic from semicordate base, deltate-apiculate, penultimate pairs in well-developed lvs 15-2(37) x (7)8-16(20) mm. Peduncle 5-18 mm; capitula and perianth of *C. surinamensis*, glabrous or puberulent calyx 1.9-3.4 x 0.7-1.4(1.7) mm, glabrous or commonly thinly brown-strigulose corolla 6.3-9(12) mm; androecium 10-14(28)-merous, tube 9-16 mm, exserted from corolla 1-7(8.5) mm; ovary of bisexual flowers linear-ellipsoid, glabrous. Pod either glabrous or thinly remotely pilosulous, valves coarsely venose lengthwise; seeds light brown speckled with darker brown, in broad view ± 10-11 x 5.5-7.5 mm, pleurogram lacking.

Distribution: Brazil, along the Amazon R. from the delta west to Peru, middle Orinoco valley in Venezuela, Guyana (Chi-Chi Mts.), Suriname (Lucie and Tapanahoni Rs.), French Guiana (Ouaqui and Inini Rs.); in savanna, second-growth woodland on terra firme, Amazonian restinga, and on rocky river banks, in Guyana up to 450 m elev.; flowering throughout the year (GU: 1; SU: 3; FG: 2).

Selected specimens: Guyana: Cuynui-Mazaruni Region, Chi-Chi Mts., Pipoly 10325 (NY, US). Suriname: Confluence of Paloemeu and Tapanahoni Rs., Wessels Boer 1327 (NY); Lucie R., Wilhelmina Gebergte, Irwin *et al.* 55414 (NY). French Guiana: Fleuve Ouaqui, forêt, rive gauche, de Granville B4909 (CAY); R. Grand Inini, en amont de Bicade, Oldeman B3531 (CAY).

Taxonomic note: *Calliandra tenuiflora* Benth. should be considered as a synonym of *C. surinamensis* following Barneby (1998: 76).

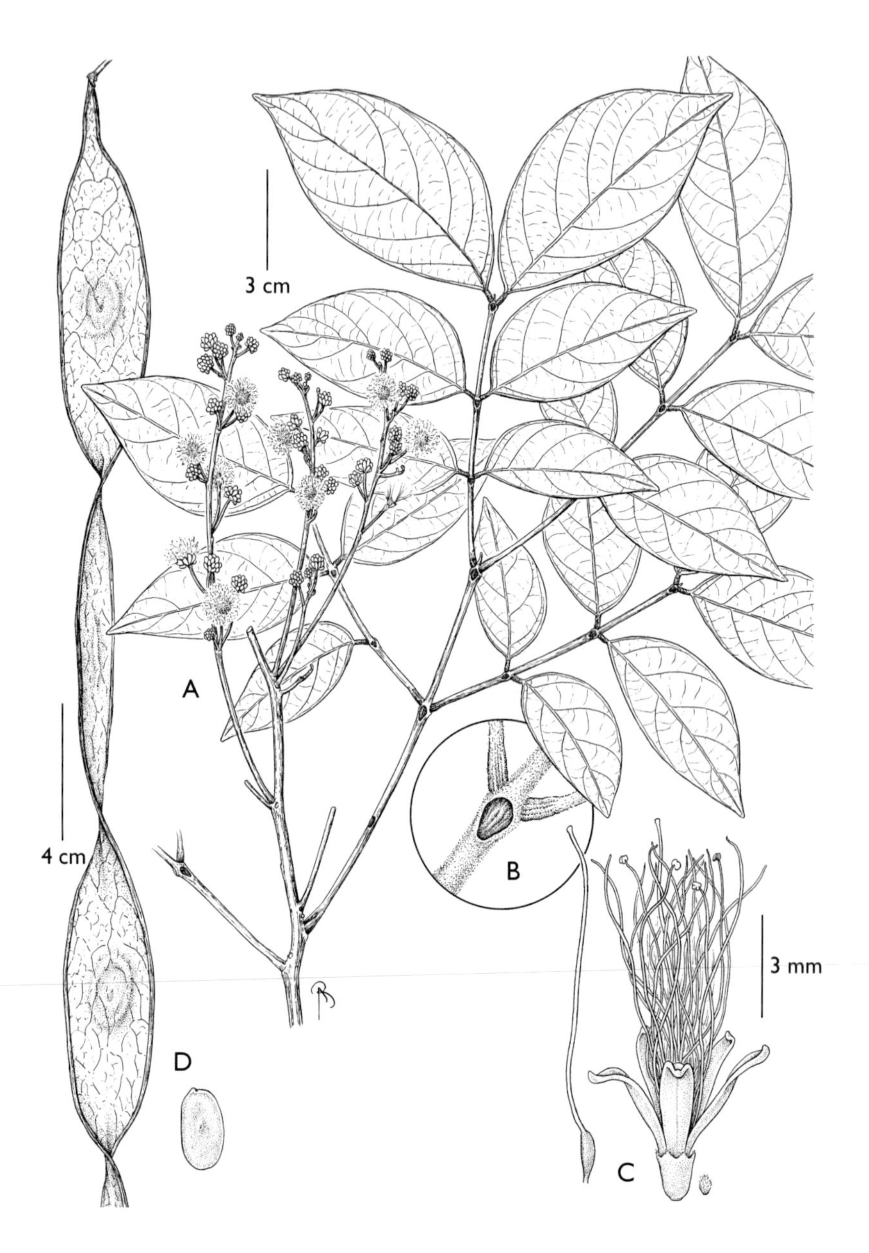

Fig. 7. *Cedrelinga cateniformis* (Ducke) Ducke: A, flowering branch; B, extrafloral nectary; C, flower; D, pod and seed (A-C, SEF 8949; D, Neill 7131; both from Ecuador). Drawing by Bobby Angell; reproduced with permission from Mori *et al.*, 2002: 490.

8. **CEDRELINGA**[12] Ducke, Arch. Jard. Bot. Rio de Janeiro 3: 70. 1922. Type: C. cateniformis (Ducke) Ducke (Piptadenia catenaeformis Ducke)

A monotypic genus of uncertain affinity. Along with its great size it may be known at anthesis by unusual asymmetry of ample, finely reticulate leaflets, primary vein at once arched backward and posteriorly displaced from mid-blade, venation pinnate; by a terminal panicle of pseudoracemose capitula, and by corolla lobes separating downward as far as stemonozone, therefore at maturity lacking a tube free from androecium. Fruit very large, papery lomentiform, twisted through about 90° at each interseminal isthmus, it is unique and unmistakable.

Distribution: SE Colombia, S Venezuela, Suriname, French Guiana, E Ecuador, E Peru, Brazil (Acre, Amazonas, Mato Grosso, Pará); 1 species.

1. **Cedrelinga cateniformis** (Ducke) Ducke, Arch. Jard. Bot. Rio de Janeiro 3: 70, pl. 6. 1922, as 'catenaeformis'. – *Piptadenia catenaeformis* Ducke, Arch. Jard. Bot. Rio de Janeiro 1: 17. 1915. Syntypes: Brazil, marshy banks of river forest near Oriximina below the Rio Trombetas, Ducke HAMP 15.704, and near Obidos, Ducke HAMP 15.710 (syntypes BM!, P!, RB!). – Fig. 7

Trees of potentially great size, fertile at 30-60(66) m, with long straight, smooth but vertically furrowed, reddish, trunk buttressed at base and up to 1-2(rarely 3) m DBH, young branchlets and axes of inflorescence densely minutely brown-puberulent, foliage glabrous, subconcolorous. Lf-formula ii-iii/2-4(5), lfts 20-36 per lf; stipules lacking; lf-stks (6.5)8-16(22) cm, petiole including firm pulvinus (3)4-8 cm, one or longer of two interpinnal segments (3)4-7.5 cm; petiolar nectaries at or close below insertion of each pinna-pair, low-convex and wrinkled, or sunk into epidermis and amorphous, similar but smaller nectaries between furthest 1-2 pairs of lfts; pinnae a little accrescent distally, rachis of longer ones 6-12(16.5) cm, one or longer of 2-3 interfoliolar segments (2)2.5-5 cm; lfts a little or distinctly accrescent distally, asymmetrically ovate or ovate-elliptic from postically cuneate, antically broader and rounded base, distal pair 7-12(13) x 3.5-5.7 cm, 1.8-2.6 times as long as wide. Primary axis of inflorescence ± 1-1.5 dm; peduncles 2-6 per node, longest of a fascicle 5-16 mm; capitula hemispherical, densely 8-20-flowered, receptacle 1-2 mm; bracts ovate or obovate-spatulate, 0.3-0.7 mm, puberulent, persistent; fls all sessile, homomorphic, 5-merous

[12] by James W. Grimes

perianth greenish-white (brunnescent when dried), glabrous except for minutely ciliolate calyx-teeth and sometimes papillate tip of corolla-lobes; calyx short-campanulate, 1-1.4 x 1.1-1.3 mm, depressed-deltate teeth 0.2-0.25 mm; corolla 4-4.6 mm, lobes ovate ± 1 x 0.9 mm at early anthesis, later separating more deeply, finally free beyond stemonozone; androecium whitish, 24-30-merous, 9-11 mm. Pod one per capitulum, pendulous, lomentiform, 20-70 cm long in outline, broad-linear but deeply constricted between each of 2-5(6) very large planocompressed 1-seeded articles, these oblong-elliptic 10-15(17) x (3.3)3.5-6 cm, and isthmi 3-18 mm wide, body twisted through ± 90° at each isthmus but plane and straight between them; indehiscent, 1-seeded segments (resembling whole pods of *Platymiscium*) breaking apart by transverse fission at isthmi and shed individually; seeds in broad outline elliptic, 25-32 x 14-18 mm (orbicular 10-11 mm diam), ± 1.5 mm thick, testa thin dry translucent, when ripe fragile, exareolate.

Distribution: Scattered through the Amazon Delta region in Brazil, Pará, S and W in the Amazon basin to Acre, Mato Grosso, to Ecuador, Peru, N into Venezuelan Guayana, Suriname, and French Guiana; in either wet or seasonally dry forest, especially along streams, sometimes locally gregarious (SU: 1; FG: 2).

Selected specimens: Suriname: Near Bigi Poika, Maripaston Road, Teunissen LBB 12202 (NY). French Guiana: Saül, near Eaux Claires, Mori *et al.* 21547 (CAY, NY); Saül, near Monts La Fumée, Boom & Mori 2300 (NY).

Vernacular name: Suriname: don-ceder.

9. **CHLOROLEUCON**[13] (Benth.) Britton & Rose, N. Amer. Fl. 23: 36. 1928. – *Pithecolobium* sect. *Chloroleucon* Benth., London J. Bot. 3: 197, 221, 1844.
Lectotype (Britton & Rose, 1928): C. mangense (Jacq.) J.F. Macbr. var. vincentis (Benth.) Barneby & J.W. Grimes (Pithecolobium vincentis Benth.)

Trees and shrubs adapted to an annual dry season, lvs then partly or wholly deciduous, annual growth differentiated into long- and short-shoots, long-shoots floriferous only their first year, annotinous branches stiffly flexuous, randomly armed with solitary or paired, axillary thorns

[13] by James W. Grimes

derived from lignescent sterile peduncles, flowering branchlets rapidly emergent from a cone of imbricate, striately veined, caducous perules, flowers at anthesis seldom coeval with mature lvs. Leaves bipinnate, subsessile lfts varying from linear to obovate. Peduncles either solitary or fasciculate at 1-4 lower, foliate or efoliate nodes of brachyblasts, ebracteolate; fls mostly ascending, either capitate or densely shortly racemose-spicate; bracts very small, deciduous long before anthesis. Flowers of each unit of inflorescence either homomorphic or commonly heteromorphic, the always sessile terminal one then a little stouter and with modified androecium; perianth normally 5-merous (random exceptions); calyx narrowly campanulate, 5-veined, short-toothed; corolla narrowly tubular with dilated limb, tube striately veined when dry; androecial tube weakly united at base to corolla, in peripheral fls varying from included to distinctly exserted, in terminal fls always exserted, strongly dilated at orifice and there randomly fimbriate by auxiliary sterile filaments; ovary sessile, either conic or truncate at apex; stigma minutely dilated. Pods sessile but sometimes basally attenuate, (Guianan species) coiled into a compressed or open helix; dehiscence tardy and inert, through one or both sutures, but these seldom gaping widely enough to release seeds; seeds compressed-lenticular, with hard pallid testa.

Distribution: Genus with 10 species; lowland and less often submontane, xeromorphic brush-woodlands, seasonally dry forests, savannas, and deserts in warm-temperate and tropical America, from NW Mexico (27°N) to the Antilles, Argentina (28°S) and SE Brazil, in equatorial latitudes found mostly in savanna enclaves within forest climax; in the Guianas 1 species.

1. **Chloroleucon acacioides** (Ducke) Barneby & J.W. Grimes, Mem. New York Bot. Gard. 74(1): 141. 1996. – *Pithecellobium acacioides* Ducke, Arch. Jard. Bot. Rio de Janeiro 3: 69. 1922. Lectotype (designated by Barneby & Grimes (1996: 141)): Brazil, labelled “Santarem (Pará), Ducke 16358” (MG 10207). – Fig. 8

Broad-crowned drought-deciduous microphyllidious tree 5-30 m tall with smooth trunk attaining 2-5 dm diam., either randomly armed at nodes with stout tapering, solitary or paired, ascending thorns 4-27 mm long, or more often unarmed, stiffly flexuous, densely lenticellate branchlets early glabrate, lf-axes and peduncles pilosulous; perulate buds axillary to mature lvs ovoid, (1.5)2-4 mm, scales dorsally castaneous, glabrous, striately veined, microscopically ciliolate. Lf-formula (vi)vii-x/30-46; stipules often obsolete, when developed either linear or linear-oblanceolate and attaining 12 mm, early caducous; lf-stks

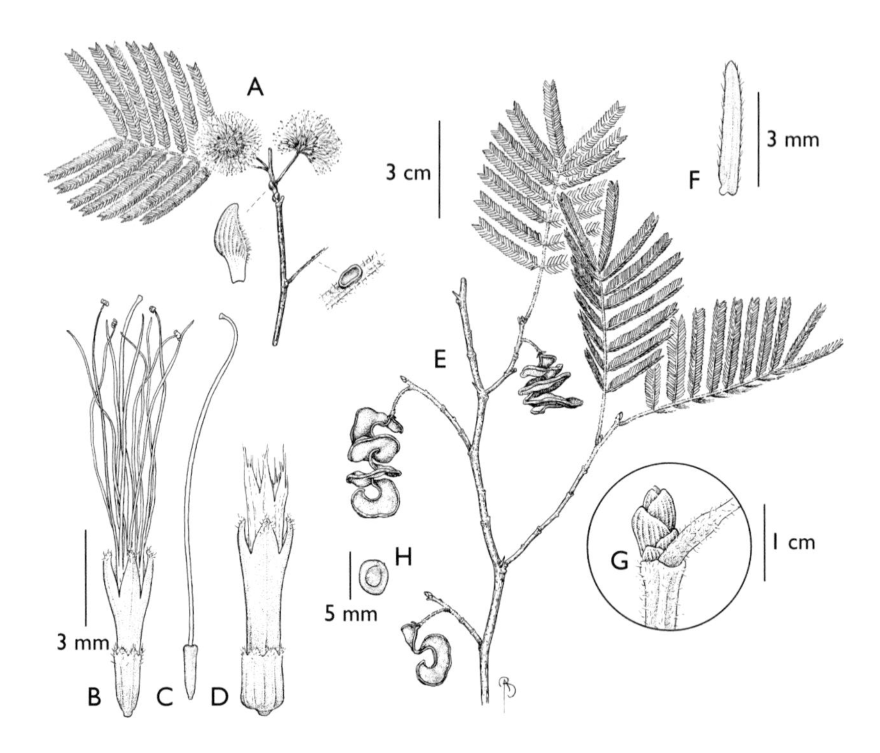

Fig. 8. *Chloroleucon acacioides* (Ducke) Barneby & J.W. Grimes: A, flowering branch with detail of bract and extrafloral nectary on petiole; B, peripheral flower; C, gynoecium; D, terminal flower; E, fruiting branch; F, leaflet; G, imbricate, striately nerved, caducous perules of a terminal bud; H, seed (A-D, Prévost 1135; E-G, Strudwick 4425, from Brazil). Drawing by Bobby Angell; reproduced with permission from NY.

4.5-8.5(12) cm, petiole proper 8-24 mm, longer interpinnal segments 4-8(11) mm; rachis of longer pinnae 2.5-4(6) cm, longer interfoliolar segments 0.5-0.8(1.2) mm; lfts subequilong except at very extremities of rachis, at first subvertically imbricate but spreading at maturity, blades linear from obliquely obtusangulate base, straight or slightly porrect at deltate or triangular acute apex, larger ones 3-6.5 x 0.5-0.9 mm, 5-7 times as long as wide, almost centric midrib 1(2)-branched above mid-blade, flanked on either side by an intramarginal primary vein produced almost to blade's apex. Peduncle solitary or geminate, 5-16 mm; capitula hemispherical 16-27-flowered, fls heteromorphic, axis of receptacle, including pedestal of sessile heteromorphic central fl, 2-3 mm; bracts minute or obsolete; peripheral fls: sessile or lowest raised on pedicel up to 0.45 mm; perianth 5-merous, calyx and corolla greenish, anthers white at early anthesis, yellowing in age; calyx campanulate, 5-veined, 1.3-2 x 0.5-0.7 mm, either glabrous or thinly pilosulous externally, deltate teeth 0.1-0.25 mm; corolla narrowly trumpet-shaped, 3.6-5.4 mm, lance-ovate lobes 1.2-2 x 0.6-0.8 mm, microscopicaly papillate on margin; androecium 10-12-merous, 11-14.5 mm long, filaments united into a tube 1.6-2.4 mm long surrounding glabrous truncate ovary; terminal fl: sessile, calyx 1.5-1.7 x 1-1.7 mm, corolla 5.5-6.5 x ± 1 mm; androecium of same length as that of peripheral fls but tube 6-8.5 mm, distally 0.8-1.5 mm diam., fimbriately dissected at orifice, but fertile stamens no more numerous; stigma slightly enlarged, ± 0.1 mm diam. Pod usually solitary, subsessile, undulately linear, 6-7 mm wide, spirally decurved and coiled through 1.5-4 circles into a compressed helix, 1.6-2 cm diam., sutures ± 0.5-0.8 mm wide, exterior (seminiferous) one sinuously constricted between seeds, thinly leathery green glabrous venulose valves becoming crustaceous fuscous, bulging over each seed; dehiscence through both sutures but tardy and irregular, seeds mostly released by weathering; seeds plumply lentiform, in broad view 3.8-4.5 x 3.6 mm, testa smooth, pale tan, fuscous over pleurogram, this nearly complete, 2.5-3 x 1.5-1.7 mm.

Distribution: Discontinuously widespread in NE Brazil, French Guiana, and E Suriname; in seasonally dry tropical woodlands and at edge of campo or on river terraces, below 200 m; flowering after rains (SU: 4 (not seen); FG: 3).

Selected specimens: French Guiana: Ile de Cayenne, Prévost 1135 (CAY, NY); Awara, Lescure 616 (CAY, NY).

10. **DESMANTHUS**[14] Willd., Sp. Pl. 4: 1044. 1806, nom. cons.
Type: D. virgatus (L.) Willd. (Mimosa virgata L.)

Unarmed herbs and subshrubs from woody taproot or xylopodium. Indument of short plain hairs, scanty in the Guianas. Stipules linear-setiform from asymmetrically auriculate, prominently veined base, persistent. Leaves bipinnate, pinnae and leaflets opposite; a sessile, cupular or almost plane nectary at insertion of first pinna-pair; leaflets (in the Guianas) subsessile, linear or linear-oblong, excentrically 1-veined, 3-10.5 mm. Flowers in small capitula borne singly in a succession of leaf-axils, heteromorphic: i) a few of lowest sterile, with 10, variable dilated whitish staminodes, usually ii) one or more functionally male, and iii) several distal ones perfect, 5(6)-merous, either iso-or diplostemonous; calyx shallowly campanulate short-toothed; petals whitish, valvate in aestivation, separate at full anthesis; anther-sacs parallel, connective

[14] by Rupert C. Barneby

glandless; pollen shed in monads; ovary sessile glabrous; style linear, stigma poriform. Pods ascending, sessile, (in the Guianas) narrowly linear 2.5-4 mm wide, nearly straight, laterally compressed but low-convex over each seed, valves stiffly chartaceous or leathery, glabrous; dehiscence inert, through both sutures; seeds obliquely basipetal, small, exocarp vesicular, pleurogrammic; endosperm present.

Distribution: American genus of 25 species, primarily of tropical and warm temperate savanna-campo habitats but several opportunistically weedy, most diverse to the N and S of the equatorial forest belt: SW United States to Uruguay and N Argentina, 1 species N over the prairies nearly to the Canadian line, 1 species naturalized in the Old World; 2 species in the Guianas.

Notes: In the Guianas *Desmanthus* can be confused only with *Neptunia*, similar in subherbaceous life-form and in heteromorphic flowers, but differing in yellow petals, bibracteate peduncles, and broader, stipitate pod. The taxonomy and nomenclature adopted herein follow Luckow (1993).

LITERATURE

Luckow, M. 1993. A monograph of Desmanthus (Leguminosae: Mimosoideae). Syst. Bot. Monogr. 38: 1-166.

KEY TO THE SPECIES

1 Petiole of mature leaves 6-16 mm long .... *1. D. pernambucanus*
Petiole of mature leaves 1-5 mm long .... *2. D. virgatus*

1. **Desmanthus pernambucanus** (L.) Thell., Mém. Soc. Sci. Nat. Cherbourg 38: 296. 1912. – *Mimosa pernambucana* L., Sp. Pl. 519. 1753. Lectotype (designated by Luckow (1993: 113)): Plate: t. 307, fig. 3. in Plukenet, Phytographia, v. 1, 1694.

Weakly suffrutescent herbs 4-20 dm, closely resembling *D. virgatus* in habit, stature, pubescence, small whitish capitula and linear pod. Stipules setiform 3-6 mm, dilated at base into a prominently veined auricle. Leafstalks ± 2-6 cm, petiole 6-16 mm; nectary between first pinna-pair sessile, elliptic, cupular or nearly plane 1-2.5 mm diam; pinnae 2-4 pairs, rachis of longer ones 1.5-5 cm; leaflets of distal pinnae 10-21 pairs, narrowly oblong or linear 5-10.5 x 1.2-3.3 mm, obtuse or minutely apiculate, ciliolate. Peduncle axillary 1-4.5 cm; capitula 8-13-flowered, lower 2-4 flowers sterile with 10 filiform or dilated, whitish staminodia 5-10 mm; perianth of upper, bisexual flowers greenish, glabrous; calyx turbinate 2-3 mm, teeth 0.3-0.5 mm; petals 3-4 mm; stamens 5 or 10, ± 5-7 mm; gynoecium glabrous, style 3-5 mm. Pod ascending, linear 45-85 x 3.2-3.8 mm, straight, plano-compressed, 13-22-seeded, valves stiff, reddish-brown nigrescent, glabrous; dehiscence and seeds of *D. virgatus*.

Distribution: Native in the West Indies and scattered in N and E South America, along and near the coast of Guyana and Suriname; naturalized in Florida, S Africa, SE Asia and Pacifica; in seasonally moist, sandy or stony open places at sea-level up to 1500 m, abundantly weedy on roadsides and in waste places; flowering in the Guianas August to January, sporadically later (GU: 1; SU: 1).

Selected specimens: Guyana: Demerara-Mahaica Region, campus of the University of Guyana, Hahn 4826 (NY, US); Georgetown, Hitchcock 16638 (GH, NY, US). Suriname: Meerzorg, Florschütz 952 (TEX).

2. **Desmanthus virgatus** (L.) Willd., Sp. Pl. 4: 1047. 1806. – *Mimosa virgata* L., Sp. Pl. 519. 1753. Type: India, the collector unknown (holotype LINN). – Fig. 9

Herbs from woody taproot 3-15 dm, stems erect, decumbent, or prostrate, thinly pubescent when young, early glabrate. Stipules setiform 2-9 mm, auriculate on side further from petiole. Leafstalks ± 1-4 cm, petiole 1-5 mm; a sessile cupular nectary 0.3-1.2 mm between first pair of pinnae; pinnae 2-5 pairs, ± 1-3 cm; leaflets 11-23 pairs, subsessile, linear-oblong 3-7 x 0.7-1.6 mm, either obtuse or acute, facially glabrous but ciliolate, excentric straight midrib simple. Peduncle axillary solitary 0.6-4 cm; bracts dimorphic, lowest sessile, sometimes united into an involucre, upper ones with slender stalk and peltate blade; receptacle of capitula 3-10 mm, up to 22-flowered; calyx obconic, of sterile flowers 0.6-1.2 mm, of bisexual flowers 1.5-3 mm, teeth 0.2-1 mm; corollas 1.5-4 mm, greenish, petals white- or red-tipped; staminodes of sterile flowers ± 2-7 mm, filiform or up to 1 mm wide; stamens of bisexual flowers 3.5-7 mm; ovary glabrous, style ± 2-4 mm. Pod ascending, in profile linear 22-88 x 2.5-4 mm, straight or almost so, abruptly apiculate, 10-27-seeded, valves leathery red-brown, glabrous, framed by often pallidly discolored sutures, low-convex immediately over each seed; dehiscence inert, through both sutures; seeds obliquely basipetal, 2.1-2.9 x 1.4-2.7 mm.

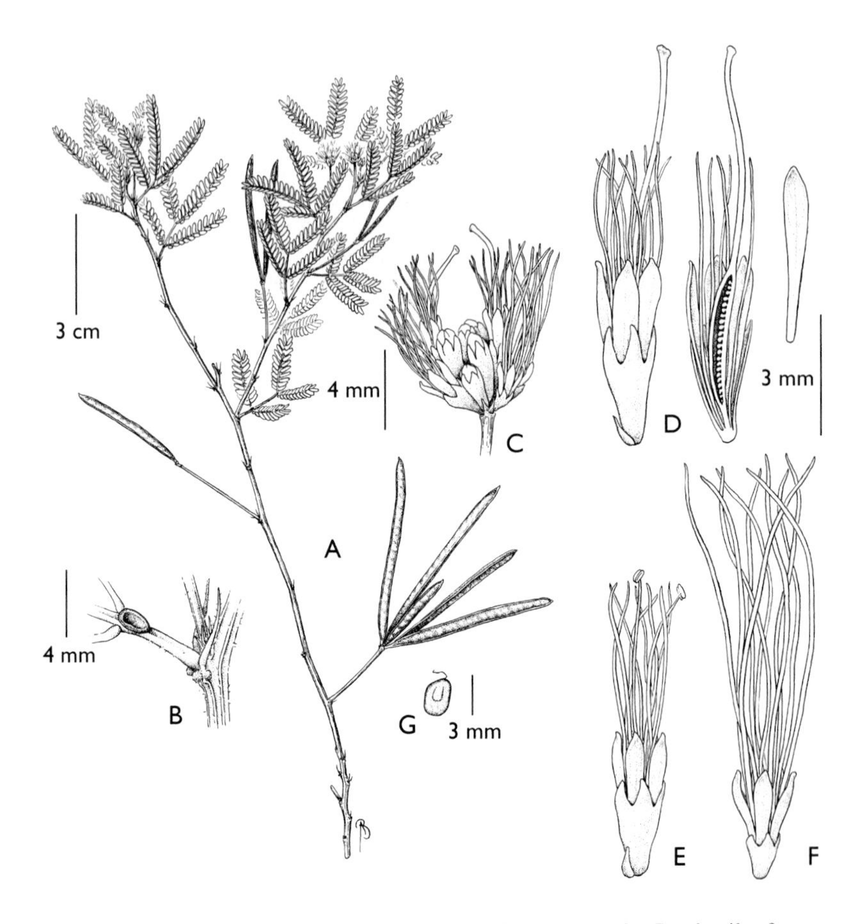

Fig. 9. *Desmanthus virgatus* (L.) Willd.: A, fertile branch; B, detail of stem leaf showing leaf petiole with gland between lower pair of pinnae and stipules at base; C, capitulum; D, bisexual flower (left), and medial section of bisexual flower (right), and petal (far right); E, staminate flower; F, sterile flower; G, seed (A, G, US Forest Dept. 6793; B-F, Acevedo 2784). Drawing by Bobby Angell; reproduced with permission from NY.

Distribution: Widespread from S Florida and Texas S through the Caribbean basin, Mexico, and C America to Uruguay, N Argentina, and Ecuador, common below 100 m on the coastal lowlands of Guyana and Suriname, to be expected in French Guiana; in disturbed grassy places, on sea beaches, at edge of mangrove swamps, and becoming prolifically weedy on roadsides and in both urban and agrarian wasteland; flowering through the year unless drought-inhibited (GU: 8; SU: 3).

Selected specimens: Guyana: Georgetown, FD 7100 (= Fanshawe 3504) (NY); Demerara-Mahaica Region, Pipoly 9041 (NY, US). Suriname: Along a ditch at plantation Jachtlust, Mennega 161 (NY); Saramacca, Wullschlägel 946 (NY).

Note: The epithet *virgatus* has often been applied specifically to the taller, more or less erect forms of this variable species, which is here, following Luckow (1993), more exactly defined by the short petioles and characteristic leaf-movements of the pinnae, and includes populations with prostrate or diffuse stems.

11. **ENTADA**[15] Adans., Fam. Pl. 318, 554. 1763, nom. cons.
Type: E. monostachya DC.

Entadopsis Britton, N. Amer. Fl. 23: 191. 1928. – *Entada* sect. *Entadopsis* (Britton) Brenan, Kew Bull. 20: 365. 1966.
Type: E. polystachya (L.) Britton [Entada polystachya (L.) DC.]

The Guianas taxa, comprising sect. *Entadopsis* (Britton) Brenan, unarmed trees and shrubs potentially sarmentose or scandent into forest canopy by means of randomly cirrhifarous leaves. Indument of short sordid hairs. Stipules small, caducous. Leaves bipinnate, pinnae either opposite or subopposite, leaflets exactly opposite, penniveined; no petiolar nectaries. Inflorescences terminal, dense, mostly plagio- or geotropic pseudoracemes of mostly geminate spiciform racemes of small, nearly or quite glabrous flowers, crowded racemes secundly ascending toward vertical; pedicels very short, persistent after disarticulation of sterile flowers. Flowers 5-merous decandrous, some functionally staminate, campanulate calyx open in bud, valvate petals at anthesis separating to base, greenish-white or reddish perianth fading and drying dark red-brown; stamans equal, not far exserted, filaments white, introrse anthers dorsifixed, connective tipped with a caducous gland; pollen shed in monads. Pods a disproportionately large craspedium, in profile broad-linear, straight or gently decurved, plano-compressed, when ripe disolved into a persistent replum and 10-17 1-seeded segments, exocarp thin, lustrous brown then peeling in flakes, endocarp separating into individually indehiscent 1-seeded anemochorous packets; seeds oblong-ellipsoid compressed, brittle testa with finely engraved pleurogram, endosperm 0.

Distribution: Circumtropical genus of ± 28 species, most numerous and diverse in Africa; in the Americas 1 species, *E. polystachya*, with 3 varieties, 2 of these varieties in the Guianas.

[15] by Rupert C. Barneby

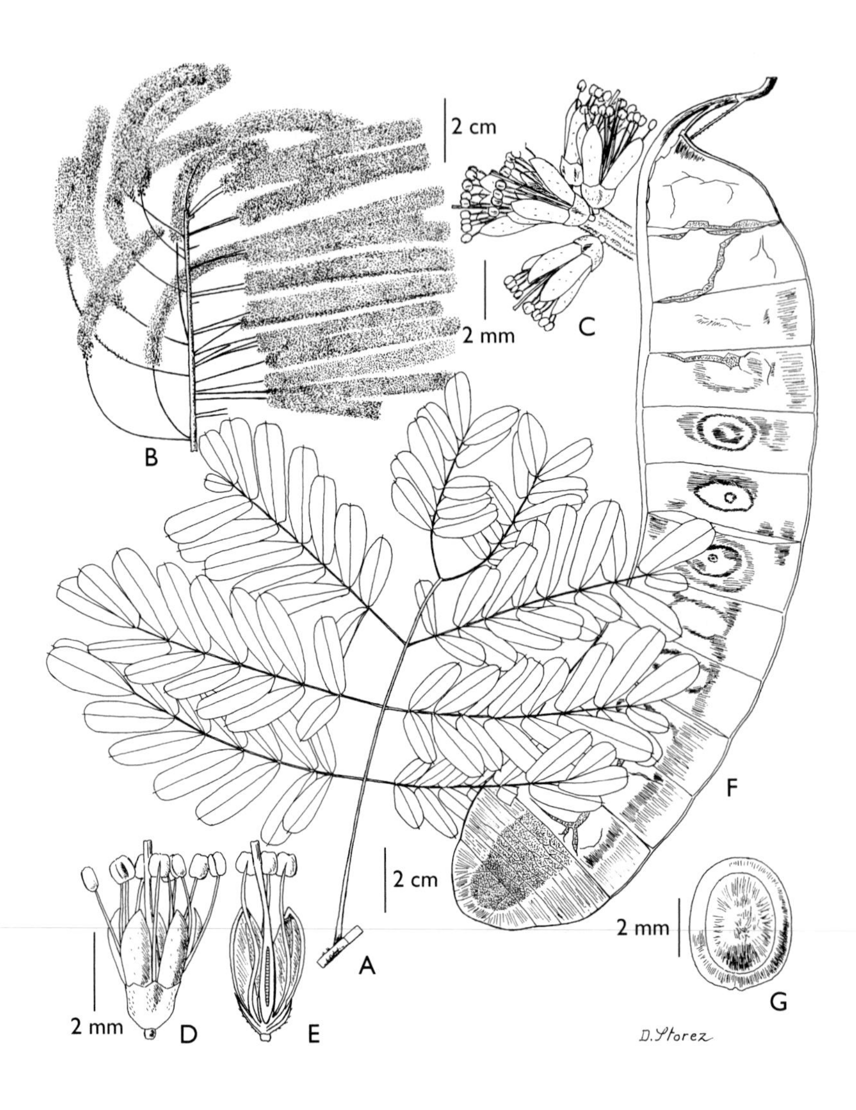

Fig. 10. *Entada polystachya* (L.) DC.: A, leaf; B, inflorescence (partial); C, inflorescence, detail; D, flower; E, longitudinal section of flower; F, pod; G, seed (A-E, Prévost 3762; F, Prance 6726; G, Hallé 1041). Drawing by Dominique Storez; reproduced with permission from P.

Notes: Clasping and then lignescent tendrils develop from the rachis of the furthest pair of pinnae, but only in random leaves. The massive one-sided, leafless terminal inflorescence is characteristic of the genus in the Guianas. A heavy mephitic fragrance attracts small coleoptera, the probable pollinators.
The stems of *E. polystachya* var. *polystachya* furnish cordage, and the roots saponin.
No vernacular names are recorded from the Guianas, but they are numerous in Latin America.

1. **Entada polystachya** (L.) DC., Prod. 2: 425. 1825 and Mem. Leg. 434, t. 61, 62. 1825. – *Mimosa polystachia* L., Sp. Pl. 520. 1753 as 'polystochia'. – *Mimosa bipinnata* Aubl., Hist. Pl. Guiane 946. 1775, a superfluous substitute. – *Mimosa chiliantha* G. Mey., Prim. Fl. Esseq. 163. 1818, a superfluous substitute. Type: "Acacia alia scandens, spica multiplici", Plumier, Cat. Pl. Amer. 17. 1703, later described and illustrated by Burm., Pl. Amer. 1: t. 12. 1755.
– Fig. 10

Mimosa caudata Vahl, Ecl. Amer. 3: 35. 1807. – *Acacia caudata* (Vahl) DC., Prod. 2: 456. 1825. Type: French Guiana, Cayenne, von Rohr s.n. (holotype C).

KEY TO THE VARIETIES IN THE GUIANAS

1 Pinnae of well-developed leaves 4-8 pairs and leaflets of longer pinnae 12-20 pairs; largest leaflets 11-23 x 3-8 mm . . *1a. E. polystachya* var. *polyphylla*
Pinnae of well-developed leaves 3-5 pairs and leaflets of longer pinnae (6)7-11(12) pairs; largest leaflets 25-50 x 9-17 mm. . . . . *1b. E. polystachya* var. *polystachya*

1a. **Entada polystachya** (L.) DC. var. **polyphylla** (Benth.) Barneby, Brittonia 48(2): 175. 1996. – *Entada polyphylla* Benth., J. Bot. (Hooker) 2: 133. 1840. Type: Guyana, on the Quitaro R. [= Guidaru, affluent of the upper Essequibo R.], Ro. Schomburgk ser. I, 604 (holotype K! = NY Neg. 1726, isotype NY!).

Small trees and opportunistically scandent bushropes 3-10 m with terete glabrate older stems, young growth puberulent or pilosulous with sordid-gray hairs, leaves bicolored, leaflets on upper face dark green and either dull or (young) lustrous, either glabrous or puberulent, beneath paler and always puberulent or pilosulous, inflorescence a massive terminal efoliate pseudoraceme of dense spiciform racemes, flowers small, strong-scented, horizontally spreading on obscure pedicels. Stipules subulate 0.5-1.5 mm, fugacious. Leaf-stalks (6.5)8-18 cm; pinnae 4-8 pairs, median ones ordinarily longest, rachis of these, including their elongate pulvinus, (4)5-9 cm, rachis of one or both distal pinnae randomly modified into a lignescent tendril (present in few specimens); a pair of ascending subulate paraphyllidia at top of each pulvinus; leaflets of longer pinnae 12-20 pairs, blades oblong from obtusely angulate base, either obtuse or more often truncate-emarginate, longer ones 11-22(23) x 3-7(8) mm; venation pinnate, midrib straight, almost centric, prominulous dorsally, secondary and further venules faint or immersed. Primary axis of inflorescence 1-3.5(4.5) dm; racemes subsessile, solitary or geminate, their axis (3)3.5-8(9) cm; bracts narrowly ovate or subulate 0.6-1 mm, caducous; pedicels 0.15-0.3 mm; perianth glabrous or calyx microscopically puberulent, petals white, yellowish, once reported as bright yellow, brunnescent when dry; calyx shallowly turbinate-campanulate, faintly 5-veined, 0.6-1.1 mm, teeth depressed-deltate, not over 0.2 mm; petals erect 2-2.6 mm; filaments white, (2.6)3-4.7 mm; ovary sessile, glabrous, style about as long as stamens. Pod in profile broad-linear, straight or almost so, plano-compressed, 25-45 x 4-8 cm, 10-15-seeded, contracted at base into a stipe 1.5-3 cm, broadly obtuse at apex, the wiry persistent replum undulately constricted between seeds, exocarp thin glabrous, fuscous, exfoliating when ripe from pale tan endocarp, this breaking up into transversely narrow-oblong, 1-seeded papery segments ± 2-3 cm wide.

Distribution: Puerto Rico, discontinuously widespread through the Amazon basin from Ecuador to NE Bolivia, Brazil (Mato Grosso, Pará and Amapá) and the Guianas; in disturbed woodland and at edge or in tree-fall openings of virgin moist forest, mostly below 300 m but ascending to ± 1000 m in Ecuador; flowering nearly throughout the year, in the Guianas most prolifically Oct.-Feb. (GU: 1; SU: 3; FG: 1).

Selected specimens: Guyana: Konaschen area, Essequibo R., Jansen-Jacobs *et al.* 1793 (NY). Suriname: Vicinity of Kwakoegroen, Maguire & Stahel 25008 (NY); N of Albina, Marowijne R., Lanjouw & Lindeman 277 (NY). French Guiana: Mont Cabassou, Ile de Cayenne, Hoff 5699 (NY).

Note: This variety is well illustrated in Mart., Fl. Bras. 15: t. 70. 1876.

1b. **Entada polystachya** (L.) DC. var. **polystachya**

Trees, shrubs and bushropes, sometimes flowering at 1.5-2 m but opportunistically sarmentose and scandent into forest canopy and potentially attaining 20 m, young growth and primary axis of inflorescence pilosulous or puberulent, leaves bicolored, leaflets bright green and usually glabrous on upper face, on lower face paler and pilosulous along midrib or overall, axis of dense flower-spikes varying from pilosulous to glabrate, but flowers themselves usually glabrous. Stipules subulate, less than 1.5 mm, caducous. Leaf-stalks (8)10-11 cm; pinnae 3-5 pairs, rachis of longer ones, including their elongate pulvinus, (5)6.5-13.5 cm, rachis of one or both of furthest pair randomly modified into a lignescent tendril; leaflets of longer pinnae (6)7-11(12) pairs, oblong from strongly inequilateral base, shallowly emarginate at apex, longer ones 2.5-5 x 0.9-1.7 cm; venation pinnate, midrib straight subcentric, prominulous only beneath, secondary venules prominulous above only in youth, immersed in mature blades. Primary axis of pseudoraceme (1)1.5-3.5 dm, that of mostly geminate racemes 5.5-11 cm; floral bracts membranous, ovate- or lance-acuminate, caducous; pedicels (0.1)0.15-0.2.5 mm, persistent, nigrescent; perianth glabrous overall or calyx sometimes puberulent; calyx hemispherical or shallowly turbinate-campanulate 1.1-1.5 mm, depressed-deltate teeth 0.15-0.4 mm; petals at anthesis white, ochroleucous, yellowish, or red, when pale turning red-brown in age or when dried nigrescent, 2.8-3.7 mm; filaments white, when fully expanded 4-5.6 mm; ovary sessile, glabrous, style nearly as long as filaments. Pod narrowly oblong or broad-linear in profile, when well fertilized 21-35 x (5)5.5-8 cm, contracted at base into a stout stipe 1.2-3 cm, broadly rounded at apex, (12)13-17-seeded, exocarp glossy, dark brown, ultimately peeling in flakes, endocarp pale tan, breaking up into transversely narrow-oblong 1-seeded papery, individually indehiscent segments (11)12-27 mm wide.

Distribution: Mexico, C America, the Antilles to Hispaniola, N Colombia, the Orinoco valley in Venezuela, and Trinidad and Tobago, widely dispersed in tropical America: Amazonian Peru, Brazil, and Bolivia, common in the lowlands of the Guianas; in swampy thickets, on river banks, in gallery forest, at edge of mangrove swamps, in seasonally flooded forest-savanna ecotone, and locally in upland brush-woodland, mostly below 300 m but attaining 1000 m in C American rainforest; flowering in the Guianas April-December, but perhaps intermittently throughout the year (GU: 5; SU: 2; FG: 3).

Selected specimens: Guyana: Kamakusa, upper Mazaruni R., de la Cruz 4100 (NY); Berbice R., S of New Dageraad, Maas *et al.* 5478 (NY). Suriname: Forest near the Ma Retraite plantation, Kegel 355 (NY); Small ridge N of the Nickerie R., opposite the fortress, Lanjouw & Lindeman 3114 (NY). French Guiana: Vicinity of Cayenne, Broadway 255, 822 (NY).

12. **ENTEROLOBIUM**[16] Mart., Flora 20(2), Beibl. 8: 116. 1837.
Type: E. timbouva Mart.

Unarmed trees, young growth ± densely pubescent with simple, pallid or brownish hairs. Leaves alternate, bipinnate; lf-formula ii-xxx/4-80; stipules small, mostly caducous; petiolar nectary either cupular-patelliform, or mounded, or sunk in petiolar groove; lft-pulvinules less than 1 mm; venation of lfts palmate-pinnate, or (in narrowest blades) reduced to midrib. Inflorescences of either homo- or heteromorphic, compact capitula borne (in Guianan species) either singly or fasciculate in contemporary lf-axils or at efoliate nodes below current year's foliage as well as axillary to some contemporary lvs of annual cycle. Fls 5-8-merous, calyx of peripheral ones campanulate, turbinate- or cylindro-campanulate; corolla gamopetalous, twice or less than twice as long as calyx; androecium ("8")10-70-merous, tube variable in length, intrastaminal disc often well developed; ovary either tapering or truncate at apex. Pods sessile, in profile oblong or broad-linear decurved through 0.5-2 circles into a reniform-auriculiform, or annular, or compressed-helicoid figure, outer (seminiferous) suture either evenly arcuate or undulately recessed between seeds, valves composed of a) thin, fuscous or blackish exocarp, b) thick, either dry mealy-fibrous or resinous-pulpy mesocarp, and c) papery pallid endocarp inflexed to form complete, often partly membranous interseminal septa; dehiscence 0, seeds released by weathering on ground or excreted by herbivores; seeds transverse, 1-3-seriate, plumply compressed-ovoid-ellipsoid, hard testa castaneous or fuscous, pleurogram complete or incomplete.

Distribution: 11 species widespread over the lowland and submontane American tropics from S Mexico and Greater Antilles S through the Magdalena Valley in Colombia, the Orinoco basin in Venezuela, and Guyana almost throughout Brazil to E Bolivia, Paraguary, N Argentina, and Uruguay; 3 species in the Guianas.

Note: Barneby & Grimes (1996: 247-248) recognize 2 sections within *Enterolobium*. *E. cyclocarpum* is placed in sect. *Enterolobium*, characterized in part by pallid indument and homomorphic flowers lacking intrastaminal discs. *E. schomburgkii* and *E. oldemanii* belong to section *Robrichia* Barneby & J.W. Grimes, characterized in part by golden-brown indument, heteromorphic flowers, and terminal flower with an intrastaminal disc.

[16] by James W. Grimes

KEY TO THE SPECIES

1 Indument of pallid hairs; flowers homomorphic, androecium 20-68-merous, intrastaminal disc lacking . . . . *1. E. cyclocarpum*
Indument of golden-brown hairs; flowers heteromorphic, androecium 10-merous (random anomalies) 2

2 Calyx of peripheral flowers 5.5-6 mm; corolla 7-8.2 mm; andreocium 20-29 mm, stemonozone 1-1.6 mm, staminal tube 3-5 mm . . . *2. E. oldemanii*
Calyx of peripheral flowers 2.5-3.3(3.5) mm; corolla 3.2-5.2 mm; androecium 10-13.5 mm, stemonozone 0.5-1 mm, staminal tube 1-2.3 mm. . . . .
. . . . *3. E. schomburgkii*

1. **Enterolobium cyclocarpum** (Jacq.) Griseb., Fl. Brit. W. Ind. 3: 225. 1860. – *Mimosa cyclocarpa* Jacq., Frag. Bot. 30, tab. 34, fig. 1. 1809. Type: Jacquin, Frag. Bot. 30. tab. 34, fig. 1. 1809.

Trees 12-50 m, young growth and corolla lobes densely yellow-white pubescent, becoming glabrate with age; infls 1-5-fasciculate at efoliate older nodes, or foliate younger ones, or on efoliate branches. Lf-formula v-xiv/15-31, stipules not seen, lf-stk 11.7-18(37.2) cm; petiole 5.0-7.0 cm, interpinnal segments 10-18 mm; pinnae 7.7-12.8 cm, sometimes proximally subopposite; lfts oblong-lanceolate 11-16 x 2.75-4, inequilateral at base, bicolored, sparingly pubescent on both surfaces, or upper surface glabrate, midvein forwardly displaced. Peduncle 2.2-6.8 cm, 30-44-flowered, bracts rhombic-elliptic, ± 1 mm, quickly deciduous; fls homomophic, on a short pedicel ± 0.5 mm, calyx tubular-campanulate 2.25-3.0 mm; corolla tubular-funnelform 4.5-7 mm, androecium 9-13 mm, tube slightly > corolla; ovary sessile, tapering at apex, glabrous, intrastaminal disc 0. Pod tightly coiled into a usually complete circle, valves 1.75-6.5 cm diam., raised over seeds and constricted between them; indehiscent, though easily breaking apart between seed-chambers; seeds plump, in profile 15-20 x 10-12 mm, with a complete pleurogram, areola darker than remainder of seed coat.

Distribution: Mexico, West Indies, Venezuela and adj. Brazil, and Guyana; in dense, usually dry, primary or gallery forest on terra firme, often cultivated; flowering February to April (GU: 1).

Specimen examined: Guyana: Takutu R., Kanuku Mts., A.C. Smith 3235 (NY).

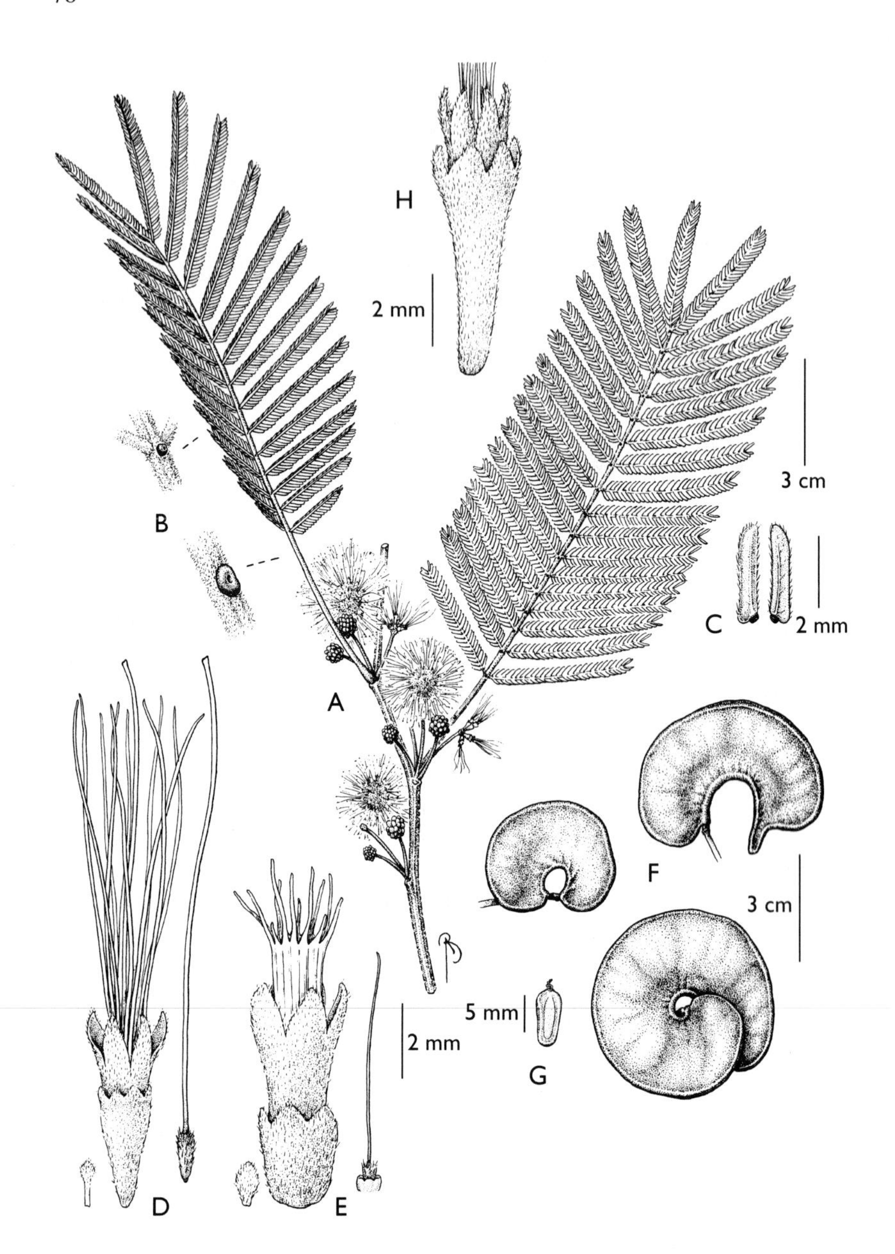

Fig. 11. A-G: *Enterolobium schomburgkii* (Benth.) Benth.: A, habit; B, leaf-nectaries; C, leaflets; D, peripheral flower; F, pods; G, seed. H: *Enterolobium oldemanii* Barneby & J.W. Grimes: H, dimorphic terminal flower (A-C, Silva 4781; D, Silva 1387; F, left pod Silva 4781, right pod Smith 41781, lower pod Silva 2474; G, Silva 4781; Silva collections all from Brazil; H, Oldeman 2242). Drawing by Bobby Angell; reproduced with permission from Mori *et al.*, 2002: 491.

2. **Enterolobium oldemanii** Barneby & J.W. Grimes, Mem. New York Bot. Gard. 74(1): 249. 1996. Type: French Guiana, ancienne réserve forestière du Matoury, Ile de Cayenne, Oldeman 2242 (holotype P!). – Fig. 11 H

Trees to 30 m, resembling *E. schomburgkii* but bark smooth and leaflets somewhat wider and flowers nearly twice as large. Lf-formula xvi-xxiii/35-49; stipules thick-coriaceous, inequilaterally triangular-lanceolate or lanceolate 4.0-4.75 x 2.0-2.25 mm, quickly deciduous; lf-stk 6.5-14 cm; petiole 9-20 mm; interpinnal segments 3-6 mm; pinnae a little decrescent at both ends of lf-stk, rachis of longer ones (2.6)3.5-5.7 cm; lfts decrescent at each end of pinna, linear to linear-elliptic from an inequilaterally truncate base, apex broadly acuminate, blades straight, plane, commonly ciliolate but glabrous on both faces, longer ones 5-6 x 0.7-1.2 mm, venation reduced to a forwardly displaced midrib. Peduncle 1-2(3) per node, 1.2-3.6 cm; capitula 32-38-flowered; bracts persistent into anthesis, that of peripheral fls linear-spatulate to oblanceolate ± 1 mm, that of peripheral fl about twice as large; peripheral fls: calyx tubular 5-6 mm, contracted at base into an obconical pedicel 0.5-1 mm; corolla 7-8 mm; androecium 10-merous, 20-29 mm, stemonzone 1-1.6 mm, tube 3-5 mm; ovary 2.6-3.5 mm, obovoid truncate, densely puberulent or puberulent only along ventral suture; terminal fl: calyx sessile, deeply campanulate 5.5-6 x 2.8-3.3 mm, teeth ± 1 mm; corolla ± 11 mm, lobes 2.5 mm; androecial tube exserted 1.5-2 mm; ovary rudimentary or similar in shape and size to that of peripheral fl, surrounded by a shallow disc. Pod, up to 3 per capitulum, curved through 0.5-1.5 circles, this 9-11 cm in diam., sutural rib slightly enlarged, finely venulose valves 3.5-4.5 cm broad; seeds (few seen mature) ovate-elliptic in outline, ± 9 x 5 mm.

Distribution: Known from French Guiana near Cayenne, and on Piste-de-St-Elie in undisturbed, primary forest; fruits of *E. oldemanii* are known from Brazil (Amazonas) (FG: 5).

Selected specimens: French Guiana: Réserve Forestière Mt. Grande Matoury, Ile de Cayenne, 8 km from Cayenne, Grimes 3267 (CAY, NY); Piste-de-St-Elie, between the Couanamama and Sinnamary Rs., Grimes *et al.*, 3310 (CAY, NY).

3. **Enterolobium schomburgkii** (Benth.) Benth., Trans. Linn. Soc. London 30: 599. 1875. – *Pithecolobium schomburgkii* Benth., London J. Bot. 3: 219. 1844. Type: Brazil, Amazonas, Pedrero, on the Rio Negro, Ro. Schomburgk ser. I, 874 (holotype K!). – Fig. 11 A-G

Trees fertile at (6)8-40(55) m, bark peeling in large flakes, young branches and all lf- and inflorescence-axes densely pilosulous, tomentulose, or subvelutinous with sinuous, brownish or bronze hairs. Lf-formula (ix) x-xxiii/(32)40-83; stipules oblanceolate 3-10 x 1-4 mm, very quickly caducous; lf-stks (6)8-19(24.5) cm, petiole (10)12-27(35) cm; longer interpinnal segments 4-9(11) mm; pinnae often a little decrescent either proximally or at both ends of lf-stk, sometimes subequilong, rachis of longer ones (2.2)3-6(6.7) cm; lfts bicolored, subequilong except at ends of rachis, linear or narrowly lance-linear from shallowly auriculate or postically rectangulate base, straight or commonly gently porrect, obtuse, longer ones 2.5-5(5.4) x 0.5-0.8 mm, blades glabrous above, silky-strigulose or -pilosulous beneath. Peduncles 3-9 per node, longer ones (13)15-30(46) mm; capitula (or capituliform racemes) (16)20-45(55)-flowered; bracts persistent into anthesis, that of peripheral fls linear-oblanceolate or -spatulate 1.2-2.5 mm, that of terminal fl ovate; peripheral fls: pedicel 0-2 x 0.2-0.4 mm; calyx narrowly vase-shaped (2.3)2.5-3.3(3.5) x 1.2-1.6 mm; corolla (3.2)3.5-5.2 mm; androecium 10.5-13.5 mm; ovary sessile, obliquely truncate ± 1.3 mm, densely strigulose overall; terminal fl: calyx broadly campanulate (2.2)2.4-3.8 x 1.8-2.6 mm; corolla (4)4.5-8 mm, staminal tube 5-9.5 mm; ovary non-functional, surrounded at base by a 5-lobed nectarial disc. Pod evenly decurved through ± 3/4-1.5 circles into a flattened spiral, compressed body, measured along middle of valves, 5-12 x 1.8-3(3.3) cm; dehiscence 0, but valves breaking under pressure transversely between seeds, these released in nature only by predators or by weathering; plump seeds in broad profile oblong-elliptic ± 6-8 x 3.2-4.7 mm, in relatively long pods 1-seriate, in shorter pods irregularly 2-seriate and so crowded as to become deformed by mutual pressure; pleurogram narrowly U-shaped 4.7-6.5 x 1.8-2.7 mm.

Distribution: S Mexico south through Colombia, Peru, much of Brazilian Amazonia, Venezuela, and the Guianas; evergreen and drier semideciduous tropical forest at low elevations but above the floodline; flowering September through April (GU: 3; SU: 4; FG: 1).

Selected specimens: Guyana: Kuyuwini R., A.C. Smith 2616 (NY); Wabawak, Kanuku Mts., FD 5862 (= WB 452) (NY). Suriname: Nature Park Brownsberg, Brokopondo Distr., Tjon Lim Sang LBB 16235 (NY). French Guiana: Route de Cayenne au km 8, côte gauche direction Cayenne, Serv. Forestière 7527 (NY).

13. **HYDROCHOREA**[17] Barneby & J.W. Grimes, Mem. New York Bot. Gard. 74(1): 23. 1996.
Type: H. corymbosa (Rich.) Barneby & J.W. Grimes (Mimosa corymbosa Rich.)

Pithecolobium sect. *Samanea* ser. *Corymbosae* ['*Corymbosa*'] Benth., London J. Bot. 3: 221. 1844.
Type: P. corymbosum Benth. [Hydrochorea corymbosa (Rich.) Barneby & J.W. Grimes]

Unarmed trees and arborescent shrubs commonly 5-27 m tall, young stems and foliage gray-or brown-pilosulous, lfts nearly always conspicuously bicolored. Lf-formula i-vii/3-35; stipules linear or subulate, caducous; petiolar nectary between or close below insertion of first or only pinna-pair, either sessile or stipitate (rarely suppressed); lft-pulvinules ± 0.5-2.5 mm; venation of lfts pinnate. Peduncles 2-11(14) cm; inflorescences compactly racemose-capitulate, axis including pedestal 1.5-6 mm; peripheral fls slenderly pedicellate, mostly 5-merous, 1-4 distal fls sessile or shortly stoutly pedicellate, 5-8-merous. Calyx campanulate or turbinate-campanulate, (1)1.4-7 mm; corolla (2.6)3-11(12) mm. Pods often several per capitulum, stiffly erect-ascending, sessile or shortly pseudostipitate, in profile linear or broad-linear, straight or almost so, compressed, mature fruit lomentiform, (4.5)5-12.5 x 0.75-1.9 cm, readily breaking transversely between seeds and less readily so through shallowly undulate sutures into square or transversely oblong, 1-seeded articles closed at each end by half of interseminal septum; seeds hard, brown, areolate, pleurogram complete.

Distribution: 3 species widespread in riparian habitats in the Orinoco and Amazon basins and in the Guianas, in Brazil extending E to Maranhão and S to the Pantanal in Mato Grosso do Sul; 2 species in the Guianas.

KEY TO THE SPECIES

1 Leaflets of longer pinnae (2)3-11(14)-jugate; pod (9)10-19 mm wide . . . . *1. H. corymbosa*
Leaflets of longer pinnae 14-35-jugate; pod 7.5-9 mm wide . . . . *2. H. gonggrijpii*

[17] by James W. Grimes

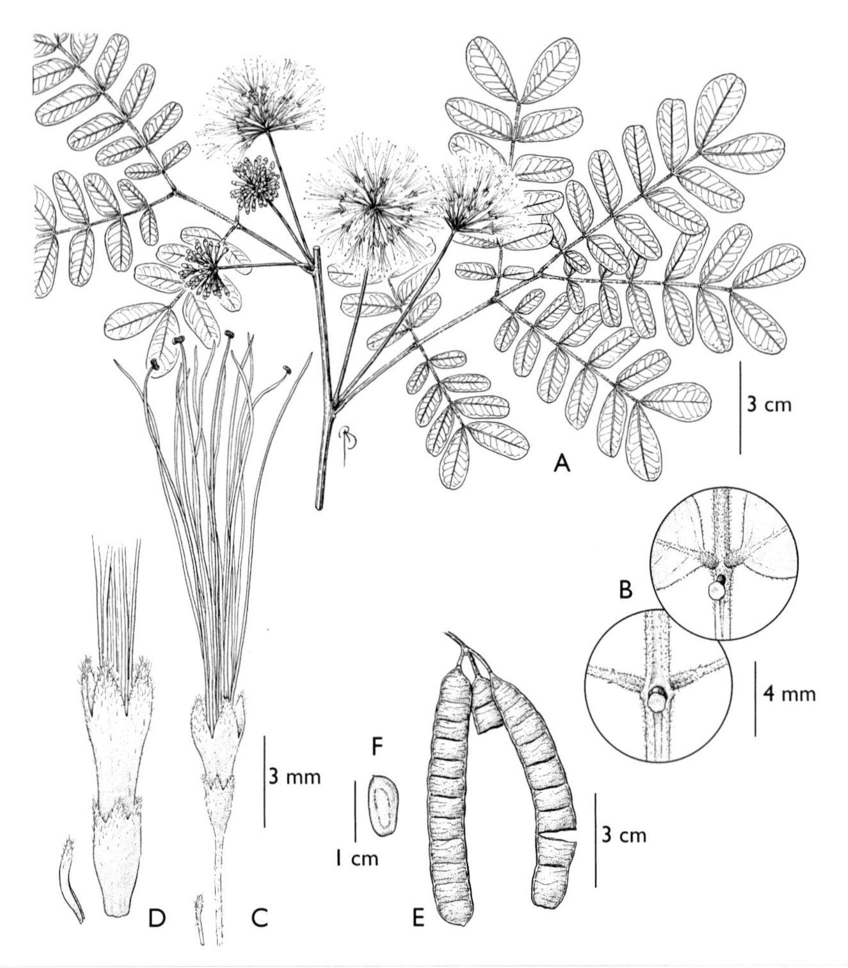

Fig. 12. *Hydrochorea corymbosa* (Rich.) Barneby & J.W. Grimes: A, flowering branch; B, extrafloral nectary on leaf rachis between a pinna pair (lower circle) and extrafloral nectary on pinna rachis between a leaflet pair (upper circle); C, peripheral flower and bract; D, terminal flower and bract; E, lomentiform pods; F, seed (A-D, Sperling 6269; E-F, Prance 10692; both from Brazil). Drawing by Bobby Angell; reproduced with permission from NY.

1. **Hydrochorea corymbosa** (Rich.) Barneby & J.W. Grimes, Mem. New York Bot. Gard. 74(1): 27. 1996. – *Mimosa corymbosa* Rich., Actes Soc. Hist. Nat. Paris 1: 113. 1792. – *Arthrosamanea corymbosa* (Rich.) Kleinhoonte in Pulle, Fl. Suriname 2(2): 327. 1940. – *Albizia corymbosa* (Rich.) G.P. Lewis & P.E. Owen, Leg. Ilha de Maracá 40, fig. 3J, pl. 4D. 1989. Type: French Guiana, Cayenne, in woods along the Kourou R., Leblond s.n. (holotype P (hb. Richard)!). – Fig. 12

Pithecolobium corymbosum Benth., London J. Bot. 3: 221. 1844. Lectotype (designated by Barneby & Grimes 1996: 28): Guyana, Ro. Schomburgk ser. II, 114 (anno 1841) (K!).

Trees attaining 5-20 m with trunk 1-3 dm DBH, young branches and all lf-axes (at least ventrally) and peduncles pilosulous with incurved, gray or sordid hairs. Lf-formula ii-v(vii)/(4)5-11(14); lf-stk of most developed lvs 4-14(18) cm, but that of depauperate lvs of some lateral flowering branchlets only 0.8-3.5 cm, petiole (0.7)1-4(5) cm, one or longest interpetiolar segment (0.7)1-3.5(4.2) cm; pinnae usually subaccrescent distally, rachis of distal or penultimate pair (3)3.5-9.5(11) cm, interfoliolar segments dilated upward, longer ones 5-13(15) mm; lfts bicolored, on upper face dark olivaceous brunnescent, on lower face dull, pale olivaceous or pale tan and almost always minutely, often remotely strigulose, accrescent upward from base of rachis or subequiform (except for broader furthest pair) upward from mid-rachis, in outline broadly or narrowly rhombic-oblong those near and above mid-rachis (11)12-31(35) x (4)4.5-14 mm, (1.7)2-3.3(4.5) times as long as wide. Peduncle (1.5)3-11(13) cm, in fruit ascending stout and lignescent, long persistent; capitula (30)35-75-flowered, fls strongly heteromorphic, pedicellate peripheral ones crowded on a narrowly clavate receptacle 2.5-5.5 mm, pedestal of 1-4(5) furthest, sessile or subsessile fls varying from drum-shaped and ± 1 mm diam. to linear and up to 5-8 mm; bracts dimorphic, those of peripheral fls linear or linear-oblanceolate, 0.7-2 mm, those of (sub)terminal ones linear-elliptic 2-4.5 mm, all early dry caducous; peripheral fls: pedicel of lower ones (4)5-11.5(13) x 0.1-0.2 mm; perianth 5(6)-merous, usually finely minutely strigulose overall, but sometimes only calyx-teeth and corolla-lobes puberulent, or whole perianth glabrous except for ciliolate corolla-lobes; calyx campanulate or turbinate-campanulate, (1)1.4-2.6 x 0.6-1.2(1.4) mm; corolla (2.6)3.5-6(6.5) mm; terminal fl(s): pedicel 0 or up to 1 mm, nearly as thick; calyx either broadly campanulate or cylindro-campanulate, 3-6 x (0.8)1.3-2.2 mm; corolla (6.5)7.5-11.5(15) mm. Pod 1-7(8) per capitulum, erect, either sessile or contracted at base into a pseudostipe 1-5 mm, in profile broad-linear, (4.5)5-10.5 x (0.9)1-1.7(1.9) cm, straight or nearly so (sometimes bent sideways, but not decurved), valves at first plano-compressed, becoming low-convex over seeds and depressed between them, mature fruit lomentiform, valves readily cracking between seeds and more reluctantly through dorsally plane, shallowly undulate sutures (0.8-1.6 mm wide), 1-seeded articles transversely oblong, as wide as pod and (5)5.5-9(10) mm long, along line of fracture 0.9-2 mm thick; seeds compressed but plump, in broad view 7.3-9.8 x 4-5.3(5.6) mm, testa firm crustaceous, tan or brown, closely investing hard greenish embryo, pleurogram complete, narrowly oblong ± 6-7.5 x 1.6-2.5 mm.

Distribution: SE Colombia, Ecuador, Peru, Bolivia, Venezuela, the Guianas, Brazil (Acre, Rondônia, Mato Grosso, Goiás, and Maranhão); in periodically or permanently inundated riparian forest and (S of the Guianas) in gallery forest, below 250 m; flowering mostly May to September, occasionally in other months (GU: 11; SU: 11; FG: 4).

Selected specimens: Guyana: Puruni R., FD 7771 (= JB 87) (NY); Rupununi R., Monkey Pond landing, SW of Mt. Makarapan, Maas *et al.* 7372 (NY). Suriname: Saramacca R., above Boschland, Maguire 24047 (NY); Para R. near Hannover, Teunissen *et al.* LBB 15314 (NY). French Guiana: S of Pedra Alice, at river margin, Irwin *et al.* 47586 (NY); riparian vegetation at Langa Tabiki, Maroni R., Prévost 1730 (CAY, NY).

2. **Hydrochorea gonggrijpii** (Kleinhoonte) Barneby & J.W. Grimes, Mem. New York Bot. Gard. 74(1): 25. 1996. – *Pithecolobium gonggrijpii* Kleinhoonte, Receuil Trav. Bot. Néerl. 22: 414. 1925. – *Arthrosamanea gonggrijpii* (Kleinhoonte) Kleinhoonte in Pulle, Fl. Suriname 2(2): 328. 1940. Type: Suriname, Forest Reserve Zanderij I, tree 141, BW 4357 (holotype U! (2 sheets)).

Pithecolobium pullei Kleinhoonte, Receuil Trav. Bot. Néerl. 22: 415. 1925.– *Arthrosamanea pullei* (Kleinhoonte) Kleinhoonte in Pulle, Fl. Suriname 2(2): 328. 1940. Type: Suriname, Zanderij I, Pulle 65 (holotype U! (2 sheets)).
Pithecolobium sabanensis Schery, Fieldiana, Bot. 28: 258. 1952. Type: Venezuela, Bolívar, Gran Sabana, S of Mt. Roraima, Steyermark 59146 (holotype F!).

Shrub or tree (1)2-20(27) m tall, young stems, all lf-axes and peduncles puberulent-tomentulose with forwardly incurved, sordid or rusty hairs. Lf-formula (ii)iii-vi(in juvenile lvs occasionally -viii)/(12)14-35 ("40"); stipules linear-lanceolate or -attenuate, 2-7.5 x 0.3-0.7 mm, early deciduous; lf-stks of lvs on most terminal fl stems (2.5)3-12 cm, of lvs on vigorous long-shoots and sapling branches up to 16-23 cm, petiole (0.8)1-2.4(5) cm, longer interpinnal segments (7)9-2(28) mm; pinnae accrescent distally, rachis of longer ones (3)4.5-8.5(10.5) cm, longer interfoliolar segments 2-4(in sapling lvs -6.5) mm; lfts conspicuously bicolored, glossy dark green (when dry brunnescent) and (except for sometimes ciliolate midrib) glabrous above, beneath pale olivaceous (drying tan), dull, either glabrous or minutely thinly strigulose, ciliolate or not, lfts except at far ends of rachis subequilong, in outline linear or narrowly oblong, those near and beyond mid-rachis (7.5)8-16(17.5) x (1.5)2-4(4.5) mm, 3.6-5(5.5) times as long as wide. Peduncle solitary or geminate, 2-5.5(8) cm; capitula subhemispherical, (11)17-40-flowered, clavate axis, including short terminal pedestal, 1.5-4 mm; bracts linear-oblanceolate, ± 1.5-2.5 mm, early caducous; fls heteromorphic, peripheral ones slenderly pedicellate, terminal 1-3 either sessile or stoutly short-pedicellate, with modified androecium; perianth 5-merous (or terminal fls randomly -8-merous), commonly strigulose overall but sometimes only distally; peripheral fls: pedicel of lowest fls (3)4-8 mm, at middle 0.2-0.3 mm diam.; calyx (1.7)2.6-4(4.8) x (1.1)1.3-2.9 mm, varying from openly turbinate-campanulate to narrowly campanulate; corolla 5.3-8.5(9) mm; terminal fl(s): calyx campanulate, sometimes deeply so, (2.8)3-6 x (1.6)1.8-3.3 mm; corolla (7.5)8-12(13.5) mm. Pod 1-7 per capitulum, not seen fully ripe but apparently quite similar in structure and dehiscence to those of *H. corymbosa*, commonly 7.5-9(11.5) mm wide, 8-13-seeded.

Distribution: Venezuela, Guyana and Suriname from the northern lowlands S to Roraima, Brazil; on river banks, at gallery margins, and in swamp-forest, at 40-1400 m; flowering October to April (GU: 14; SU: 6).

Selected specimens: Guyana: Along Ireng R., between Waipa and Sand Hill Rapids, Maguire *et al.* 46238 (NY); Makauria Cr., FD 3313 (= Fanshawe 577) (NY). Suriname: Poikakreek, Gonggrijp & Stahel 4211 (NY); Coppename R. headwaters, vicinity of camp 5, Maguire 25060 (NY).

14. **INGA**[18] Mill., Gard. Dict. Abr. ed. 4. 1754.
Type: I. vera Willd., Sp. Pl. 4: 1010. 1806.

Spreading crowned trees exhibiting most often Troll's architectural model with plagiotropic branches, trunk often irregular, compressed, sometimes slightly grooved, base sometimes buttressed or at least swollen, with rounded roots spreading outwards from trunk. Inner bark producing in several species a light, not sticky, whitish, red or yellow sap. Leaves alternate, once pinnate with 1-6 pairs of leaflets (see note below); stipules herbaceous or scaly, lanceolate, rarely foliaceous and broad, caducous or persistent; leaflets opposite; petiole and rachis terete and naked to dorso-ventrally compressed, marginate to more or less widely winged, provided with a cup-shaped or disc-shaped nectary between each pair of leaflets; leaflets foliaceous, elliptic, generally weakly asymmetrical at base, apex acuminate or not; venation pinnate, secondary veins parallel, often straight at base, then curved. Inflorescences of axillary heads or spikes, solitary or fascicled (3-6), 15-50(+)-flowered, in axil of either young developing or mature leaves, sometimes gathered on short shoots, in axil of undeveloped,

[18] by Odile Poncy

atrophied leaves and then simulating compound inflorescences, less often ramiflorous or cauliflorous; bracts more often scaly, oblanceolate or linear, seldom foliaceous, often deciduous; flowers homomorphic, often white or whitish; calyx and corolla 5-lobed, campanulate or tubular, calyx often irregular, generally glabrous or sparsely pubescent; corolla tubular or funnel-shaped, glabrous to densely pilose, 5 to 50 mm long, two to five times as long as calyx (see note below); stamens united in a tube either longer or shorter than corolla (i.e., "exserted" or "not exserted"); anthers dorsifixed, connective enlarged and circular; pollen 16-36-celled polyads; ovary 1, rarely 2, with 15-30 ovules; stigma tubular or funnel-shaped. Pods indehiscent, generally straight or slightly curved, twisted in a few species, coiled in one, sutures narrow to markedly widened; seeds covered with a bright white, edible pulp (seedcoat = sarcotesta), cotyledons colored (green, brownish or purplish), shiny.

Distribution: Ca. 300 species (Pennington, 1997) throughout C and S America, from Mexico to N Argentina, West Indies, with a great diversity in Amazonia; in rain forest, secondary open vegetation on wet soil; a few species have a mesic habitat; in the Guianas 58 species.

Vernacular names: Guyana: warakosa, waitey. Suriname: warakusa, swietie-boontje (Creole). French Guiana: pois-sucré (Créole); inga (Wayapi); waki (Galibi).

Phenology: Some species are likely to be caducifoliate (e.g., *I. alba*, *I. graciliflora*) in French Guiana (D. Loubry obs.), with no well-defined flowering season, but most flower in the dry season. Flower opening time (as observed in a few species) is specific, as there are day-blooming species, night-blooming species, and afternoon-blooming species, each flower lasting one day.

Economic uses: No use known in the Guianas, where Inga-trees are not cultivated. The pulpy seedcoat of wild trees is reportedly edible.

Notes: The descriptions given here are exclusively based on the materials collected in the Guianas; for the characters not available from Guianese specimens, the description is completed using extra-area specimens or literature, and the references mentioned.
The following conventions are used:

- Leaves: "jugate" refers to the number of pairs of leaflets: "leaf 2-jugate" = leaf with 2 pairs of leaflets, "leaf 5-jugate" = leaf with 5 pairs of leaflets.
- Flowers: an approximate ratio calyx length/corolla length is given for each species, e.g. 1/5 means corolla five times longer than the calyx.

LITERATURE

Leon, J. 1966. Central american and West Indian species of Inga (Leguminosae). Ann. Missouri Bot. Gard. 53: 265.

Pennington, T.D. 1997. The genus Inga. Botany. Royal Botanic Gardens, Kew, 844 p.

Pittier, H. 1916. A preliminary revision of the genus Inga. Contr. U.S. Natl. Herb. 18: 173.

Poncy, O. 1984. Graines, germinations et plantules dans le genre Inga (Légumineuses, Mimosoideae): étude morphologique chez quelques espèces de Guyane française. Bull. Soc. Hist. Nat. Toulouse 120: 36-42.

Poncy, O. 1985. Le genre Inga (Légumineuses, Mimosoideae) en Guyane française. Mém. Mus. Natl. Hist. Nat., B, Bot. 31: 1-126, 12 pl.

Poncy, O. 1991. Deux nouvelles espèces de Inga (Mimosaceae) et notes nomenclaturales et taxonomiques sur trois autres espèces. Bull. Mus. Natl. Hist. Nat., B, Adansonia 13(3-4):147-154.

Souza, M. 1993. El genero Inga (Leguminosae: Mimosoideae) del Sur de Mexico y Centroamerica, estudio previo para la Flora Mesoamericana. Ann. Missouri Bot. Gard. 80: 223-269.

KEY TO THE SPECIES

(The key relies as much as possible on vegetative characters, to aid in identification of sterile material. Where the key is based on variable characters, such as leaflet-number, the observed variation is included in the key, unless it occurs very infrequently).

1 Plant strongly hairy, mainly on young twigs, flowers, leaf-rachis, veins and pods 2
Plant not strongly hairy 9

2 Petiole and leaf-rachis not winged; pod tetragonal (sutures > 1 cm wide) . . . . *48. I. rubiginosa*
Petiole winged or not, leaf-rachis winged at least partially; pod not as above 3

3 Leaflets (2)3 pairs; high dense rainforest 4
Leaflets 4-5 pairs; low, mesophytic forest or near of it 5

4 Petiole and rachis segments winged at least part of them, hairs long, hirsute, upper leaflets < 15 cm long; calyx > 1 cm long, glabrous or if hair very sparse; pod often twisted, covered with dense brown hirsute hairs . . . . *17. I. fastuosa*
Petiole not winged, rachis segments winged, hairs short, velvety, upper leaflets up to 30 cm long; calyx < 1 cm long, velvety pubescent; pod not as above . . . . *30. I. lomatophylla*

5 Leaflets narrow, ca. 4 times as long as broad; calyx and mature pod sparsely pubescent; pod straight and flat . . . . *31. I. longiflora*
Leaflets broad, ca. 2 as long as broad; calyx and mature pod densely velvety or pilose 6

6 Mature pod almost cylindrical and grooved, velvety . . . . *24. I. ingoides*
Mature pod flat, often twisted, densely rusty pilose 7

7 Calyx tubular > 2 cm long . . . . *21. I. grandiflora*
Calyx not tubular < 2 cm long 8

8 Leaflets strictly 4 pairs; calyx funnel-shaped > 1 cm long . . . . *10. I. calanthoides*
Leaflets 4-5 pairs; calyx campanulate < 1 cm long . . . . . *12. I. cayennensis*

9 Leaflets never longer than 5 cm (plant entirely glabrous) 10
Leaflets longer than 5 cm, at least the distal ones (plant glabrous or pubescent). 12

10 Inflorescence subcapitate; foliar nectaries stipitate, leaflets 2-4 pairs . . . . *22. I. heterophylla*
Inflorescence capitate; foliar nectaries not stipitate, leaflets 3-6 pairs 11

11 Tall tree, dense high rainforest; leaflets 3-5 pairs, maximum length 5 cm; pod 10-20 cm long. . . . . *20. I. gracilifolia*
Treelet or shrub, low mesophytic bush; leaflets 4-6 pairs, maximum length 3 cm; pod < 10 cm long . . . . *58. I. virgultosa*

12 Stipules persistent very developed, foliaceous, rounded, wider than long, up to 2 cm wide . . . . *52. I. stipularis*
Stipules not as above 13

13 Leaflets 1 or 2 pairs. 14
Leaflets 3 pairs or more 40

14 Leaflets 2 pairs 15
Leaflets 1 pair (= leaves bifoliolate). 37

15 Foliar rachis narrowly to broadly winged. 16
Foliar petiole and rachis not winged, terete or flattened or marginate 26

16 Inflorescence spicate 17
Inflorescence umbellate or nearly so 21

17 Flowers minute, calyx < 2 mm long, regular 18
Flowers not minute, calyx > 2 mm long, irregular (i.e. markedly cleft on one or two sides, or the 5 teeth not equal); 20

18 Leaves velvety pubescent on veins and inferior surface; rachis of inflorescence < 2 cm long .... *8. I. brachystachys*
Leaves glabrous; rachis of inflorescence > 2 cm long 19

19 Petiole and rachis broadly winged (4-6 mm wide), wing triangular distally; calyx ≥ 1.5 mm long .... *13. I. cordatoalata*
Petiole and rachis narrowly winged (up to 4 mm wide), not triangular distally; calyx ≤ 1 mm long .... *35. I. marginata*

20 Flowers bright yellow; mature pod yellow, looking fleshy < 15 cm long ... . *44. I. pilosula*
Flowers white or yellowish-white;mature pod dark green, rigid > 15 cm long .... *51. I. splendens*

21 Inflorescence subcespitose, rachis clavate ca. 5 mm long .... *29. I. leptingoides*
Inflorescence cespitose, flowers on a spherical head 22

22 Leaflet winged only distally; pod > 20 cm long .... *19. I. graciliflora*
Leaflet winged on entire length; pod < 20 cm long 23

23 Pod markedly asymetrical at base; corolla ≤ 5 mm long ....*41. I. obidensis*
Pod symmetrical at base; corolla > 5 mm long...................... 24

24 Larger leaflet < 10 cm long, nectaries inconspicuous ca. 1mm .... *25. I. jenmanii*
Larger leaflet > 10 cm long, nectaries conspicuous 2-3 mm 25

25 Leaflets (2)3 pairs; calyx ≥ 5 mm........................ *9. I. brevipes*
Leaflets (1)2 pairs; calyx < 5 mm.................... *56. I. umbellifera*

26 Inflorescence spicate.. 27
Inflorescence umbellate 34

27 Flowers minute, calyx ≤ 2 mm long, regular 28
Flowers not minute, calyx > 2 mm long, irregular................... 30

28 Leaflets 2-3 pairs, foliar nectary raised to stipitate; total length of inflorescence < 4 cm long, calyx ≥ 1,5 mm .... *37. I. mitaraka*
Leaflets 2 pairs, foliar nectary sessile; inflorescence length > 4 cm long, calyx < 1.5 mm long.. 29

29 Mature pod plump, fleshy-like, yellow (2-3 cm wide); seasonal forest and scrubland .... *27. I. laurina*
Mature pod slender (ca. 1 cm wide), envelope thin, papery; undisturbed and secondary rainforest .... *35. I. marginata*

30 Flowers pubescent, corolla white-yellowish or greenish 31
Flowers glabrous, corolla white pinkish or reddish 33

31 Leaf coriaceous; calyx velvety pubescent, > 2 mm diam.; pod > 2.5 cm wide, pericarp thick and rigid, seeds not prominent . . . . *51. I. splendens*
Leaf not coriaceous; calyx narrow (> 2 mm diam.), glabrous or with sparse hairs; pod < 2.5 cm wide, pericarp thin and supple, seeds prominent . . . 32

32 Leaf petiole plus rachis < 6 cm long; rachis of inflorescence < 2 cm long, calyx < 5 mm long . . . . *28. I. leiocalycina*
Leaf petiole plus rachis > 6 cm long; rachis of inflorescence > 2 cm long, calyx > 5 mm long . . . . *32. I. longipedunculata*

33 Petiole and leaf rachis woody; mature pod woody, verrucose, brownish, > 2.5 cm wide; leaflets 2-3 pairs . . . . *49. I. sarmentosa*
Petiole leaf rachis not woody; mature pod not woody, yellow, < 2.5 cm wide; leaflets always 2 pairs. . . . . *11. I. capitata*

34 Leaflets strictly 2 pairs, dark green (living and herbarium specimen either), pulvini not different in color, terminal leaflets markedly twice as large as proximal ones, nectaries prominent, conspicuous, > 1 mm wide. 35
Leaflets (2)3 pairs, often discolor, pulvini blackish, terminal leaflets not twice as large as proximal ones, nectaries < 1 mm wide 36

35 Calyx pubescent; pod brown-yellowish and velvety, verrucose with oblique ribs; non flooded forest. . . . . *23. I. huberi*
Calyx glabrous; pod smooth, bright yellow at maturity; riverine forest. . . . .
. . . . *50. I. sertulifera*

36 Calyx < 2 mm, corolla < 7 mm; mature pod up to 40 x 4 cm, margins undulating and thickened, seeds not contiguous and prominent . . . .
. . . . *33. I. loubryana*
Calyx > 2 mm, corolla > 8mm; mature pod up to 28 x 2.5 cm, straight, seeds contiguous and not prominent . . . . *.42. I. paraensis*

37 Petiole > 15 mm long and conspicuously winged, wing triangular . . . .
. . . . *40. I. nubium*
Petiole < 15 mm long, poorly winged, channelled 38

38 Inflorescence spicate; leaflets always 1 pair . . . . *39. I. nouragensis*
Inflorescence umbellate; leaflets 1-2(3) pairs 39

39 Peduncle of inflorescence > 1 cm, corolla > 5 cm long . . . . . *25. I. jenmanii*
Peduncle of inflorescence < 1 cm, corolla < 5 cm long . . . . *.41. I. obidensis*

40 Leaflets 3 pairs 41
Leaflets 4 pairs or more 63

41 Foliar rachis conspicuously winged (i.e. > 3 mm wide), either for entire length of each segment, or only for a part of it, or only the distal segment winged 42
Foliar rachis not winged or occasionally poorly so (i.e. < 3 mm wide) . . . 53

42 Stipules conspicuous > 10 x 3 mm; calyx > 1.5 cm long, corolla > 3 cm long. 43
Stipules < 10 x 3 mm; calyx < 1.5 cm long, corolla < 3 cm long. 44

43 Total leaf petiole + rachis length < 7 cm; leaflets (2)3 pairs; total inflorescence peduncle + rachis length < 4 cm. . . . . *45. I. poeppigiana*
Total leaf petiole + rachis length > 10 cm; leaflets 3(4) pairs; total inflorescence peduncle + rachis length > 8 cm. . . . . *34. I. macrophylla*

44 Foliar rachis-segments winged for entire length; flowers not minute, calyx > 1.5 mm long 45
Foliar rachis segments markedly winged only just below insertion of leaflets, or only distal rachis segment winged; flowers minute, calyx ± 1 mm long 50

45 Inflorescence umbellate (rachis clavate). . . . . *9. I. brevipes*
Inflorescences spicate 46

46 Leaves pubescent 47
Leaves glabrous 48

47 Leaves with sparse but conspicuous hairs on rachis and inferior surface of leaflets, leaflets 3-4 pairs; flowers arranged distically, floral bracts perpidencular to rachis and markedly sheathing flower; calyx < 7 mm long. . . . . *14. I. disticha*
Leaves velvety or scabrous, leaflets 3(4) pairs; flower arrangement not as above; calyx > 7 mm long . . . . *47. I. rhynchocalyx*

48 Leaves coriaceous; rachis of inflorescence > 2 cm long; calyx up to 8 mm long, calyx and corolla densely pubescent; pod thick, woody, > 2.5 cm wide . . . . *51. I. splendens*
Leaves not coriaceous; rachis of inflorescence < 2 cm; calyx < 5 mm long, calyx and corolla glabrous or unconspicuously pubescent; pod not woody, < 2.5 cm wide 49

49 Small understorey tree; leaflets maximum length 10 cm; calyx ca. 2 mm . . . . *6. I. auristellae*
Middle-sized to large tree; leaflets minimum length cm; calyx 4-5 mm . . . . *1. I. acreana*

50 Inflorescence umbellate, most often at defoliated nodes; pod length 25-35 cm; petiole and rachis woody (leaflets 2-3 pairs) . . . . *19. I. graciliflora*
Inflorescence spicate, axillary to leaves; maximum pod length 20 cm . . . 51

51 Inflorescences most often on axillary short shoots simulating compound inflorescences; pod maximum 2 cm wide (leaflets 3(4) pairs). . . . *4. I. alba*
Inflorescences most often fascicled in leaf axis; pod > 2 cm wide 52

52 Petiole < 1 cm length, wing on foliar rachis very short just below insertion of leaflets and raising around nectaries, nectaries < 2 mm diam.; pod straight ca. 2.5 cm wide(leaflets exactly 3 pairs) . . . . *7. I. bourgonii*
Petiole > 1 cm long; wing of rachis not as above, nectaries well-developed 3 mm diam. or more; pod curved minimum 3 cm wide (leaflets 3-4 pairs) . . . . *43. I. pezizifera*

53 Petiole and rachis marginate, or grooved, or flattened 54
Petiole and rachis terete (section circular or nearly so) 57

54 Upper leaflets > 13 cm long, petiole and rachis flattened to marginate, sometimes with a very short triangular wing below leaflet insertion (inflorescences spicate or umbellate) 55
Upper leaflets < 11 cm long, petiole and rachis marginate to grooved (inflorescences umbellate) 56

55 Petiole and rachis more or less woody, leaflets 2-3 pairs; inflorescence capitate mostly at defoliated nodes . . . . *19. I. graciliflora*
Petiole and rachis not woody, leaflets 3-4 pairs; inflorescence spicate mostly on axillary short shoots. . . . . *4. I. alba*

56 Leaflets 3 to 5 pairs, no leaflet longer than 7 cm; dense high rainforest, large tree (see couplet 11) . . . . *20. I. gracilifolia*
Leaflets 2 to 4 pairs, upper leaflets up to 10 cm; mixed mesophytic or low forest. . . . . *26. I. lateriflora*

57 Leaves velvety-tomentose on petiole, rachis, veins; flowers minute (calyx ± 1 mm long) (inflorescence spicate) . . . . *5. I. albicoria*
Leaves entirely glabrous; flowers not minute (calyx > 1.5 mm long) 58

58 Petiole and rachis woody, stocky, lenticellate, leaflets coriaceous (stipules very conspicuous and persistent) 54
Petiole and rachis not as above (stipules unconspicuous, or caducous) . . . 55

59 Petiole 1 cm long or less; pod ca. 3 cm wide, verrucose, not cauliflorous; canopy tree; (leaflets (2)3 pairs). *49. I. sarmentosa*
Petiole > 5 cm long; pod ca. 2 cm wide, reticulate, often cauliflorous; understorey small tree (leaflets (3)4 pairs . . . . *46. I. retinocarpa*

60 Inflorescence spicate; foliar nectaries stalked. . . . . *37. I. mitaraka*
Inflorescence umbellate; foliar nectaries sessile. 61

61 Stipules large (up to 15 x 5 mm), caducous, maximum leaflet length 20 cm; flowers sessile. . . . . *18. I. flagelliformis*
Stipules much smaller, maximum leaflet length 13 cm; flowers pedicellate 62

62 Calyx < 2 mm, corolla < 7 mm; mature pod up to 40 x 4 cm, margins undulating and thickened, seeds not contiguous and prominent .... *33. I. loubryana*
Calyx > 2 mm, corolla > 8mm; mature pod up to 28 x 2.5 cm, straight, seeds contiguous and not prominent .... *42. I. paraensis*

63 Leaflets maximum 4 pairs 64
Leaflets 4 pairs or more 74

64 Leaflets stricly 4 pairs 65
Leaflets (2)3-4 pairs 67

65 Leaflets glabrous on both faces .... *2. I. acrocephala*
Leaflets velvety-tomentose beneath 66

66 Calyx ≥ 4 mm; pod straight, markedly ribbed, velvety, quandrangular in section .... *48. I. rubiginosa*
Calyx ≤ 3 mm; pod flat, glabrous, curved .... *36. I. melinonis*

67 Leaf rachis not conspicuously winged (terete or marginate) 68
Leaf rachis conspicuously winged 72

68 Petiole > 1 cm long, reduced to pulvinus, proximal nectary markedly wider than others (a strictly riverine species) .... *38. I. nobilis*
Petiole > 1 cm long, nectaries not different in size 69

69 Petiole and rachis robust and woody, petiole > 10 cm long (a small understory species) .... *46. I. retinocarpa*
Petiole not woody, < 10 cm long 70

70 Leaves pubescent (rachis and inferior surface of leaflet); flowers > 1 mm long .... *54. I. suaveolens*
Leaves glabrous; flowers minute, ca. 1 mm long 71

71 Foliar nectaries very prominent, pezize-like; pod curved, thick, > 2 cm wide .... *43. I. pezizifera*
Foliar nectaries irregular, not conspicuously prominent; pod straight, thin, < 2 cm wide .... *4. I. alba*

72 Stipules > 6 mm wide; corolla > 3 cm long .... *34. I. macrophylla*
Stipules < 6 mm wide; corolla < 3 cm long 73

73 Calyx pubescent; pod flat < 13 cm long, hairy .... *14. I. disticha*
Calyx glabrous, striate; pod quadrangular, > 13 cm long, glabrous .... *53. I. striata*

74 Foliar rachis not winged or very narrowly so, or only distally on each segment 75
Foliar rachis conspicuously winged on entire length of each segment 77

75 Leaves pubescent (rachis and inferior surface of leaflets) . . . . 55. *I. thibaudiana*
Leaves glabrous 76

76 Leaflets < 7 cm long; mature pod straight and dark green . . . . 20. *I. gracilifolia*
Leaflets > 7 cm long; mature pod coiled and orange . . . . 16. *I. fanchoniana*

77 Leaves glabrous . . . . 3. *I. alata*
Leaves pubescent or velvety (rachis and inferior surface of leaflets). 78

78 Foliar nectaries not circular, kidney shape (pod sulcate) . . . . . . 15. *I. edulis*
Foliar nectaries circular 79

79 Pod flat and glabrous . . . . 54. *I. suaveolens*
Pod sulcate (sutures hypertrophied) and velvety 80

80 Leaves small (distal leaflets < 12 cm); bud ovoid, calyx campanulate (not riverine) . . . . 24. *I. ingoides*
Leaves large (distal leaflets > 12 cm long); bud cylindrical, calyx tubular (riverine). . . . . 57. *I. vera* subsp. *affinis*

1. **Inga acreana** Harms, Notizbl. Königl. Bot. Gart. Berlin 6: 298. 1915. Type: Peru, Rio Acre, Ule 9425 (holotype B, destroyed, isotypes G, K, MG, photo US).

Small tree up to 10 m high; plant glabrous except on flowers, and very young leaves; branchlets with elongated lenticels. Leaves 3-4-jugate; stipules linear, 7-8 x 1-2 mm, caducous; petiole and rachis marginate to winged; petiole 2.5-3 cm long, segments of rachis up to 4 cm long, wing obtriangular up to 10 mm wide just below insertion of leaflets; nectaries orbicular, sessile, 1 mm diam.; leaflets elliptic, proximal ones 5-7 x 2-3 cm, markedly smaller than medium and distal ones 10-13(18) x 3.5-5(6) cm; base acute, apex with a 1-1.5(2)cm long, mucronate acumen; blades crimped, glabrous on both faces, or puberulate when young. Inflorescence of 2-3 spikes axillary to developed or undeveloped leaves (in the latter case, inflorescences simulate a terminal panicle), or on short lateral shoots, 4-8 cm long; peduncle slender, 3-4 cm long; rachis ca. 1 cm long, 30-flowered; bracts scaly, spatulate, 1-2 mm long; flowers sessile, bud cylindrical; calyx narrow, tubular, ca. 5 mm long, slightly pubescent, regular, teeth ca. 1 mm; corolla tubular, 7-8 mm, pubescent; ratio calyx/ corolla length ca. 1/2; staminal tube not exserted. Pod 11-14 x 1.5-2(2.5) cm, flat, glabrous, rigid, greenish brown when ripe, asymmetrical at base, valves shallowly rippled, sometimes lenticelled, seeds prominent, endocarp abundant, stringy, expanding in between seeds; seeds not contiguous, cotyledons 2 x 1-1.3 cm, dark blue-green.

Distribution: An uncommon species of the dense primary rainforest, either riverine or non-flooded; known from western Amazonia and Venezuela, Suriname and French Guiana (SU: 2; FG: 8).

Selected specimens: Suriname: Emmaketen, Daniëls & Jonker 878 (U); Juliana Top, Irwin *et al.* 54771 (G, NY, U). French Guiana: Arataye R., Poncy 474 (P), Sastre 5597 (P); Trail Belizon-Saül, Aubréville 315 (CAY, P, U); Saül, Beekman 63 (CAY, P), Feuillet 482 (CAY, P), Sabatier 1052 (CAY, P); Sinnamary, Piste-de-St-Elie, Sabatier & Prévost 3876 (CAY, P); Upper Maroni Basin, Atachi-Bacca Mts., de Granville 10617 (CAY, NY, P, U, US).

2. **Inga acrocephala** Steud., Flora 26: 759. 1843. Type: Suriname, Suriname R., Hostmann 1067 (holotype U, isotypes K, G, P, NY).
– Fig. 20 E-H

Medium-sized tree; bark greyish to whitish, with transversally elongated irregular lenticels; branchlets slightly sulcate, greenish grey. Leaves 4-jugate, entirely glabrous, even when young, except on stipules; stipules laciniate or oblanceolate, 4-5 x 2-3 mm, rounded at apex, pubescent, caducous; petiole and rachis unwinged, sometimes narrowly marginate; petiole 2.5-4 cm long; rachis segments 2-5 cm long; nectaries cup-shaped, sessile or shortly stipitate, 1-1.5 mm diam.; leaflets narrowly elliptic, proximal leaflets 6-10(12) x 2.5-4(5)cm, distal ones 10-16(22) x 2.5-4(5)cm); base acute, asymmetrical; acumen 1-2 cm long, with a mucronate apex; blades smooth and dark green above; young leaves hanging, bright dark purple, with whitish nectaries. Inflorescence spikes never solitary but 2-5 together in axils of developed or undeveloped leaves (in the latter case, the inflorescences simulate a terminal panicle), or on short lateral shoots and then simulating a lateral panicle; peduncle variable in length (2 to 7 cm), slender, sparsely tomentose; rachis 1-2 cm, bearing 20-30 closely appressed, sessile flowers; bracts 1-2 mm long, scaly, spatulate and curved, persistent; bud tubular; calyx and corolla sparsely pubescent; calyx campanulate 2-4 cm long, the 5 teeth unequal; corolla funnel-shaped, 5-7 mm long; ratio calyx/corolla length 1/3-1/2; staminal tube not or shortly exserted. Pod on a thick peduncle (up to 5 mm diam.), large (20-25 x 3-4 cm), ligneous-fibrous with many lenticels, glabrous, bright and greenish-brown when ripe, sutures poorly developed; seeds 10-15, cotyledons up to 2.5 x 1.5 cm, green. Seedling: epicotyl up to 25 cm long, first two leaves alternate, bifoliolate with a marginate, 1-2 cm long petiole and narrowly elliptic leaflets (7-10 x 3-4 cm).

Distribution: A species of high dense non-flooded rainforest, occurring in the Guianas and Brazil (Pará); the unique record from Guyana refers to a sterile specimen placed here with doubt (GU: 1?; SU: 11; FG: 9).

Selected specimens: Guyana: Takutu Mts., Smith 3178 (K, NY, P, US). Suriname: Mennega 508 (U, US); Clevelandia, Maguire *et al.* 47112 (U); Coppename R., BW (Gonggrijp & Stahel) 6242 (U); Jodensavanne, BW 5115 (U), Schulz 8271 (U), 7492 (U); Kabalebo Dam Project, Lindeman & Görts *et al.* 345 (BBS, U); Litany R., de Granville *et al.* 11948 (BBS, CAY, P, U, US); Maratakka R., Snake Cr., Maas & Tawjoeran LBB 10826 (BBS, P, U); Tafelberg, Maguire *et al.* 24121 (G, K, P, NY, RB, US); Wilhelmina Gebergte, Lucie R., Schulz LBB 10379 (BBS, U). French Guiana: Piste de Kaw, Poncy 592 (CAY, P); Saül, Mori & Pipoly 15567 (CAY, P, NY); Sinnamary, Bordenave 864 (CAY, K, NY, P, U); St-Laurent-du-Maroni, Godebert, Wachenheim 204 p.p. (P), 312 (P, U), 384 (P); St-Laurent-Mana Rd., SF 31M (P), 7332 (CAY, P); Upper Oyapock R., Trois-Sauts, Grenand 989 (CAY, P).

Vernacular name: French Guiana: inga pini (Wayapi).

3. **Inga alata** Benoist, Bull. Mus. Natl. Hist. Nat. 27: 198. 1921. Type: French Guiana, Mélinon 8a (1845) (holotype P).

Middle-sized tree, up to 30 m x 60 cm; bark light grey-brown, with horizontal rows of dense, red-brown lenticels, branchlets angular, velvety. Leaves (3)4(5)-jugate; stipules either lanceolate, spatulate or rhomboid, 6-8 x 2-4 mm, glabrous or sparsely pubescent, longitudinally scored, caducous; petiole and rachis segments 2-4 cm long, petiole thick, flattened to marginate; rachis segments broadly marginate to winged, 3 to 6 mm wide; nectaries circular to elliptic, conspicuous, sessile or shortly stipitate, 1.5-2 mm diam.; leaflets acuminate, base cuneate to acute; blades chartaceous and crimped, glabrous or sparsely pubescent beneath; young leaves often reddish with a white tomentum, nectaries darker red. Inflorescence fascicles of 3-5 spikes either in axil of mature leaves or axillary to short shoots, 4-5 cm long; peduncle slender and short, 1-3.5 cm long; rachis 1-1.5 cm long, 30-50-flowered; bracts inconspicuous (< 1 mm long), scaly, recurved, persistent; flowers sessile, glabrous or very slightly tomentose; bud obtrullate; calyx tubular 1-2 mm long, regular; corolla funnel-shaped 5-7 mm, with 1-2 mm long, sharp lobes; ratio calyx/corolla length ca 1/4; staminal tube shorter than the corolla. Pod 15-25 x 2.5-3.5 cm, glabrous, flat with the seeds often prominent, dark green when ripe, mucronate at apex; margins poorly developed, valves shallowly rippled, eventually verrucose; seeds 8-14, not contiguous, cotyledons green, 1.5 x 1 cm. Seedling: first two leaves opposite, unijugate, stipules ca 5 mm long, laciniate, petiole winged, ca 2 cm long, leaflets with undulating margins (note: seedlings in early development are identical to those of *I. pezizifera*).

Distribution: An uncommon species of undisturbed dense non-flooded rain forest; the Guianas, Amazonia, westward to Ecuador, known from Suriname only from 2 uncertain sterile collections (GU: 1; SU: 2?; FG: 11).

Selected specimens: Guyana: Surama, McDowell *et al.* 2080 (P, U, US). French Guiana: Arataye R., Saut-Pararé camp, Poncy 124 (CAY, P), 220 (CAY, P), Sastre 4830 (CAY, P), Villiers 3730 (CAY, NY, P, U); Nouragues camp, Prévost & Sabatier 2713 (CAY, P); Ile Cayenne, Fort Diamant, Poncy 256 (CAY, P, U); Inini Mts., Bellevue, de Granville *et al.* 7795 (CAY, P, U); Haut Tampoc, Cremers 4586 (CAY, P); Upper Oyapock R., Trois-Sauts, Grenand 716 (CAY, P), Jacquemin 1864 (CAY, P), Ouroucourou, Oldeman B3199 (CAY, P, U).

Vernacular name: French Guiana: masulapa (Wayapi).

4. **Inga alba** (Sw.) Willd., Sp. Pl. 4: 1013. 1806. – *Mimosa alba* Sw., Fl. Ind. Occid. 2: 976. 1800. Type: French Guiana, "Cayenne", von Rohr s.n. (holotype BM, isotype C). – Fig. 18 E-G

Inga thyrsoidea Desv., Journ. Bot. 3: 71.1814. Type: "Hab. in Gujana", Desvaux s.n. (holotype P).

Large tree, up to 40 m x 1.20 m; trunk irregular, with heavy, rounded buttresses up to 1.50 m high; bark brown-reddish to orange-red, dotted with many small lenticels, slash exudes a reddish-translucent sap; branchlets with velvet-like, rust-colored pubescence. Leaves 3(4)-jugate, very variable in size; stipules linear, ca. 7 mm long, often persistent; petiole and rachis velvety beneath; petiole terete or marginate, 2-4.5 cm long; rachis segments 2-3 cm long, marginate or winged, only distal part of rachis often winged just below insertion of leaflets; nectaries 1-2 mm in diam., sessile, variable in shape but prominent and often wider than rachis; leaflets glabrous, elliptic, variable in size and shape, distal ones up to 10 x 5 cm, apex acute or acuminate, base obtuse to cuneate; blades greyish beneath. Inflorescence of axillary spikes generally on short lateral shoots up to 6(8)cm long, 2-3 in axil of abortive leaves; peduncle tomentose, 5-10 mm long; rachis ca. 1 cm long, with 15-25 close flowers; bracts inconspicuous, scaly, spatulate; bud clavate, calyx minute (ca. 1 mm long) with sparse hirsute hairs, teeth short; corolla tomentose, funnel-shaped, 3-4 mm long; ratio calyx/corolla length 1/4; staminal tube 10 mm or more, exserted at length. Pod (10)15-20 x 1-2 cm, margins thin, valves leathery, thin, greyish-green, smooth to velvet-like, with reticulate thin shallow ripples; seeds 9-14, prominent, not contiguous, small with thin pulp, cotyledons dark green to brown, ca. 13 x 7 mm. Seedling with two first leaves opposite, stipules linear, 3 mm, petiole marginate, 1.5 cm, nectary red, leaflets reddish then greenish-grey, dull, glabrous.

Distribution: A widely distributed species, rather common, sometimes cultivated; in lowland dense rainforest, also in disturbed secondary forests, bushy savanna areas (GU: 11; SU: 7; FG: 19).

Selected specimens: Guyana: de la Cruz 4440 (NY, US); Tillett & Tillett 45861 (NY, US); Sandwith 1085 (K, U); Kamazuka, de la Cruz 2763 (NY, US), 2832 (NY, US); Kanuku Mts., Massif W, A.C. Smith 3351 (NY, P, U), Moco-Moco, Maas & Westra 3901 (P, U); Mazaruni Station, FD (Davis) 2292 (FDG, K), Upper Mazaruni, Gleason 2832 (K); Rupununi, Jansen-Jacobs *et al.* 2286 (BRG, P, U); Tumatumari, Gleason 287 (K, US). Suriname: Brownsberg, Tawjoeran LBB 12576 (BBS, U), Vreden LBB 13728 (BBS, K, U), Sectie O, BBS 39a (P, U); Coppename, Stahel & Gonggrijp BW 6174 (K, U); Jodensavanne, Mapane Cr., Roberts LBB 9872 (BBS, P, U); Lely Mts., Mori & Bolten 8489 (NY, P, U); Voltzberg, Raleighfalls, Roberts LBB 14041 (P, U). French Guiana: sin loc., Desvaux s.n. (P); Cayenne, Benoist s.n. (P); Cabassou, Prévost 440 (CAY, P); Matoury, Leblond s.n. (P, US); Saül, Boom & Mori 1730 (CAY, NY, P), Approuague-Arataye R., Station Nouragues, Sabatier 3499 (CAY, NY, P, U, US); Kourou, Alexandre 81; Oyapock R., Irwin *et al.* 48067 (K, NY, U, US); Saül, Boom & Mori 1730 (CAY, NY, P, U); Savane Dorothée, Hallé 4029 (MPU, P); Maroni R., Mélinon 79 (P), 250 (K, P); Saint-Laurent-du-Maroni, Acarouani, SF 7667 (P, U), Sagot 977 (K, P, U); Mana Rd., SF 160 M (P, U), 206 M (BBS, P, U), 6128 (P) ; Regina-Rd., Loubry 30 (CAY, P, U); Sinnamary, Piste-de-St-Elie, Loubry 1267 (CAY, NY, P, U); Trois-Sauts, Grenand 1412 (CAY, P).

Vernacular names: Guyana: mapourokuni (or "maporokon") (Arawak), whikie. Suriname: rodebast prokonie, prokonie. French Guiana: sisi-pilan (Wayapi), lebi-weko (Aluku), abonkini (Saramac.).

Use: Canoe (Djukas).

Note: *Inga alba* is the tallest *Inga* species.

5. **Inga albicoria** Poncy, Bull. Mus. Natl. Hist. Nat., B, Adansonia 18(1-2): 67. 1996. Type: Guyana, Essequibo R., Groete Cr., Fanshawe 1252 = FD 3988 (holotype K, isotypes BRG, FDG, NY). – Fig. 13

Medium sized tree to 25 m tall; bark cream-colored, slash green then whitish; branchlets light brown, puberulous, lenticels beige to whitish. Leaves 3-jugate; stipules triangular, more or less falcate, 3-4 x 1 mm, caducous; petiole and rachis slightly flattened to marginate, brownish pubescent; petiole 1-2(+) cm, segments of rachis 1.5-3 cm; nectaries circular, prominent, deep cup-shaped or funnelform, ca. 1 mm in diam.,

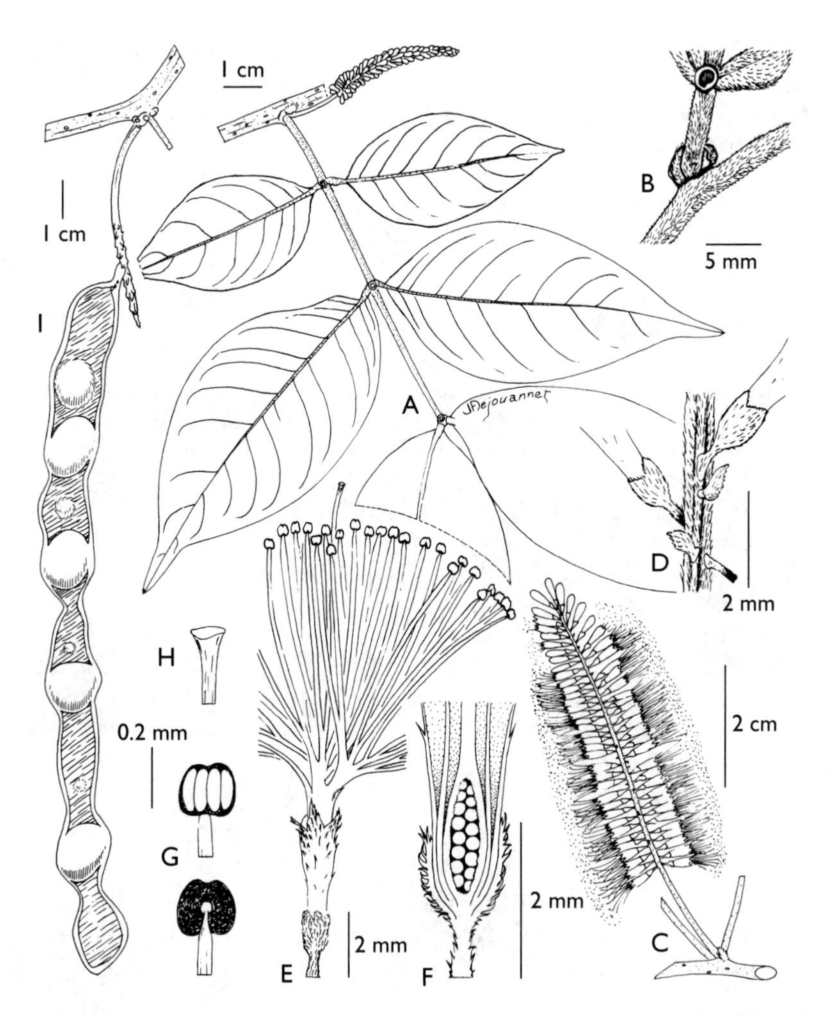

Fig. 13. *Inga albicoria* Poncy: A, habit: leaf and young inflorescence; B, base of a young leaf, nectary, stipules; C, inflorescence; D, inflorescence rachis, detail; E, flower; F, flower, longitudinal section; G, anther, ventral and dorsal views; H, stigma; I, pod (A-H, Sastre 5709, 5880; I, de Granville *et al.* 10961). Drawing by Jean-François Dejouannet; reproduced with permission from Poncy, Adansonia 18: 69. 1996.

1 mm high; leaflets elliptic, proximal ones 5-7 x 1.5-2.5 cm, distal ones 7-12 x 2.5-5 cm, base acute, almost symmetrical, apex acuminate and apiculate, surface glabrous above, fine velvet-like pubescence beneath, this denser on veins. Inflorescence of 1-3 spikes in axil of either old or contemporary leaves; peduncle 1.5-2.5 cm, generally with a bract in the middle; rachis 5-7 cm long, bearing over 50 well separated, sessile or

very shortly pedicellate flowers; bracts inconspicuous, hooked, spatulate, ca. 1 mm long; flower bud clavate, apiculate; calyx and corolla with very tiny and sparse inconspicuous hairs, calyx campanulate, minute, 1 mm long, corolla funnel-shaped, 4-5 mm long; ratio calyx/corolla length ca 1/4; staminal tube slightly exserted. Pod flattened, glabrous, to 20 x 1.5 cm, valves shallowly rugose, sutures straight or undulating, not enlarged; seeds 9-14 not contiguous, prominent.

Distribution: Dense primary undisturbed rainforest; Guyana, French Guiana, and Brazil (Amapá and Pará) (GU: 1; FG: 12).

Selected specimens: French Guiana: Inini Mts., Mt. Atachi Bacca, de Granville 10961 (CAY, G, NY, P, US, VEN); Kaw, Billiet & Jadin 6418 (CAY); Rémire, Poncy 1360 (CAY, P); R. Mana, Saut-Ananas, Cremers 7541 (CAY, P); R. Approuague, R. Arataye, Saut-Pararé, Sastre 5709 (CAY, P, U), 5717 (CAY, P), 5880 (CAY, P, U), Villiers 3729 (CAY, P), Station Nouragues, Poncy 1039 (CAY, NY, P, U), Sabatier & Prévost 2709 (CAY, K, NY, P, U); Saül, Mori *et al.* 18766 (CAY, NY, P, U), 24762 (CAY, NY, P).

Note: A species belonging to section Bourgonia (spiky inflorescences, small flowers, minute calyx), which includes several closely related species in the Guianas. It also shows strong affinities with *Inga cylindrica* Mart. (sect. *Bourgonia*) from S and NE Brazil, Pennington (1997) treats it as a synonym of the latter. Curiously not recollected in Guyana since the type collection.

6. **Inga auristellae** Harms, Notizbl. Königl. Bot. Gart. Berlin 6: 298.1915. Type: Peru, Alto Acre, Seringal Auristella, Ule 9426 (isotypes G, U, US). – Fig. 26 G-K

Small understorey tree to 10 m tall; trunk slender, sometimes with small narrow buttresses; bark thin, pale brownish to greyish, no lenticels; very young parts of branchlets, pulvini and rachises of leaves tomentose, otherwise glabrous. Leaves 3-jugate; stipules linear, ca. 5 mm long, caducous; petiole marginate or narrowly winged, ca. 1 cm long; rachis segments 2-3(4) cm long, obtriangularly winged up to 4 mm wide; nectaries 1 mm diam.; leaflets elliptic, base acute or cuneate, apex acuminate and mucronate, proximal ones 4-6 x 1.5-2.5 cm, medial and distal ones 6-9 x 2-3.5 cm; blades greyish green, glabrous on both surfaces except prominent primary vein above sometimes tomentose; young leaves glabrous, whitish to pale red, with red nectaries. Inflorescence of 1-4 spikes, slightly pubescent on all parts; peduncle slender, 2-3 cm long, one sterile bract at 2/3 along its length, rachis 1.5-2 cm long, bracts 1.5 mm long, scaly and recurved, persistent; flowers 20-30 very close, sessile or very shortly pedicellate; bud obovate or obtrullate; calyx tubular, ca. 2 mm long, regular, lobes acute, ca. 1 mm long; corolla 6-8 mm long, glabrous except on teeth margins; ratio calyx/corolla length ca. 1/3; staminal tube as long as corolla. Pod 12-20 x ca. 1.5 cm, valves thin, leathery, glabrous, shallowly rippled, margins not thickened, mature fruit green, flexible; seeds 12-14(15), contiguous, somewhat prominent.

Distribution: A species distributed throughout the Amazon Basin and the Guianas; locally abundant in the understorey of tall dense rainforest, flooded or not (GU: 2; SU: 3; FG: 19).

Selected specimens: Guyana: Upper Essequibo-Upper Takutu Region, Clarke 8139 (BRG, P, US), Spider Mt., Jansen-Jacobs *et al.* 6132 (BRG, P, U). Suriname: Gran Rio R., Tresling 458 (U); Kabalebo Dam Project, Lindeman & Görts *et al.* 579 (BBS, K, P, U); Voltzberg, Lanjouw 915 (U). French Guiana: Approuague R., Oldeman T85 (CAY, P), Cr. Calebasse, Oldeman B1787 (CAY, P, U); Cr. Ipoucin, Oldeman B2214 (CAY, P), Cr. Parépou, Oldeman 2836 (CAY, P, U); Arataye R., Station Nouragues, de Granville *et al.* 10367 (CAY, P); Haut Mana, Bellevue, Cremers 7562 (CAY, P, U); Haut Oyapock, Euleupousin, de Granville T1142 (CAY, P), Yaroupi, Oldeman T577 (CAY, P, U); Inini R. Basin, Bellevue Mts., Cremers 7562 (CAY, NY, P, U, US); Saül, Hahn 3604 (CAY, US), Airport Rd., Mori *et al.* 15379 (CAY, NY, P), 15373 (CAY, NY, P), Bélizon Rd., Aubréville 308 (CAY, P), Carbet Maïs, de Granville 3264 (CAY, P, U), Eaux-Claires, Mori *et al.* 22183 (CAY, NY, P), 23046 (CAY, NY, P), Pennington *et al.* 13844 (CAY, K), Grand Boeuf Mort, Oldeman B4166 (CAY, NY, P, U); Sinnamary, de Granville 131 (CAY, P).

Vernacular name: French Guiana: inga sili (Wayapi).

7. **Inga bourgonii** (Aubl.) DC., Prod. 2: 434. 1823. – *Mimosa bourgonii* Aubl., Hist. Pl. Guiane 2: 941, pl. 358. 1775, as "bourgoni". Type: French Guiana, Aublet s.n. (holotype BM). – Fig. 26 L-O

Inga assimilis Miq., Linnaea 19: 130.1847. Type: Suriname, Hostmann & Kappler 1690 (holotype U, isotype P).

Medium-sized tree; bark greyish to pale brown, smooth with horizontally oriented rows of reddish lenticels; branchlets glabrous except very young parts. Leaves 3(4)-jugate, glabrous; stipules conspicuous, linear or falcate, enlarged and rounded at apex, often persistent; petiole short or reduced to pulvinus, 0.5-1(1.5) cm long, flattened; rachis segments 2.5-5 cm long each, marginate or with distally developed wing, sometimes ridged around foliar nectary; leaflets narrowly to broadly elliptic (proximal pair 6-8 x 3-6 cm, medial and distal pair 10-14(18) x 5-8 cm), base acute or cuneate, apex acuminate; nectaries sessile, ca. 1 mm diam.; young leaves glabrous, reddish-brown; leaves of juvenile individuals often 2-jugate, longer and narrower than those of adults. Inflorescence axillary spikes, solitary or 2-3(4) in axil of old mature leaves, or at defoliated nodes; peduncle 0.5-1.5 cm long, rachis 2.5-5 cm, 30-50-flowered; bracts scaly, recurved, spatulate, persistent; flowers very dense, sessile; calyx and corolla sparsely pubescent; calyx 1 mm long, corolla funnel-shaped 5-6 mm, teeth 1 mm; ratio calyx/corolla length 1/5; staminal tube as long as corolla or exserted. Pod 10-12 x 2(+) cm, cylindrical and looking fleshy when ripe, glabrous, sutures 4-5 mm wide, faces leathery, with transversal close shallow ribs; seeds 11-14, contiguous.

Distribution: Not a very common species, although distributed through the Guianas and Amazonia, westwards to Peru, northwards to Venezuela; dense primary forest, non flooded or riverine, a few collections recorded from secondary forest (GU:3; SU: 9; FG: 15).

Selected specimens: Guyana: Jenman 3898 (NY); NW Distr., Barima-Waini Region, Powis Cr., McDowell 4430 (CAY, P, U, US); Port Kaituma, Davis/FD 244 (FDG, K, NY). Suriname: Hostmann 77 (G, K, NY, P), 967 (G, K, NY, P); Splitgerber s.n. (P, U); Cadrissi R., BBS 25 (BBS, U); Corantyne R., Kanarie Kr., Stahel & Gonggrijp BW 3021 (U); Kaboerie, BW 5962 (U); Saramaca R., Jacob Kondre, Maguire 23855 (NY, P, U). French Guiana: Approuague R., Arataye Cr., Pararé camp, Sastre 5683 (CAY, P, U); Nouragues Station, Sabatier & Prévost 2715 (CAY, NY, P, U); Couata Cr., Oldeman T54 (CAY, P, US); Comté R., Oldeman 1034 (CAY, P, U), 1089 (CAY, P); Haut Oyapock, Trois-Sauts, Grenand 1048 (CAY, P), Oldeman T775 (CAY, P, U); Kaw, Mt. Favard, de Granville 6879 (CAY, NY, P, U, US); Kaw Rd., Camp Caïman, Prévost 206 (CAY, P, U); Mana, Mélinon 129 (P), Sagot 166 (P, US); Maroni R., Sastre 4135 (CAY, P, U, US); Oyapock R., Cr. Armontabo, Sabatier 1088 (CAY, P); Saül, Eaux-Claires, Mori *et al.* 22219 (CAY, NY, P), Prance *et al.* 30666 (CAY, NY, P).

Vernacular name: French Guiana: bougouni (Créole), a name used for several closely related species belonging to the series Bourgonia (*Inga alba, I. bourgoni, I. marginata*).

8. **Inga brachystachys** Ducke, Trop. Woods 90: 12. 1947. – *Inga brachystachya* Ducke, Arch. Jard. Bot. Rio de Janeiro 3: 54. 1922, non DC., 1825, nom. illeg. Lectotype (designated by Pennington 1997: 308): Brazil, Pará, near Belem do Pará, Huber MG 3857 (hololectotype RB, isolectotypes B, G, P).

Small to middle-sized understorey trees; branchlets slightly grooved, slightly pubescent. Leaves (1)2-jugate; petiole ca. 0.5 cm long, winged, rachis 1.5-2.5 cm long, narrowly to broadly winged; nectaries orbicular; stipules linear, 2-3 mm long, caducous; leaflets sessile, elliptic, proximal ones (4)6-8(9) x (2)3-4(5) cm, distal ones 7-10(15) x (3)5-7 cm; blades glabrous above (except primary vein often tomentose), brownish tomentose, velvety beneath, apex acute, or with a short, rounded acumen, base acute, markedly asymmetrical; young leaves pinkish brown, velvety. Inflorescence of 2-5 spikes axillary to mature leaves or at defoliated nodes, or on lateral short shoots; peduncle slender, 1-2.5 cm long; rachis 1-1.5 cm, 15-20-flowered; bracts inconspicuous, linear, ca. 0.5 mm long, pubescent; flowers sessile, sparsely pubescent; bud clavate; calyx tubular, ca. 1.5 mm long, teeth inconspicuous; corolla funnel-shaped, 5-7(10) mm long; ratio calyx/corolla length ca. 1/4 or lower; staminal tube exserted; ovary glabrous with 16-20 ovules, stigma tubular. Pod flattened, 7-11 x 1.7-2(2.7) cm, valves thin, shallowly but conspicuously reticulate when immature, smooth and swollen when ripe, margins not thickened.

Distribution: An understorey species of dense undisturbed non-flooded rainforest; known from the Guianas and Central and Lower Amazon Basin (GU: 6; SU: 3; FG: 7).

Selected specimens: Guyana: Acarai Mts., Clarke 7127 (BRG, P, U), Henkel 5224 (BRG, P, US); Kanuku Mts., Hoffman 382 (BRG, P, US), 449 (BRG, P, US), Sipu R., Clarke 7747 (BRG, P, US); Ro. Schomburgk ser. I, 595 (G, K). Suriname: Brokopondo, van Donselaar 1710 (BBS, U), 1755 (BBS, U); Wilhelmina Mts., Lucie x Oost R., Irwin *et al.* 55701 (NY, P, U). French Guiana: Mt. Inini, de Granville *et al.* 7999 (CAY, NY, P, US); Mt. Bakra, Emerillons, de Granville & Cremers 11765 (CAY, NY, P, U, US); Saül, Mori *et al.* 8725 (CAY, NY, P), Eaux-Claires, Mori *et al.* 24687 (CAY, NY, P); Piste-de-St-Elie, Molino 1107 (CAY, MO, MPU, P), Sabatier & Prévost 3835 (B, CAY, G, K, NY, U, US), 4363 (CAY, K, MO, NY, P).

Note: *Inga brachystachys* can easily be confused with the very common and widely distributed *I. marginata*, from which it differs by its tomentose leaves, less conspicuous foliar nectaries, shorter inflorescences and larger flowers, as well as by its habit.

9. **Inga brevipes** Benth., J. Bot. (Hooker) 2: 144. 1840. Type: Guyana, Pirara, Ro. Schomburgk ser. I, 740 (Jun 1838) (holotype K, isotype G, P, photo US).

Small tree; branchlets terete, lenticellate, slightly pubescent when young. Leaves (2)3-jugate, velvety at least when young; stipules linear to falcate, 3(+) x 1 mm, pubescent; petiole and rachis winged; petiole ca. 1 cm long or more on 2-jugate leaves, segments of rachis 2.5-3.5 cm, wing oblanceolate up to 1 cm wide; nectaries sessile, circular, proximal one often wider (up to 2 mm in diam.); leaflets elliptic, proximal pair 6-8 x 3-4 cm, medial and distal pairs 12-15 x 5-6 cm, apex rostrate, base acute to rounded, asymmetrical; blades papyraceous, glabrous on both faces. Inflorescence globose, axillary to mature leaves; peduncle very short (<5 mm long), ca. 30-flowered; bracts scaly, spatulate, ca. 2 mm long; pedicel 3-4 mm long; flower bud clavate; calyx and corolla pubescent, calyx obconical, 5-7 mm long, teeth inconspicuous; corolla 10(+) mm long, lobes 3 mm; ratio calyx/corolla length ca. 1/2. Pod on a peduncle 7 mm long, flat, 7 x 2 cm, base asymmetrical, apex apiculate (mature pod unknown).

Distribution: Endemic of S Guyana, on forested slopes of Kanuku Mts. and nearby bushy areas of the Rupununi Savanna, although a few Brazilian collections were attributed to this taxon by Pennington (1997) (GU: 7).

Selected specimens: Guyana: Kanuku Mts., Iramaipang, Wilson-Browne 483 (K, NY, U, US), 5889 (BRG, FDG), western slopes, A.C. Smith 3310 (K, NY, P, U, US); Upper Takutu R., Clarke *et al.* 6796 (P, US); Jansen-Jacobs *et al.* 659 (BBS, NY, P, U); Karanambo, Poncy 927 (P).

Note: *Inga brevipes* is very closely related to *I. umbellifera* of which it might be only a variety; pending more material, it is treated here as a separate species and considered as an endemic of S Guyana.

10. **Inga calanthoides** Amshoff, Bull. Torrey Bot. Club 75: 385. 1948. Type: Suriname, Tafelberg, Maguire 24547 (holotype NY, isotypes G, K, MO, P, U, RB, US).

Tree 15 m high; branchlets terete, dark greyish, young parts hirsute. Leaves (3)4-jugate; pulvinus, midrib above and all parts beneath densely pubescent; stipules scaly, triangular, 3 mm wide at base, 2 mm long, caducous; petiole marginate to narrowly winged, ca. 1 cm long; rachis winged, each segment 1-2(3) cm long, wings semi-elliptic to rhomboid, distal one distinctly obtriangular; nectaries orbicular, very small, < 1 mm diam., sessile or shortly stipitate; leaflets elliptic, proximal ones small, up to 4 x 2 cm, distal ones up to 10 x 4.5 cm, base acute and asymmetrical, apex acuminate and apiculate; blades leathery, scabrous above, tawny and tomentose beneath. Inflorescence spicate, single or geminate, in axil of mature leaves; peduncle less than 1 cm long; rachis ca. 2 cm with 5-6 lax flowers; bracts channelled, ca. 3 mm long, caducous; pedicels up to 5 mm; flower bud obconical; calyx and corolla covered with dense appressed white hairs; calyx funnel-shaped, 1.5-2 cm long, teeth acute, corolla 4(+) cm long; ratio calyx/corolla length 1/2 or higher; staminal tube exserted; stigma cup-shaped, ca. 1 mm diam. Pod densely hirsute, reddish-brown, flat, 15 x 2.5 cm, sutures 0.5 cm thick.

Distribution: *Inga calanthoides* is known only from the type collection, therefore it appears to be endemic to the Tafelberg Mountain in Suriname (SU:1).

Note: Pennington (1997: 641) suggests that "proper collecting will reveal [it] to be conspecific" with *Inga calantha* Ducke and *I. micradenia* Spruce ex Benth.

11. **Inga capitata** Desv., J. Bot. 3 : 71. 1814. Type: French Guiana, Cayenne, Desvaux s.n. (holotype P).

Inga albicans Walp., Linnaea 14: 298.1840. Type: Brazil, Blanchet 389 (holotype G).
Inga falcistipula Ducke, Arch. Jard. Bot. Rio de Janeiro 3: 56. 1922. Lectotype (designated by Pennington 1997: 458): Brazil, Pará, Ducke 16326 (hololectotype MG, isolectotypes BM, P, RB, US).
Inga capuchoi Standl., Trop. Woods 33: 12. 1933. Type: Brazil, Pará, Capucho 398 (isotype G).

Medium-sized tree; bark whitish or greyish with conspicuous lenticels; plant glabrous on all parts. Leaves 2(3)-jugate; stipules linear to falcate up to 10 mm long, with a beaked apex, persistent; petiole and rachis terete or slightly channelled distally; petiole 1(+) cm long, rachis 2.5-4 cm long, lenticellate; nectaries ostiolate, very small (1 mm diam.); leaflets elliptic, proximal ones 7-10 x 3.5-5 cm, distal ones generally up to 15 x 7 cm; base attenuate, acumen triangular, acute or mucronate; blades coriaceous, bright above. Inflorescence in fascicles of 2-6 spikes axillary to mature leaves; peduncle thick, angular, lenticellate, up to 6(8) cm long; rachis ca. 1 cm, 30-flowered; bracts 3-4 x 1 mm, lanceolate to spatulate, rostrate, caducous; bud closed (corolla and calyx opening simultaneously) cylindrical to obtrullate, apex acute; calyx narrowly campanulate ca. 7-10 mm long, 2-3 mm wide, teeth inconspicuous, regular; corolla campanulate 10-12 mm long; flowers white with corolla teeth sometimes pinkish; ratio calyx/corolla length 3/4 to 4/5; staminal tube hardly longer than corolla; intrastaminal ring present, conspicuous, ovary elliptic, ovules 13-15. Pod 7-12(15) x 2-3 cm, straight and supple, subcylindrical and bright yellow when ripe, calyx more or less persistent at base, apex mucronate, sutures neither thickened nor prominent, faces slightly rugose and rippled; seeds 9-12, contiguous or not.

Distribution: A species widely distributed in S America, from Panama, the Amazon Basin, to SE Brazil; in dense non-flooded rainforest (GU: 7; SU: 8; FG: 16).

Selected specimens: Guyana: UpperTakutu-Upper Essequibo Region, Kwitaro landing, Clarke 6201 (P, US); Mazaruni Station, Fanshawe 538 = FD 3274 (BRG, FDG, K, US); Jenman 7131 (BRG, K); Pomeroon R., Stockdale 14 (BRG, K); Berbice-Corentyne Region, Canje R., Pipoly *et al.* 11436 (CAY, FDG, NY, P, US), 11463 (CAY, FDG, NY, P, US); Iwokrama, Clarke *et al.* 4305 (BRG, NY, P, US). Suriname: sin. loc., Hostmann 258 (K, P as "Hortman", U); Brownsberg, Mori & Bolten 8368 (NY, U); Capoeuca R., Lindeman 5401; Coppename R., Stahel & Gonggrijp BW 6190; Raleighfalls, Stahel & Gonggrijp BW 6152 (BBS, P, U, US), Kaboerie, BW 5021 (U); Tumuc-Humac Mts., Acevedo Rodr. *et al.* 6070 (P, US); Wilhelmina Mts., Lucie R., Irwin *et al.* 54458 (NY, P, U, US); Perica R., Lindeman 5293 (BBS, P, U). French Guiana: Approuague R., Arataye Cr., Poncy 198 (CAY, P, U), 473 (CAY, P), Nouragues Station, Sabatier & Prévost 2836 (CAY, NY, P, U, US); Comté R., Aubréville 358 (CAY, P), Oldeman B1110 (CAY, P), 1447 (CAY, P); Gourdonville, Benoist 1520 (P); Ht-Mana, Cr. Baboune, de Granville 4791 (CAY, P,NY); Bas-Oyapock, Cr. Gabaret, Grenand & Guillaumet 3216 (CAY, NY, P); Ht-Oyapock, Roche Touatou, Cremers *et al.* 14107 (B, CAY, MO, NY, P, U, US); Trois-Sauts, Grenand 971 (CAY, P); Inini Mts., de Granville 7606 (CAY, P); Saül, Boom & Mori 2054 (CAY, NY, P), Mori & Pipoly 15616 (CAY, NY, P), Rte Bélizon, Mori *et al.* 24911 (CAY, NY); Sinnamary, Piste-de-St-Elie, Prévost 4098 (CAY, P); Saint-Laurent-du-Maroni, Benoist 1114 (P).

Vernacular name: French Guiana: inga muluaya (Wayapi).

12. **Inga cayennensis** Sagot ex Benth., Trans. Linn. Soc. London 30: 636. 1875. Lectotype (designated by Pennington 1997: 631): French Guiana, Cayenne, Sagot 164 (hololectotype K, isolectotypes NY, P, U). – Fig. 14 E-G

Inga cayennensis Sagot ex Benth. var. *sessiliflora* Ducke, Arch. Jard. Bot. Rio de Janeiro 3: 60.1922. Lectotype (designated by Pennington 1997: 633): Brazil, Pará, R. Xingu, Ducke 17172 (hololectotype MG, isolectotypes G, P, US).
Inga dysantha Benth., Trans. Linn. Soc. London 30: 626. 1875. Type: Brazil, Manaus, Spruce 1816 (holotype K, isotypes NY, P).

Small to medium-sized tree, up to 15 m x 20 cm, bark whitish outside, yellowish inside; plant hairy on all parts, very densely so on young twigs, leaf rachis and petiole, and pods; branchlets reddish-brown to dark brown velvety hairy. Leaves 4-6-jugate; stipules triangular, ca. 2.5 x 2.5 mm, coriaceous, sometimes persistent; petiole (1)1.5-3 cm long, not winged; rachis segments 1.5-3.5(+) cm long, winged, 5 to 15 mm wide, oblong to obtriangular; nectaries ca. 1 mm diam., orbicular, pale green, medial ones sometimes wanting; leaflets elliptic, basal ones not much smaller than distal ones, (3.5)6-10 x (1.5)2-5 cm, acumen acute, slightly curved, base cuneate to acute; blades soft, tomentose on both faces, veins densely pubescent. Inflorescence solitary or geminate, axillary, lax, hairy spikes of 10-15 flowers; peduncle 1-2 cm long; rachis 3-4 cm long; bracts linear up to 2 mm, caducous; pedicel either wanting or up to 4 mm long; bud obconical to obovate, corolla rippled; calyx tubular to campanulate, 4-7 mm long, with sparse indument, teeth reduced, cuspidate; corolla up to 2(+) cm long, covered with dense appressed glossy hairs, funnelform, basal part of tube sometimes very narrow; ratio calyx/corolla length ca. 1/4; staminal tube exserted or not; stamens off-white; stigma cup-shaped. Pod 10-20 x 2-2.5 cm, flattened, straight or often twisted, entirely covered with dense reddish-brown hirsute hairs, rigid when ripe, sutures scarcely developed, up to 5 mm wide; seeds (8-10)20-25 contiguous, cotyledons ca. 1.5 x 0.7 cm, dark green, eventually violaceous. Seedling: two first leaves opposite, bifoliolate, lacking nectary, densely whitish tomentose, number of leaflets increasing rapidly on next leaves (up to 7 pairs), sometimes with stipitate nectaries only on distal pairs; hairs whitish turning reddish-brown on young leaves, crowded on buds and veins.

Distribution: An Amazonian species; not found in primary lowland forest but associated with open areas, seasonal forest, or mesophytic low forest (inselbergs, savanna edges), although never riverine; in French Guiana frequent in disturbed forests and in secondary vegetation in the coastal area (GU: 2; SU: 6; FG: 24).

Selected specimens: Guyana: Kuyuwini Landing, Jansen-Jacobs *et al.* 2823, 3164 (BRG, CAY, P, U). Suriname: Brokopondo, van Donselaar 3758 (BBS, U); Brownsberg, sectie O, BW 4206 (U); Fallawatra, Jimenez-Saa LBB 14411 (U); Sipaliwini Savanna, Oldenburger *et al.* 345 (BBS,

K, P, U, US), 564 (K, NY, U); Tafelberg, Maguire 24709 (G, K, NY, P, U, US). French Guiana: sin. loc., Perrottet s.n. (P), Poiteau s.n. (K, P), Mélinon s.n. (P), Soubirou s.n. (P); "Cayenne", Leprieur s.n.; Approuague R., Arataye Cr., Station Nouragues, Poncy 859 (P), Sabatier & Prévost 2918 (CAY, K, NY, P, U), Sabatier 3537 (CAY, K, NY, P, U, US); Route de Kaw, Camp Caïman, de Granville 2951 (CAY, P); Iracoubo, Poncy 297 (CAY, P, U); Kourou, Oldeman B1341 (CAY, P, U); Mana, Acarouany, Sagot 1164 (G, K, P, U), SF 297M (CAY, P, U); Maroni, Gandoger 68 (P); Savane Matiti, Sastre 6157 (CAY, P); Saül, Pic Matecho, Mori & Smith 25046 (CAY, P, NY); Sinnamary, Piste-de-St-Elie, Lescure 720 (CAY, P), Sastre 6018 (CAY, P, U); Trinité Mts., Cremers 12793 (CAY, P), 12584 (CAY, P); Saint-Laurent-du-Maroni, Godebert, Wachenheim 136 (P); Gourdonville, Benoist 1609 (P); Route de Cayenne, SF 7551 (P).

13. **Inga cordatoalata** Ducke, Arch. Jard. Bot. Rio de Janeiro 3: 53. 1922. Type: Brazil, Pará, Peixeboi, Siquiera HJBR 8270 (holotype MG, isotypes P, RB, US).

Tree to 18 m x 40 cm, trunk with low buttresses, bark whitish; terminal twigs terete, slightly puberulous, lenticellate. Leaves 2-jugate, glabrous; stipules conspicuous on young shoots, ovate, up to 8 x 4 mm, caducous; petiolule terete, thick, 3-5 mm long, petiole winged, triangular, 5-9 x 3-4(6) mm; rachis winged, (0.9)1.5-2(4) cm long, distal portion triangular, up to 5-6 mm wide; nectaries circular, sessile, ca. 1 mm diam.; leaflets elliptic, base acute to obtuse, apex attenuate to shortly acuminate, proximal ones (2.5)4-5.5(7) x (1)2-2.5(4) cm, distal ones (4)7-10(14) cm; chartaceous, glossy above; primary vein depressed above, markedly prominent below, secondary veins 6-8 pairs. Inflorescence spicate, axillary, solitary; peduncle slender, (1)2-3(+) cm long; rachis (4)6-8(10) cm long, flowers very densely packed, ca. 100 per spike; bract spatulate, 1.5 mm long, pedicel 0.5 mm, calyx (1)1.5-2 mm long, 1.5-1.8 mm wide at base of lobes, lobes subequal, 0.3 mm long, sparsely haired, corolla tubular, ca. 4 mm long, lobes 1-1.5 mm; ratio calyx/corolla length ca. 1/3; staminal tube inserted or hardly as long as corolla. Pod unknown within our area (12-17 x 1.5-1.8 cm, glabrous, when ripe swollen around seeds and constricted between them, fide Pennington, 1997).

Distribution: An Amazonian species, recorded from Ecuador, Peru, Brazil (Amazonas, Pará), Guyana and French Guiana (GU: 2; FG: 2).

Selected specimens: Guyana: Upper Essequibo, Acarai Mts., Clarke *et al.* 7636 (BRG, P, US), Sepu R., Clarke 7772 (BRG, P, US). French Guiana: Saül, Mori *et al.* 23276 (CAY, K, NY, P); R. Oyapock, savane-roche Bâton Pilon, Oldeman B2539 (CAY, NY, P, U).

Note: The two available specimens from French Guiana are included here, in agreement with Pennington's treatment (1997). Pending additional, especially fruiting, collections they are, however, likely to represent a distinct species, as they differ from the type collection by several characters including bi-jugate leaves, larger stipules, and somewhat larger flowers.

14. **Inga disticha** Benth., J. Bot. (Hooker) 2: 143. 1840. Type: Guyana, Ro. Schomburgk ser. I, 25 (holotype K, isotypes G, K, P, U, US).
– Fig. 14 A-D

Inga crevauxii Sagot, Ann. Sci. Nat., Bot. 6 : 331. 1882. Type: French Guiana, Maroni R., Crevaux s.n. (holotype P).

Treelet to medium-sized tree; bark greyish; terminal twigs often staggered, irregular in section and covered by hirsutulous reddish-brown hairs, lenticels verrucose, brown. Leaves 3-4-jugate; stipules lanceolate, beaked, up to 10 mm long, tardily falling, pubescent beneath; petiole usually winged (seldom marginate), ca. 1.5(+) cm long; rachis winged, each segment ca. 2-2.5 cm, up to 7 mm wide, sparsely pubescent; nectaries orbicular, subsessile to stipitate, up to 1 mm diam.; leaflets elliptic to lanceolate, often falcate, proximal pair (2.5)4-8 x (1)2.5-3 cm, distal pair (6.5)9-14 x (2)3.5-5 cm, acumen long, sometimes mucronate, base obtuse; blades papyraceous, scabrous to sparsely tomentose above, tomentose beneath; juvenile leaves reddish-pink with white hairs. Inflorescence spicate, in fascicles of 2-5 axillary to contemporary leaves; peduncle slender and short (up to 2 cm long); rachis 4-5 cm long with 12-20 flowers loosely and distichously arranged; bracts perpendicular to rachis when flower is open, channelled, ca. 10 mm long, with a slender subulate apex, externally tomentose; flowers sessile; bud cylindrical; calyx tubular ca. 4-6 mm long, pubescent, irregular (one of five grooves between lobes deeply cleft), 5 lobes equal, acute, ca. 2 mm long; corolla variable in length, usually 15-20 mm, lobes ca. 2 mm long, pubescent; ratio calyx/corolla length ca. 1/4; staminal tube often shorter than corolla. Pod flat, hairy when young, 6-9(12) cm long, ca. 2(3) cm wide, sutures thickened up to 1 cm wide, non prominent, valves slightly convex, with very shallow transverse ripples, calyx persistent; seeds (7-8)15-17, ca. 1.2 x 0.6 cm. Seedling: juvenile leaves usually have up to 5 pairs of leaflets that are narrower and with a longer acumen than leaflets of adult tree.

Distribution: Amazonian species, Brazil (Amazonas and Pará); in the Guianas present on river banks at the edge of dense lowland rainforest, occasionally in non-flooded forest (GU: 18; SU: 9; FG: 21).

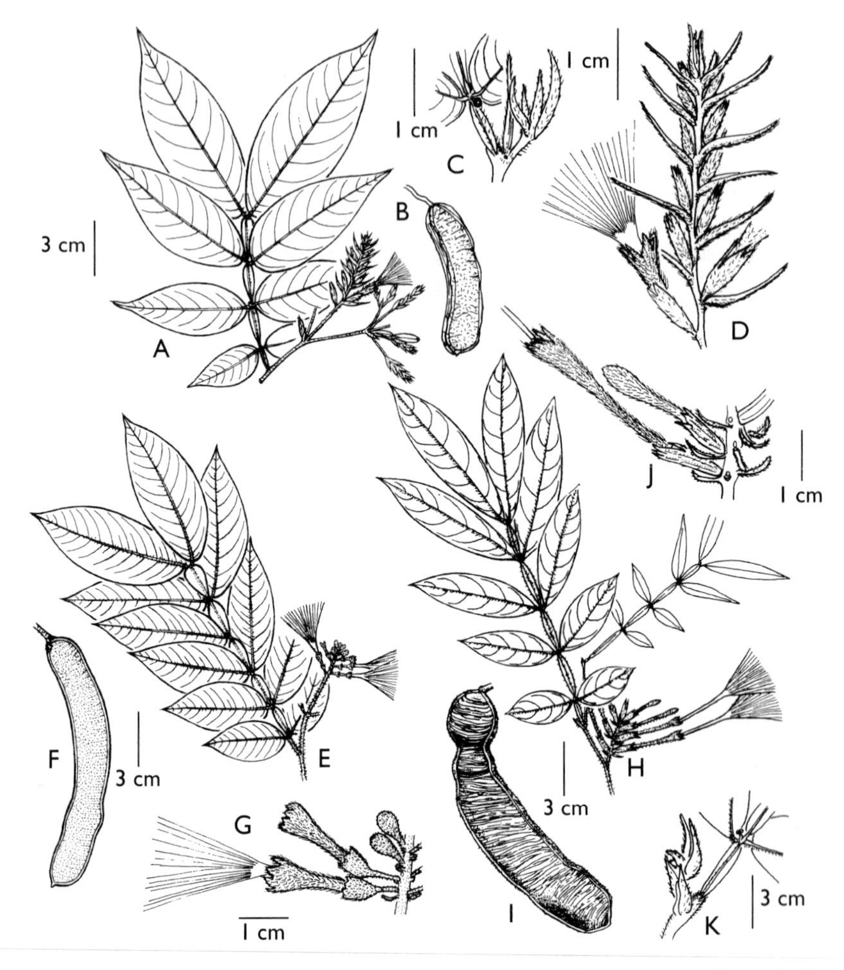

Fig. 14. A-D: *Inga disticha* Benth.: A, habit, leaf and inflorescence; B, pod; C, extremity of leaf axis, stipules, nectary; D, inflorescences, bracts, flower. E-G: *Inga cayennensis* Sagot ex Benth.: E, habit, leaf and inflorescence; F, pod; G, partial inflorescence, flower. H-K: *Inga longiflora* Spruce ex Benth.: H, habit, leaves and inflorescence; I, pod; J, partial inflorescence, bracts, flowers; K, extremity of leaf axis, stipules, petiole, nectary. Drawing by Odile Poncy; reproduced from Poncy, 1985: 58.

Selected specimens: Guyana: Kamoa R., Clarke 3029 (BRG, P, US), Jansen-Jacobs *et al.* 1677 (BRG, P, U); Berbice, McDowell 2301 (BRG, P, U, US), Mutchnik 1026 (BRG, P, US); Demerara, Hariwa Cr., Fanshawe FD 3164 (FDG); Essequibo R., Guppy FD 6221 (FDG, K,

NY, U, US); Kuyuwini R., Smith 2567 (G, NY, U, US), Sandwith 583 (K, NY, U), Kassikaityu R., Clarke *et al.* 4787 (BRG, P, US); Potaro R., McDowell 3328 (BRG, P, US); Waraputa Falls, Jansen-Jacobs *et al.* 2029 (P, U); Upper Demerara, Jenman 4038 (BRG, K, NY); Berbice R., McDowell 3268 (BRG, P, US); Powis Cr., Persaud FD 6778 (FDG, U); Rupununi, Kuyuwini R., Henkel 3204 (BRG, P, US), Jansen-Jacobs 2324 (BRG, P, U); Rewa R., Clarke 6604 (BRG, P, US). Suriname: Brownsberg, Dawson LBB 14613 (BBS, U); Coppename R., Boon 1042 (BBS, U), Lanjouw 958 (K, NY, U); Corentyne R., Wonotobo, Stahel & Gonggrijp BW 2544 (P, U); Litany R., de Granville *et al.* 11884 (B, BBS, CAY, NY, P, U, US), 12579 (B, CAY, K, NY, P, U), Rombouts 827 (U); Marowijne R., Lanjouw & Lindeman 2039 (BBS, K, NY, U), Man Kaba, Sauvain 458 (BBS, CAY, P, U); Saramacca R., Daniëls & Jonkers (U). French Guiana: Maroni, Crevaux s.n. (P), Haut-Maroni, R. Lawa, Dagou-édé, Sastre & Bell 8168 (CAY, P, U, US), Langa Soula, de Granville *et al.* 9539 (CAY, NY, P, U), Maripasoula, Fleury 214 (CAY, P), Lawa R., Schnell 11716 (P); Ht-Oyapock, Trois-Sauts, Grenand 1504 (CAY, P), Jacquemin 1731 (CAY, P), Oldeman T966 (CAY, P); Grand-Inini, Fourmi, Oldeman B3525 (CAY, P, U), Saut-Liki, de Granville B3670 (CAY, P), B3692 (CAY, P); Litani, Saut-Lavaud, de Granville 12511 (CAY, K, P, U, US); Mts. Inini, Bellevue, de Granville *et al.* 8170 (CAY, P, U); St-Georges Oyapock, Cr. Gabaret, Cremers 9865 (B, CAY, NY, P, U, US), Cr. Armontabo, Prévost 2025 (CAY, P), Saut-Maripa, Jacquemin 2099 (CAY, P); Saint-Laurent-du-Maroni, SF 7890 (P, U); Orapu R., Martin & Sanip 158 (CAY, P); Saül, Savane roche Dachine, Chareyre 64 (CAY); Sinnamay R., Cr. Tigre, Hoff *et al.* 6602 (CAY, P), Petit-Saut, Prévost 1304 (CAY, P).

Vernacular names: Guyana: karoto (Arow). Suriname: baboen-weko (Saramac.). French Guiana: inga tupewi, inga takwa-u (Wayapi)

15. **Inga edulis** Mart., Flora 20(2), Beibl. 8: 113. 1837. – *Mimosa inga* Vell., Fl. Flumin. 11: 431. 1827, non L. Type: Brazil, t. 3 in Vellozo, Icones Fl. Flumin. 11, 1831 ("1827"), as *Mimosa ynga*. – Fig. 15 A-D

Inga scabriuscula Benth., London J. Bot. 4: 606. 1845. Lectotype (designated by Pennington 1997: 742): Suriname, Hostmann 887 (hololectotype K, isolectotypes G, K, NY, P, U, US).
Inga benthamiana Meisn., Linnaea 21: 253. 1848. Type: Suriname, Jodensavanne, Kegel 1206 (holotype GOET, isotype NY, P).
Inga scabriuscula Benth. var. *villosior* Benth., Fl. Bras. 497. 1876. – *Inga complanata* Amshoff, Natuurwetensch. Stud. Suriname & Curaçao 2: 39. 1948, nom. nov. Type: Brazil, Amazonas, Manaus, Spruce 1750 (holotype K, isotypes G, P).

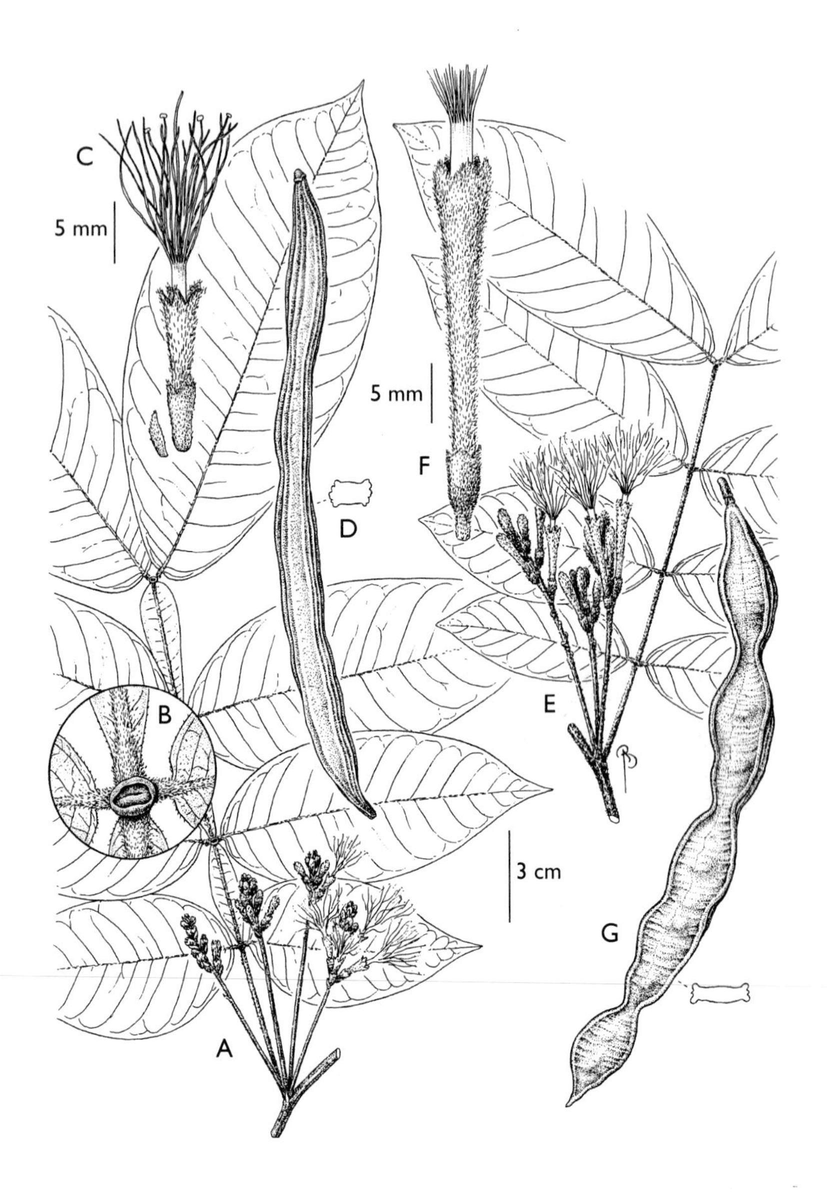

Fig. 15. A-D: *Inga edulis* Mart.: A, habit, leaf and inflorescences; B, foliar nectary; C, flower and bract; D, pod. E-G: *Inga rubiginosa* (Rich.) DC.: E, habit, leaf and inflorescences; F, flower; G, pod (A-C, Mori *et al*. 20844; D, Irwin 2699; E-F, Mori *et al*. 20722; G, Oldeman 2243). Drawing by Bobbi Angell; reproduced with permission from Mori *et al*., 2002: 497.

Medium-sized canopy tree, to 25 m x 40 cm; bark light greyish or brownish with dense regular lenticels; branchlets brown tomentose, angular, ribbed. Leaves (3)4(5)-jugate; stipules linear, 5-7 mm long, usually caducous, pubescent; petiole 2-3 cm long, unwinged (winged on seedlings and young individuals), angular to flattened; segments of rachis up to 3.5 cm long and winged, elliptic to obovate, usually ca. 1 cm wide; leaflets elliptic, proximal ones up to 9 x 4 cm, distal ones up to 13 x 6 cm, acuminate, base obtuse; blades tomentose on both faces, more densely so on veins; young leaves dark red to reddish-brown, veins and acumen green. Inflorescence spicate, mostly 3-6-fascicled in axil of young or mature leaves; peduncle 2-3 cm long; rachis up to 4 cm long, 30-40-flowered; bracts caducous, channelled, 5-7 x 1-3 mm, sheathing calyx at base, apex acicular; flowers sessile; bud ovoid to clavate; calyx 5-6(+) mm long, striate, tubular, scarcely tomentose, regular, lobes ca. 1.5 mm; corolla variable in length, usually 12-14 mm long, densely velutinous; ratio calyx/corolla length ca. 1/2; staminal tube not exserted; stigma funnel-shaped. Pod terete, 20-30 cm long and up to 2 cm diam., with longitudinal grooves (formed by hypertrophied sutures, while valves are hardly visible), covered by tightly appressed, light greenish-brown pubescence, apex acuminate; seeds ca. 15-20, longitudinally arranged; cotyledons shiny, black 1-1.5 cm long. Seedling: pubescent, with two first leaves bifoliolate and opposite, petiole widely winged, nectaries undifferentiated, orbicular.

Distribution: A wide S American extension from Colombia to S Brazil, but not in the West Indies; dense primary or disturbed, riverine, flooded or non-flooded rainforest; this fast-growing species, of great variation in habitat and plasticity in morphology, often regenerates very soon after cutting or in forest gaps (GU: 6; SU: 6; FG: 17).

Selected specimens: Guyana: Essequibo R., Bartica, Sandwith 516 = FD 1249 (FDG, K); Berbice R., New Dageraad, Maas *et al.* 5461 (BRG, P, U); Kanuku Mts., Moku-moku Cr., A.C. Smith 3440 (K, NY, P, U), Wabuwak, Wilson Browne FD 5700 (BRG, FDG); Mazaruni Station, Fanshawe FD 2952 (FDG, K); Spider Mt., Jansen-Jacobs *et al.* 5873 (P, U). Suriname: Maroni, Man Kaba, Sauvain 521 (BBS, CAY, P); Kabalebo Dam Project, Lindeman & Görts *et al.* 410 (BBS, U); Paramaribo, Jodensavanne, Kegel 1206 (P, U); Wilhelmina Mts., Juliana top, Irwin *et al.* 54900 (K, NY, P, U), 55721 (NY, U). French Guiana: sin loc., Mélinon s.n. (P); "Cayenne", Martin s.n. (P); Approuague R., Arataye Cr., Sastre 5632 (CAY, P), Poncy 119 (CAY, P); Comté R., Aubréville 378 (P); Ht-Maroni, Tampoc R., SF 7892 (P); Maripasoula, Boulard s.n. (P); Inini Mts., de Granville *et al.* 8041 (CAY, P); Saül, La Fumée Mts, Mori *et al.* 19209 (CAY, NY, P), 20844 (CAY, NY, P), 20901 (CAY,NY, P), Bélizon Rd., Mori *et al.* 21113 (CAY, NY, P), Galbao Mts., de Granville *et al.*

8740 (CAY, NY, P); Saint-Laurent-du-Maroni, Acarouany, Sagot 965 (P); Mana R., SF 269M (P, U); Sinnamary, Piste-de-St-Elie, Sabatier & Prévost 3761 (CAY, K, NY, P, U, US); Sommet Tabulaire, Cremers 6504 (CAY, P); Trois-Sauts, Lescure 367 (CAY, locality recorded as Brazil).

Vernacular names: Suriname: tetei watji (Saramac.). French Guiana: inga yowapuku (Wayapi), tetei weko (Aluku).

Use: In C America and Amazonian Brazil, *I. edulis* is protected or cultivated as a shade-tree or fruit-tree; the pods often reach 1 m long (W Amazonia). Only the wild form with small pods is known in the Guianas, where no uses are recorded.

16. **Inga fanchoniana** Poncy, Bull. Mus. Natl. Hist. Nat., B, Adansonia 5: 103. 1983. Type: French Guiana, Sinnamary, Piste-de-St-Elie, Prévost 478 (holotype P, isotypes CAY, NY, U).

Large tree up to 30 m x 40 cm; trunk subcylindrical or fluted, bark brownish-grey, looking like dry mud, exfoliating in small sheets or scales, lenticels irregular; branchlets stocky, angular; all parts glabrous, except slightly tomentose perianth. Leaves (3)4(5)-jugate; stipules triangular to oblong, subulate, up to 8 x 2.5 mm, often lignified and persistent after leaf-abscission; petiole and rachis unwinged, both terete or shallowly grooved below insertion of leaflets and up to 4 mm thick; petiole stocky, up to 6 cm long, often lignified and lenticellate; segments of rachis 3-4 cm long; nectaries sessile, orbicular, sometimes located a bit above insertion of leaflets, distal one often lacking; apex of rachis above insertion of last pair of leaflets often persistent; leaflets broadly elliptic, base often attenuate, apex obtuse, acumen rounded and mucronate; leaflets of proximal pair 7-11(13) x 3-6(7) cm, those of next pairs larger, up to 18-9 cm; blades often irregular in outline, thick, coriaceous, surface above dark green, smooth, below scabrous or very sparsely tomentose with very prominent veins. Inflorescence mostly at defoliated nodes, either in axil of leaf scars or of abortive leaves on very short (1 cm) shoots, often sheathing twigs; peduncle 1-2 cm long, flowers 25-30, crowded on subcapitate or clavate heads; receptacle 3-5 mm; long bracts scaly, grooved and spatulate, ca. 1 mm long, persistent; bud tubular to clavate; flowers sessile, calyx ca. 2 mm long, teeth inconspicuous, regular; corolla up to 1 cm long, funnel-shaped; ratio calyx/corolla length 1/4-1/5; staminal tube up to 1.5 cm, markedly exserted; stamens 30-40; ovary 1, shortly stipitate, ovules 10-15. Pod bulky, spirally coiled in 1-3 spirals, sometimes irregularly curved, pericarp rigid, subligneous, valves convex, thick, looking fleshy, orange at maturity, 2.5-4 cm wide, margins 0.5-1 cm wide, not prominent, lenticellate, brownish; seeds large with white pulp, 3-5 mm thick; embryo ca. 3 x 1.5 cm, cotyledons green. Seedling: glabrous, two first leaves bifoliolate and opposite, leaflets showing already characteristic attenuate base.

Distribution: Known from three localities in the primary dense rainforest in French Guiana, doubtfully from W Guyana (Mori & Bolten 8210 appears close to *I. fanchoniana* enough to mention it), and Pennington (1997) attributes one specimen from Amapá, Brazil (GU: 1?; FG: 9).

Selected specimens: Guyana: Matthew's Ridge, Mori & Bolten 8210 (NY, P). French Guiana: Sinnamary, Piste-de-St-Elie, Prévost 1631 (from same tree as type coll.) (CAY, NY, P), Prévost 1648 (CAY, P, U), Loubry 15 (CAY, P), Sabatier 1756 (CAY, K), Sabatier & Riera 2058 (CAY, P); Haut-Sinnamary, Cr. Plomb, Loubry 1857 (CAY, P); Approuague R., Nouragues Station, Poncy 850 (CAY, P).

Note: The aspect of the trunk and bark, not typical for *Inga*, makes the species difficult to assign to the genus in the field; the coiled pod, orange at maturity and looking fleshy, is distinctive enough to avoid confusion with any other genus as well as with any other *Inga* species.

17. **Inga fastuosa** (Jacq.) Willd., Sp. Pl. 4: 1014. 1806. – *Mimosa fastuosa* Jacq., Fragm. bot.: 15, t. 10. 1800. Type: Venezuela, Caracas, plate of Jacquin. – Fig. 16

Inga venosa Griseb. ex Benth., Trans. Linn. Soc. London 30: 623. 1875. Type: Trinidad, Sieber 104 (isotype G).
Inga guaremalensis Pittier, Contr. Fl. Venez. 5. 1921. Type: Venezuela, Upper Guaremales, Pittier 9105 (isotypes G, NY).

Small to medium-sized tree, 15-20 m tall; bark dark reddish brown, lenticels irregular in size and distribution; all vegetative and reproductive parts covered with dense hirsute rusty to brown hairs. Leaves (2)3-4 jugate; stipules conspicuous, membranaceous, channeled, triangular sometimes falcate, 12-15 x 3-6 mm, apex acute to subulate, externally densely pubescent, often persistent; petiole and leaf-rachis winged, 1-1.5(2) cm wide, elliptic to obtriangular; petiole (1)1.5-2 cm; segments of rachis (2)3-5(7) cm long; nectaries orbicular, up to 1 mm in diam., usually stipitate, sometimes proliferating (2 or 3 at same insertion of leaflets); leaflets elliptic, highly variable in size, proximal pair 3 x 1.5 to 9 x 5 cm; distal pair 12 x 7 cm to 28 x 13 cm; base obtuse to rounded, apex acute and shortly acuminate, both surfaces with long reddish sparse hairs.

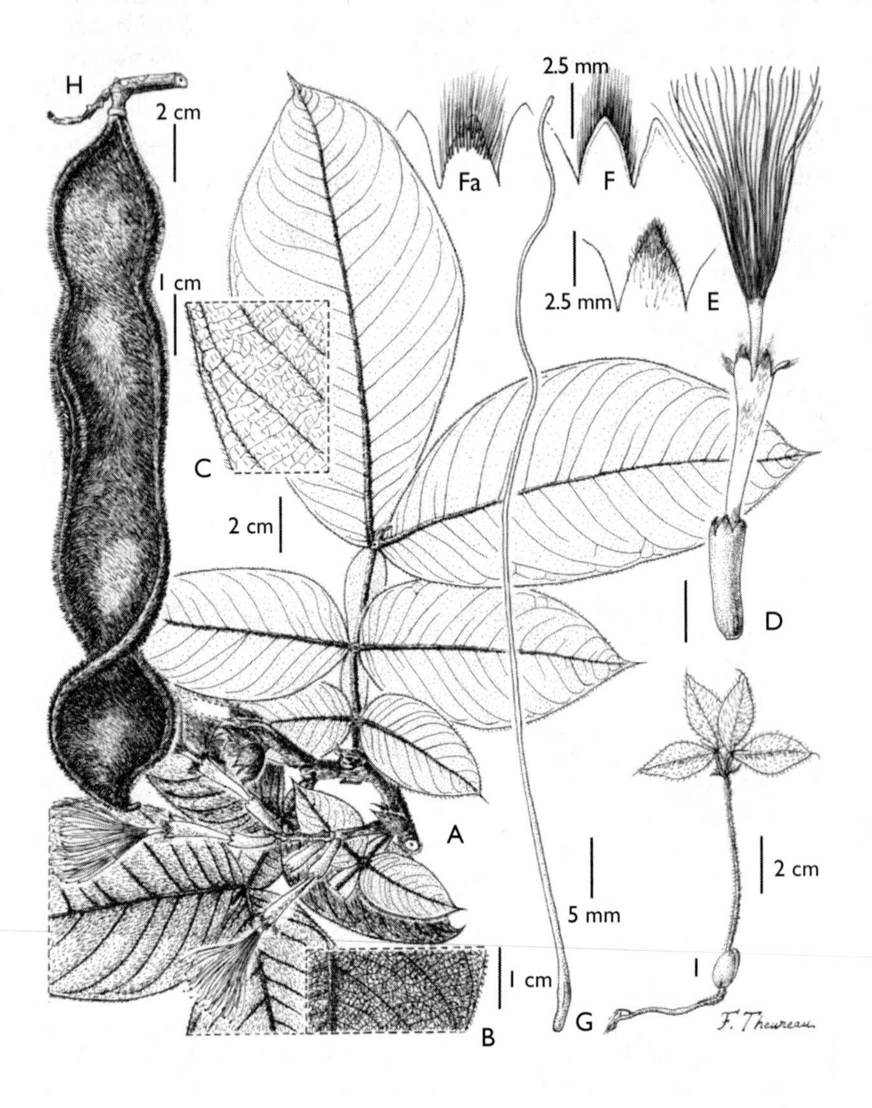

Fig. 16. *Inga fastuosa* (Jacq.) Willd.: A, habit, leaves and inflorescences; B, detail of leaf venation (young leaf); C, detail of leaf venation (adult leaf); D, flower; E, lobe of calyx, detail; F, Fa, lobe of corolla, internal (F) and external (Fa) views; G, gynoecium; H, pod; I, seedling (A-G, Mori *et al*. 22208; H-I, Poncy 610). Drawing by Françoise Theureau; reproduced with permission from P.

Inflorescence of 1-3 spikes axillary to mature leaves; peduncle 1-3 cm long; rachis 2-3 cm, 15-20-flowered, scars conspicuous and salient after abscission of flowers; bracts subulate, triangular to oblong, up to 10 x 8 mm, with sharp acicular apex; flowers sessile; bud cylindrical; calyx tubular, ca. 1.5 cm long, with hairs sparse except for pubescent borders of teeth; corolla tubular ca. 4 cm long, with dense appressed silky hair; ratio calyx/corolla length 2/5; ovary slightly pubescent; stigma cup-shaped; staminal tube shortly exserted. Pod flattened, 20-30(40) x 4(+) cm, often twisted and covered by dense hirsute rusty hair, sutures ribbed, ligneous, up to 6 mm wide; seeds up to 25 per pod, subrectangular, narrow and contiguous inside pod, pulpy seed-coat poorly developed, cotyledons shiny black. Seedling: hairy even on hypocotyl, first two leaves bifoliolate and opposite.

Distribution: Restricted to northern S America (Venezuela and the Guianas) and the West Indies; poorly collected fertile, but sterile collections from forest inventories in French Guiana (not included here) suggest that this species is not that rare in undisturbed dense rainforest communities in non-flooded areas (GU: 1; FG: 5).

Selected specimens: Guyana: Kanuku Mts., Moku-moku Cr., A.C. Smith 3446 (G, K, NY, P). French Guiana: Approuague R., Arataye Cr., Oldeman 2495 (CAY, P), Saut-Pararé, Poncy 610 (CAY, P); Saül, Foresta 663 (CAY, P, U), Eaux-Claires, Mori *et al.* 22208 (CAY, NY, P), Galbao Mts., de Granville *et al.* 8866 (CAY, K, NY, P, U, US).

18. **Inga flagelliformis** (Vell.) Mart., Flora 20(2), Beibl. 8: 112. 1837. – *Mimosa flagelliformis* Vell., Fl. Flumin. 11: t. 27. 1831. Type: plate of Vellozo.

Medium-sized tree to 20 m tall; trunk regular, unbuttressed, bark pale grey; plant glabrous on all parts except calyx and corolla; twigs angular, often chanelled above. Leaves 2-3-jugate; stipules conspicuous, elliptic to rhomboidal or falcate, 13-20 x 5-10 mm, green, caducous; petiole and rachis terete to slightly canaliculate, woody when old; petiole 0.6-2.5 cm long, rachis segments 1-3 cm long; foliar nectaries sessile, ostiolate, narrower than rachis; leaflets (2-3 pairs) elliptic, proximal pair 6-10(12) x 3-4.4(6) cm, distal pair 12-17 x 4.5-6 cm, acuminate, base acute to cuneiform; blades coriaceous and smooth on both faces, above dark green, below pale green, primary veins prominent, secondary veins 6-8 pairs. Inflorescence capitate, "axillary, solitary or paired and sometimes on short axillary shoots" (fide Pennington, 1997; not seen on

Guianese material); peduncle slender, 4-5 cm long, with sparse short hairs; rachis globose, 1.5-2 mm in diam.; bracts spatulate, 2 mm long, adaxially pubescent; flowers pedicellate, pedicel 7-8 mm, with sparse short hairs; calyx tubular, 2.5 mm long, longitudinally striate, hairs sparse, denser on teeth, those acute and very short (0.3 mm); corolla 8-9 mm long, base narrowly tubular, distal part funnel-shaped, with sparse hairs, denser on lobes, those narrowly triangular, acute, 2.5 mm long; ratio calyx/corolla length 2/5; staminal tube as long as corolla. Pod oblong, straight or slightly curved, flat, 12-18 x 3-3.5 cm, pedicel markedly acentric, apex mucronate, 1.5 cm thick when ripe, sutures 5 mm, valves bright green, sometimes with pale green dots, ornamented with slight transversal ripples; seeds 12-16, cotyledons ca. 2 cm long, bright green.

Distribution: A Brazilian species (Amazonian and Atlantic forest), recently recorded from non-flooded lowland rain forest in French Guiana (FG: 3).

Selected specimens: French Guiana: Approuague R., Nouragues Station, Poncy 874 (CAY, NY, P, U), 961 (CAY, P, U); Montagne de Kaw, Réserve Naturelle Trésor, Poncy 1348 (CAY, P).

19. **Inga graciliflora** Benth., London J. Bot. 4: 582. 1845. Type: Guyana, Roraima, Ro. Schomburgk ser. II, 756 = Ri. Schomburgk 1396 (holotype K, isotypes P).

Small to medium-sized understorey tree to 15 m tall; bark whitish, lenticellate, inner bark yellow; twigs covered with a very short tomentum. Leaves (1)2-3-jugate, variable in aspect and size (see note below); stipules up to 3 mm long, linear, caducous; petiole and rachis glabrous or sparsely puberulous, winged or not, or shortly winged below insertions of leaflets; petiole variable in length, generally about one third of that of rachis segments, latter 3-8 cm long; nectaries sessile, flattened, up to 2 mm diam.; leaflets elliptic, proximal ones 7-12 x 2-5 cm, distal ones up to 20(23) cm long; base cuneate, apex acute or inconspicuously acuminate; blades chartaceous, with scattered tiny hairs, base of veins puberulous. Inflorescence capitate, 1-4 in axil of mature old leaves or on defoliated twigs; peduncle, bract, calyx and corolla covered with short indument; peduncle short, ca. 5 mm or less; head spherical or ovoid, up to 20-flowered; bracts persistent, scaly, ca. 1 mm long, narrowly triangular (basal bracts somewhat wider); flowers pedicellate, pedicels 2-3 mm; calyx campanulate, regular, 1 mm long; corolla funnel-shaped, 4-5 mm long; ratio calyx/corolla length 1/4 or lower; staminal tube exserted, twice as long as corolla. Pods in bundles of up to 8 per capitula, glabrous, 20-35 x ca. 2 cm, convex around seeds which are not contiguous, valves joining between seeds, with a reticulate ornamentation, sutures ribbed, not thickened; seeds 13-14, embryo dark green.

Distribution: The Guianas, Amazonian Peru and Brazil; in the Guianas it appears to be a rare species of the undisturbed lowland non-flooded rainforest (GU: 5; SU: 1; FG: 5).

Selected specimens: Guyana: Barima R., Koriabo, van Andel 1322 (P, U); Mazaruni R., Tillett & Boyan 45875 (K, NY, U, US); Moruka, Fanshawe 2682 = FD 5475 (BRG, FDG, K, NY); Pakaraima Mts., Maas *et al.* 7550 (P, U). Suriname: Nickerie Distr., Kabalebo Dam Proj., Lindeman & Görts *et al.* 295 (BBS, U). French Guiana: Maroni, Grand-Inini R., Sabatier & Prévost 3346 (CAY, K, P), Haut-Marouini R., Roche Koutou, de Granville *et al.* 9546, (B, CAY, MG, NY, P, PORT, U); Saül, Oldeman 3208, (CAY, P), 3236 (CAY, P); Saint-Laurent-du-Maroni, route de Mana, SF 7233 (CAY, P).

Vernacular names: Guyana: karoto, waremesuri.

Note: This species is clearly delimited by its characteristic flowers and fruits, however the leaves are remarkably variable. While materials from Guyana are similar to the type ((2)-3-jugate leaves and unwinged petiole and rachis), those collected in French Guiana have (1)2-jugate leaves with a more or less winged rachis (variable on a single collection – de Granville *et al.* 9546), resembling the leaves of *I. obidensis*.

20. **Inga gracilifolia** Ducke, Arch. Jard. Bot. Rio de Janeiro 3: 52. 1922. Type: Brazil, Pará, Peixeboi, Goeldi HAMP 8324 (holotype MG, isotypes G, K, P, RB, US).

Medium-sized to tall tree (to 30 m x 50 cm); trunk cylindrical, not buttressed, bark reddish to orangish pale brown, inconspicuously lenticellate; whole plant glabrous, except on very young twigs and leaves. Leaves (3)4-5-jugate and small; stipules scaly, 1 mm long, generally persistent; petiole + rachis up to 7 cm long, marginate to narrowly winged, channelled; petiole and rachis segments 1-1.5(2) cm; nectaries small, ca. 1 mm diam., orbicular, sessile to slightly stipitate; leaflets elliptical to rhomboidal, proximal ones very small, 1.5-2.5 x 0.8 1.3 cm, distal ones 4-5(6) x 1.5-2.5(3) cm, base variable, often markedly asymmetrical particularly on distal leaflets, superior edge parallel and inferior one perpendicular to rachis; apex acuminate, acumen triangular ca. 5 x 5 mm, inconspicuously apiculate; blades smooth, glossy; young leaves dark red. Inflorescence (described from Brazilian specimens: see note below) capitate, solitary in axil of mature leaves; peduncle slender, 3-4 mm long; head 25-35-flowered, bracts minute (0.5 mm), caducous; flowers inconspicuously pedicellate; calyx 1 mm long, corolla 5 mm long, narrowly tubular, distally funnel-shaped; ratio calyx/corolla length 1/5; staminal tube exserted, to 8 mm long. Pod flattened, 18-25(over 30) cm long x 2-3 cm wide, bright green, valves with slightly prominent tiny reticulate ripples, sutures not thickened, somewhat undulating; seeds 13-18, not contiguous, prominent. Seedling and juvenile plant up to 8-jugate, very similar to that of *Zygia sabatieri* Barneby & J.W. Grimes in its early once-pinnate stage.

Distribution: The Guianas, Amazonian Venezuela, Peru and Brazil; dense undisturbed lowland, non-flooded, rainforest in Amazonia; this species is likely to be deciduous (pers. obs. and D. Loubry's pers. comm.) (GU: 2; FG: 7).

Selected specimens: Guyana: NW Distr., Matthews' Ridge, Maguire 40496 (K, NY, P, U); Essequibo, Keriti Cr., Fanshawe 860 = FD 3596 (FDG, K). French Guiana: Approuague R., Arataye Cr., Station Nouragues, Poncy 870 (CAY, NY, P, U), 1854 (CAY, K, NY, P, U); Kaw Mts., Acevedo & Mori 12317 (CAY); Sinnamary, Piste-de-St-Elie, Riera 227 (CAY, P), 1501 (CAY, P), Loubry 1199 (CAY, P); Passoura, Sabatier & Prévost 3611 (CAY, P).

Vernacular name: Guyana: tureli.

Note: No flowering collection is available from the Guianas. The description of the inflorescences is based on a specimen from Pará, Brazil, which differs slightly from the Guianese collections, by 5-6-jugate leaves and somewhat narrower leaflets.

21. **Inga grandiflora** Ducke, Arch. Jard. Bot. Rio de Janeiro 3: 59. 1922. Lectotype (designated by Pennington 1997: 649): Brazil, Pará, near Gurupá, Ducke MG 17180 = HJBR 10102 (hololectotype MG, isolectotypes P, RB).

Understorey tree, 10 m tall; trunk irregular; bark pinkish-brown, densely covered with prominent reddish-brown lenticels; twigs robust, angular in section or terete, lenticellate (lenticels longitudinally lenticular and yellowish), tomentose. Leaves large, 4(5)-jugate; stipules scale like, ovate, apex acute, 5 x 4 mm, tomentose; petiole ca. 5 cm, narrowly winged (5-7 mm wide), tomentose; rachis 13-14 cm, each segment 3.5-5 cm, winged (up to 10 mm wide), tomentose on midvein above and on veins and wings below; nectaries ostiolate, 1 mm diam., mostly sessile or shortly stalked; leaflets sessile, elliptic, base acute to rounded, apex acute to rounded, mucronate, proximal leaflets 8-11 x 4-5 cm, distal leaflets 15-22 x 7-9 cm; secondary veins 13-15 pairs; blades chartaceous, villose on veins and margins. Inflorescence axillary, solitary, spicate; peduncle thick, 4-6 cm long, tomentose; rachis 5-8 cm, bearing 12-20 large, sessile flowers; bracts (from Ducke 10103, Gurupá, Brazil) scale like, triangular, villose, 3 mm long, caducous; calyx 3-3.5 x 1 cm, tubular, slightly asymmetrical, longitudinally striate, tomentose, lobes subequal, 3-5 mm; corolla 5-6 cm, tube narrow, hirsute, lobes 4-6 mm long, acute; staminal tube exserted to 2 cm out of corolla. Pod 20-30(35) x 2.5-3 cm, straight or curved or twisted, velutinous (hairs reddish), base attenuate, apex acuminate, margins to 1 cm thick, raised.

Distribution: Amazonian Brazil and Peru, and French Guiana (FG: 4).

Selected specimens: French Guiana: Approuague R., Arataye Cr., Station Nouragues, Poncy 958 (NY, P, U), 994 (P), 1306 (CAY, K, NY, P), 1831 (CAY, K, NY, P, U).

Note: The specimens from Nouragues, French Guiana, represent the first records of the species in the Guianas; detailed tree inventories in the Nouragues area have shown that this species is rather common in medium-height forest on granitic soil on the foothills of an inselberg, but does not occur in the other plots with dense high lowland forest on lateritic soil.

22. **Inga heterophylla** Willd., Sp. Pl. 4: 1020. 1806. Type: Brazil, Sieber s.n., comm. Hoffmannsegg (holotype B-Willd. 19043, photo P).
– Figs. 17; 19 D-G

Inga protracta Steud., Flora 26: 758. 1843. Type: Suriname, Hostmann 1194 (holotype U, isotypes G, K, P).

Shrub or small tree, sometimes bushy, bark greyish with small lenticels, twigs slender; all parts of plant glabrous. Leaves (2)3-jugate; stipules linear, up to 3 mm long, generally persistent; petiole up to 1 cm long, flattened, grooved or marginate, segments of rachis ca. 1 cm long, marginate, channelled; foliar nectaries ostiolate, diam. < 1 mm, and stipitate (stipe

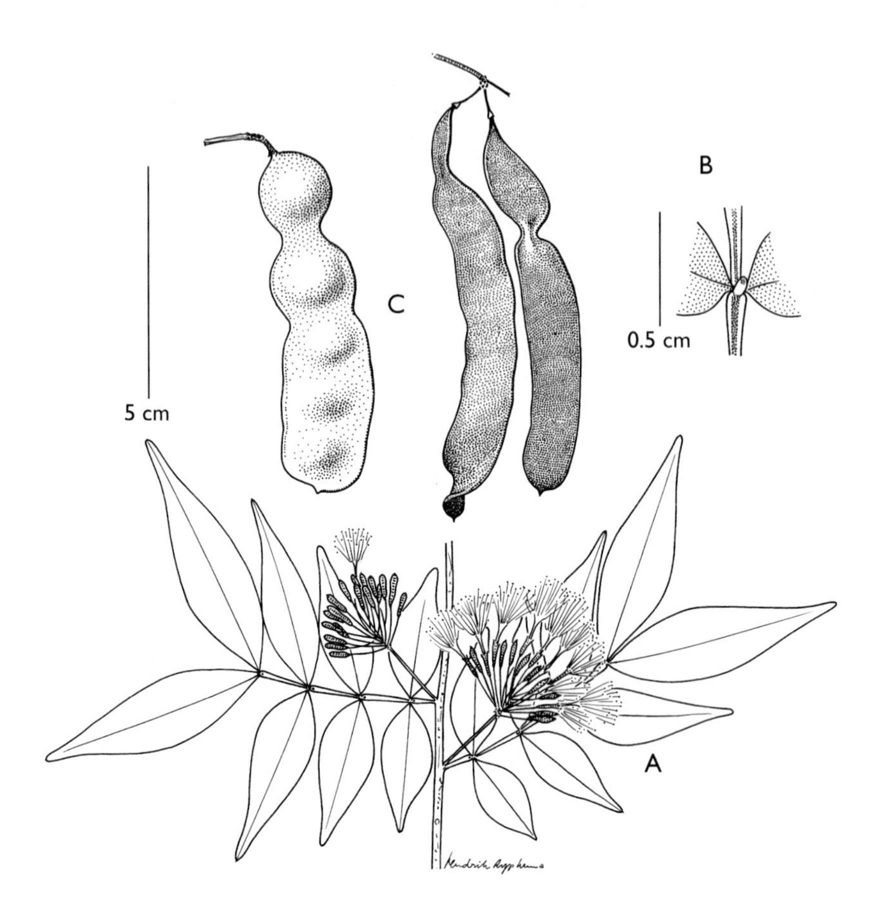

Fig. 17. *Inga heterophylla* Willd.: A, flowering branch; B, extrafloral nectary; C, pods (A-B, FD 6452; C, BW 3832). Drawing by Hendrik Rypkema; reproduced with permission from U.

up to 1 mm long and very slender) seldom sessile; leaflets elliptic, small (distal pair rarely larger than 5 x 2), base markedly asymmetrical, apex acuminate; blades smooth and glabrous on both faces, veins not prominent. Inflorescence of solitary subglobose heads axillary to old leaves; peduncle slender, 1-2 cm long; rachis 3-5 mm long often simulating a capitulum, 30-40-flowered; bracts scale like, curved, up to 1 mm long; pedicels thin, 5-7 mm long; bud obovate; calyx campanulate, regular, ca. 1 mm long; corolla ca. 5 mm long, longitudinally striate, lobes up to 1 mm long; ratio calyx/corolla length 1/5; staminal tube slightly longer than corolla. Pod flattened, glabrous, up to 12 x 1.5 cm, mucronate, valves leathery, finely reticulate, 6-12 seeds prominent, sutures not thickened.

Distribution: Widely distributed in mixed or seasonally dry forests in C America, Venezuela, Trinidad, the Guianas and Brazil; in French Guiana it is recorded only from the low forest on white sands along the western coast, St-Laurent-du-Maroni area (GU: 7; SU: 15; FG: 11).

Selected specimens: Guyana: Berbice Savanna, Henkel 2478 (BRG, P, US); Corentyne R., Mapenna Cr., Fanshawe 2959 = FD 6289 (BRG, FDG, K, U); Issano Rd., Fanshawe FD 7114 (FDG, K, U); Mazaruni, Kartabo Rd., Persaud ("CAP") 71 = FD 6452 (FDG, K, P, U); Pakaraima Mts., Hoffman 2325 (BRG, P, US); Puruni R., Mazaruni R., Boyan 102 = FD 7786 (FDG, K, NY, P), Jenman 5317 (BRG). Suriname: sin. loc., Hostmann 1194 (G, K, U, P), BW 2798 (K, U), BW 3832 (BBS, U), BW 3115, 3119 (U), BW 4680 (BBS, U), Budelman 1296 (BBS), van Donselaar 1356 (BBS), Maas & Tawjoeran LBB 10831 (BBS, U), Schulz LBB 7367 (BBS, U), 7458 (U); Jodensavanne, Mapane Cr., Boerboom LBB 9506 (BBS, U); Paramaribo, Wane weg, Kramer & Hekking 2855 (BBS, U); Zanderij, BW 4364 (BBS, P, U), Mennega 21 (U). French Guiana: sin. loc., Richard s.n. (P); Saint-Laurent-du-Maroni, Mélinon 485 (P), Wachenheim 52 (P), 295 (U), 312 (P); Saint-Elie Village, Poncy & Crozier 1151 (CAY, P); Saint-Jean, Benoist 1081 (P, U); Plateau des Malgaches, SF 4005 (CAY, P, U); Route Mana, SF 7430 (P), 7433 (P); R. Tampoc, SF 7894 (P, U).

Vernacular name: Guyana: shirada.

23. **Inga huberi** Ducke, Arch. Jard. Bot. Rio de Janeiro 3: 49. 1922. Type: Brazil, Belem do Pará, Huber 2050 = HJBR 10006 (holotype MG, isotype G, RB). – Figs. 18 A-D; 19 A-C

Medium-sized to large tree, up to 30 m x 45 cm; bole irregular, compressed or fluted, outer bark dark reddish-brown to blackish, inner bark pinkish to yellowish brown; lenticels paler, dense and irregular in size; youngest parts of branchlets with a velvet-like greenish brown indument. Leaves glabrous, almost always 2-jugate; stipules linear, 6-8 mm long, persistent; petiole 1-1.5 cm long, rachis 3-4 cm long, both thick, ligneous and lenticellate, not winged, flattened and triangular just below leaflet-insertion; nectaries cup-shaped, conspicuous, proximal one often elliptic and larger than distal one; leaflets elliptic to oblong, proximal ones 9-12 x 3-5 cm, distal ones 15-20 x 5-8 cm, base attenuate, apex acute and apiculate; blades leathery, smooth, very dark above, slightly paler and greyish beneath; young leaves pubescent, yellowish brown beneath. Inflorescence globose, 1-3 in axils of mature or young leaves, peduncle thick, tomentose, 3-6 cm long, 20-30-flowered; bracts scale like, curved upwards; pedicel 1-1.5 mm

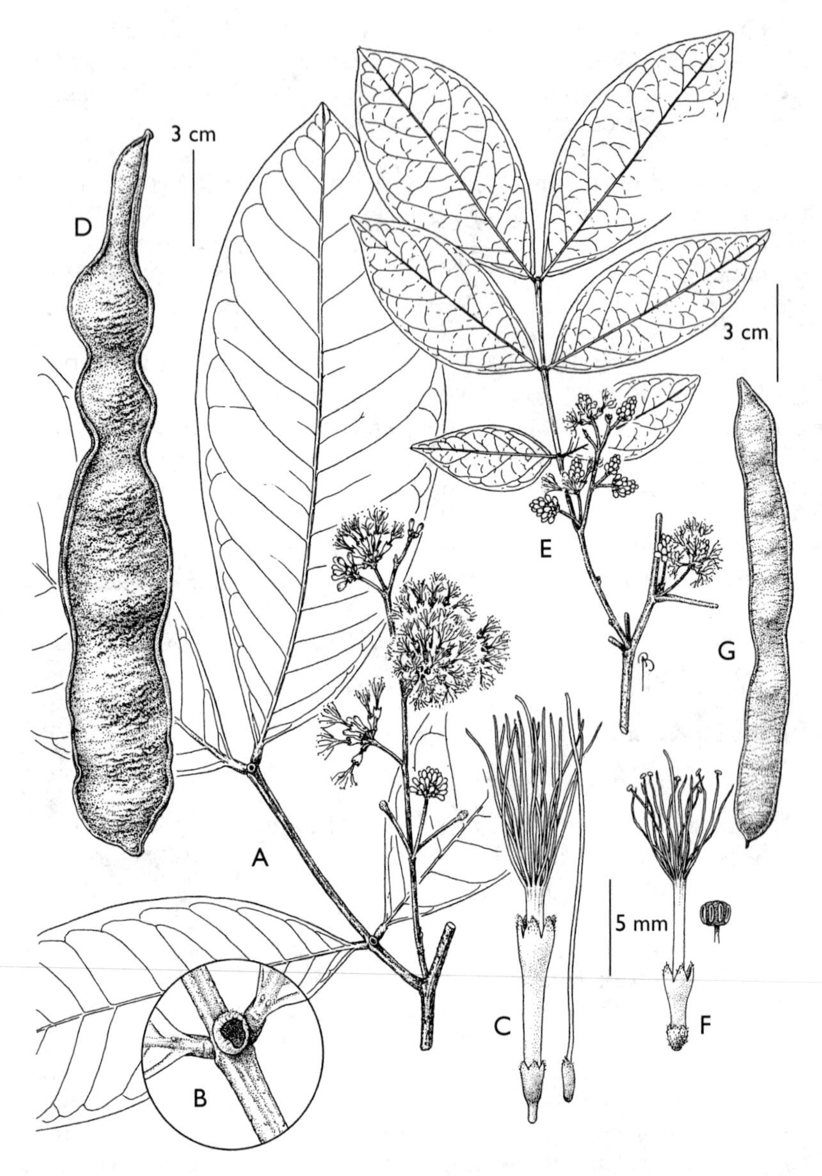

Fig. 18. A-D: *Inga huberi* Ducke: A, habit, leaf and inflorescences; B, foliar nectary; C, flower and pistil; D, pod. E-G: *Inga alba* (Sw.) Willd.: E, habit, leaf and inflorescences; F, flower and anther (ventral view); G, pod (A-C, Maguire 54444; D, Mori & Gracie 18903; E-F, Marshall 191; G, SF (Service Forestier) 7667). Drawing by Bobbi Angell; reproduced with permission from Mori *et al.*, 2002: 495.

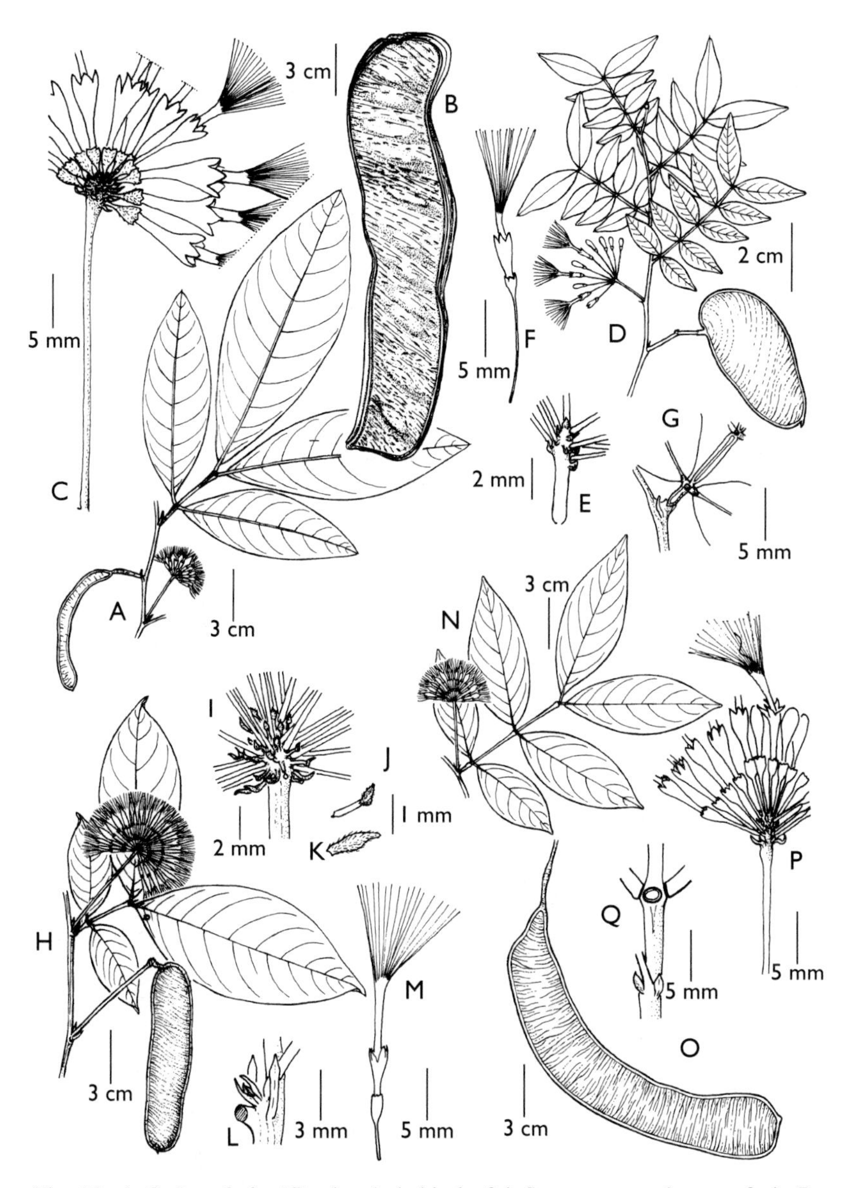

Fig. 19. A-C: *Inga huberi* Ducke: A, habit, leaf, inflorescence and young fruit; B, pod; C, inflorescence. D-G: *Inga heterophylla* Willd.: D, habit, leaves, inflorescence and pod; E, rachis of inflorescence, bracts; F, flower; G, leaf rachis, nectaries, stipule. H-M: *Inga sertulifera* DC.: H, habit, leaf, inflorescence, pod; I, detail of the flower head, bracts; J, flower bract; K, bract at base of flower head; L, base of leaf and inflorescence, stipules; M, flower. N-Q: *Inga paraensis* Ducke: N, habit, leaf and inflorescence; O, pod; P, inflorescence; Q, base of the leaf rachis, stipules. Drawings by Odile Poncy; reproduced from Poncy, 1985: 39.

long; bud clavate; calyx 2-2.5 mm long, campanulate, tomentose, regular, teeth inconspicuous; corolla 8-10 mm long, funnel-shaped, tube glabrous, lobes triangular, 1-2 mm long, pubescent; ratio calyx/corolla length 1/4; staminal tube 10-12 mm long, inconspicuously exserted. Pod flattened, up to 20 cm long, 3.5-5 cm wide, often irregular in outline, base and apex asymmetrical, valves greyish green to brownish, rigid, with, irregular oblique thick salient veins, margins thickened, enlarged, up to 1 cm wide; seeds 10-15, not contiguous, oblique, cotyledons ca. 1.7 x 1 cm, whitish to pale violet. Seedling: slender, first leaves opposite, bifoliolate, glabrous, petiole marginate, leaflets narrow, 3-5 x 1-2 cm, leathery, dark green, stipules as on mature trees.

Distribution: Eastern Amazonia, the Guianas and eastern Venezuela; dense evergreen non-flooded rainforest, not found in secondary vegetation (GU: 4; SU: 3; FG: 10).

Selected specimens: Guyana: Demerara, Mabura Hill, Polak 347 (U), 349 (seedling, U); Essequibo R., Moraballi Cr., Sandwith 119 (K, U), 526 (K). Suriname: Suriname R., Jodensavanne, LBB 9863 (BBS, NY, U); Wilhelmina Mts., Juliana Top, LBB 10325 (BBS, U), Maguire *et al.* 54444 (K, NY, P, U). French Guiana: Saül, Mori *et al.* 15593 (CAY, NY, P, U); Approuague R., Arataye Cr., Pararé Camp, Poncy 234 (CAY, P), 586 (CAY, P); Nouragues Station, Poncy 913 (P), Sabatier & Prévost 2718 (CAY, K, NY, P, U), 2752 (CAY, K, NY, P), 2830 (CAY, K, NY, P); Régina, Mt. Tortue, Feuillet 3626 (CAY, P, U, US), 10309 (CAY, P, U, US); Sinnamary, Petit-Saut, Blanc s.n. (CAY, P).

24. **Inga ingoides** (Rich.) Willd., Sp. Pl. 4 : 1012. 1806. – *Mimosa ingoides* Rich., Actes Soc. Hist. Nat. Paris 1: 113. 1792. Type: French Guiana, Cayenne, Leblond s.n. (holotype P-Juss.).

Inga merianae Splitg., Pl. Nov. Surinam 19. 1842. Type: Suriname, Splitgerber 337 (holotype L, isotypes K, U).
Inga galibica Duchass. & Walp., Linnaea 23: 747. 1850. Type: Guadeloupe, Duchassaing s.n. (holotype P).

Tree up to 10 m high; bark greyish with thick whitish lenticels; branchlets irregular, angular, young parts tomentose. Leaves 4-5-jugate; stipules tomentose, ca. 7 x 1.5 mm, caducous; petiole and rachis covered with dense pale reddish hairs; petiole to 3 cm long, terete to marginate; segments of rachis 2-4 cm, winged, elliptic to obovate, to 15 mm wide; nectaries ostiolate or cup-shaped, circular to elliptic, ca. 1.5 mm diam.; leaflets elliptic, proximal ones up to 10 x 4 cm, distal pair up to 17 x 7 cm, base rounded and symmetrical to markedly asymmetrical, apex acute, apiculate; blades papery, bullate, scabrous above, tomentose beneath. Inflorescence of 1-3 spikes in axils of mature or young leaves; peduncle 3-6 cm long, thick, angular, longitudinally grooved, tomentose; rachis up to 4 cm, 15-20-flowered; bracts narrow, subulate, 3-5 mm long, caducous; pedicel 3-5 mm; bud ovoid, apiculate; calyx campanulate, yellowish green to rusty brown, up to 10 x 4 mm, regular, teeth up to 3 mm long; corolla yellowish green, funnel-shaped, ca. 15 mm long, hairy, lobes 2-3 mm; ratio calyx/corolla length ca. 2/3; staminal tube not exserted from corolla, stamens white; stigma cup-shaped. Pod subcylindrical, deeply grooved, (-)5-20 cm long, ca. 1.5 cm diam., calyx accrescent around base, apex attenuate and rostrate; pericarp semi-rigid, velvety, yellowish brown when ripe, formed by hypertrophied sutures of pod, while valves are narrow and inconspicuous; seeds 15-20, longitudinally orientated; cotyledons black, asymmetrical. Seedling: pubescent, first two leaves unijugate, opposite, petiole winged.

Distribution: The geographical extent is wide, including the Lesser Antilles and tropical S America as far south as S Brazil; in low forests, gallery forest, on white sands; in the Guianas abundant along the coast, in secondary forests, by roadsides and rivers, and far from the coast, eg. in the Rupununi Savanna (S Guyana), in humid habitats such as forested river banks (GU: 19; SU: 19; FG: 27).

Selected specimens: Guyana: Berbice R., FD 6394 (FD, K, U); Demerara, Jenman 1524 (BRG, U), 5074 (BRG); Garden of Eden, Persaud & Omawale 90 (BRG); Bot. Garden Georgetown, Harrison & Persaud 1552 (BRG); Parika, Gillespie 1037 (P, US), 1224 (BRG, P, U, US), Pipoly 11279 (BRG, CAY, P, US); Rupununi, Kanuku Mts., Takutu R., FD 2188 (FDG, K), A.C. Smith 3277 (L, NY, P, U), 3524, Jansen-Jacobs *et al.* 15 (U); Iserton, A.C. Smith 2506 (K, NY, P, U); Rupununi, Shiriri Mt., Jansen-Jacobs *et al.* 529 (BBS, BRG, CAY, P, U, US), Sand Cr., Wilson-Browne FD 5631 (BRG, FDG), Witaru Falls, Jansen-Jacobs *et al.* 155 (BBS, CAY, P, U, US); Up. Essequibo, Makawatta Mt., Hoffman & Gopaul 312, Gran Sabana, Harris *et al.* 1093 (P, US). Suriname: sin. loc., de Vriese s.n. (K), Splitgerber s.n. (P, U), Hostmann 78a (P, U), 685 (P, U), 884 (P,U), Kegel 290 (P, U); Coppename R., Lanjouw & Lindeman 1383 (U); Coronie Distr., Jenny, Teunissen M. & P. LBB 15132 (BBS, U); Domburg, BW 4305 (U); Nickerie, Teunissen M. & P. LBB 13221 (BBS); Paramaribo, Celos, den Outer 863 (BBS, U, WAG); Parwa R., Braams Point, Stahel 30 (BBS); Saramaca R., Hamburg, BBS 107 (BBS, U); Suriname R.x Para R., Lindeman 3686 (U); Oceaan Weg, Teunissen M. & P. LBB 15047 (BBS); Pl. Liberté, Florschütz 867 (U); Wilhelmina Mts., Lucie x Oost R., Irwin *et al.* 55355 (NY, U), 55550 (NY, U). French Guiana: sin.

loc., Richard s.n. (P), Leblond 282 (P), Mélinon 125a (P), Leprieur s.n. (P); Ilet La Mère, Lemée s.n. (P); Ile de Cayenne, Leguillou s.n. (P), Jacquemin 1869 (CAY, P), Lemée s.n. (P), Centre Orstom, Jardin, friches, David *et al.* 98 (CAY, P), Prévost 3756 (CAY, K, NY, P), 4076 (CAY, MO, P), Chaton, Jacquemin 2213 (CAY, P), Montabo, Prévost 792 (CAY, P), Rorota, Labat 67 (CAY, P), Rte Madeleine, Prévost 480 (CAY, P); Iracoubo, Poncy 298 (P), Prévost 608 (CAY, P); Kourou, Crevaux s.n. (P), Benoist 1464 (P); Savane Fracas, Billiet & Jadin 4766 (CAY,); Sinnamary, Malmanoury, Skog & Feuillet 290 (CAY, P, US); Sinnamary, Feuillet 603 (CAY, P), Prévost 799 (CAY, P); Mana, Les Hattes, Feuillet 754 (CAY, P), Lescure 714 (CAY, P); Tonate, Bordenave 433 (CAY, P), Poncy 279 (CAY, P).

Note: A few specimens from southern Guyana have somewhat larger flowers than described above (eg. Wilson-Browne 25 = FD 5631).

25. **Inga jenmanii** Sandwith, Bull. Miscell. Inform. Kew 1931: 368. 1931. (as "jenmani"). Type: Guyana, Essequibo R., Moraballi Cr., Sandwith 187 (holotype K, isotypes BRG, NY).

Inga sertulifera DC. var. *minor* Benth., London J. Bot. 4: 581. 1845. Type: Guyana, Pomeroon R., Ro. Schomburgk ser. II, 810 = Ri. Schomburgk 1427 (BM, K, MEL, NY, P).

Small to large tree up to 30 m tall; bark dark red-brown, densely lenticellate, inner bark yellow; branchlets slender, lenticellate, with very short, thin hairs, especially on young parts. Leaves 1-2(3)-jugate; stipules inconspicuous, linear 2-3 mm long, caducous, pubescent; petiole and rachis not or rarely very shallowly winged, marginate, channelled; petiole slender, 5-10(+) mm long; rachis when present 1-2 cm long; nectaries inconspicuous, sessile or shortly stipitate, less than 1 mm diam.; leaflets elliptic, base acute, apex acute or acuminate, (3)5-8(10) x (1.5)2-4(5) cm (1-jugate leaves and distal pair of 2-jugate leaves); proximal leaflets of 2-jugate leaves never longer than 6 cm. Inflorescence capitate, solitary in axil of contemporary or old leaves; peduncle 1-2 cm long, slender, generally < 30-flowered; bracts scale like, boat-shaped, curved, pubescent; flowers pedicellate, pedicel 4-6 mm long; calyx 1-1.5 mm long, campanulate, teeth puberulous, regular; corolla 7-8 mm long, funnel-shaped; ratio calyx/corolla length 1/6-1/7; staminal tube included in, or exserted from corolla; stigma funnel-shaped. Pod 7-12 x 2 cm, glabrous, dark green and up to 1.5 cm thick at maturity, valves smooth, sutures longitudinally striate, not prominent, rounded, 3-6 mm wide; seeds 6-12, contiguous.

Distribution: In the present taxonomic delimitation (see note below) known only from the Guianas; dense rainforest, sciophilous (GU: 17; SU: 2; FG: 11).

Selected specimens: Guyana: sin. loc., Ro. Schomburgk ser. II, 782 (K, P); Barima-Waini, Port Kaituma, Hoffman *et al.* 557 (BRG, P, US); Bartica, Moraballi Cr., Sandwith 171 (K, NY, P), 1139 (FDG, K); Siba Cr., Fanshawe FD 3410 (FDG, K); Demerara, Jenman 3908 (K, NY, P); Mazaruni R., de la Cruz 2245 (K, U); Mt. Ayanganna, Tillett & Tillett 45155 (K, NY, P, U); Potaro-Konawaruk Road, McDowell 3540 (BRG, P, US); Potaro-Konawa Road, McDowell 3427 (BRG, P, US, U), 3436 (BRG, US); Supunani Cr., Jenman 6587 (BRG, K). Attributed here with doubt: Cuyuni-Mazaruni, Paruima R., Clarke 1124 (BRG, P, US); Barama R., Kokerite, van Andel 1113 (P, U), Yakishuru, van Andel 1162 (P, U). Suriname: Wilhelmina Mts., Lucie R., Irwin *et al.* 54512 (BBS, NY, U), 55780 (NY, P, U). French Guiana: Sinnamary, Piste-de-St-Elie, Loubry 841 (CAY, P), Sabatier & Prévost 4055 (CAY, NY, P); Cr. Plomb, Loubry 1374 (CAY, P); Passoura Cr., Sabatier & Prévost 4075 (CAY, P); Approuague R., Nouragues Station, Poncy 873 (CAY, NY, P, U), 1038 (CAY, NY, P, U), Sabatier & Prévost 2689 (CAY, K, NY, P, U, US); Maroni, Wachenheim 54 (P), 313 (P); Saint-Laurent-du-Maroni, Route de Mana, SF 7337 (P).

Note: *Inga jenmanii* has previously been considered a synonym of *I. sertulifera* (Poncy, 1985), and was later included in *I. sertulifera* subsp. *leptopus* T.D. Penn. (Pennington, 1997). Recent collections mainly from French Guiana, often with bifoliolate leaves, support the status of *I. jenmanii* as a distinct species; the species inhabits non-flooded rainforests, eventually at top of hills, while *I. sertulifera* is essentially a riverine, strictly heliophilous species.

26. **Inga lateriflora** Miq., Linnaea 19: 131. 1847. Type: Suriname, Hostmann 1691 (holotype U, isotypes G, NY, P).

Tree to 25 m high, trunk not buttressed; bark smooth, greyish to pinkish, with horizontal rows of lenticels and exfoliation rings; branchlets terete. Leaves glabrous at every stage, but highly variable in number, shape and size of leaflets, depending on age and localities; leaflets numerous (up to 8 pairs) and very narrow in juvenile stages, decreasing in number and increasing in width as tree grows; stipules lanceolate, acicular, 2 mm long, often persistent; petiole and rachis marginate, channelled; petiole 1-2 cm long; nectaries orbicular, ca. 1 mm in diam.; fertile specimens with 2-4 pairs of leaflets, elliptic to lanceolate, apex rostrate, base acute, proximal pair (3)3.5-5 x (1)1.7-2.7 cm, distal pair(s) (4.5)6-10 x (2)2.5-3 cm; blades above smooth and dull due to a very tiny granulation on surface. Inflorescence capitate, axillary to mature leaves, more often at defoliated nodes, and often axillary to abortive leaves of a very short (ca. 1 cm) inconspicuous shoot; peduncle up to 1 cm long; heads 12-15-flowered, bracts spatulate and curved, pubescent; flowers tiny, calyx regular, ca. 1 mm, with sparse hairs; corolla up to 4 mm long, entirely glabrous, teeth linear-triangular 1 mm long; ratio calyx/corolla length ca. 1/4; staminal tube exserted and twice as long as corolla; stigma tubular, slightly enlarged. Pod straight, glabrous, 10-15 x 1-1.5 cm (French Guiana, Suriname) to 20 x 1.8 cm (Guyana); sutures not thickened, valves leathery, when ripe yellow and moniliform (seeds not contiguous, prominent); cotyledons light brown ca. 1 cm long. Seedling: two first leaves opposite and 2-jugate, leaflets linear; leaves of juvenile plant up to 8-jugate (see above).

Distribution: Recorded throughout Amazonia; restricted to xerophytic or seasonal forests (at the Nouragues Field Station (French Guiana) it is found only in low forest on slopes or at the top of the granitic outcrop, as noted for *Inga virgultosa* and *I. cayennensis*), *I. lateriflora* and its relative *I. heterophylla* grow in similar habitats, and can be taxonomically confused (GU: 18; SU: 18; FG: 5).

Selected specimens (including few sterile and juveniles): Guyana: sin. loc. Gleason 298 (K, NY), Jenman 2428 (BRG, K, NY), de la Cruz 1080 (NY), 2209 (NY), 2428 (NY), 2558 (NY), 2560 (NY), 4228 (NY); Kanuku Mts., Wabuwak, Wilson-Browne FD 5737 (BRG, FDG, K); Mabura Hill, Poncy 922 (P); Mazaruni Station, Davis 488 = FD 444 (FDG, K, U), Davis FD 2483 (FDG, K, NY, U), Fanshawe FD 4334 (FDG, K); Tapacooma Cr., Jenman 6608 (BRG, K, NY, U); Timehri, Sta-Mission, Poncy 914 (P); Upper Mazaruni, de la Cruz 2233 (K, NY, U); Upper Demerara R., Jenman 4247 (BRG, K). Pomeroon-Supenaam., Akawini R., Hoffmann *et al.* 2555 (CAY, US). Suriname: Brownsberg, BW 4511 (P, U), 6651 (K, U), Sectie O, BW 454 (U), 1633 (K, U), 1793 (U), 2052 (U), 2378 (U), 3898 (U), 4784 (U); Jodensavanne, Heyligers 212 (BBS, U), 586 (BBS, U), Sabajo & Roberts LBB 11193 (P, U), Schulz 9325 (BBS, U); Moengo, Lanjouw & Lindeman 682 (BBS, K, U); Nickerie, Kabalebo Dam Proj., Lindeman & Görts *et al.* 479 (U); Zanderij, Samuels 545 (K, U), Stahel BW 149 (BBS, CAY, U). French Guiana: Approuague R., Station Nouragues, Larpin 810 (CAY, P); Inini Mts., de Granville 7338 (CAY); Mt. St-Marcel, de Granville *et al.* 15382 (CAY); Emerillons, Feuillet 1238 (CAY); Saül, Feuillet 439 (CAY).

Vernacular name: Guyana: shirada.

27. **Inga laurina** (Sw.) Willd., Sp. Pl. 4: 1018. 1806. – *Mimosa laurina* Sw., Prodr. 85. 1788. Type: St. Kitts, Masson s.n. (holotype, BM).

Mimosa fagifolia L., Spec. Pl. 516. 1753. Based on Plukenet, Almagestum Botanicum, p. 44, tab. 241, fig. 2. 1696. – *Inga fagifolia* (L.) Willd. ex Benth., Trans. Linn. Soc. London 30: 607. 1875, non G. Don, 1832. Type: "habitat in Barbados", tab. 241, fig. 2 of Plukenet.

Small tree, often bushy, to tall tree up to 30 m; whole plant glabrous. Leaves 2-jugate, stipules persistent, falcate, ca. 5 x 2 mm; petiole and rachis stocky, flattened or marginate; petiole 0.5-1 cm long, rachis 1.5-3 cm long; nectaries conspicuous, sessile, 1-2 mm diam.; leaflets broadly elliptic, proximal ones up to 9 x 5 cm, distal ones up to 15 x 7.5 cm; base and apex acute; blades leathery, veins prominent below. Inflorescence of axillary spikes, solitary or geminate in axils of contemporaneous leaves; peduncle 1-2 cm long; rachis up to 12 cm long, bearing 50 or more minute sessile flowers similar to those of *I. marginata*. Pod up to 10 x 2 cm, fleshy when ripe, yellow and smooth, subcylindrical, margins ca. 5 mm wide; seeds contiguous.

Distribution: Widespread in S and C America and the Caribbean; morphologically very close to *I. marginata*, however, *I. marginata* occurs in dense rainforest while *I. laurina* is to be found in drier habitats or close to them (e.g. in gallery forests), including marsh forests, scrubland on white sands, secondary forest; Pennington (1997) considered that "it is one of the few [*Inga*] species which will tolerate relatively dry areas..."; in the Guianas known only from northern Guyana (one collection of Leprieur (G), supposedly from French Guiana, is of doubtfull origin), where it seems to be mainly cultivated raising the question whether the few specimens collected "from the wild" might be adventitious (GU: 9; FG: 1?).

Selected specimens: Guyana: Berbice, FD 766 (=Hohenkerk 4) (BRG, FDG, K); Demerara, Jenman 4859 (BRG, K), 4862 (BRG, K); Persaud & Omawale 101 (BRG), McDowell *et al.* 2279 (BRG, P, U); Garden of Eden, Persaud 206 (BRG); Pomeroon R., Hoffman 2486, Kairimap R., Hohenkerk 729 (BRG, FDG, K); Georgetown (cult.), Kortright 8861 (BRG, K). French Guiana: Leprieur s.n. (G).

28. **Inga leiocalycina** Benth., London J. Bot. 4: 598. 1845. Lectotype (designated by Poncy 1985): Guyana, Pomeroon R., Ro. Schomburgk ser. II, 829 = Ri. Schomburgk 1391 (hololectotype P, isolectotypes K, MO, NY, P, US). – Fig. 20 A-D

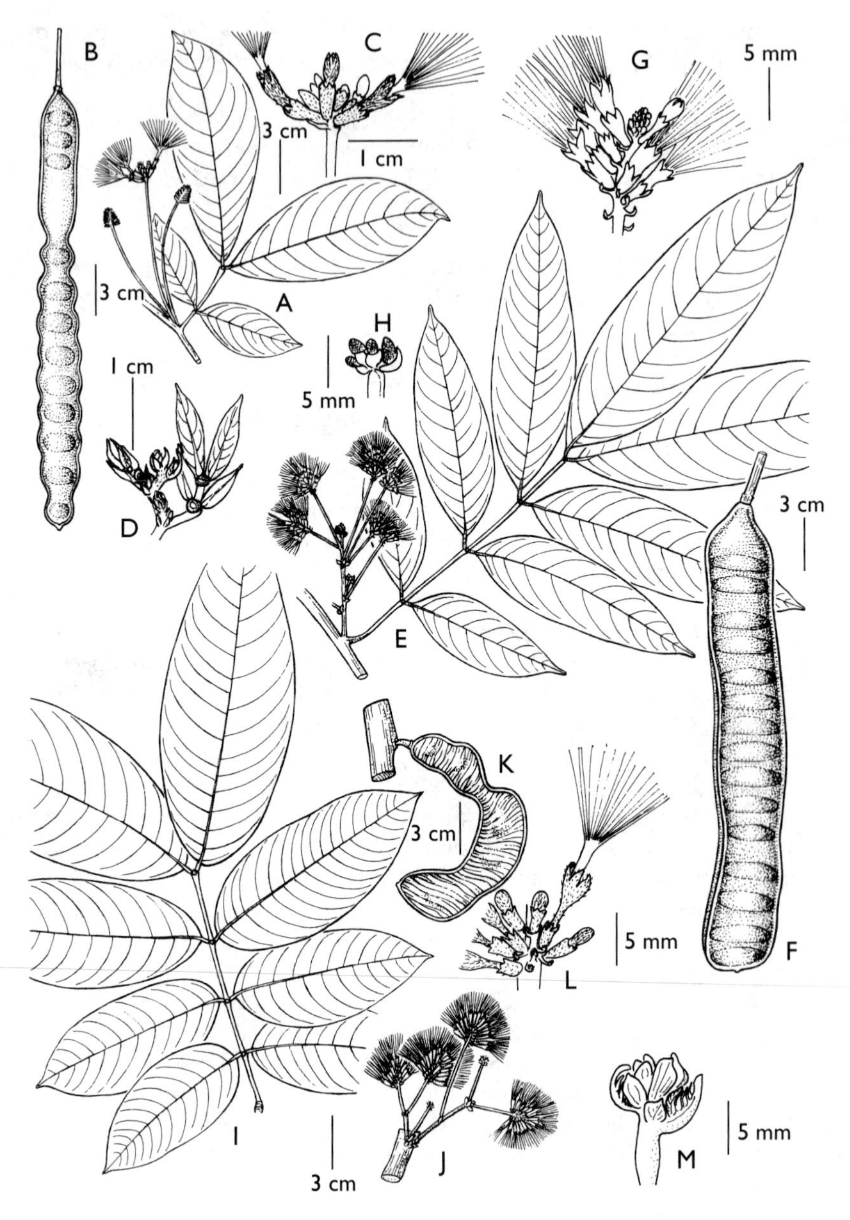

Fig. 20. A-D: *Inga leiocalycina* Benth.: A, habit, leaf and inflorescences; B, pod; C, flowers; D, extremity of leaf axis, young leaves, stipules. E-H: *Inga acrocephala* Steud.: E, habit, leaf and inflorescences; F, pod; G, inflorescence, flowers; H, extremity of inflorescence axis, young inflorescences. I-M: *Inga melinonis* Sagot: I, leaf; J, cauliflorous inflorescences; K, pod; L, partial inflorescence, flower; M, extremity of leaf axis, stipules. Drawing by Odile Poncy; reproduced from Poncy, 1985: 49.

Tree to 30 m x 60 cm; bark reddish brown, rough with tiny whitish lenticels; branchlets pubescent when young. Leaves 2-jugate, with silky, reddish or whitish tomentum on petiole, rachis and veins, more dense on young parts; stipules linear to rhomboidal, ca. 4 mm long, pubescent, caducous; petiole and rachis unwinged, petiole terete up to 2 cm long, rachis terete to flattened, less often marginate 2-4 cm long; nectaries conspicuous, prominent, generally wider than rachis, circular to transversally elliptic, pale green; leaflets elliptic, proximal pair up to 10 x 4 cm, distal pair up to 18 x 7 cm; base acute, apex acute, apiculate; surface above glabrous, veins prominent beneath, with more or less conspicuous reddish hairs. Inflorescence of 2-5 spikes in axils of juvenile or adult leaves; peduncle slender, up to 8 cm long; rachis 1-2 cm, bearing 20-30 densely aggregated, sessile flowers; bracts inconspicuous, 1-2 mm long, linear; flower bud cylindrical; flowers very narrow, calyx tubular < 1 diam., 2-3 mm long, glabrous, striate, with irregular lobes; corolla tubular, ca. 8 mm long, covered with dense, appressed, whitish hairs; ratio calyx/corolla length 1/3-1/2; staminal tube not or slightly exserted. Pod straight, 15-30 x (1.5)2-2.5 cm, when ripe bright green, glabrous, with narrow (up to 3 mm wide) ribbed sutures, valves constricted between prominent seeds; cotyledons green, up to 1.5 x 1 cm. Seedling: whitish tomentose, two first leaves opposite, bifoliolate, petiole winged.

Distribution: In the Guianas and central Amazonian Brazil; lowland dense rainforest, in non-flooded areas (GU: 16; SU: 9; FG: 11).

Selected specimens: Guyana: Barama R., Kariako, van Andel 1261 (P, U), Kokerite, van Andel 1161 (P, U); Barima R., Jenman 6964 (K, U), Pipoly 8416 (BRG, P, U, US); Essequibo R., Macouria R., Henkel 1991 (BRG, P, US); Moraballi Cr., Sandwith 515 (K, P, U); Upper Takutu, Kwitaro R., Clarke 6220 (BRG, P, US); Iwokrama, Clarke *et al.* 4304 (BRG, P, US); Mazaruni R., Fanshawe FD 3359 (FDG, K, NY, U); Puruni R., Boyan 107= FD 7791 (FDG, K, P, U); Pomeroon R., Ri. Schomburgk 1391 (P); Rupununi Savanna, Isherton, A.C. Smith 2480 (K, NY, P, U), Kuyuwini R., Jansen-Jacobs *et al.* 2917 (BRG, NY, P, U), 3008 (BRG, P, U), Wakadanawa, Jansen-Jacobs *et al.* 5482 (BRG, P, U). Suriname: Nickerie, Wessels Boer 315 (U); Voltzberg, Lanjouw 912 (U); Raleighfalls, van Troon LBB 16174 (BBS); Wilhelmina Mts., Lucie R., Irwin *et al.* 55353 (U), 55381 (NY, P), 55791 (U); Jodensavanne, Schulz LBB 8942a (BBS, U); Tumuc-Humac, Talouakem, Acevedo-Rodriguez *et al.* 6043 (BRG, CAY, P, US), 6091 (BRG, CAY, P, US). French Guiana: Approuague R., Arataye R., St-Pararé, Poncy 440 (CAY, P, U, US); Haut-Oyapock R., Trois-Sauts, Grenand 1503 (CAY, P), Prévost & Sabatier 2818 (CAY, K, NY, P, U); Maroni R., Mélinon 348 (P, U), Cr. Sparwine, SF 195M (CAY, P, U), SF 6136 (CAY, P); Saül, Eaux-Claires, Mori *et al*. 23059 (CAY, NY, P), Galbao trail, Mori *et al*. 20879 (CAY, NY, P); Saint-Laurent, Acarouany, Sagot 780 (P), Rte Mana, Feuillet 780 (CAY, P); Sinnamary, Piste-de-St-Elie, Prévost 1786 (CAY, P).

Vernacular name: French Guiana: masulapa pila (Wayapi).

Note: Pennington (1997) proposed changes in the taxonomy of the *Inga punctata* group (including *I. leiocalycina, I. stenoptera, I. longipedunculata*, and their synonyms). Based on the knowledge of the Guianese materials, and after a re-examination of the type specimens, the distinction between *I. longipedunculata* and *I. leiocalycina* is maintained here.

29. **Inga leptingoides** Amshoff, Bull. Torrey Bot. Club 75: 384. 1948. Type: Suriname, Tafelberg, Maguire 24264 (holotype NY, isotypes K, P, U).

Small tree; branchlets lenticellate; whole plant glabrous. Leaves (1)2-jugate; stipules linear, 3 x 1 mm, caducous; petiole and rachis winged, obtriangular; petiole ca. 1 cm long; rachis ca. 2 cm (3.5 cm); nectaries conspicuous, prominent, proximal one transversally elliptic, ca. 2.5 mm wide, distal one smaller, circular; leaflets elliptic, proximal ones ca. 7 x 3.5 cm, distal ones up to 10 x 7 cm, apex acute, not or slightly acuminate, and carinate, base cuneate to acute; blades leathery, shiny, veins forming a thin, salient network above and beneath. Inflorescence subcapitate, solitary in axils of mature leaves, 25-35-flowered; peduncle thin, 2-3.5 cm long; rachis clavate ca. 7 mm long; bracts inconspicuous 1 mm long or less, spatulate; pedicel slender, ca. 2 mm long; flower bud clavate; calyx campanulate, minute (1 x 1 mm); corolla funnel-shaped, up to 5 mm long; ratio calyx/corolla length ca. 1/4; staminal tube non-exserted. Fruit unknown.

Distribution: Known only from the Tafelberg in Suriname; in low mesophytic "*Clusia* bush", growing on rocky substrate (SU: 2).

Specimen examined: Suriname: Tafelberg, Maguire 24640 (US, NY, P, RB, U).

30. **Inga lomatophylla** (Benth.) Pittier, Contr. U.S. Natl. Herb. 18: 195. 1916. – *Inga speciosa* Spruce ex Benth. var. *lomatophylla* Benth., Trans. Linn. Soc. London 30: 620. 1875. Type: Venezuela, San Carlos de Rio Negro, Spruce 3097 (as Brazil) (holotype K, isotypes G, NY, P). – Fig. 21

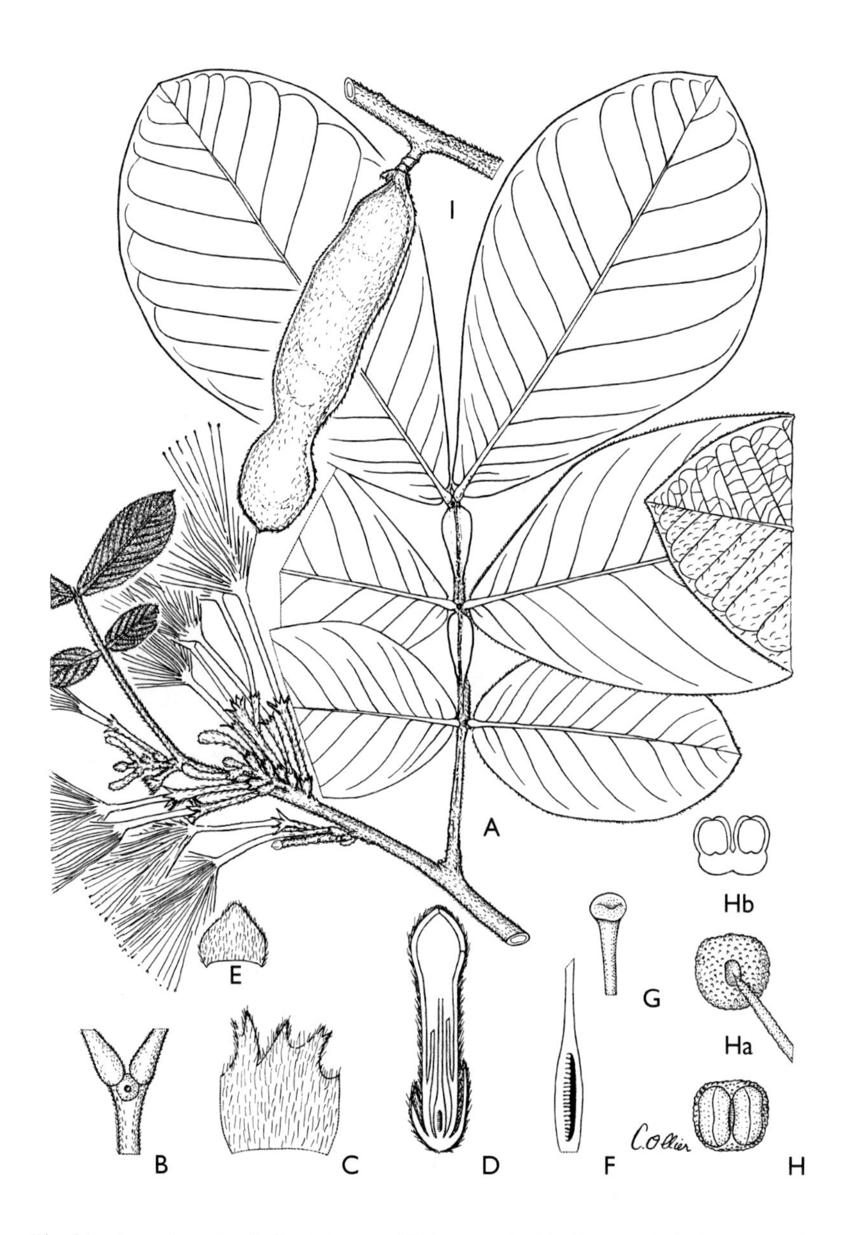

Fig.21. *Inga lomatophylla* (Benth.) Pittier: A, habit (leaf and inflorescence); B, extremity of leaf-stalk with foliar nectary; C, flower bract; D, flower bud, longitudinal section; E, calyx (display); F, ovary; G, stigma; H, Ha, Hb, anther: dorsal view (H), ventral view (Ha), section (Hb); I, pod (A-H, Poncy 929, Prévost 1148; I, Poncy 929). No scales known for this drawing. Drawing by Corinne Ollier; reproduced with permission from P.

Inga speciosa Spruce ex Benth., ibid., non Martens & Galeotti, 1843. Type: Brazil, Santarem, Spruce 973 (holotype K, isotype G, P).
Inga speciosa Spruce ex Benth. var. *bracteifera* Ducke, Arch. Jard. Bot. Rio de Janeiro 4: 17. 1925. Lectotype (designated by Poncy 1991: 152): Ducke HJBR 10104 (hololectotype P, isolectotypes G, K, RB, US).

Small to medium-sized tree; bark pale, greyish to whitish; branchlets grooved, densely tomentose, reddish brown when young, lenticels sparse, irregular, impressed. Leaves (2)3-jugate; stipules scale like, triangular, 2-3 x 2-3 mm, pubescent; petiole and rachis densely pubescent; petiole up to 6 cm long, thick, slightly channelled above, not winged or only distally; rachis segments 5-7 cm long, generally winged, if so elliptic, up to 12 mm wide; nectaries inconspicuous, 1 mm diam. or less; leaflets elliptic, large, proximal pair up to 14 x 8 cm, distal pair up to 28 x 14 cm; apex acute, base acute to rounded; blades thick, discolored, surface above dark green and rough, below paler, pubescent, veins of both faces hairy. Inflorescence of 2-4 lax spikes axillary to undeveloped or reduced leaves; either on terminal or lateral axes, these axes thick, pubescent, angular; peduncle up to 2 cm long; rachis 3-5 cm, with 30-40 flowers; bracts pubescent, elliptical to rhomboidal, ca. 3 x 3 mm, caducous; bud cylindrical; calyx tubular, 7-9 mm long, tomentose, short lobes inconspicuous; corolla tubular up to 35 mm long, narrow, with dense appressed hairs, lobes 3-4 mm, subulate; ratio calyx/corolla length 1/4; staminal tube exserted; ovary 1 or 2 per flower, glabrous; stigma cup-shaped. Pod (not known from Guianese collections) ca. 15 x 2 cm, at maturity subcylindrical, sparsely pubescent, sutures not prominent, 3-5 mm wide. Seedling (observed at Station Nouragues, French Guiana): first two leaves opposite and unijugate, petiole widely winged, surface of blades discolored above (greyish green and shiny along veins, green elsewhere).

Distribution: Uncommon species of dense lowland primary forest in Amazonian Brazil, Peru and French Guiana (several sterile records and field observations show that the species is widely distributed in French Guiana) (GU: 1; FG: 10).

Selected specimens: Guyana: Upper Takutu-Upper Essequibo, Sipu R., Clarke *et al.* 7826 (BRG, P, US). French Guiana: Approuague R., Station Nouragues, Poncy 929 (CAY, K, P, U, US), 929 b,c (CAY, P); Comté R., Cr. Rupert, Oldeman B1117 (CAY, P); Route Cayenne-Régina, Prévost 1148 (CAY, P), 1794 (CAY, P); Sinnamary, Piste-de-St-Elie, Sabatier & Prévost 1312 (CAY, P), 3219 (CAY, INPA, K, MO, NY, P, U), 3769 (CAY, NY, P, U); Ouaqui R., Grigel, de Granville B5053 (CAY, P, U).

31. **Inga longiflora** Spruce ex Benth., Trans. Linn. Soc. London 30: 620. 1875. Type: Brazil, Pará, Obidos, Spruce 484 (holotype K, isotype P). – Fig. 14 H-K

Inga tubaeformis Benoist, Bull. Soc. Bot. France 66: 390. 1920. Type: French Guiana, Saint-Laurent-du-Maroni, Gourdonville, Benoist 1491 (holotype P).

Small tree up to 12 m; bark reddish brown. branchlets cylindrical, striate, ferrugineous hirsute. Leaves (3)4(5)-jugate; stipules linear or lanceolate, up to 10 x 1.5 mm long, persistent, pubescent; petiole and rachis winged, petiole base and primary vein densely pilose; petiole ca. 2.5(+) cm long, rachis segments 2-3 cm long, wing elliptic, up to 7 mm wide; nectaries often pedicellate, to 3-4 mm long, orbicular and reduced in diam. (1 mm or less); leaflets narrowly elliptic, proximal ones 3-8 x ca. 3 cm or smaller, distal ones 6-13 x 1.5-3.5 cm; base acute, apex apiculate; blades papery, glabrous and smooth above, sparsely pubescent beneath; veins pubescent on both faces. Inflorescence solitary or less often geminate spikes in axils of adult leaves; peduncle ca. 1 cm long, hirsute; rachis 1.5-2.5 cm long, 6-12-flowered; bracts linear 3-6 mm long, pubescent, persistent; flower bud obovate; calyx tubular, acuminate before opening, 10-13 mm long, striate, hairy, markedly irregular (so as to appear almost bilabiate); corolla tubular up to 4.5 cm long, covered with dense appressed glossy white hairs, lobes narrow, 4-5 mm long; ratio calyx/corolla length 1/4; staminal tube exserted and very long (up to 6 cm). Pod 12-22 x 3 cm, flattened, slightly curved, rigid, hairs falling off when ripe, sutures 3-4 mm wide, only very slightly prominent; valves flattened, seeds not conspicuously prominent. Seedling: juvenile form as observed from numerous seedlings surrounding parent tree (Station Nouragues, French Guiana): stem, petiole, rachis and veins reddish brown, densely pubescent; stipules linear, 5 mm long; leaves 2-4-jugate; features of adult leaves already present, but leaflets much narrower, lanceolate.

Distribution: Amazonian species, uncommon in the Guianas where it is known from the northern half of French Guiana, poorly collected in Suriname, recently recorded from Guyana (several collections of the very recognizable juvenile forms attest to the presence of the species in quite a number of localities) (GU: 2; SU: 2; FG: 6).

Selected specimens: Guyana: Upper Essequibo-Upper Takutu, Acarai Mts., Clarke 7509; Sipu R., Clarke 7715. Suriname: Corantyne R., Lindeman 6658 (U), Upper Tapanahoni, Schulz 8150 + seedling 8156 (U). French Guiana: Ile de Cayenne, R. Tonégrande, Oldeman B882; St-Laurent, Gourdonville, Benoist 1491 (P); Oyapock, Miriaflor, Oldeman 1294 (CAY, P), Saut-Armontabo, Oldeman B2417 (CAY, P); Sinnamary, Piste-de-St-Elie, Sastre 5484 (CAY, P); Maroni R., Wachenheim 204 (p.p.).

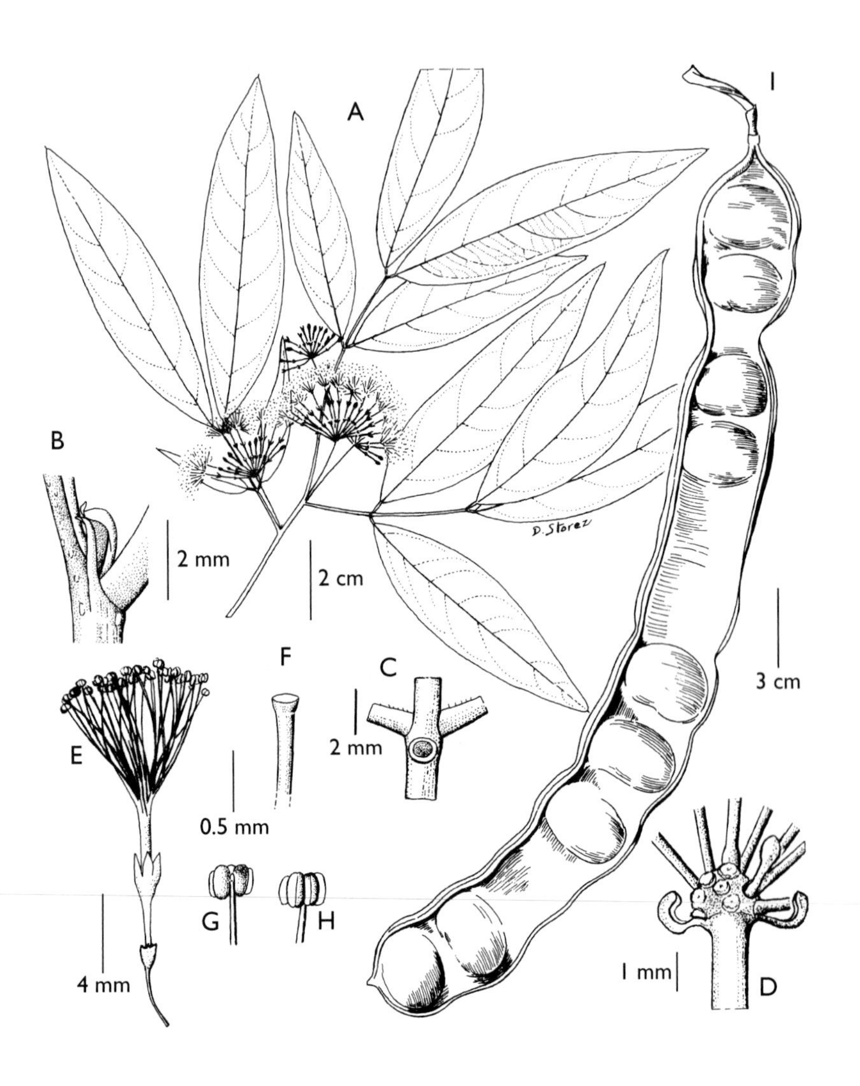

Fig. 22. *Inga loubryana* Poncy: A, habit; B, stipules; C, foliar nectary; D, base of umbel; E, flower; F, stigma; G, H, stamens; I, pod (A-H, Loubry 1135; I, Béna 1105). Drawing by Dominique Storez; reproduced with permission from Poncy, Adansonia 29: 251. 2007.

32. **Inga longipedunculata** Ducke, Arch. Jard. Bot. Rio de Janeiro 3: 56.1922. Type: Brazil, Pará, R. Tapajoz, Mangabal, Ducke HAMP 16453 = HJBR 10074 (isotypes P, RB, US).

Small to medium-sized tree, up to 18 m tall; branchlets slightly ferrugineous-pubescent. Leaves 2(3)-jugate; stipules triangular, 3 mm long, caducous; petiole and rachis segment(s) 2.5-5 cm long, unwinged, terete, velvety pubescent; nectaries orbicular, prominent, ca. 1 mm in diam., distal one sometimes lacking; leaflets elliptic, distal ones ca. 15 x 6 cm (up to 20 cm long), base cuneate to attenuate, apex acute, blades above glabrous or with very sparse long hairs, below glabrous, secondary veins regularly parallel. Inflorescence of long spikes, solitary or geminate in axils of mature leaf; peduncle 7-10(+) cm long; rachis 3-6 cm long, 30-40-flowered; flowers sessile, bracts 1 mm long, curved; bud cylindrical, apiculate; calyx tubular, 5-8 mm long, tomentose, lobes irregular; corolla 12-15 mm, tubular, covered with glossy appressed hairs; ratio calyx/corolla length ca. 2/5; staminal tube not exserted; stigma discoid. Pod woody, up to 35 cm long, ca. 2 cm wide; seeds prominent (collected from ground).

Distribution: Known from eastern Amazonia (Pará, Amapá) and French Guiana (Saül and Sinnamary areas), and one doubtful specimen from Guyana; in high dense primary non-flooded rainforest, blooming at night (S. Mori observ.) (GU: 1?; FG: 11).

Selected specimens: Guyana: Waini R., Baramanni Cr., Fanshawe 2348 = FD 5084. French Guiana: sin. loc., Richard s.n. (P); Kaw Mts., Réserve Trésor, Poncy 1347(CAY, P); Approuague R., Station Nouragues, Poncy 963 (CAY, P); Saül, Mori *et al.* 21015 (CAY, NY, P), 22023 (CAY, NY, P, US), 23739 (CAY, NY, P), 23892 (CAY, NY, P, US), Mts. Galbao (est), de Granville *et al.* 8681 (CAY, P, U, US), Rte Bélizon, Mori *et al.* 24743 (CAY, P, US); Oyapock R., Cr. Gabaret, Blanc 113 (B, CAY, K, NY, P, U, US); Sinnamary Basin, Petit-Saut, Sabatier & Prévost 2185 (CAY, NY, P, U).

33. **Inga loubryana** Poncy, Adansonia 29(2): 249-254. 2007. Type: French Guiana, Sinnamary, Piste-de-St-Elie, PK 19, Loubry 1135 (holotype P, isotypes CAY, US). – Fig. 22

Medium-sized to large tree, up to 35 m x 70 cm. Leaves 2-3-jugate, stipules linear, 2-3 mm long, caducous; petiole and rachis not winged, cylindrical or flattened and slightly canaliculate, lignified and lenticellate on old leaves; petiole 1-2(+) cm long, rachis segments approximately twice as long (2-4 cm); nectaries generally conspicuous, orbicular, border

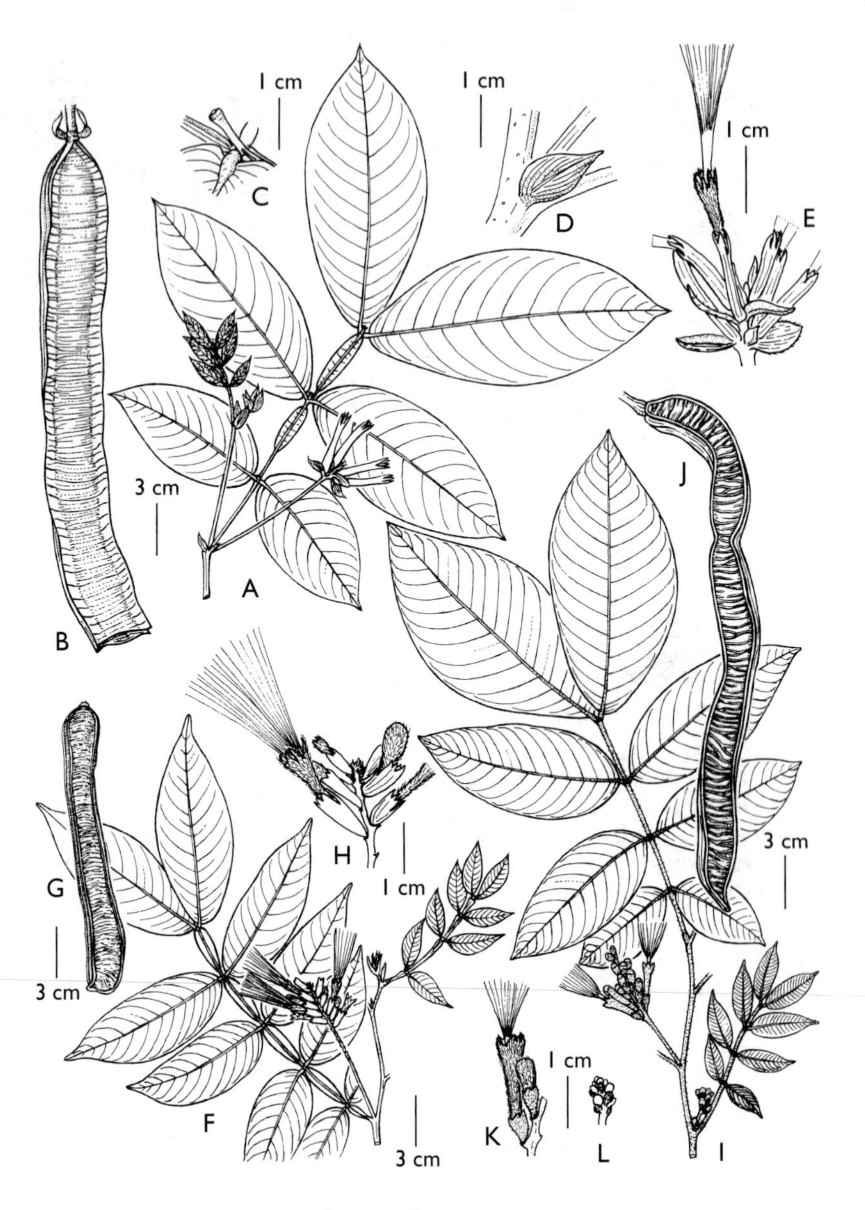

Fig.23. A-E: *Inga macrophylla* Humb. & Bonpl. ex Willd.: A, habit, leaves and young inflorescence; B, pod; C, foliar nectary; D, stipule; E, inflorescence, flower. F-H: *Inga sapindoides* Willd. (not in the Guianas): F, habit, leaves and inflorescence; G, pod; H, flowers. I-L: *Inga rubiginosa* (Rich.) DC.: I, habit, leaves and inflorescence; J, pod; K, flowers; L, young inflorescence, bracts. Drawing by Odile Poncy; reproduced from Poncy, 1985: 63.

thick, sessile or shortly stipitate (markedly stipitate in a few cases), distal nectary sometimes lacking; leaflets elliptic, proximal ones 5.5-8.5 x 2-3.5 cm, distal ones (7.5)9-13 x (2.5)3-5(6) cm; base cuneate to attenuate, apex acute, sometimes with a short, wide acumen (ca. 6 x 5 mm), mucronate; blades variable in aspect, typically chartaceous, glossy, often discolored, dark green above, paler and greyish below, primary vein prominent on both faces; venation camptodromous; inferior surface often very smooth, glossy, densely covered with tiny dots, secondary veins not prominent. Inflorescence umbellate, 1-3 axillary to adult leaves or young leaves of new growth; peduncle (1)2-4(5) cm long; rachis globose or shortly clavate, ca. 2 mm diam., 15-20-flowered; bracts scale like, curved, ca. 1 mm long, puberulent, caducous; sometimes a larger bract at base of umbel; pedicel slender, 3-6 mm long, calyx campanulate, ca. 1 mm long, teeth inconspicuous and slightly puberulent, corolla 5-6 mm long, glabrous, funnelform; ratio calyx/corolla length 1/5; staminal tube not exserted from corolla; ovary glabrous, ovules 15-20, stigma cup-shaped. Pod large, up to 40 x 4 cm, ca. 15-seeded, valves shallowly transversally wrinkled, sutures undulating around seeds, subligneous and thickened up to 5 mm at maturity; seeds not contiguous, prominent at maturity.

Distribution: Guyana and northern French Guiana; dense evergreen rainforest on white sand or laterite (GU: 3; FG: 9).

Selected specimens: Guyana: Bartica, Moraballi Cr., Fanshawe 259 = FD 2995 (FDG, K, U); Mazaruni, Puruni R., Takutu Cr., Fanshawe 2136 = FD 4872 (BRG, K, NY); Potaro R., Mahdia R., Fanshawe 1038 = FD 3774 (FDG, K). French Guiana: Mts. de la Mirande, SF 3766 (P); Mt. Grand Matoury, Poncy 993 (P); "Cayenne", Béna, SF 1105 (P, U); Paracou, Pétronelli 221 (CAY, P); Sinnamary, Piste-de-St-Elie, Sabatier 3570 (CAY, NY, P); St-Laurent-du-Maroni, Route de l'Acarouany, km 1.1, SF 7650 (NY, P, U); St-Laurent-du-Maroni, Route de Mana, SF 7145 (CAY, NY, P, U), SF 7437 (CAY, NY, P, U).

34. **Inga macrophylla** Humb. & Bonpl. ex Willd., Sp. Pl. 4: 1015. 1806. Type: Venezuela, Orinoco, Humboldt & Bonpland 915 (isotype P).
– Fig. 23 A-E

Inga bracteosa Benth., London J. Bot. 4: 609.1845. Type: Guyana, Roraima, Ro. Schomburgk ser. II, 695 = Ri. Schomburgk 1080 (holotype K, isotypes G, P, US).

Small to medium-sized tree (up to 10-12 m x 30 cm); bark dark brown to dark red, inner bark yellow; branchlets glabrous, angular, with very few inconspicuous lenticels, slightly grooved to longitudinally channelled.

Leaves (2)3(4)-jugate; stipules foliaceous, 15 x 8-12 mm, elliptic to triangular, persistent, glabrous, with parallel prominent veins; petiole ca. 4-5 cm long, terete or angular, glabrous or sparsely pubescent; rachis segments 4.5-7cm long, winged either from base or only on its distal part, wing up to 10(+) mm wide; leaf-apex often persistent beyond last pair of leaflets, up to 1.5 cm long; nectaries sometimes orbicular, sessile cups ca. 1 mm diam. or less, most often stipitate up to 8 mm high; leaflets broadly elliptic, proximal ones 10-12 x 5-6 cm, distal ones up to 20 x 12 cm, apex acute to acuminate, terminating in a sharp tip, base rounded to slightly cordate, pulvini and primary vein tomentose; blades papery, rough above, sparsely pubescent on both faces; young leaves pale green and densely whitish tomentose. Inflorescence of 2-4 spikes in axils of adult or juvenile leaves; peduncle (5)7-10 cm long, thick, glabrous to slightly pubescent, slightly grooved; rachis 3-5 cm long, 15-25-flowered; bracts lanceolate, 2(3)cm x 4 mm, foliaceous and persistent, 2-3 basal ones wider, together simulating an involucre; flowers sessile; bud cylindrical; calyx ca. 3 cm long, tubular, striate, glabrous or with sparse hairs, lobes irregular, 5-8 mm long, narrow, acute; corolla up to 6 cm long, covered with dense glossy appressed hairs; ratio calyx/corolla length 1/2; staminal tube long-exserted, free filaments 3 cm long; ovary with ca. 30 ovules, stigma widely cup-shaped. Pod large, up to 40 x 4 cm, quadrangular in section when ripe, sutures up to 1.5 cm wide, salient, winged, faces shallowly rippled, glabrous, or pubescent when very young. Seedling: first two leaves 1-jugate, opposite, smooth and tomentose, petiole winged.

Distribution: As currently circumscribed (Pennington, 1997), widely distributed throughout Amazonia; in French Guiana, recorded only in the forests of the interior, not from riverine areas; its sciophilous habit was observed in dense evergreen forest where it flowers in the understorey, but heliophilous behavior also occurs, as around Saül, where it is abundant in secondary vegetation (roadside, village, airstrip) (GU: 1; SU: 3; FG: 13).

Selected specimens: Suriname: Mapane, Akintosoela, Elburg UVS 17075 (BBS); Blakawatra, 60 km SW Paramaribo, den Outer 938 (BBS, P, U). French Guiana: Approuague R., Arataye Cr., Nouragues Station, Poncy 1034 (CAY, K, NY, U); Saül, Feuillet 498 (CAY, P), Foresta 670 (CAY, P, U), de Granville B5379 (CAY, NY, P, US), Mori *et al.* 14846 (CAY, NY, P, U), 14882 (CAY, NY, P), Prévost 761 (CAY, P), Oldeman B4085 (CAY, P); Haut-Oyapock, Trois-Sauts, Grenand 564 (CAY, P), 1141 (CAY, P), Lescure 269 (CAY, P), 493 (CAY, P), 761 (CAY, P).

Vernacular name: French Guiana: inga yowa (Wayapi).

Note: The species is better known in the Guianas under the synonymous binomial *Inga bracteosa* (Pennington, 1997). Curiously the species has not been re-collected in Guyana since Schomburgk collected the type specimen of *I. bracteosa*.

35. **Inga marginata** Willd., Sp. Pl. 4: 1015. 1806. (nom. cons. prop. Benth., Mim. 608. 1875). Type: Venezuela, Caracas, Bredemeyer s.n. (holotype B-Willd. no. 19031, photo P), typ. cons.

Mimosa semialata Vell., Flora Flum. 11: 1831. – *Inga semialata* (Vell.) Mart., Flora 20(2), Beibl. 8: 111. 1837. Type: Plate of Vellozo in Flora Flum. 11: t. 5, fig. 7. 1831.

Medium-sized tree (20-25 m tall); bark reddish brown, with tiny lenticels; branchlets irregular, glabrous or slightly tomentose. Leaves 2-jugate, glabrous; stipules linear, 4(+) x 1.5 mm, caducous; petiole up to 2 cm long, marginate; rachis up to 4 cm long, marginate to slightly winged, wing obtriangular, up to 4 mm wide distally; nectaries circular, sessile, ca. 1 mm diam.; leaflets elliptic to oblong, proximal ones small (up to 8 x 3 cm), distal ones twice larger, base acute, asymmetrical, apex acuminate and apiculate; blades papyraceous, glabrous on both surfaces; juvenile leaves pinkish brown to red, shiny, with dark red nectaries. Inflorescence axillary spikes, solitary to 3 in axils of mature leaves, peduncle ca. 1 cm long; rachis variable in length (up to 8 cm) with 30-50 loosely clustered, glabrous flowers, either sessile or shortly pedicellate; bracts linear, 1-2 mm long, conspicuously overtopping buds on young spikes, bud obovate or bulb-like; calyx 1 mm long or less, campanulate, lobes inconspicuous, corolla funnel-shaped, up to 4 mm long, lobes triangular, 1 mm; ratio calyx/corolla length 1/4; staminal tube included in corolla or exserted. Pod sessile, glabrous, subcylindrical, 6-12(18) cm long, ca. 1 cm wide, sutures not thickened, when ripe yellowish, valves convex, not rigid, finely transversally ribbed, seeds slightly prominent; seeds contiguous or not, with poorly developed pulp, embryo green, ca. 5 x 7 mm.

Distribution: Widely distributed through Amazonia and C America, the Guianas, rather common in the lowland forests of French Guiana; primary rainforest, and dense forest on white sand (GU: 10; SU: 3; FG: 18).

Selected specimens: Guyana: Sandwith 120 (K); Marudi Mts., Locust Cr., Stoffers *et al.* 280 (BRG, U, P); Essequibo x Mazaruni R., Mazaruni Station near Bartica, Fanshawe FD 3536 (FDG, K, U); Fanshawe 800 = FD 6310 (FDG, K), Tutin 460 (BM, K, U), Sandwith FD 1126 (FDG), 468 (K, P, U); Upper Essequibo, Rewa R., Jansen-Jacobs *et al.* 6062 (BRG, P, U); Roraima, Ro. Schomburgk ser. II, 918 = Ri. Schomburgk 1443 (K, P); Upper Potaro, Kopinang, Boom & Samuels 9300 (NY, P). Suriname: Kabalebo Dam Project, Lindeman & de Roon 802 (BBS, U), Lindeman & Cowan 7028 (U); Brownsberg, BW 3223 (K, U). French Guiana: Approuague R., Arataye Cr., Saut-Pararé, Barrier 3888 (CAY, P), Poncy 471 (CAY, P, U); Comté R., Cr. Gabrielle, Grenand 1864 (CAY, P); Maroni R., Pompidou-Papaïchton, Sastre 6474 (CAY, P); Saül, Mori *et al.* 8701 (CAY, NY, P), 18219 (CAY, NY, P, U), Maas *et al.* 2285 (K, U), de Granville B4179 (CAY, P, U), Philippe 27012 (CAY, ILLS, P), Prévost 768 (CAY, P); Boeuf Mort, Oldeman B4179 (CAY, P), de Granville B4591 (CAY, P), La Fumée Mts, Mori & Boom 15029 (CAY, NY, P), Rte Bélizon, Mori *et al.* 23978 (CAY, NY, P), 23984 (CAY, NY, P), 24806 (CAY, NY, P); Sinnamary, Piste-de-St-Elie, Prévost & Sabatier 3793 (CAY, G, K, MO, NY, P), Sabatier & Prévost 4622 (CAY, P).

Vernacular names: Guyana: hickuritoro (FD 1126), maparakun (Tutin 460), warakosa.

Note: This highly variable species generally presents quite constant features in the Guianas area, e.g. the leaves always 2-jugate; variations in the Guianas mostly involve the flowers being sessile or shortly pedicellate, and variation in the length of the staminal tube; however, an interesting variation is shown in a set of very similar specimens from Guyana (Bartica-Mazaruni area) and a specimen from E Suriname: the leaves have minute stipules, a thicker rachis, leathery, glossy leaflets, and a longer pod.

36. **Inga melinonis** Sagot, Ann. Sci. Nat., Bot. 6(13): 335. 1882. Lectotype (designated by Poncy 1985: 48): French Guiana, Mélinon 453 (hololectotype P). – Fig. 20 I-M

Inga cyclocarpa Ducke, Arch. Jard. Bot. Rio de Janeiro 4: 14. 1925, non Willd. 1806. Lectotype (designated by Pennington 1997: 369): Brazil, Pará, Ducke HJBR 10031 (hololectotype RB, isolectotype U).

Medium-sized tree; bark light brown, inner bark yellowish, producing a yellow sap; branchlets terete, pubescent, with wide lenticels. Leaves 4-jugate, brownish tomentose on all parts except upper surface of blades; stipules rectangular-linear up to 3 x 12 mm, longitudinally striate, caducous; petiole and rachis terete to marginate; petiole 5-10 cm long, rachis segments 5-8 cm long; nectaries sessile, orbicular with thick margin, ca. 2 mm diam.; leaflets narrowly elliptic, proximal ones up to 15 x 6 cm, distal ones up to 25 x 9 cm; apex acute to acuminate; base acute; blades green, smooth, glabrous above, brownish to greyish, velvet-like in texture beneath. Inflorescence either axillary to old leaves or at defoliated nodes; one fertile node produces 2-4 spikes and 1-2 short shoots, with several spikes each in axils of undeveloped leaves; peduncle ca. 2 cm long; rachis ca. 1.5 cm long, over 40-flowered; flowers sessile, entirely pubescent; bud cylindrical; bracts up to 2 mm, scale like, curved, persistent; calyx tubular, 2-3 mm long, lobes inconspicuous; corolla tubular ca. 1 cm long, lobes subulate, 2-3 mm long; ratio calyx/corolla length ca. 1/4; staminal tube not exserted from corolla. Pod ligneous, flat, curved, asymmetrical, glabrous, 15-25 x 3-3.5 cm, when ripe yellowish brown, sutures ridged, up to 1 cm wide, valves yellowish brown, wrinkled, seeds slightly prominent.

Distribution: Amazonia, uncommon in French Guiana, albeit fast-growing in the clearings of dense primary forest, also recorded from northern Guyana, where it grows either in high dense humid forest or in low forest on sandy soil; a few sterile collections from Suriname are placed here (GU: 5; SU: 2; FG: 9).

Selected specimens: Guyana: sin. loc., Graham 184 (NY), Gleason 148 (NY), Barama R., van Andel *et al.* 1132 (P, U); Mazaruni Station, Davis 500 = FD 2495 (FDG, K); Lower Essequibo, Groete Cr., Fanshawe 1253 = FD 3989 (FDG, K). Suriname: Jodensavanne, Lindeman 3788 (U); van Troon, LBB 16173 (BBS). French Guiana: Maroni, Mélinon 370 (P), 443 (P), 456 (P), Wachenheim 304 (NY, P); Sinnamary, Piste-de-St-Elie, Prévost 848 (CAY, P), 3839 (CAY, P); Saint-Laurent-du-Maroni, Route de Mana, SF 7287 (CAY, NY, P, U), 7446 (P, U).

Vernacular names: Guyana: karoto. Suriname and French Guiana: kodiaweko (Paramaka).

37. **Inga mitaraka** Poncy, Bull. Mus. Natl. Hist. Nat., B, Adansonia 18(1-2): 70. 1996. Type: French Guiana, Tumuc-Humac Mts., dense forest on foot of Mitaraka granite outcrop, de Granville 1197 (holotype P, isotype CAY). – Fig. 24

Medium-sized to large tree (up to 30 m x 50 cm); bark blackish, dotted with greyish spots, lenticels pale; branchlets and leaves glabrous, very young parts slightly tomentose. Leaves (2)3-jugate; stipules linear, up to 7 mm long, often persistent; petiole and rachis not winged, slender, flattened; petiole 1.5(2)-3.5 cm long; rachis segments 2.5-4 cm long; nectaries shortly stipitate, cup-shaped, 1-1.5 mm diam.; leaflets elliptic, often irregular in outline, proximal ones (4.5)5-7 x (1.5)2-3 cm, distal ones 7-11(13) x 3-4(6) cm, attenuate at base, apex acuminate (acumen

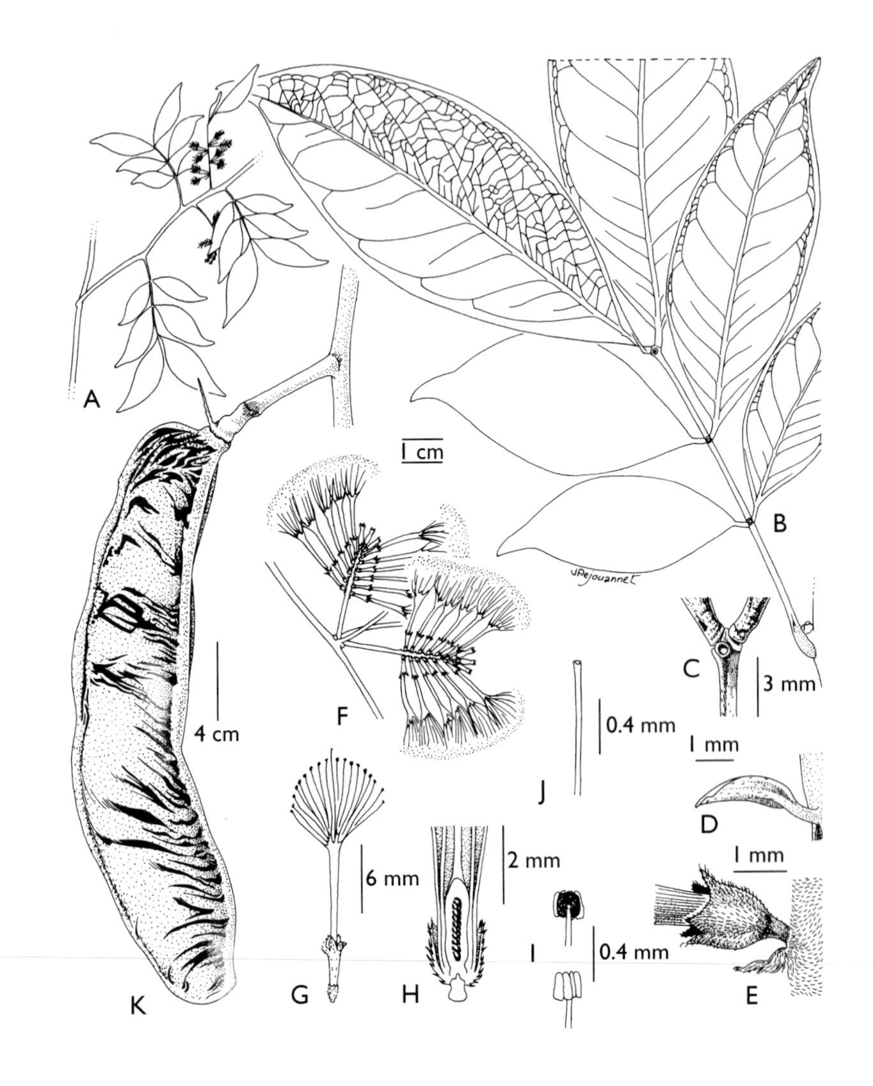

Fig. 24. *Inga mitaraka* Poncy: A, habit; B, leaf; C, extremity of foliar rachis, nectary; D, stipule; E, bract and base of flower, detail; F, inflorescences; G, flower; H, flower, longitudinal section; I, anther, dorsal and ventral views; J, stigma; K, pod (A-J, de Granville 1197; K, Grenand 143). Drawing by Jean-François Dejouannet; reproduced with permission from Poncy, Adansonia 18: 71. 1996.

5-12 mm long); blades chartaceous, veins prominent on both faces. Inflorescence of 2-4 spikes in axils of adult leaves or of undeveloped leaves on shortened shoots; these axes, as well as peduncles, rachis and flowers very short pubescent; peduncle and rachis 1-2 cm long, rachis 25-30-flowered; flowers sessile; bracts scale like, curved, 1 mm long, caducous; calyx campanulate, 1.5-2 mm long; corolla funnel-shaped, 5-6 mm long, glabrous except for teeth margins; ratio calyx/corolla length 2/3; staminal tube exserted, anthers and stigma not seen. Pod rigid, ligneous, 10-15 x ca. 3 cm, glabrous, sutures ridged, 5 mm thick, irregular, valves smooth, dark green when ripe; seeds ca. 10-13.

Distribution: Southern French Guiana and Brazil, Amapá, Jari R. drainage; high dense primary rain-forest (GU: 1?; FG: 7).

Selected specimens: Guyana: Kanuku Mts., A.C. Smith 3479 (K, P, U). French Guiana: Approuague R., Oldeman B2278A; Mts. Trinité, Poncy *et al.* 1418 (CAY, K, NY, P, U); Trois-Sauts, Grenand 143 (CAY), Jacquemin 1517 (CAY), Prévost & Grenand 973 (CAY, P), Lescure 528 (CAY, NY, P).

Vernacular name: French Guiana: inga-u (Wayapi).

Notes: The collection from Kanuku Mts. (Guyana) is doubtfully placed here, the pod being verrucose. *Inga mitaraka* was synonymized under *I. alba* by Pennington (1997). It is here retained as a distinct species which differs from *I. alba* by various characters including its pubescence, inflorescence, and especially the pod.
I. mitaraka seems locally abundant in the upper Oyapock R. area of amerindian wayapi settlements; the wayapis feed on the seeds of this species and are thus likely to protect it, just as they do for *I. edulis*; they also use the sap as a dye.

38. **Inga nobilis** Willd., Enum. Pl. Hort. Berol. 1047. 1809. Type: Brazil, Hoffmannsegg (holotype B-Willd. no. 19036).

Inga corymbifera Benth., J. Bot. (Hooker) 2: 144. 1840. Type: Guyana, Ro. Schomburgk ser. I, 226 p.p. (holotype K, isotype G, NY, P, US).
Inga riedeliana Benth., London J. Bot. 4: 595. 1845. Type: Brazil, Riedel s.n. (holotype K, isotype G).
Inga riedeliana Benth. var. *surinamensis* Benth., London J. Bot. 4: 595.1845. Lectotype (designated by Pennington 1997: 382): Suriname, Hostmann 830 (hololectotype K, isolectotypes P, U).
Inga sericantha Miq., Linnaea 19: 132. 1847. Type: French Guiana, “ad Fl. Carouany”, Kappler 1692 (holotype U).
Inga conglomerata Benoist, Bull. Mus. Hist. Nat. Paris 27: 115. 1921. Type: French Guiana, Maroni, Wachenheim 149 (holotype P, isotype K).

Small riparian, often bushy tree; bark dark brown or greyish, lenticels very small; branchlets finely tomentose when young. Leaves (3)4-jugate; stipules linear, up to 10 x 2 mm, apiculate, often persistent; petiole and rachis terete, puberulous, petiole 0.5-1 cm long, rachis segments 2.5-4 cm long; nectaries orbicular, sessile, up to 1.5 mm diam., proximal one more conspicuous, larger, discoid, irregular in outline; leaflets elliptic, sometimes slightly falcate, variable in size, proximal ones generally 4-7 x 2-3 cm, distal ones 10-15 x 4-5 cm, base acute, apex shortly acuminate; blades glabrous, veins more or less conspicuously tomentose beneath; young leaves green with whitish hairs, and white nectaries. Inflorescence spicate, 2-3 in axils of either mature or young leaves, or on short shoots; distal part of a vigorous flowering branch simulates a terminal panicle; peduncle length variable (2-5 cm); rachis ca. 1.5 cm long, 15-25 close flowers; flowers subsessile to pedicellate, pedicel up to 3 mm long; bracts persistent, up to 2.5 mm long, spatulate with a subulate apex; flower bud clavate; calyx tubular, 3-5(+) mm, pubescent, lobes regular, inconspicuous; corolla funnel-shaped, to 10 mm long; ratio calyx/corolla length 1/3 to 1/2; staminal tube generally not exserted. Pod 8-12(+) x 2-2.5 cm, smooth, glabrous, apiculate, subcylindrical and yellow to pale brown, fleshy when ripe, sutures up to 1 cm wide, valves convex, shallowly wrinkled; seeds 10-15, closely packed, embryo green ca. 2 x 1 cm, contact surface between cotyledons uneven. Seedling: first two leaves opposite, 1-jugate, petiole ca. 2 cm long and winged; petiole length increasing up to 5 cm and wing width decreasing to zero as seedling develops; adult leaf features are visible on first 3-jugate leaves.

Distribution: Throughout the Guianas, Amazonian Venezuela, Brazil; a riparian species, very common along rivers, unknown from non-flooded primary forest, rare in non-riparian, secondary vegetation; ca. 100 collections seen (GU: 14; SU: 8; FG: 18).

Selected specimens: Guyana: Ro. Schomburgk ser. II, 31; Barima-Waini R., Pipoly *et al.* 8245 (BRG, NY, P, US); Berbice R., Canje R., Gillespie *et al.* 2416 (P, US); Corentyne R., Jenman 6 (K, P), 230 (BRG, K); Upper Demerara R., Jenman 4261 (BRG, K), 5030 (BRG, K); Essequibo R., Arakaka, Gillespie 1424 (BRG, P, US); Kuyuwini R., A.C. Smith 2572 (NY, P, U); Kwitaro R., Clarke *et al.* 6506 (BRG, P, US); Pomeroon R., Ri. Schomburgk 1419 = Ro. Schomburgk ser. II, 839 (K, P); Potaro R., Kaieteur Falls, Hahn 4756 (P, U); Saut Aracai Mt., Chodikar, Guppy 437 = FD 7452 (FDG, K, NY, U). Suriname: Brownsberg, Brokopondo, van Donselaar 1477 (BBS, U); Paramaribo, Uitkijk, den Outer 892 (BBS, U, WAG); Litany R., de Granville *et al.* 12020 (CAY, P, U); Saramaca R., Maguire *et al.* 24952 (NY, P, U); Wilhelmina Mts., Lucie R., Irwin *et al.* 55996 (NY, P, U), Schulz LBB

10053 (BBS, U). French Guiana: Approuague R., Arataye Cr., Poncy 134 (CAY, P), 208 (CAY, P, U), Emerillon Cr., Oldeman B1887 (CAY, P); Grand-Inini R., Saut-Liki, de Granville B3345 (CAY, P); Haut-Maroni R., Itany R., Cremers 5112 (CAY, P); Mana R., Arouany Cr., Hallé 639 (CAY, P), Baboune Cr., Cremers 7530 (CAY, P); Bas-Oyapock R., Armontabo Cr., Prévost & Grenand 2014 (CAY, P), Gabaret Cr., Lescure 782; Haut-Oyapock R., Camopi, Oldeman 2536 (CAY, P); Trois-Sauts, Grenand 525 (CAY, P); Maroni, Grand-Soula, Schnell 12293 (P), Marouini R., Cremers 5027 (CAY, P); Sinnamary R., Saut Vata, Bordenave 485 (CAY, P), Saut Takari-Kante, Hoff 5894 (CAY, NY, P, U); Saint-Laurent-du Maroni, SF 5110 (CAY, P).

Vernacular names: Guyana: warakosa. French Guiana: mikotawa (Wayana), yawi-inga (Wayapi), miumiu maksi axibine (Palikur).

Note: In spite of the leaf polymorphism, *Inga nobilis* is easy to recognize by its very short or even absent petiole, and by the proximal nectary larger than the others.

39. **Inga nouragensis** Poncy, Bull. Mus. Natl. Hist. Nat., B, Adansonia 18(1-2): 67. 1996. Type: French Guiana, Nouragues Station, Sabatier & Prévost 2569 (holotype P, isotypes CAY, K, NY). – Fig. 25

Tree of medium to large size, trunk up to 55 cm diam.; bark creamish to brownish, smooth, "with tightly spaced hoop marks" (Mori & Boom 14734); branchlets greyish, irregular, bark flaking off, internodes short, young parts pubescent. Leaves 1-jugate, bifoliolate; stipules inconspicuous, 1.5 mm long and caducous; petiole marginate and grooved to narrowly winged, obtriangular, to 2 mm wide; nectary orbicular 0.5 mm diam.; leaflets elliptic, 5-10(12) x 2.5-4(5) cm, apex generally acuminate, base attenuate or acute, often markedly asymmetrical; blades glabrous or with very sparse short hairs, chartaceous to coriaceous, often folded along primary vein, at least apical part, primary vein conspicuously prominent above. Inflorescence of 1-3 short spikes in axils of either young or mature leaves; peduncle slender, 1.5 to 3 cm long; rachis 6-10 mm long, bearing ca. 25 close, sessile flowers; bracts scale like, 1-2 mm long; flower bud ovoid; calyx campanulate, 1-1.5 mm, teeth not always conspicuous, pubescent at apex; corolla funnel-shaped, 4-6 mm long, glabrous; ratio calyx/corolla length 1/4; staminal tube exserted; ovary glabrous, 14-15 ovules, stigma tubular. Pod (-)10-16 cm x 1.5-1.8 cm, glabrous, sutures not thickened, valves shallowly wrinkled (mature pod unknown); seeds 8-11.

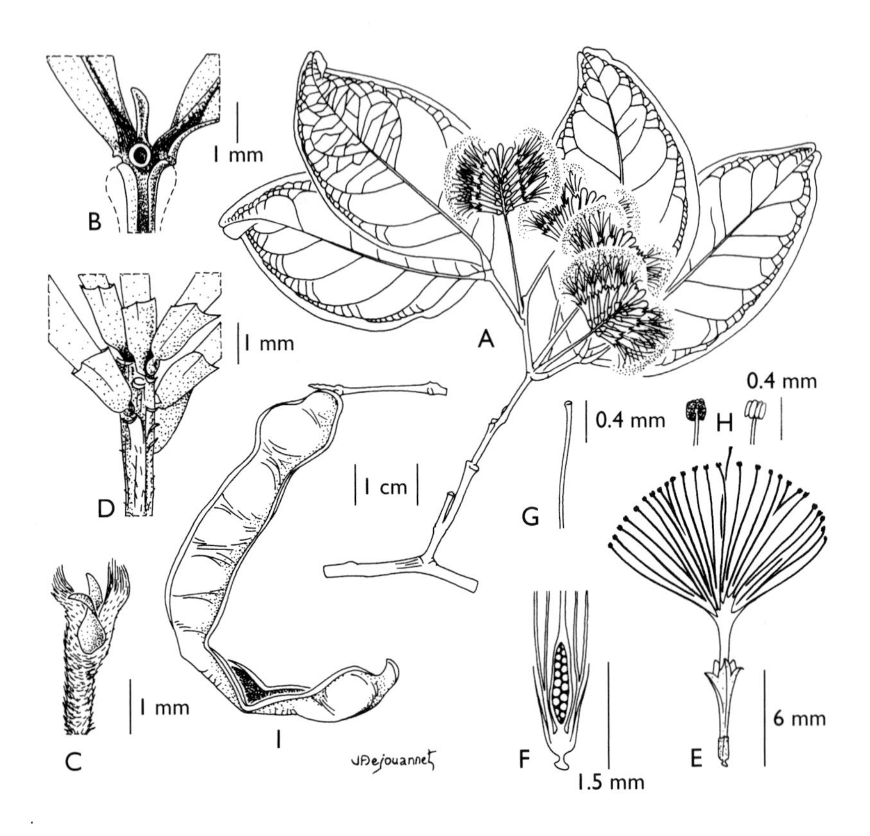

Fig. 25. *Inga nouragensis* Poncy: A, habit; B, rachis and foliar nectary; C, stipules; D, detail of inflorescence rachis; E, flower; F, flower, longitudinal section; G, stigma; H, anther, ventral and dorsal views; I, pod (A-H, Sabatier *et al*. 2569; I, Mori *et al*. 15530). Drawing by Jean-François Dejouannet; reproduced with permission from Poncy, Adansonia 18: 72. 1996.

Distribution: French Guiana; apparently a sciophilous and slow growing species, in dense primary non-flooded forest (FG: 10).

Selected specimens: French Guiana: Approuague R., Arataye Cr., Nouragues Station, Poncy 992 (CAY, P, U), Sabatier & Prévost 2624 (CAY, K, NY, P, U); Saül, Mori *et al*. 14734, 15530 (CAY, NY, P); Sinnamary, Piste-de-St-Elie, Sabatier & Prévost 3643 (CAY, P), 3787, 3801 (CAY, NY, P), 4321 (CAY), 4365 (CAY, K, NY).

40. **Inga nubium** Poncy, Bull. Mus. Natl. Hist. Nat., B, Adansonia 13(3-4): 148. 1991. Type: French Guiana, Tumuc Humac, de Granville 1475 (holotype P, isotype CAY).

Small tree, up to 15 m tall; branchlets slender, terete, with pale lenticels. Leaves 1-jugate, glabrous; stipules linear ca. 3 mm long, caducous; petiole ca. 2 cm long, winged, obtriangular up to 8 mm below insertion of leaflets; nectary circular ca. 1 mm in diam.; leaflets elliptic, (6)8-12(13) x (2)3-4.5(6) cm, base acute to cuneate, apex acute to obtuse, often markedly acuminate, apiculate; blades thin, soft, somewhat bullate, glabrous, greyish below. Inflorescence capitate, generally solitary in axils of mature leaves, 20-30-flowered; peduncle slender, ca. 1 cm long; bracts pubescent, spatulate, up to 2 mm long; flowers pedicellate, pedicel ca. 5 mm long; flower bud cylindrical, narrow; calyx and corolla tubular, tomentose, calyx 2(+) mm long, teeth inconspicuous; corolla with irregular lobes; ratio calyx/corolla length 1/3; staminal tube not or hardly exserted; stigma tubular. Pod (mature pod unknown) flattened, 6-8 x ca. 2.5 cm, glabrous, sutures 2-3 mm wide, valves shallowly ribbed transversally; seeds 8-10.

Distribution: Suriname, French Guiana, Amapá (Brazil); hyper-humid submontane forest (SU: 1; FG: 5).

Selected specimens: French Guiana: Mt. Galbao, de Granville *et al.* 8621 (B, CAY, NY, P, U, US); Sommet Tabulaire, de Granville 3500 (CAY, P); Mt. Bellevue, de Granville *et al.* 7649 (B, BR, CAY, G, K, MG, MO, P, U, VEN); Saül, Mori *et al.* 24964 (CAY, NY). Suriname: Wilhelmina Mts., Irwin *et al.* 55127 (NY, U).

Note: By its floral morphology, *Inga nubium* is closely related to the common riparian species *I. sertulifera*, and to the less common *I. jenmanii*, from which it can be distinguished mainly by its constantly 1-jugate leaves and triangularly winged petiole.

41. **Inga obidensis** Ducke, Arch. Jard. Bot. Rio de Janeiro 3: 49. 1922. Type: Brazil, Pará, Obidos, Ducke MG11826 (holotype MG, isotypes BM, G, P, US).

Small tree; branchlets slender, terete, puberulous, lenticellate. Leaves 1-2-jugate; stipules linear, puberulous, 1-2 mm long; petiole 5-15 mm, marginate or poorly winged, rachis 2-3.5 mm, winged (wing triangular, 3-5 mm wide distally); nectaries sessile, flat, inconspicuous (< 1 mm diam.); leaflets (1)2 pairs, in 1-jugate leaves (7)9-15 x (2)4-5 cm, in

2-jugate leaves proximal ones 5-8 x 2.5-3.5 cm, distal ones 9-14 x 3.5-6 cm, base cuneate to rounded, apex narrowly attenuate to acuminate and mucronate; blades smooth, glabrous, secondary veins 5-6 pairs. Inflorescence (not available on Guianese specimens, here described from type collection and following Pennington, 1997) in axils of old leaves or below, umbellate; shortly pedunculate; bracts persistent, minute; flowers shortly pedicellate; calyx campanulate, 1-1.5 mm; corolla funnelform, 4-5 mm long; staminal tube exserted. Pod glabrous 7-11 x 2.5-3 cm, straight, base asymmetrical, valves with conspicuous transverse to oblique veins, yellow at maturity; 2-6 seeds.

Distribution: S Guyana and Amazonian Brazil (GU: 2).

Selected specimens: Guyana: W Kanuku Mts., A.C. Smith 3177 (NY, P), Clarke 1815 (BRG, P, US).

42. **Inga paraensis** Ducke, Arch. Jard. Bot. Rio de Janeiro 4: 12. 1925. Lectotype (designated by Poncy 1985): Brazil, Pará, Tapajoz R., Villa Braga, silva non inundabili, Ducke RB 16698 (hololectotype RB, isolectotypes K, P, U). – Fig. 19 N-Q

Tree up to 35 m high; bark variable, beige to red-brown, lenticels conspicuous in horizontal lines; branchlets glabrous, angular, red-brown with dense paler lenticels. Leaves (2)3-jugate, entirely glabrous; stipules falling late, linear to slightly falcate, 4-8 mm long; petiole (1-2 cm long), rachis (each segment 2-3.5 cm long) flattened, shallowly grooved adaxially; nectaries sessile, orbicular; pulvinus black, ca. 0.7 mm; leaflets chartaceous, narrowly elliptic to elliptic, almost symmetrical, base cuneate to attenuate, apex most often markedly acuminate (acumen 1(+) cm long), proximal ones 5-8 x 2.5-4 cm, distal ones 7.5-14 x 3-7 cm. Inflorescence capitate; peduncle slender, up to 8(10) cm long; bracts linear, acicular, 2-3 mm long; flowers ca. 30 per head, glabrous; pedicel 7-9 mm long; calyx tubular, narrow, 3-4 mm long, teeth insconspicuous; corolla tubular, 9-11 mm long, lobes acute, sharp, 2 mm long; staminal tube not exserted; stigma tubular. Pod 18-25(28) x 2-2.5 cm, glabrous; pedicel 5-10 mm long, is continued by attenuate base of pod; sutures poorly developed; seeds up to 18(22) per pod, close to each other but not appressed.

Distribution: Described from Brazilian Amazonia, now clearly recorded from C French Guiana, and from one locality in Guyana; the single specimen from Suriname is placed here with hesitation (GU: 1; SU: 1?; FG:8).

Selected specimens: Guyana: Upper Essequibo R., Gunn's, Jansen-Jacobs *et al.* 1911 (BRG, CAY, P, U). Suriname: Mori & Bolten 8518 (NY). French Guiana: Nouragues Station, Poncy 938 (CAY, P), 954 (CAY, P), 988 (CAY, NY, P, U, US), 990 (BBS, CAY, K, MO, NY, P, U, US), 1050 (CAY, K, MO, NY, P, U, US), Sabatier & Prévost 2742 (CAY, K, NY, P, U); Route N 2 Cayenne-Régina, pk 79, Prévost & Sabatier 4934 (CAY, G, MO, P, U, US); Saül, Mori *et al.* 24706 (CAY, NY, P).

43. **Inga pezizifera** Benth., London J. Bot. 4: 587. 1845. Type: Guyana, Ro. Schomburgk ser. II, 124 = Ri. Schomburgk 50 (holotype K, isotypes BM, NY, P). – Fig. 26 A-F

Inga subsericantha Ducke, Arch. Jard. Bot. Rio de Janeiro 3: 55. 1922. Type: Brazil, Pará, Tapajoz R., Pimental, Ducke HAMP 16732 (isotype K).
Inga urnifera Kleinhoonte, Recueil Trav. Bot. Néerl. 22: 414. 1925. Lectotype (designated by Pennington 1997: 216): Suriname, Sectie O, tree no. 625, BW 5264 (hololectotype U).

Tree, up to 30 m x 40 cm; bark greyish to light red-brown, lenticels paler, dense, small; all parts of plant glabrous except slightly pubescent flowers. Leaves (2)3-4-jugate; stipules conspicuous, linear to rhomboidal, up to 8 x 2 mm, caducous; petiole and rachis terete or marginate, sometimes (see note below) narrowly winged; petiole up to 3 cm long, rachis segments 2.5-4 cm long; nectaries usually very well-developed, cup-shaped, up to 5 mm diam., wider than rachis especially on young trees; leaflets elliptic, distal pair up to 17 x 7 cm; base acute, less often cuneate, apex acute to slightly acuminate; blades glabrous, shiny; juvenile leaves pale green with inconspicuous white hairs. Inflorescence spicate, spikes either in axils of mature leaves or of undeveloped leaves of short shoots; peduncle up to 25 cm; rachis ca. 1 cm, with 25-40 very crowded flowers; bracts scale like, inconspicuous, 1 mm long, persistent; flowers sessile to shortly pedicellate; calyx (ca. 1 mm long) and corolla (ca. 5-6 mm long) tubular, slightly pubescent; ratio calyx/corolla length 1/5; staminal tube not or little exserted; stigma tubular. Pod flat, glabrous, 15-25 x 3-4 cm, slightly to markedly curved, subligneous, rigid and dark green when ripe, sutures scarcely enlarged to 5 mm wide, not salient; insertion of peduncle markedly asymmetrical; valves smooth or shallowly obliquely ribbed; seeds ca. 3 x 1 cm, cotyledons green. Seedling: two first leaves opposite, 1-jugate, petiole winged; adult vegetative features do not develop before young seedling has 3-jugate leaves and is at least 70 cm tall.

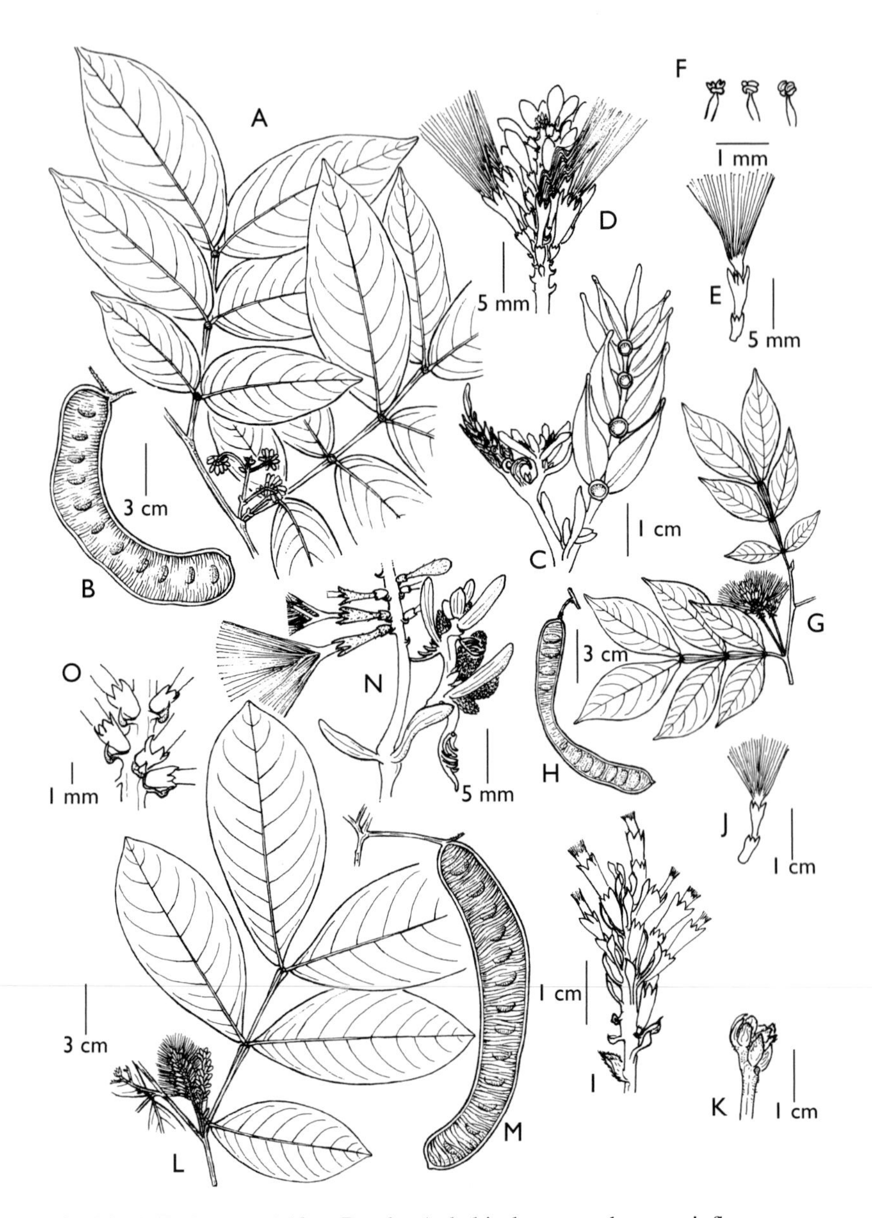

Fig. 26. A-F: *Inga pezizifera* Benth.: A, habit, leaves and young inflorescences; B, pod; C, extremity of leaf axis: foliar nectaries, stipules; D, inflorescence; E, flower; F, stamens. G-K: *Inga auristellae* Harms: G, habit, leaves and inflorescence; H, pod; I, inflorescence; J, flower; K, vegetative bud, stipules. L-O: *Inga bourgonii* (Aubl.) DC.: L, habit, leaves and inflorescences; M, pod; N, extremity of leaf axis, stipules, young inflorescences and leaves, flower; O, flower base, bracts. Drawing by Odile Poncy; reproduced from Poncy, 1985: 32.

Distribution: Widely distributed in southern C America, Amazonia and northern S America; in the Guianas it shows wide ecological tolerance as well as morphological plasticity, as it can be found either in the primary undisturbed forest, in secondary vegetation and in riverine forests; it is a very fast growing species in open areas (GU: 10; SU: 4; FG: 22).

Selected specimens: Guyana: Barama R., van Andel *et al.* 776 (P, U), 1120 (P, U); Chimapu, Maguire *et al.* 46135 (NY, U); Iwokrama, Clarke *et al.* 4296 (BRG, P, US); Mazaruni Station, Davis FD 2972 (FDG, K, U), FD 3269 (K, U); Potaro-Siparuni, Konaward, McDowell 4812 (BRG, P, US); Rupununi, Lethem-Surama Rd, Acevedo *et al.* 3450 (BRG, CAY, P, US); Upper Essequibo, Sipu R., Clarke *et al.* 7100 (BRG, P, US). Suriname: Roberts & van Troon LBB 14808 (BBS, U); Jodensavanne, Lindeman 6925 (U); Lelydorp, Teunissen LBB 15208 (BBS, U). French Guiana: Approuague R., Arataye Cr., Oldeman 3012 (CAY, P), Nouragues Station, Poncy 930 (B, CAY, K, NY, P, U, US); Cabassou, Prévost 412 (CAY, P), Sauvain 196 (CAY, P); Regina, Cayenne-Regina Road, Feuillet 1381 (B, CAY, NY, P, U, US), Prévost 1271 (CAY, P), Mt. Tortue, Feuillet *et al.* 10368 (CAY, P, U, US); Roura, Jacquemin 2563 (CAY, P); Inini, Mt. Atachi-Bacca, de Granville *et al.* 10904 (B, CAY, G, NY, P, U, US); Saül, Mori & Gracie 18278 (CAY, NY, P, U), 22180 (CAY, NY, P), 24886 (CAY, NY); Mana, Saut-Sabbat, de Granville B5330 (CAY, P, U); Maroni R., Bena SF 12N (CAY, P, U), Mélinon 81 (P); Mts. de la Trinité, Poncy 1419 (CAY, P); Saint-Laurent-du-Maroni, Poncy 300 (CAY, P, U); St-Jean, Benoist 1263 (P); Savane Dorothée, Hallé 4030 (CAY, MPU, P); Sinnamary, Piste-de-St-Elie, Loubry 1126 (CAY, P), Prévost 3814 (CAY, G, K, MO, NY, P, U, US), Sabatier 3052 (CAY, INPA, NY, P).

Note: The leaf of *Inga pezizifera* is usually easy to recognize with 3-4 pairs of wide leaflets, an unwinged rachis and especially by the large cup-shaped nectaries after which the species is named. Nevertheless, variations are frequent (e.g. 2-jugate leaves, narrow leaflets, poorly-developed nectary, rachis somewhat winged) and somewhat limit quick and accurate identification.

44. **Inga pilosula** (Rich.) J.F. Macbr., Publ. Field Mus. Nat. Hist., Bot. Ser. 13(3): 41. 1943. – *Mimosa pilosula* Rich., Actes Soc. Hist. Nat. Paris 1: 113. 1792. Type: French Guiana, Cayenne, Leblond s.n. (holotype P-Juss., isotype P).

Inga nitida Willd., Sp. Pl. 4: 1013. 1806. Type: Brazil, Pará, Sieber s.n. comm. Hoffmannsegg (holotype B-Willd. no. 19027).
Inga pilosiuscula Desv., J. Bot. 3: 71. 1814. nom. nud.

Inga setifera DC., Prod. 2: 432. 1825. Type: sin. loc., anon., s.n. "Muséum Paris 1821" (holotype G-DC).
Inga platycarpa Benth., J. Bot. (Hooker) 2: 142. 1840. Type: Ro. Schomburgk ser. I, 534 (holotype K).
Inga affinis Steud., Flora 26: 758. 1843. Type: Hostmann 1157 (holotype U, isotypes K, NY).

Shrub, treelet or seldom medium-sized tree to 20 m tall; bark light grey, with brownish stripes and sparse orbicular lenticels; branchlets glabrous or slightly pilose. Leaves 2-jugate; stipules linear, oblanceolate or falcate, to 7 x 2 mm, persistent; petiole and rachis pilose mainly on veins, petiole up to 1(+) cm long, most often with obtriangular wing up to 8 mm wide; rachis 2.5-4 cm long, generally winged only on distal half, winged part obovate to obtriangular, to 1 cm wide; terminal, linear rachis appendage often persistent beyond superior leaflets; nectaries very conspicuous, up to 3 mm diam., cup-shaped, proximal one irregular, larger than distal one; leaflets elliptic, very variable in size, proximal pair (average size 7 x 3.5 cm) about half size of distal one (13 x 6 cm); apex acute, acuminate or not, apiculate, base acute; surface rough above and beneath, slightly pubescent beneath; young leaves pinkish to purplish and whitish tomentose. Inflorescence spicate, solitary or 2-3 in axils of mature leaves; peduncle length variable, up to 7 cm; rachis 1-2 cm, flowers 20-30, closely clustered, sessile; bracts 3-4 mm long, narrowly spatulate, caducous, sparsely pubescent; flower bud cylindrical; calyx tubular, 6-8(+) mm long, densely pubescent at base, lobes very irregular, eventually simulating a 2-labiate calyx; corolla tubular, bright yellow, up to 15 mm long, densely tomentose, lobes regular, 2-3 mm long; ratio calyx/corolla length ca. 1/2; staminal tube shorter than corolla; stigma cup-shaped. Pod glabrous, straight, to 12(15) x 2.5 cm, apiculate, calyx persistent, valves rigid, fibrous, yellow when ripe, sutures 3-4 mm wide, valves flat or convex, with a network of transverse shallow ribs; seeds 5-10(15), not contiguous, embryo flat ca. 1.5 x 1 cm, cotyledons black. Seedling: pubescent, two first leaves opposite, 1-jugate, petiole winged.

Distribution: Widespread in Amazonia, and also present in Venezuela and Trinidad; strictly heliophilous species, most able to survive in open or savanna vegetation in the Guianas, in forested galleries along savanna creeks, at the margin of disturbed forests in coastal areas, and sometimes in dense rainforest, on rocky river banks near rapids, probably requiring humid soils, a reason why it was not recorded from areas with mesophytic vegetation such as that associated with granite outcrops (GU: 14; SU: 3; FG: 16).

Selected specimens: Guyana: Barama R., van Andel 978 (BRG, P, U); Berbice R., Ebini Station, Harrison & Persaud 1177 (BRG, K); Capsey Lake, Stockdale & Jenman 8794 (BRG, K); Demerara R., Jenman

5102 (BRG), Mazaruni Station, Jenman 5266 (BRG), Tapercooma Cr., Jenman 6558 (BRG, K); Pomeroon R., Anderson 335 (BRG, CAY, FDG, K); Pomeroon R.-Suppename R., Hoffman 2477 (BRG, P, US); Roraima, Ro. Schomburgk ser. II, 586 = Ri. Schomburgk 959 (U); Toucan Mts., Kamoa R., Jansen-Jacobs *et al.* 1600 (BRG, K, P, U); Rupununi R., de la Cruz 1802 (K); Upper Essequibo R., Gunn's, Jansen-Jacobs *et al.* 1454 (BRG, CAY, P, U), Henkel 3234 (BRG, CAY, P, US). Suriname: Wilhelmina Mts., Zuid R., Irwin *et al.* 57611 (NY, P); Lely Mts., Lindeman & Stoffers *et al.* 819 (BBS, NY, P, U). French Guiana: Approuague R., Arataye Cr., Sastre 5934 (CAY, P, U); Cayenne, Bena SF 1035 (CAY, P, U), Mélinon 77 (P); Montjoly, SF 3486 (CAY, P); Route du Tour de l'Ile, Feuillet 2230 (B, CAY, NY, P, U, US); Iracoubo, Bellevue Village, Sastre 6111 (CAY, P); Regina, Cayenne-Régina Road, Sastre & Bell 8030 (CAY, P, U, US); Saint-Laurent, Mana Rd., Feuillet 3583 (CAY, P, US), Concession Gosset, SF 7528 (CAY, P, U), Godebert, Wachenheim 261 (P), Gourdonville, Benoist 1508; Sinnamary, village, Sastre 6105 (CAY, P), Piste-de-St-Elie, Loubry 18 (CAY, P), Prévost 3244 A & B(CAY, K, MO, NY, P); Sabatier & Prévost 4868 (CAY, K, MO, P).

Vernacular name: Guyana: home-waiki.

Use: In Guyana it is appreciated as both an ornamental and medicinal plant, and often grown in villages.

Note: Bright yellow flowers are a unique feature within the Guianan *Inga's*, it makes *I. pilosula* very easy to identify when in flower.

45. **Inga poeppigiana** Benth., London J. Bot. 4: 602. 1845. Type: Peru, Loreto, "Maynas, in sylvis ad Yurimaguas", Poeppig 2436 (holotype K, isotype P, photo NY.)

Small tree; bark pale; branchlets ribbed, irregular, young parts pubescent; stipules conspicuous, lanceolate, 8-12 mm long, 4 mm wide, striate, often persistent; vegetative parts as well as axes of floral spikes provided with sparse hairs 1-2 mm long. Leaves (2)3-jugate; petiole and rachis winged; on 3-jugate leaves petiole up to 1 cm long, segments of rachis 2-2.5 cm long, elliptic, 4-6 mm wide; nectaries regular, circular, sessile or very shortly stipitate, less than 1 mm in diam.; persistent terminal appendage ca. 8 mm long, prolonging rachis beyond distal leaflets; leaflets elliptic, proximal ones 3-4 x 1.5-2.5 cm, distal ones to 11 x 4.5 cm; base strongly asymmetrical (internal edge cuneate, straight, external edge rounded), apex acuminate (acumen 1-1.5 cm long) and apiculate; blades supple, indument somewhat denser beneath. Inflorescence spicate and solitary

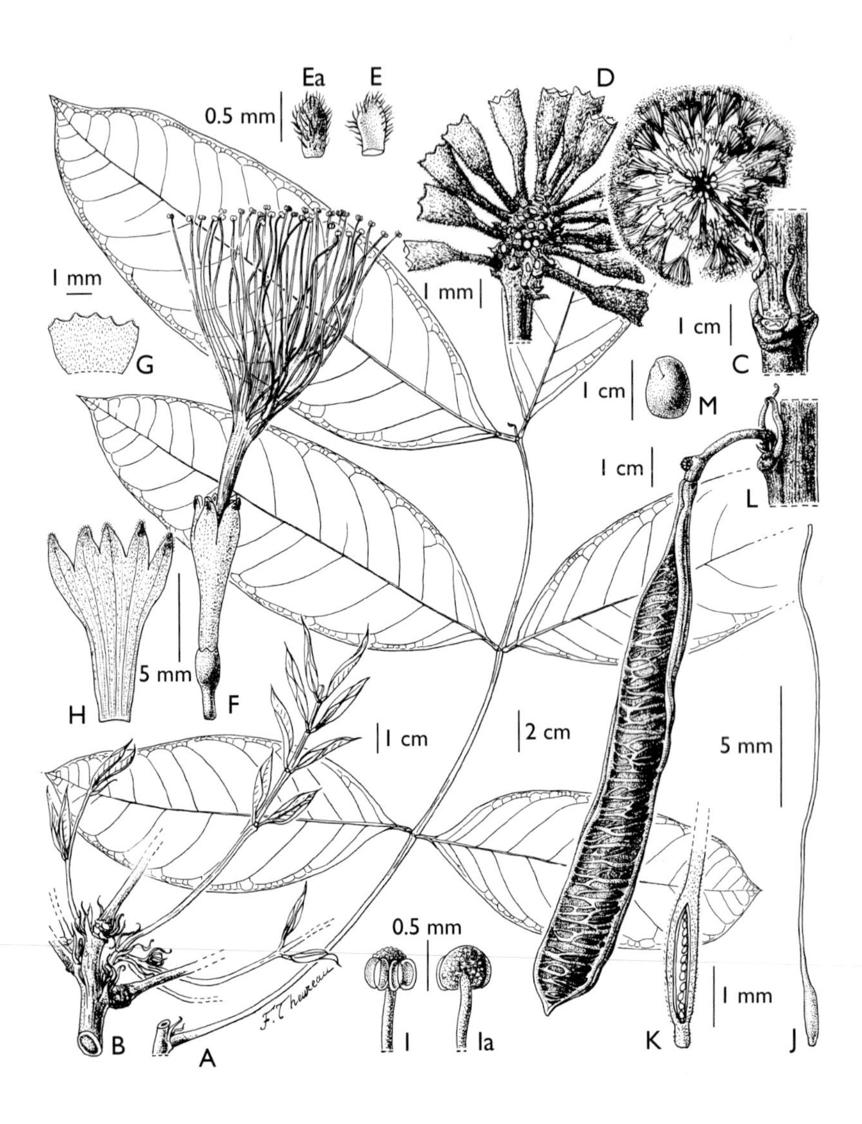

Fig.27. *Inga retinocarpa* Poncy: A, leaf; B, woody twig with young leaves; C, woody twig, detail of leaf scar and stipules; D, flower head, profile (rachis) and from above; E, Ea, flower bract, adaxial (E) and abaxial (Ea) views; F, flower; G, calyx (display); H, corolla (display); I, Ia, anther, abaxial (I) and adaxial (Ia) views; J, gynoecium; K, ovary; L, pod; M, seed (A-K, Villiers *et al*. 2086; L-M, de Granville B4761). Drawing by Françoise Theureau; reproduced with permission from Poncy, Adansonia 13: 149. 1991.

in axils of adult leaves; peduncle pubescent, ca. 1.2 cm; rachis ca. 2 cm, 15-flowered or less; bracts resembling stipules but smaller (ca. 5 mm long), channelled; flowers sessile; flower bud clavate; calyx tubular, ca. 2 cm long, green, a little bulging at base, glabrous, striate, teeth irregular; corolla tubular, up to 5 cm long, greenish white, sparsely pubescent; ratio calyx/corolla length 2/5; staminal tube exserted, 7 cm long: whole androecium 11 cm; style 12 cm, stigma cup-shaped. Pod straight, 20-30 x 1.5 cm, sutures not thickened, valves thin, leathery, densely rusty velutinous, seeds not contiguous, prominent (seeds not seen).

Distribution: Western Amazonia, and recorded in Amapá (Brazil) and French Guiana; small understorey tree of dense undisturbed rainforest, it seems to be a nocturnal bloomer (S. Mori observ., 1992) (FG: 3).

Selected specimens: French Guiana: Approuague, Station Nouragues, Poncy 950, 959 (CAY, NY, P, U); Saül, Eaux Claires, Mori *et al.* 22177 (CAY, NY, P).

Note: Sterile material of this species could be confused with the very common understorey species *Inga auristellae*, or with *I. disticha*. The very large, conspicuous flowers suggest an affinity with the Amazonian *I. longiflora*, also belonging to series Longiflorae Benth.

46. **Inga retinocarpa** Poncy, Bull. Mus. Natl. Hist. Nat., B, Adansonia 13(3-4): 147. 1991. Type: French Guiana, Arataye R., Villiers & Feuillet 2086 (holotype P, isotypes P, CAY, NY, U). – Fig. 27

Small understorey tree up to 8 m high; bark light brown to grey, verrucose, inner bark yellow to orange; branchlets fistulose and covered with a thin downy brown tomentum (also seen on stipules, petiole and foliar rachis). Leaves very large, 3-4-jugate, entirely glabrous; stipules linear ca. 12 x 2 mm, persistent; rachis and petiole thick, woody, terete; petiole up to 15 cm long x 4 mm diam., rachis segments 4-6 cm long, longitudinally grooved; nectaries inconspicuous, ca. 1.5 mm diam.; terminal appendage prolonging rachis beyond distal leaflets often persistent; leaflets oblong, proximal ones 7-12 x 4-5 cm, distal ones up to 28 x 10 cm; base cuneate to attenuate, apex acute, shortly acuminate; blades coriaceous, dark green, veins prominent beneath; young leaves greyish green to violet, whitish tomentose. Inflorescence subcapitate, ramiflorous, arising just below foliated part of twigs or less often on old branches or on trunk; axes downy, 2-3 at one node and very contracted; peduncle ca. 1.5 cm; rachis subspherical or ovoid ca. 3 mm long, ca. 30-flowered; bracts 1 mm long, inconspicuous, pubescent below; flowers greenish white, shortly pedicellate, pedicel up to 1.5 mm long, very finely tomentose; calyx 1.5 x 1 mm, regular; corolla funnel-shaped, ca. 12 mm long, lobes sharp, 2 mm long; ratio calyx/corolla length 1/8; staminal tube ca. 15 mm, conspicuously exserted; stigma tubular. Pod 8-14(18) cm x 1.7-2.5 cm, often twisted, semi-rigid, olive-green to brownish when ripe, velvety pubescent, valves convex, with a conspicuous network of prominent ribs, sutures up to 5 mm wide, salient and rounded; seeds 12-15, embryo flattened, dark-brown, ca. 10 x 8 mm. Seedling: first two leaves opposite, rachis terete, leaflets elliptic and large, with a brownish indument on veins, at early stage very similar to the seedlings of *I. rubiginosa*.

Distribution: Known from French Guiana and Pará (Brazil), one sterile collection from Guyana is doubtfully placed here; an uncommon sciophilous species of the understorey of dense rainforest (GU: 1?; FG: 13).

Selected specimens: Guyana: NW Distr., Matthew's Ridge, Maguire & Cowan 39348 (NY). French Guiana: Bassin de l'Oyapock, Cr. Gabaret, Cremers 9956 (CAY, P, US); Saül, de Granville 4638 (CAY, NY, P); de Granville B 4761 (CAY, P), Eaux Claires, Acevedo *et al.* 5002 (CAY, P, US); Approuague R., Arataye Cr., Villiers 2130 (CAY, NY, P), Nouragues Station, Poncy 893 (CAY, NY, P, U, US), 1328 (CAY, K, NY, P, U), Riera 1557 (CAY, P, U); Grand Matoury, Cremers *et al.* 13893 (CAY, P), Grimes 3268 (CAY, NY), Mts. de Kaw, Bordenave 262 (CAY, P), Réserve Naturelle Trésor, Poncy *et al.* 1370 (CAY, P, U).

Note: With all axes apparently orthotropic, and poorly branched, it exhibits unusual architectural features within the genus, most species following Troll's model.

47. **Inga rhynchocalyx** Sandwith, Kew Bull. 1948: 318. 1948. Type: Guyana, Essequibo R., Keriti Cr., Fanshawe 886 = FD 3622 (holotype K).

Tree to 30 m x 40 cm, buttressed to 1 m; young twigs dark brown, densely lenticellate, glabrous. Leaves 3(4)-jugate, glabrous or inconspicuously pubescent when young; stipules linear, 3-5 mm, caducous; petiole and rachis segments winged, 2.5-3.5(+) cm long, obovate to obtriangular, up to 8-13 mm wide, edges rounded; nectaries orbicular, 1 mm diam.; leaflets narrowly elliptic, glossy, proximal ones 7-8 x ca. 3 cm, distal ones 13-18 x 6-7 cm, base acute, apex cuneate, apiculate to acuminate; both surfaces glabrous. Inflorescence spicate, solitary or geminate in axils of mature leaves; peduncle grooved, 3-5 cm long, with inconspicuous sparse hairs; rachis appearing subcapitate when young, up to 4 cm long at end of flowering, 20-50(+)-flowered, scars of the flowers prominent and conspicuous; bracts linear, to 10 mm long, overtopping young buds; flowers sessile, crowded; flower buds elliptic, topped by a sharp acumen 2-4 mm long; calyx acuminate, greenish to yellowish or cream, tubular, ca. 1 cm long excluding acumen, irregularly lobed, either only once cleft and then looking 1-labiate, or twice cleft; corolla cream to white, pubescent, 17-20 mm long, tubular, 5-lobed; ratio calyx/corolla length 1/2 or higher; staminal tube exserted; stigma cup-shaped. Pod (description based on 2 collections from the ground) flat, 20-35 (45) x 2.8-4 cm, twisted, margins undulated between seeds which are not contiguous, valves striate.

Distribution: Rare species, known only from the Guianas and adjacent Venezuela; the ecology is unclear; the few samples studied were collected either in dense greenheart-*Mora* rainforest inGuyana, in dense rainforest of C French Guiana or in xeromorphic forest on stony soil in Suriname (GU: 3; SU: 2; FG: 3).

Selected specimens: Guyana: NW Distr., Waini R., Fanshawe 2348 = FD 5084 (FDG, K, NY, P); Demerara, Mabura Hill, Pipoly 8615 (BRG, NY, P, US). Suriname: Lely Mts., Plateau SW, Lindeman & Stoffers *et al*. 262 (BBS, U), 538 (BBS, CAY, P, U). French Guiana: Saül, Mori *et al*. 23262 (CAY, NY, P, U); Sinnamary, Saut Takari-Kanté, Prévost *et al*. 3190 (CAY, P); Mts. Trinité, Plateau Tabulaire, Poncy *et al*. 1238 (CAY, P).

Note: *Inga rhynchocalyx* belongs to the series Acuminatae (a group of species with an acuminate calyx, included in Pilosulae sensu Pennington 1997); it has close affinities with *I. acuminata* Benth. of C America, Venezuela and Trinidad, which has much smaller flowers and leaves. The specimen Fanshawe 2348 = FD 5084 from Guyana is placed here with hesitation because of its different foliar rachis (not winged, the nectaries larger and flattened).

48. **Inga rubiginosa** (Rich.) DC., Prod. 2: 434. 1825. – *Mimosa rubiginosa* Rich., Actes Soc. Hist. Nat. Paris 1: 113. 1792. Type: French Guiana, Cayenne, Leblond (holotype P, not traced).

– Figs. 15 E-G; 23 I-L

Medium-sized to large tree, up to 35 m x 50 cm; bark pale grey to yellowish, with orangish spots, lenticels irregular; branchlets pubescent and reddish brown, lenticellate. Leaves 4-jugate, downy on all parts except upper surface of leaflets; stipules triangular, 2 mm wide at base, 1.5 mm long, very caducous; petiole terete or grooved, ca. 3 cm long; rachis slightly flattened to marginate, segments 2.5-4 cm long; nectaries sessile, discoid, up to 2 mm diam.; leaflets oblong to elliptic, proximal ones up to 11 x

5 cm, distal ones up to 17(20) x 8(10) cm; base generally obtuse, apex acute, apiculate; blades thick, dark green, rough with veins puberulous above, tomentose, pale reddish and slightly shiny below. Inflorescence spicate, generally 2-4 in axil of mature leaves, occasionally on short lateral shoots; peduncle shallowly grooved to flattened, pubescent, short (1-3 cm long); rachis 2-4 cm, 10-20-flowered; bracts triangular to ovate, channelled, caducous; flowers lax, sessile; flower bud obovate, upper corolla ribbed; calyx and corolla pilose; calyx campanulate, 4-6 mm long, greenish to pale-brown, inconspicuously toothed; corolla tubular, 2-2.8 cm long, covered with dense appressed shiny hairs, paler than calyx, lobes 2-3 mm long; ratio calyx/corolla length 1/5-1/6; staminal tube not conspicuously exserted, stamens white. Pod ca. 20-30 cm long, tetragonal, straight or curved, greyish to brownish green and velvety, valves 2-3 cm wide, shallowly ribbed transversely, sutures enlarged up to 1.5 cm, with prominent, rounded longitudinal ribs; seeds up to 20, closely spaced. Seedling: two first leaves opposite, 1-jugate; subsequent leaves showing characteristics of adult leaves (petiole, hairs, nectaries).

Distribution: Through Amazonia and the Guianas; a lowland primary rainforest species, probably an evening-blooming species, according to S. Mori's observations in Saül (GU: 6; SU: 7; FG: 22).

Selected specimens: Guyana: Moruca R., Kumaka Rd., van Andel 1761 (BRG, P, U); Essequibo x Mazaruni R., Bartica, Sandwith 605 (G, K, NY, P, U); Potaro-Siparuni, Iwokrama, Clarke 210 (BRG, P, US); Upper Essequibo, Akarai Mts., Chodikar, A.C. Smith 2909 (K, NY, P, U); Kanuku Mts., Wabuwak, FD 5811 (BRG, K); Mazaruni Station, FD 2512 (K, U), FD 3737 (K). Suriname: Brownsberg, Tawjoeran LBB 14643 (BBS, P, U); Sectie O, BW 6074 (U), 6141 (K, U); Jodensavanne, Mapane Cr., Elburg & Roberts LBB 12635 (BBS, U); Schulz 7280 (BBS, P, U); Kabalebo Dam Project, Lindeman & Görts *et al.* 63 (BBS, U); Blakawatra, den Outer 867 (BBS, U). French Guiana: Approuague R., Arataye Cr., Nouragues Station, Poncy 1016 (CAY, P, U, US); Cayenne, Martin s.n. (P), Perrottet s.n. (P), Poiteau s.n. (P); Grand Matoury, Oldeman 2243 (CAY, P), Cremers *et al.* 13855 (CAY); Kaw Mts., Camp Caïman, Poncy 591 (CAY, P); Saint-Laurent-du-Maroni, Acarouany, Sagot s.n. (P), SF 216M (BBS, CAY, P), Camp Lorrain, SF 52M (CAY, P, U), Route de Mana, SF 7411 (CAY, P, U), 7534 (CAY, P, U), Godebert, Wachenheim 386 (P); Saint-Georges de l'Oyapock, Grenand 2133 (CAY); Saül, Route de Bélizon, Mori & Ek 20722 (CAY, NY, P), Eaux Claires, Mori *et al.* 23099, 23102 (CAY, NY, P), La Fumée W, Mori *et al.* 20787 (CAY, NY, P); Route de Bélizon, Mori *et al.* 24950 (CAY, NY); Sinnamary, Piste-de-St-Elie, Sabatier & Prévost 2215 (CAY, NY, P, U), Sastre 6121 (CAY, P).

Vernacular name: French Guiana: akiki-inga (Wayapi).

49. **Inga sarmentosa** Glaziou ex Harms, Notizbl. Königl. Bot. Gart. Berlin 6: 303. 1915. Type: Brazil, Rio de Janeiro, Glaziou 12629 (isotypes G, K, P). – Fig. 28

Medium-sized, low buttressed tree; glabrous on all parts; bark pale brown; branchlets stocky, with conspicuous, very dense, verrucose lenticels. Leaves (2)3(4)-jugate; stipules falcate, ca. 10 x 3 mm, often persistent; petiole and rachis terete, woody and lenticellate like branchlets; petiole very short (seldom longer than 0.5 cm), segments of rachis 2-4 cm long; leaflets elliptic, proximal ones 7-12 x 3.5-6.5 cm, distal ones 13-22 x 5-8.5 cm, base acute, apex acute ending in a triangular to falcate acumen, blades leathery, smooth and shiny above, beneath tertiary veins form a conspicuous, dense network. Inflorescence subspicate, axillary, often solitary in axils of mature leaves; peduncle 2-6 cm long, thick (ca. 3 mm in diam.); rachis 1.5-2.5 cm long, flowers 20-30, very densely clustered, sessile or shortly pedicellate; bracts linear to spatulate, up to 3 mm long; calyx tubular, ca. 10(+) mm long, greenish, irregular (one groove between two of the 2 mm long lobes deeply cleft), opening at same time as corolla; corolla tubular, 13-17 mm long, whitish, lobes ca. 2 mm, pinkish to purplish; ratio calyx/corolla length ca. 3/4; staminal tube not or hardy exserted, stamens and style pinkish, stigma tubular or narrowly funnel-shaped; ovary 12-20-ovulate; intrastaminal ring surrounding ovary at its base. Pod woody, flattened, variable in length but generally ca. 13-16 cm, up to 4 cm wide, light brown with pale or reddish lenticels, denser along border of valves and sutures, sutures rounded, not prominent, ca. 1.5 cm wide; endocarp fibrous, brownish; seeds not contiguous, pulpy seedcoat poorly developed, embryo dark, brownish purple to violet or black, ca. 15 x 7 mm. Seedling: first two leaves 1-jugate and opposite, blades glabrous and coriaceous.

Distribution: Amazonia and the Guianas, disjunct in Peru and Venezuela, it seems to be more common through the Lower Amazon and French Guiana; primary dense lowland rainforest; only a few (sterile, and thus of doubtful identification) collections from Suriname and a single record from Guyana (GU: 1?; SU: 2; FG: 11).

Selected specimens: Guyana: Demerara, Mabura Hill, Jansen-Jacobs *et al.* 1860 (BRG, CAY, P, U). Suriname: Hannover Distr., Para, Teunissen & Lindeman LBB 15305 (BBS, U); Tumuc-Humac Mts., Talouakem, Acevedo-Rodriguez *et al.* 6153 (BRG, P, US). French Guiana: Comté R., SF 3735 (CAY, P); Sinnamary, Piste-de-St-Elie, Sabatier 845 (CAY, P, U, US), 2372 (CAY, P), Prévost & Sabatier 3006, 3915 (CAY, NY, P, U), Sabatier & Prévost 4135 (CAY, NY, P), 4140 (CAY, P); Petit-Saut, Loubry 243 (CAY, P); Ht-Oyapock R., Trois-Sauts, Grenand 1272 (CAY, P); Saint-Georges de l'Oyapock, Grenand 3042 (CAY, INPA, K, MO, P); Saül, Eaux Claires, Mori *et al.* 22286 (CAY, NY, P).

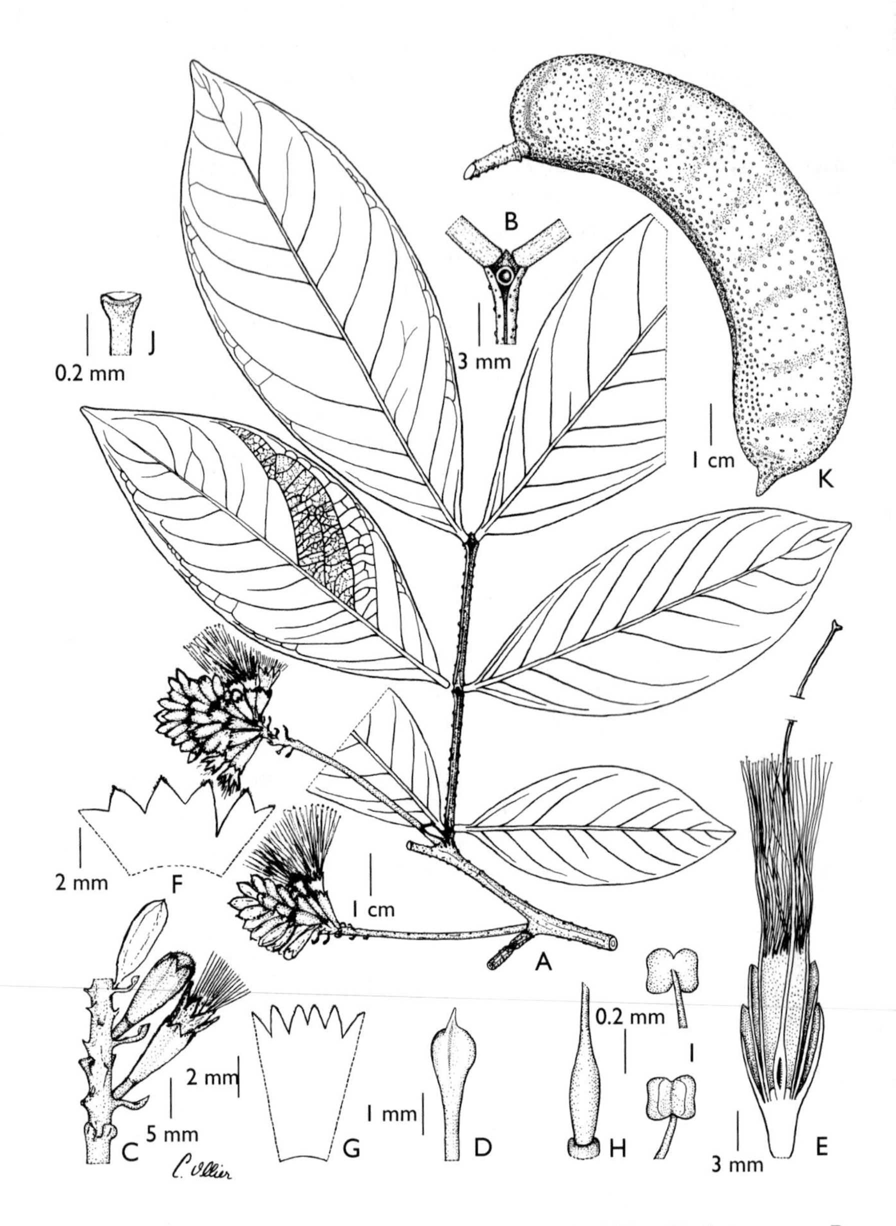

Fig. 28. *Inga sarmentosa* Glaziou ex Harms: A, habit, leaf and inflorescences; B, extremity of leaf-stalk with foliar nectary; C, inflorescence, detail; D, flower bract; E, flower, longitudinal section; F, calyx (display); G, corolla (display); H, ovary; I, anther, dorsal and ventral view; J, stigma; K, pod (A-J, Sabatier *et al*. 4140; K, Sabatier 845). Drawing by Corinne Ollier; reproduced with permission from P.

Vernacular name: French Guiana: inga-manio (Wayapi).

Note: In this treatment *Inga sarmentosa* is considered distinct from *I. capitata*, contrary to Pennington (1997: 458), who placed the former as a synonym of the latter, based on the assumed existence of intermediate forms. In French Guiana at least, a clear delimitation is observed, however, between both species occurring in the same forests. The uncertainty about the true origin of the type collection of *I. sarmentosa* is discussed in Poncy, 1991.

50. **Inga sertulifera** DC., Prod. 2: 436. 1825. – *Mimosa coriacea* Pers., Syn. Pl. 2: 262. 1806. – *Inga coriacea* (Pers.) Desv., J. Bot. Agric. 3: 71. 1814, non Willd.. Type: French Guiana, Cayenne, Stoupy s.n. (holotype P-Juss.). – Fig. 19 H-M

Small tree, often bushy, bark grey to whitish, lenticels numerous, pale, branchlets thick, angular, whole plant glabrous except very young branchlets puberulous. Leaves 2-jugate; stipules up to 15 mm long, linear or falcate, persistent; petiole (up to 1 cm long) and rachis (1.5-3 cm long) marginate, persistent linear appendage beyond superior leaflets; nectaries sessile, prominent, proximal one triangular or irregular, distal one circular up to 2.5 mm diam.; leaflets narrowly elliptic, distal ones (10-16(20) x 5-7 cm) twice as large as proximal ones (4-7(10) x ca. 3 cm), base cuneate, apex acute, apiculate or not, carinate; blades coriaceous, smooth and bright on both surfaces, below paler, veins not prominent; young twigs and leaves deep purple on all parts, nectaries whitish inside. Inflorescence umbellate, 1-3 in axils of mature leaves; peduncle slender, 5-7 cm long, receptacle spherical or clavate, with 50(+) pedicellate flowers; bracts 1.5-2 mm long, spoon-like, apical enlarged part pubescent, a few larger bracts at base of capitulum; flower bud oblong; pedicel 4-5 mm long; calyx greenish, tubular, ca. 3 mm long, regular, lobes inconspicuous; corolla white, funnel-shaped, up to 10 mm long; ratio calyx/corolla length ca. 1/3; staminal tube markedly exserted. Pod up to 15 cm long, but often short (2-3 cm) and then 1- or 2-seeded, 2.5-3 cm wide, apiculate, cylindrical, looking fleshy and bright yellow when ripe, valves supple, leathery, finely wrinkled or smooth, sutures up to 5 mm wide, not prominent; seeds contiguous and appressed, embryo black, up to 15 x 8 mm. Seedling: with first two leaves opposite and 1-jugate, leaflets ca. 5 x 2 cm; petiole obtriangularly winged.

Distribution: Venezuela, the Guianas, Brazilian and Peruvian Amazonia; a riverine species common along rivers, sporadically growing in disturbed places in the coastal area of French Guiana (GU: 14; SU: 6; FG: 24).

Selected specimens: Guyana: Fanshawe FD 3166 (FDG), Ro. Schomburgk 137.S (K, U); Demerara, Mamali, Fanshawe FD 4165 (FDG); Essequibo R., Cuyuni-Mazaruni, Gillespie 1449 (P, U, US), Kako R., Hahn 5536 (BRG, P, U, US), 5540 (BRG, P, U, US), Morabali Cr., Sandwith FD 1160 (FDG, K), Sandwith 284 (K, U), Supurani Cr., Jenman 6588 (BRG, NY), Sipu R., Clarke *et al.* 7729 (BRG, P, U); Potaro R., Kaieteur Park, Cowan & Soderstom 2233 (K, US), Gillespie 1361 (BRG, P, US, U); Ireng R., Maguire *et al.* 46246 (NY, U); Rupununi, Kuyuwini Landing, Jansen-Jacobs *et al.* 2503 (BRG, P, U). Suriname: Hostmann 237 (K, NY, U); Commewijne R., Perica R., Schulz 7258 (P, U); Marowijne R., Hugh-Jones 52 (K, U); Hermina village, Lanjouw & Lindeman 2024 (K, U); Saramacca R., Maguire *et al.* 24054 (K, NY, P, U); Wilhelmina Mts., Lucie R., Irwin *et al.* 55362 (P, U). French Guiana: Approuague R., Arataye Cr., Poncy 192 (CAY, P, U), Saut Machicou, de Granville 8252 (CAY, NY, P, U, US); Comté R., Aubréville 311 (P), Oldeman 1246 (CAY, P), 1213 (CAY, P, US); Gabrielle Cr., Hallé 2699 (CAY, MPU, P); Ht Marouini R., Langa Soula, de Granville 9637 (CAY, NY, P, U); Ht Oyapock R., Oldeman T754 (CAY, P); Iracoubo, Village Bellevue, Poncy 287 (Cay, P, U); Mana, Mélinon 152 (P); Mana R., Tamanoir Cr., Hallé 484 (CAY, P), Saut Ananas, Cremers 7538 (CAY, P); Maroni, Ouaqui Cr., SF 7805 (P, U), Schnell 12145 (P, U), Goodou Campou, Sastre & Bell 8210 (CAY, P, US), Papaichton, Sastre 6471 (CAY, P), Saut Gaa Caba, Sastre 6505 (CAY, P); Saint-Laurent, Portal R., SF 4509 (P); Sinnamary R., Petit-Saut, Prévost 1302 (CAY, P, U), Tigre Cr., Hoff 6589 (CAY), 6613 (CAY, P), Saut Dalles, Loubry 1671 (CAY, P), Saut Mouche, Bordenave 425 (CAY, P).

Vernacular names: Guyana: waitey. French Guiana: inga-tawa (Wayapi).

Notes: This species is morphologically very close to *Inga umbellifera* (see note under that species).
Pennington (1997) distinguishes 2 subspecies; the materials from the Guianas all refer to the typical *I. sertulifera* DC. subsp. *sertulifera* (see also *I. jenmanii*).

51. **Inga splendens** Willd., Sp. Pl. 4: 1017. 1806. Type: Brazil, Pará, Hoffmannsegg 18 (holotype B-Willd. no 19037, photo P, G).

Inga floribunda Benth., J. Bot. (Hooker) 2: 143: 1840. Type: Guyana, Ro. Schomburgk ser. I, 364 (holotype K, isotype P).
Inga hostmannii Pittier, Contr. U.S. Natl. Herb. 18 (5): 188. 1916. – *Inga splendens* Willd. var. *hostmannii* (Pittier) Ducke, Arch. Jard. Bot. Rio de Janeiro 4: 15. 1925. Type: Suriname, Hostmann s.n. (holotype GH).

Medium-sized tree, to 25 m tall, outer bark greyish, with small, closely spaced brownish lenticels arranged in horizontal rows, inner bark orange-yellow to whitish; branchlets thick, irregular, glabrous or sparsely pubescent when young. Leaves 2(3)-jugate; stipules triangular, ca. 3 x 1-2 mm, generally caducous but persistent on short shoots; petiole stocky (up to 4 mm diam.), marginate to winged, distally up to 1 cm wide; rachis of 2-jugate leaves up to 4 cm long, winged at least distally, up to 8 mm wide below insertion of leaflets; nectaries sessile, orbicular, 1.5-2 mm diam., with a thick, prominent border; leaflets elliptic-obvovate, proximal ones (6.5)8-9.5(12) x (2)3.5-4.5(6) cm, distal ones twice as large or more, base markedly asymmetrical, acute or obtuse, apex obtuse, cuspidate, blades leathery, edges often rolled downwards, glabrous and smooth above, more or less pubescent, veins prominent beneath, lateral veins parallel and straight for proximal two thirds. Inflorescence spicate, 2(+) spikes in axils of mature leaves, more often in axils of atrophied leaves of lateral short shoots (4-6 cm long); peduncle slender, somewhat angular, slightly pubescent, up to 4 cm long; rachis 1-3 cm long, 20-30 closely clustered sessile flowers; bracts spatulate, channelled, surrounding buds, caducous; flower buds ovoid to clavate with acute apex; calyx greenish, tubular, 4-8 mm long, hairy, irregularly cleft, sometimes appearing bilabiate, corolla yellowish to white, with glossy hairs, funnel-shaped, 1.3-20(25) mm long, lobes ca. 2 mm long, ratio calyx/corolla length 1/3-2/3; staminal tube slightly exserted. Pod on a thick ligneous peduncle, flattened, ligneous and rigid, slightly to markedly curved, up to 20 x 3.5 cm, glabrous (pubescent when young), strongly apiculate, valves very dark green when ripe, rugose, sutures brownish, often lenticellate; seeds not prominent, contiguous, embryo ca. 2.5 x 0.8 cm, greenish brown. Seedling: pubescent, two first leaves opposite, 1-jugate, sessile, stipules conspicuous, up to 13 mm long, leaflets oblanceolate, very narrow at base; juvenile plant may present only one pair of leaflets when up to 2 m tall, then 2 pairs, these oblong to narrowly obovate, winged petiole increasing up to 5 cm.

Distribution: Venezuela, the Guianas, Brazil (Pará); dense primary lowland rainforest, mostly on flooded areas along river banks, seldom on “terra firme” (GU: 15; SU: 17; FG: 16).

Selected specimens: Guyana: FD 6032 (K), 6033 (K), Gleason 916 (K); Demerara R., Jenman 4339 (BRG, K); Barama R., Kariako, van Andel & Samuels 831 (P, U); Barima-Waini Reg., Ayambara Fall, Pipoly 8196 (BRG, NY, P, U, US); Cuyuni-Mazaruni, Waramaden, McDowell 3188 (P, U, US), Oko Cr., Tutin 332 (K, U); Matthews Ridge, Mori *et al.* 8165 (NY); Pakaraima Mts., Kurupung R., Hoffman *et al.* B 2393 (BRG, P, US); Potaro-Siparuni, Kato, Hahn 5752 (P, U, US); UpperTakutu-Upper Essequibo, Kassikaytiu, Clarke *et al.* 4991 (BRG, P, US), Rewa

R., Clarke 3981 (BRG, P, US), Kuyuwini R., Clarke 2211 (BRG, P, US). Suriname: BW 3631, 4717 (BBS, U), BW 5266 (U), 6307 (BBS, U), Schulz 10055 (BBS, U); Lanjouw & Lindeman 3240 (NY, U), Stahel 314 (BBS, U); Brownsberg, Brokopondo, van Donselaar 1441 (BBS, U), Commewijne, Perica R., Schulz 7165 (BBS, U, P); Kabalebo Dam Project, Lindeman & Görts *et al.* 665 (BBS, U); Kaboerie, BW 5095, 4293 (K, U), 5918 (P, U); Saramacca R., Cr. Dungeoman, Maguire *et al.* 24059 (K, P, U); Wilhelmina Mts., Lucie R., Maguire *et al.* 54031 (NY, U); Soderstrom & Holmgren 54148 (NY, P, U). French Guiana: Cr. Jupiter, Loubry 777 (CAY, P); Approuague R., Arataye Cr., Poncy 207, 282 (CAY, P), Oldeman 2492 (CAY, P), La Folie Cr., Oldeman 1494 (CAY, P); Camopi R., Tamouri Cr., Lescure 143 (CAY, P); Maroni, Saut Lavaud, de Granville 12510 (CAY, P); Haut Oyapock, Trois-Sauts, Grenand 956 (CAY, P); Ouaqui R., Grigel, de Granville B4892 (CAY, P), St Macaque, de Granville 1721 (CAY, P); Sinnamary R., Cr. Plomb, Bordenave 814 (CAY, K, NY, P, U, US), Saut-Tigre, Bordenave 841 (CAY, P, US), Petit-Saut, Molino 896 (CAY, MPU, P); Piste-de-St-Elie, Sabatier & Prévost 3640 (CAY, NY, P, U, US); Saint-Laurent, SF 4208 (CAY, P).

Vernacular names: Suriname: phoedoe. French Guiana: a-ipopita (Wayapi).

Note: *Inga splendens* exhibits an important variation in the size of the flowers, mainly the corolla length, and the calyx/corolla length ratio is not constant.

52. **Inga stipularis** DC., Prod. 2: 435. 1825. Type: French Guiana, Cayenne, Patris s.n. (holotype G-Del).

Small to medium-sized tree; bark pale greyish to whitish, lenticels verrucose, slash white; branchlets irregular; whole plant glabrous. Leaves (2)3-4-jugate; stipules conspicuously developed, foliaceous, persistent, sessile or shortly petiolate, triangular to rounded, wider than long, ca. 1.5 x 2 cm or more; petiole and rachis terete, petiole up to 2 cm long, rachis segments 2-5 cm, terminal appendage prolonging rachis beyond leaflets often persistent; nectaries sessile, prominent, circular, up to 2 mm diam., with a thick border; leaflets elliptic to oblong, proximal ones generally 5-8 x ca. 3 cm, distal ones 8-12 x 4-5 cm, sometimes larger, base cuneate, symmetrical or nearly so, apex shortly acuminate, blades coriaceous, tertiary veins netted and slightly prominent beneath; juvenile twigs and leaves uniformly bright purple. Inflorescence spicate, (1)2-5 spikes in axils of adult or young leaves; peduncle thick, variable in length, generally 4-8 cm; rachis 1-2 cm long, 40(+)-flowered, conspicuously scarred after flower abscission, flowers very closely clustered, sessile; bracts lanceolate,

6-10 x 2 mm, caducous; flower bud cylindrical, often pinkish at apex; calyx tubular, ca. 4 mm long, 5 teeth acute, regular; corolla funnel-shaped, 7-8 mm long, white, lobes ca. 1 mm, slightly pinkish; ratio calyx/corolla length ca. 1/2; staminal tube generally not exserted; stigma tubular. Pod moniliform, (-)10-16(+) x 1.5-2 cm, dark green when ripe, valves leathery, smooth; sutures not enlarged; seeds 8-15, embryo dark brown to black, ca. 10 x 6 mm. Seedling: first two leaves alternate, 1-jugate, glabrous, stipules inconspicuous, leaflets narrowly elliptic, 4 x 1.5 cm.

Distribution: Venezuela, the Guianas and Eastern Amazonia (Brazil, Pará), very common in French Guiana, apparently less so in Guyana; dense rainforest, tolerating the understorey shade, as well as fast-growing and especially conspicuous and common in clearings or disturbed areas, likely to be short-lived (GU: 4; SU: 4; FG: 19).

Selected specimens: Guyana: Essequibo, Bartica-Potaro Road, Fanshawe FD 3481, FD 4173 (FDG, K); Kamoa Mts., Clarke 4927 (BRG, P, US); Mabura Hill, Ek 1147 (BRG, P, U). Suriname: Arrowhead Basin, Maguire *et al.* 24450 (NY, P, U); Brownsberg, BW 6572 (P, U), 6898 (BBS, U); Wilhelmina Mts., Lucie R., Maguire *et al.* 54099 (P, U). French Guiana: Cayenne, Route du Galion, Descoings & Luu 20147 (CAY, P), Matoury, Hallé 1011 (CAY, P, U); Bas-Oyapock, Chemin Maripa, Oldeman T837 (CAY, P); Haut-Oyapock, Mts. St Marcel, de Granville 2652 (CAY, P); Maroni R., Paul Isnard, Cremers & Hoff 11300 (B, CAY, NY, P, U, US); Maroni R., Haut-Marouini, Langa Soula, de Granville *et al.* 9237 (CAY, P); Mts. Bakra, Emerillon, Cremers & de Granville 13175 (CAY, NY, P, U); Paracou, Pétronelli 11 (CAY, NY, P); Piste Tibourou, Dutrève & Crozier 35 (CAY, K, NY, P); Mts. Inini, Bellevue, Sabatier & Prévost 3223 (CAY, P, U); Mt. Saint-Marcel, de Granville *et al.* 15513 (CAY, K, P, US); Réserve Nat. Mts. Trinité, Poncy 1470 (CAY, P); Régina, Mt. Tortue, Feuillet 10286 (CAY, P, U, US); Saül, Mts. La Fumée, Mori & Pennington 18013 (CAY, K, NY, P); Sinnamary, Cr. Plomb, Sabatier & Molino 4843 (CAY, MO, P), Piste-de-St-Elie, Gentry & Zardini 50338 (CAY, MO, P), Puig 10246 (CAY, P), Parcelle Ecerex, Prévost & Sabatier 4495 (CAY, MO, P).

Vernacular name: French Guiana: sisipay (Wayapi).

53. **Inga striata** Benth., London J. Bot. 4: 608. 1845. Type: Brazil, Sello s.n. (holotype K).

Inga nuda Salzm. ex Benth., London J. Bot. 4: 607. 1845. Type: Brazil, Bahia, Salzmann s.n. (holotype K, isotype G, MO, P).
Inga commewynensis Miq., Stirp. Surinam. Select. 1. 1851. Type: Suriname, Commewijne R., Focke 392 (holotype U).

Inga prieurei Sagot, Ann. Sci. Nat., Bot. ser. 6, 13: 332. 1882. Type: French Guiana, Leprieur s.n. (holotype P, isotype K).
Inga perrottetii Sagot, ibid.: 334. 1882. Type: French Guiana, Perrottet s.n. (holotype P).
Inga nuda Salzm. ex Benth. var. *longiflora* Benth., Trans Linn. Soc. London 30: 618. 1875. Type: French Guiana, Poiteau s.n. (holotype B destroyed, isotype RB fragments).

Medium-sized tree; branchlets pale brown, lenticellate, angular. Leaves (3)4-jugate, stipules lanceolate, ca. 8 x 2 mm, pubescent, often persistent; petiole up to 3.5 cm long, terete to winged; rachis segments ca. 3 cm long, winged; nectaries orbicular, 1 mm diam., very prominent to stipitate; leaflets elliptic, proximal pair (5)8-10(12) x (2)3.5-5.5(6.5) cm, distal pair up to 16(20) x 7(9) cm; base obtuse to slightly cordate, apex acuminate, surface rough above, veins beneath conspicuously prominent and pubescent; young leaves densely yellowish brown pubescent, nectaries green. Inflorescence spicate, solitary or geminate in axils of mature leaves; peduncle stocky, angular, up to 3 mm thick and very variable in length (1 to 7 cm); rachis 2-5 cm, 40-50-flowered, flowers crowded, sessile (shortly pedicellate on Prance *et al*. 15025); bracts linear 5-10 mm long, generally persistent; calyx tubular, 10-15 mm long, striate, glabrous, markedly cleft on one side; corolla tubular, variable in length, to 20 mm, densely hairy, lobes 2-3 mm; ratio calyx/corolla length 2/3; staminal tube shorter than corolla; stigma cup-shaped. Pod 15-20 x ca 1.5 cm, rigid, tetragonal, dark green and glabrous when ripe, sutures up to 1 cm wide and salient, ribbed, valves smooth or very shallowly wrinkled, endocarp pithy, pulpy; seedcoat poorly developed, embryo brown ca. 1 x 0.5 cm. Seedling: two first leaves opposite, 1-jugate, petiole widely winged, blades bullate with sparse white hairs.

Distribution: In its recent new delimitation, *Inga striata* has a peri-Amazonian distribution; not a very common species, although recorded in several places in coastal (including the sea shore) or open areas, and secondary forest; also recorded from the forest of the interior in French Guiana (GU: 3; SU: 3; FG: 12).

Selected specimens: Guyana: Marudi Mts., Locust Cr., Stoffers *et al*. 291 (BBS, BRG, CAY, NY, P, U); Potaro-Siparuni, Paramakatoi, Clarke *et al*. 1252 (BRG, P, US); Rupununi, Kuyuwini R., Jansen-Jacobs *et al*. 2958 (BRG, CAY, NY, P, U). Suriname: Mouth of Coppename R., Lindeman 5694 (U); Voltzberg, de Jong & Holthuyzen LBB 15783 (BBS, NY). French Guiana: Leprieur 347 (P); Ouanary, Geay 1891 (P); Ile de Cayenne, Fort Diamant, Feuillet 55 (CAY, P, U), Lac Lalouette, David *et al*. 74 (CAY), Mahury, de Granville 3449 (CAY), Prévost 810 (CAY, P), Vidal, Feuillet 1546 (CAY); Ilet La Mère, Forget 479 (CAY); Approuague R., Arataye Cr., Poncy 470 (CAY, P).

54. **Inga suaveolens** Ducke, Bol. Técn. Inst. Agron. N. 2: 5. 1944. Type: Brazil, Amazonas, São Paulo de Olivença, Santa Maria, Ducke 1522 (holotype IAN, isotypes NY, P, US).

Small to medium-sized tree, 15-20 m high, bark smooth; branchlets striate or ridged, young shoots tomentose, lenticels numerous. Leaves 4(5)-jugate; stipules linear, 3-5 mm long, with sharp apex, pubescent, caducous; petiole and rachis marginate to winged, primary vein covered with short ferrugineous hairs above, almost glabrous beneath; petiole 0.5-1.5 cm long, segments of rachis (1.5)2-3.5(5) cm long; nectaries cup-shaped, prominent, sometimes short-stalked, 1-2 mm diam., proximal ones often transversally elliptical; pulvini pubescent, leaflets elliptic to oblanceolate, proximal ones (3)4.5-8 x (1.5)2.5-3 cm, distal ones 9-13(16) x ca. 3.5(6) cm, base cuneate to acute, apex apiculate and mucronate, veins 7-10 pairs, salient and sparsely pubescent beneath; young leaves densely tomentose. Inflorescence of 2-3 spikes in axils of mature leaves; peduncle tomentose, (2)3-6 cm long, rachis (1.5)2-3 cm, up to 6 cm when old, 30-40-flowered; bracts linear 1.5-3 mm long, caducous, puberulent; flowers sessile; flower bud cylindrical to ovoid, apex acute or shortly acuminate; calyx tubular 4-6(8) mm long, striate, with long scattered hairs, denser at base, irregular, shortest lobes 1mm, deepest cleft 4 mm; corolla 11-12(15) mm long, tubular, with shiny appressed dense hairs, 5 equal teeth 1-2 mm; ratio calyx/corolla length 1/2-2/3; staminal tube as long as corolla or exserted (variable on same specimen); stigma discoid. Pod 6-13 x 2 cm (18 x 3 cm on Smith 3178 from Guyana), almost cylindrical when ripe, glabrous, smooth, valves plump, very shallowly wrinkled, sutures not thickened.

Distribution: Amazonian Ecuador to French Guiana with a single record in Amazonian Brazil (the type specimen) (GU: 10; FG: 19).

Selected specimens: Guyana: Corentyne R., Jenman 85 (BRG, K); Kanuku Mts., Hoffman & Gopaul 350, 459 (BRG, P, US), Wabuwak, Wilson-Browne FD 5861 (BRG, U); Upper Essequibo, Gunn's, Jansen-Jacobs *et al.* 1446 (BRG, P, U), Rupununi, Kuyuwini R., Henkel *et al.* 3194 (BRG, P, US), Jansen-Jacobs *et al.* 2838, 3204 (BRG, P, U) ; Upper Takutu, Kassikaityu, Clarke *et al.* 7916, 8808 (BRG, P, US). French Guiana: Approuague R., Arataye Cr., Poncy 364 (P, CAY, U), Sastre 5854 (CAY, P, U); Maroni R., Aloïké, Fleury 543 (CAY), Boniville, Schnell 11478 (CAY, K, NY, P, U, US); Grand-Inini R., de Granville C34 (CAY, P, U); Ht Marouini, de Granville *et al.* 9264 (B, CAY, MG, NY, P, U, US); Oyapock R., Eureupoucigne, Oldeman B3271 (CAY, P), Zidockville, Prévost & Grenand 913 (CAY, P, U); Saül, Grenand 970 (CAY, P), Eaux-Claires, Mori & Ek 20736 (CAY, NY, P), Mori *et al.* 22024, 23114, 23344 (CAY, NY, P), La Fumée, Mori *et al.* 20736 (CAY, NY, P), Rte Bélizon,

Mori & Gracie 24231 (CAY, P, NY), Philippe *et al.* 26948 (CAY, NY, P), Sentier Botanique, Mori *et al.* 24151, 24180 (CAY, NY, P); Trois-Sauts, Prévost & Grenand 913 (CAY, P); Approuague R., Oldeman B3271 (CAY, NY, P, U).

Vernacular name: French Guiana: paku-inga (Wayapi).

Note: Pennington (1997) cited a few specimens from non-flooded forest in C French Guiana. In addition, we attribute here a set of specimens collected from several riparian forests in French Guiana and Guyana, whose leaves differ in having the foliar rachis mostly winged, and a denser pubescence. Characters overlap between the two forms which are thus difficult to distinguish clearly. Nevertheless, this inclusive approach probably requires the re-evaluation of *Inga suaveolens*, and might lead to the recognition of the Guianese materials as distinct species. It is also necessary to reconsider the status of *I. java* Pittier, treated by Pennington as a synonym of *I. densiflora* Benth., because the type specimen of the former is very close to our Guianese collections of *I. suaveolens*; in fact Pennington placed a few of them in *I. densiflora*.

55. **Inga thibaudiana** DC., Prod. 2: 434. 1825. Type: French Guiana, Cayenne, Thibaud s.n. (holotype G-DC, microfiche P).

Inga gladiata Desv., Ann. Sci. Nat. Paris 1, 9: 427. 1826. Type: French Guiana, Cayenne, Desvaux s.n. (holotype P).

Treelet often bushy in secondary or riparian environments, to large tree up to 25 m x 50 cm in high dense forest, with rounded low buttresses; bark light greyish to beige, verrucose, lenticels many, closely spaced, irregular. Leaves 4-5(6)-jugate; stipules scale like, triangular (1 x 1.5 mm), caducous; petiole terete, 1-2 cm long; rachis angular, sometimes distally winged (often winged on young trees), pubescent, segments ca. 2(+) cm; nectaries conspicuous, 2-3 m diam. and generally wider than rachis, transversally elliptic or almost triangular; leaflets elliptic to narrowly elliptic, proximal ones 3-6(8) x 2-4(5) cm, distal ones 8-15 x 4-5(7) cm; base acute, often strongly asymmetrical, apex often falcate, acute to conspicuously acuminate (acumen up to 2 cm long); surface above rough, sparsely tomentose, dark greyish green, beneath silky, pale yellowish brown, secondary veins prominent, strictly parallel and straight in their proximal half; young leaves pale grayish green, pubescent, nectaries whitish, bright, large. Inflorescence spicate, (1)2-5 spikes in axils either of adult or young developing leaves; spikes 20-30-flowered; peduncle pubescent, variable in length, up to 4 cm; rachis 1.5-2 cm; bracts scale like, triangular, 1 x 1 mm, persistent; flower bud cylindrical; calyx and corolla greenish white; calyx tubular 4-5 mm long, pubescent, corolla tubular, ca. 2 cm long, narrow, densely hairy, teeth acute 1-2 mm; ratio calyx/corolla length ca. 1/4; staminal tube not conspicuously exserted; stigma cup-shaped. Pod 15-25(+) x 2.5 cm, 1.5 cm thick, straight or distally curved, semi-rigid, leathery, covered with velvet-like tomentum, yellowish to brownish green to reddish brown when ripe, valves smooth, somewhat convex, sutures 3-7 mm wide, not ribbed, lenticellate; seeds up to 30, very closely spaced; embryo brownish green, 8-12 x 1 cm, cotyledons irregular in shape. Seedling: pubescent, first two leaves opposite, 1-jugate, petiole winged, leaflets lanceolate, markedly asymmetrical, apex falcate; leaflet number increasing up to 7 pairs, petiole and rachis winged on juvenile plant.

Distribution: C America, Venezuela, Colombia, the Guianas, Brazil (Amazonia and southwards to Rio de Janeiro); one of the most widely distributed species in the genus; either in dense lowland primary rainforest where it is quite common, or along rivers, or in disturbed areas of secondary vegetation where it proves especially successful (GU: 25; SU: 10; FG: 23).

Selected specimens: Guyana: Ro. Schomburgk ser. II, 74 (K, P), Ri. Schomburgk 172 (K, P); Boerasirie, Little 16765 (U); Cuyuni-Mazaruni, Kako R., Hahn 5542 (CAY, P, US), Imbamadai, Gillespie 2774 (P, U, US), Pakaraima Mts., Henkel & Hoffman 234 (BRG, P, US), Hoffman *et al.* 1678 (CAY, P, US), Paruima R., Clarke *et al.* 6158 (BRG, P, US), Utshe R., McDowell 2782 (BRG, P, U, US); Ebini, Exper. Station, Persaud 1216 (K); Mabura Hill, Jansen-Jacobs *et al.* 1974 (CAY, P, U, US); NW District, Moruca R., Acquero, van Andel 1747 (P, U), Barama R., Kariako, van Andel 980 (P, U); Pomeroon R., Mainstay Lake, Mc Dowell 4113 (CAY, P, U, US); Upper Demerara, Jenman 4134 (BRG), Linden Rd., Pipoly 9686 (NY, P, US); Essequibo R., Moraballi Cr., Sandwith 386 (K), YaYa Landing, Ek *et al.* 794 (BRG, P, U); Lower Essequibo, Suddie, de Granville & Poncy 11724 (CAY, P); Upper Essequibo, Gunn's, Jansen-Jacobs *et al.* 1451 (CAY, P, U, US), 1539 (BRG, P, U, US), Kuyuwini R., Clarke 4499 (BRG, P, US); Wassarai Mts., Clarke 8011 (BRG, P, US); Rupununi, foot of Kanuku Mts., FD 2207 (K), Surama, Acevedo 3366 (P, U, US). Suriname: Commewijne, Perica R., Schulz LBB 3637, LBB 7219, LBB 7259 (BBS, P, U), LBB 13055 (BBS, CAY); Cottica R., Moengo, Lanjouw 545 (P, U); Jodensavanne, Lindeman 3677 (BBS, U), Kabalebo Dam Project, Lindeman & Görts *et al.* 549 (BBS, U); Tafelberg, Maguire 24549 (NY, U), Watramiri, BW 5978 (P, U); Zanderij, BW 6038 (U). French Guiana: Ile de Cayenne, Baduel, Cremers & Hoff 12946 (CAY, P), Cabassou, de Granville *et al.* 15641 (B, CAY, NY, P), Risquetout, Sabatier & Prévost 4405 (CAY); Approuague R., Arataye Cr., Nouragues Inselberg, Larpin 978 (CAY, P); Mahury, Gentry & Zardini 50318 (CAY, MO); Rorota, Hallé 1078 (CAY,

P); Mts. Kaw, Piste Degrad Lalanne, Feuillet 3597 (CAY, P); Route de l'Est (Cayenne-Régina), Comté R., Sastre & Bell 8232 (CAY, P, U, US); Iracoubo, Village Bellevue, Sastre 6112 (CAY, P); Réserve Mts. La Trinité, Munzinger & Poncy 1376 (CAY, P); Saül, Carbet Maïs, de Granville 2005 (CAY, P), village, Hahn 3690 (CAY, P, U, US), Galbao Mt., de Granville *et al.* 8879 (B, CAY, NY, P); Maroni R., Sauts Providence, Sastre *et al.* 8037 (CAY, P); Saut-Sabbat, Feuillet 598 (CAY, P); Savane Matiti, Pérez & Crozier 814 (CAY, QCA); St-Laurent, Rte Mana, SF 4010, 7220, 7328 (CAY, P); Sinnamary, Parcelle Ecerex, Prévost 4753 (CAY, G, K, MO, P); Haute-Wanapi, Inselberg, de Granville *et al.* 16186 (CAY, P).

Vernacular name: French Guiana: alakwa-inga (Wayapi).

Note: *Inga thibaudiana* is a very fast growing heliophilous tree, and one of the best adapted species for colonizing disturbed areas; it shows a great morphological plasticity and wide ecological tolerance. Pennington (1997) distinguished 3 subspecies, with only subsp. *thibaudiana* occurring in the Guianas. In the present treatment this distinction is not followed, because the apparent distinctive characters used to separate the infraspecific taxa do not seem to be relevant in our area: e.g. the winged foliar rachis seems related to the age of the tree based on our observations in French Guiana.

56. **Inga umbellifera** (Vahl) Steud. ex DC., Prod. 2: 432. 1825. – *Mimosa umbellifera* Vahl, Eclog. Amer. 3: 30. 1807. Type: French Guiana, Cayenne, von Rohr s.n. (holotype C, isotype BM, microfiche P).

Inga sciadion Steud., Flora 26: 758. 1843. Type: Suriname, Hostmann 170 (holotype U).

Bushy shrub or small tree to 10 m tall; bark brownish with paler lenticels; branchlets glabrous. Leaves (1)2-jugate, glabrous; stipules lanceolate, ca. 3 x 1 mm, pubescent, caducous; petiole and rachis winged, wings obtriangular to auriculate, spreading around nectary, up to 6 mm (petiole) or 10 mm (rachis) wide just below leaflet insertion; petiole 1.5-2 cm long, rachis ca. 2.5(+) cm long; nectaries sessile, cup-shaped, 2-3 mm diam.; leaflets elliptic or oblong, variable in size, proximal pair (-)5-9 x 3-4 cm, distal pair 10-18 x 4-6 cm, sometimes smaller; base cuneate, apex cuspidate to acuminate; blades papery to leathery, smooth and bright above, paler beneath; young leaves green, with sparse white hairs, nectaries white inside. Inflorescence umbellate, generally solitary in axils of adult or young leaves, rarely on short shoots; capitula 20-25-flowered; peduncle slender, pubescent, ca. 1 cm long (but 2.5 cm on one collection); bracts scale like, spatulate, ca. 2 mm long; flower bud cylindric; flowers pedicellate, pedicel up to 8-10 mm long, pubescent; calyx tubular, regular, variable in length (2-4 mm), teeth inconspicuous, pubescent; corolla tubular 7-10 cm long, very sparsely tomentose or glabrous, lobes 1-2 mm; ratio calyx/corolla length ca. 1/3; staminal tube not conspicuously exserted. Pod cylindrical, fleshy, smooth or very shallowly wrinkled and bright yellow when ripe, 7-10 x ca. 2 cm, conspicuously cuspidate at apex, sutures 5-8 mm wide, not prominent; seeds contiguous, cotyledons dark-brown or black.

Distribution: The Guianas, Amazonia Brazil and Peru; strictly heliophilous, but tolerant of various edaphic conditions, growing mostly in mesophytic low forest on poor or thin soils, either on sand (coastal areas of the Guianas where it can be seen along savannas) or granite (low forest at the foot of rocky outcrops), less often riparian (and then perhaps on rocky or sandy areas), 2 collections seen from high dense rainforest (GU: 11; SU: 5; FG: 19).

Selected specimens: (a selection out of more than 100 collections). Guyana: Barima-Waini R., Powis Cr., McDowell *et al.* 4432 (BRG, CAY, P, U, US); Pomeroon R., Ro. Schomburgk ser. II, 751 (= Ri. 1400) (K, P); Corentyne R., Jenman 369 (BRG, K, P); Kanuku Mts., Rupununi R., Jansen-Jacobs *et al.* 164 (BBS, P, U), Wabuwak, Wilson-Browne FD 5682 (BRG, FDG); Lower Essequibo, Groete Cr., Fanshawe 1771 = FD 4507 (FDG, K); Mazaruni Station, FD 3339 (FDG, K); Upper Essequibo, Acarai Mts., Clarke *et al.* 7601 (BGR, P, US), Kwitaro R., Clarke *et al.* 6505 (BRG, P, US); Rewa R., Clarke *et al.* 6651 (BRG, P, US), Wassarai Mts., Clarke *et al.* 8496 (BRG, P, US). Suriname: Jodensavanne, Schulz 7725 (U); Kabelstation, Stahel 414 (U); Lely Mts., SW Plateau, Lindeman & Stoffers *et al.* 146 (BBS, K, U); Saramacca R., Maguire *et al.* 24044 (NY, P, U). French Guiana: Approuague R., Arataye Cr., Camp Arataï, de Granville *et al.* 16629 (CAY, P), Inselberg Nouragues, Larpin 875 (CAY, P), Poncy 147 (CAY, P); Ile de Cayenne, La Mirande, Oldeman B1282 (CAY, P); Grand-Ouaqui R., Saut-Macaque, Schnell 11938 (CAY, P, U); Hte-Mana, Cr. Baboune, Cremers 7465 (CAY, P); Kourou, Savane Matiti, Poncy 95 (CAY, P), Sastre 5986 (CAY, P); Iracoubo, Bellevue, Sastre & Moretti 4189 (CAY, P); Mts. Atachi Bacca, de Granville *et al.* 10943 (CAY, P); Saül, Eaux Claires, Pennington 13863 (CAY, K), Rte Bélizon, Mori *et al.* 24788 (CAY, NY, P); Saint-Laurent-du-Maroni, Gandoger 142 (P), Godebert, Wachenheim 271, 384 (P), Gourdonville, Benoist 1552 (P), Karouany, Sagot 959 (K, P); Maripasoula, SF 6030 (CAY, P).

Note: Although *Inga umbellifera* and *I. sertulifera* are very closely related, especially as both occur in riverine habitats, a few characters allow them to be easily distinguished from each other, as pointed out by Pennington (1997: 271): "*I. umbellifera* differs from *I. sertulifera* in the broader wing on the rachis, in the larger, flatter foliar nectaries and presence of indument on the inflorescence and flowers. The calyx of *I. umbellifera* is often larger than that of *I. sertulifera*".

57. **Inga vera** Willd., Sp. Pl. 4: 1010. 1806. Lectotype (designated by Bässler 1992): Sloane, Voy. Jamaica 2: 58, t. 183, f. 1. 1725.

In the Guianas only: subsp. **affinis** (DC.) T.D. Penn., The genus Inga, Botany 716. 1997. – *Inga affinis* DC., Prod. 2: 433. 1825. Type: Brazil, Raddi s.n. (holotype G-DC). – Fig. 29

Inga meissneriana Miq., Stirp. Surinam. Select. 2. 1851. Type: Suriname, Suriname R., Focke 1143 (holotype U, isotype P).

Shrub or small tree, often bushy; bark greyish; branchlets angular, longitudinally striate and poorly lenticellate, tomentose when young. Leaves (4)5-6-jugate; densely hairy on stipules, petiole, rachis and veins on both surfaces of leaflets; stipules linear, ca. 5 mm long, caducous; petiole and segments of rachis 1-3 cm long, winged; nectaries sessile, orbicular, to 1 mm diam.; leaflets narrowly elliptic, proximal pair to 8 x 2.5 cm, distal pair to 11 x 4 cm; base acute, apex acute to shortly acuminate, apiculate; blades supple, hairy on both faces (more densely on veins), on young leaves hairs bright reddish brown. Inflorescence of 1-3 spikes axillary to mature or juvenile leaves; peduncle ca. 2 cm long, hairy, rachis up to 3 cm long, provided with 15-20 well separated, sessile flowers; bracts channelled, ca. 4 x 2 mm, caducous; flower bud cylindrical to clavate; calyx tubular, 6-9 mm long, pubescent, lobes 1 mm long; corolla tubular, 15-20 mm long, densely hairy with up to 3 mm long lobes; ratio calyx/corolla length ca. 2/5; staminal tube not exserted from corolla. Pod sessile, up to 8 cm long, ca. 1 cm wide, entirely covered with greenish brown hairs forming a velvety indument, surface of valves only 3 mm wide between hypertrophied, sulcate, up to 8 mm thick sutures.

Distribution: C America, widely distributed through S America; in our area strictly riparian, often growing colonially on sandy river banks (GU: 11; SU: 7; FG: 13).

Selected specimens: Guyana: Pomeroon R., Ro. Schomburgk ser. II, 831 (= Ri. 1423) (K, MEL, P); Corentyne R., Jenman 340 (BRG, K), Berbice, McDowell *et al*. 2640 (BRG, CAY, P, US); Cuyuni R., Triana Fall, Loekie FD 2038 (FDG, K); Pomeroon R., Ocoshi Cr., Fanshawe FD 6402 (FDG, K, P, U); Mazaruni R., Cuyuni R., Gillespie 2212 (BRG, P, U, US), 2232, 2245 (BRG, P, US); Rupununi, Dadanawa, Jansen-Jacobs *et al*. 2042 (BRG, P, U); Upper Essequibo-UpperTakatu R., Gillespie 1763 (BRG, P, US), McDowell *et al*. 1920 (BRG, CAY, P, US). Suriname: Coppename R., Hekking 1007 (U); Gonini, Versteeg 61 (U); Kabalebo R., Florschütz & Maas 2610 (U); Wilhelmina Mts., Lucie R.-Oost R.,

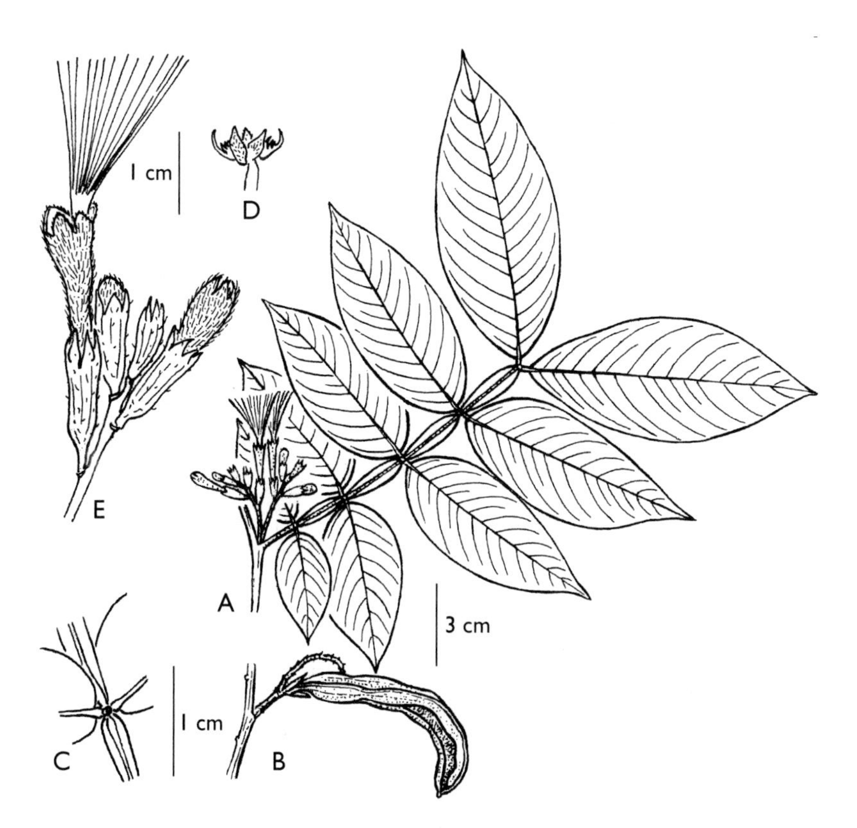

Fig.29. *Inga vera* Willd. subsp. *affinis* (DC.) T.D. Penn.: A, habit, leaf and inflorescence; B, pod; C, detail of leaf rachis, nectary; D, extremity of leaf axis, stipules; E, flower and flowerbuds. Drawing by Odile Poncy; reproduced from Poncy, 1985: 55, as *I. meissneriana*.

Irwin *et al.* 55400 (NY, P, U); Tapanahoni R.-Paloemeu R., Wessels Boer 1299 (U); UpperTapanahoni, Rombouts 680 (K, U). French Guiana: Approuague R., Pierrette, Poncy 248 (CAY, P, U); Camopi, Mt. Alikéné, Moretti 1356 (CAY); Cayenne, Expl. For. „Compagnons Réunis", CTFT 951 (CAY); Grand Inini R., Degrad Fourmi, de Granville *et al.* 8171 (CAY, P, U); Mana R., Sagot 926 (CAY, P), Saut-Dalles, Loubry 1150 (CAY, P); Maroni, Béligui, Sastre & Moretti 3814 (CAY, P, U), Cr. Ouaqui, SF 7778 (CAY, P, U), Cr. Sparwine, SF 6161 (CAY, P); Taluwen-Antecume, Fleury 1152 (CAY, P); Sinnamary R., Cr. Tigre, Hoff *et al.* 6590 (CAY, NY, P, U), Petit Saut, Sabatier 845 (CAY), Saut Stéphanie, Hoff 6039 (CAY, P).

Vernacular name: French Guiana: katapali (Wayapi).

Note: This widely distributed species was divided into 3 subspecies by Leon (1966); Pennington (1997) revised the delimitation of the infraspecific taxa and established *I. vera* subsp. *affinis*.

58. **Inga virgultosa** (Vahl) Desv., Ann. Sci. Nat. Paris 9: 426. 1826. – *Mimosa virgultosa* Vahl, Eclog. Amer. 3: 32. 1807. Type: French Guiana, "Cayenne", von Rohr s.n. (holotype C, isotype BM).

Shrub or treelet, often bushy, to 6 m tall; bark greyish; branchlets angular, verrucose, with brownish lenticels; whole plant glabrous. Leaves (3)4-6(10)-jugate and very small; stipules linear, ca. 1.5 mm long, persistent; petiole and rachis marginate, channelled; petiole ca. 0.5 cm long, rachis segments up to 1 cm long; nectaries orbicular or ovate, diam. < 1 mm; leaflets elliptic, up to 3 x 1 cm, apex acuminate, base rhomboidal or nearly so, cuneate to rounded; young twigs and leaves pale pinkish purple. Inflorescence umbellate, axillary, generally solitary; peduncle very slender, short (ca. 5 mm); capitula 10-15-flowered; bracts scale like, 0.8-1 mm long; pedicels slender, 8-10 mm long; calyx minute, ca. 1 mm long, corolla ca. 4 mm long, funnel-shaped, teeth sharp, 0.5 mm; ratio calyx/corolla length 1/4; staminal tube long-exserted. Pod glabrous, variable in length but often less that 6 cm, 2-2.5 cm wide, nearly cylindrical and yellow when ripe, base markedly asymmetrical, apex acuminate, sutures 2-3 mm wide, not prominent, valves smooth to finely wrinkled. Seedling: first two leaves opposite, 3-jugate, with a marginate, 1 cm long petiole; in the juvenile plant the number of leaflets increases to 7 pairs.

Distribution: Endemic to the eastern Guianas (Suriname, French Guiana, and adjacent areas of Brazil (Amapá)); commonly found in mesophytic scrubby low forest on granite, on the foothills or at the top of inselbergs, as well as in some areas of the coastal zone with well-drained but infertile soils (SU: 3; FG: 18).

Selected specimens: Suriname: Bakhuis Mts., Bordenave *et al.* 8226 (BBS, CAY, K, P, U); Nassau Mts., Lanjouw & Lindeman 2567 (BBS, K, P, U); Brokopondo, Tjon Lim Sang LBB 14831 (BBS, U). French Guiana: Approuague R., Nouragues Station, Larpin 599 (CAY, P), Loubry 103 (CAY, P); Bas-Oyapock, Inselberg Armontabo, Cremers 7040, 7085 (CAY, P, U), 7085A (seedling, CAY, P); Cayenne, Leblond 438 (P), Sabatier 3047 (CAY, P), Prévost 3049, 4435 (CAY, P); Matoury,

Lemée s.n. (P); Maripasoula, Fleury 107 (CAY); Massif Emerillon, de Granville 3787 (CAY, P, U); Massif de l'Inini, Mt. Bellevue, de Granville 8014 (CAY, P); Mt. St Marcel, Sastre 4443 (CAY, P), de Granville *et al.* 15395 (CAY, P, US); Mts. Tumuc-Humac, Kalimantan, de Granville 1120 (CAY, P), Talouakem, de Granville 12208 (CAY, P); Saül, Pic Matecho, de Granville 3352 (CAY, P).

15. **LEUCAENA**[19] Benth., J. Bot. (Hooker) 4: 416. 1842.
    Lectotype (Taxon 13: 300. 1964): L. diversifolia (Schlechtend.) Benth. (Acacia diversifolia Schlechtend.)

Distribution: A tropical and warm-temperate genus of 22 species, most diverse and numerous in tropical Mexico and C America, 2 extending into SW United States, 1 into Venezuela and NW Peru, 1 species (*L. leucocephala*) extensively planted for shade and forage and thence weedy and fully naturalized in both New and Old Worlds; in the Guianas 1 species.

Note: *L. leucocephala*, may be distinguished from other Guianan MIMOSOIDEAE by this syndrome of characters: flowers densely capitulate and subtended by peltate bracts; perianth 5-merous decandrous, filaments free to base, and anthers relatively large (± 1 mm), connective lacking a gland; pod plano-compressed, tapering backward into a stipe, inertly dehiscent through both sutures.

1. **Leucaena leucocephala** (Lam.) de Wit, Taxon 10: 54. 1961. – *Mimosa leucocephala* Lam., Encycl. 1: 12. 1783. – *Acacia leucocephala* (Lam.) Link, Enum. Hort. Berol. Alt. 2: 444. 1822. Type: cultivated at Paris, of American origin (holotype P-LA).

   Leucaena glauca sensu Benth., J. Bot. (Hooker) 4: 416. 1842 & Trans. Linn. Soc. London 30: 443. 1875, type (*Mimosa glauca* L.) excluded; *L. glauca* of almost all Guianan literature.

Unarmed trees of rapid growth and precocious maturity, sometimes flowering when subherbaceous and only 1-2 m tall but potentially attaining ± 10 m with trunk up to 1 dm DBH or more, young growth either densely puberulent with short, appressed or incumbent, pallid hairs or sometimes subglabrous, thin-textured olivaceous leaflets facially glabrous, usually ciliolate. Stipules triangular-ovate up to 1.5 mm,

[19] by Rupert C. Barneby

caducous. Leaves bipinnate, pinnae and leaflets in opposite pairs; leaf-stalk of longer leaves 1-2 dm, petiole ± 2-4 cm; a sessile, round or elliptic, cupular or almost plane nectary ± 1-2.5 mm diam. at insertion of first pinna-pair, and similar but smaller ones at tip of pinnae; pinnae (3)4-9(10) pairs, rachis of longer ones (4.5)5.5-10.5 cm; leaflets of longer pinnae (9)12-20 pairs, shortly pulvinulate, blades lanceolate from inequilaterally cuneate or shallowly semicordate base, acute or acuminate, larger ones 9-16(20) x 1.7-4.5 mm, mostly 3-veined from pulvinule, the almost straight midrib dividing blade ± 1:2, weakly pinnate distally, 2 posterior primary veins weak and short. Inflorescence of globose, densely many-flowered capitula borne solitary and geminate in axil of coeval or immediately hysteranthous leaves; involucre of 2-3 broad short confluent bracts at or close below base of capitula, receptacle globose or pyriform; bracts peltate, small round ciliolate blade raised on a filiform stalk, covering young flower buds, subpersistent; flowers homomorphic, 5-merous, 10-androus, hermaphrodite and some functionally staminate, greenish perianth glabrous except for minutely puberulent calyx-teeth and tip of petals; calyx deeply campanulate, slightly dilated distally, 2.7-4 x 1-1.7 mm, depressed-deltate teeth 0.2-0.5 mm; petals valvate in bud, early free to base, erect, linear-oblanceolate, 4.5-6 mm; filaments white, free to base, 9-10.5 mm, anthers dorsifixed, locules parallel, thinly pilosulous, connective not gland-tipped; pollen shed in monads; ovary subsessile, ± 3 x 1 mm, puberulent, style 7-8.5 mm, truncate and membranous at stigma. Pods usually several (up to ± 20) per capitulum, in profile broad-linear, straight, tapering into a stipe ± 5-16 mm, more abruptly acute or apiculate distally, when well fertilized 10-21 x 1.7-2.1 cm, (16)18-24(28)-seeded, body plano-compressed, greenish-brown or fuscous, thinly puberulent valves framed by straight, narrowly prominulous sutures, slightly elevated and paler over each seed, exocarp faintly venulose, endocarp pallid smooth, dehiscence inert, through both sutures; seeds subhorizontal on filiform funicle, compressed-obovoid, hard testa lustrous brown, areolate, endosperm thin.

Distribution : Supposedly native to Mexico or C America but now dispersed around the globe in tropical and warm-temperate climates; a prolifically weedy, arborescent subshrub of disturbed, often barren places at low elevations; valued as a nursery shade tree and widely planted for cattle-forage, for fuel, and as a hedge, tolerant of coppicing; flowering nearly throughout the year, the pods long persistent; common along and near the coast of the Guianas (GU: 2; SU: 4; FG: 7).

Selected specimens: Guyana: New Amsterdam, Hitchcock 16830 (NY); Craing Village, Demerara R., FD 6786 (= Fanshawe 3326) (NY, U). Suriname: Experimental farm Kabo, Distr. Saramacca, Everaarts 1155 (NY, U); lower Commewijne R., near plantation Mon Souci, Hekking 1116 (NY). French Guiana: Iles du Salut, Ile Saint Joseph, de Granville 8229 (NY); Iles du Salut, Ile Royale, Cremers 8662 (NY).

Vernacular names: Guyana: lead-tree. West Indies: jumby-bean.

Uses: The foliage provides rich nourishment for cattle and other ruminants, and agronomic breeding is reducing the mimosine in leaves and pods, which is toxic to horses and swine. Green pods are sold in Mexican markets as a vegetable. The shiny seeds are strung into necklaces.

16. **MACROSAMANEA**[20] Britton & Rose ex Britton & Killip, Ann. New York Acad. Sci. 35: 131. 1936.
Type: M. discolor (Willd.) Britton & Rose ex Britton & Killip (Inga discolor Willd.)

Pithecolobium sect. *Samanea* ser. *Coriaceae* [sic] Benth., Trans. Linn. Soc. London 30: 589, sp. ult. exclusa. 1875 & in Mart., Fl. Bras. 15(2): 443 ("*Coriacea*"). 1876.
Lectotype: P. adiantifolium (Kunth) Benth. [Macrosamanea discolor (Willd.) Britton & Rose ex Britton & Killip]

Unarmed, either free-standing or when crowded sarmentose, arborescent shrubs. Leaves alternate, bipinnate; lf-formula iii-ix(xv)/(9)10-26(31); stipules firm, deciduous; lfts pulvinate; venation palmate and pinnate. Inflorescences of axillary capitula or pseudoracemose; bracts at base of capitulum either subtending the lowest 2-5 fls or immediately below them, reflexed and charged ventrally with a nectary, or all bracts lacking nectary. Flowers normally 5-merous but teeth and lobes unequally cleft, sometimes confluent; calyx cylindric or cylindro-campanulate; corolla narrowly trumpet-shaped, usually white and externally silky-pubescent; androecial tube adnate at base to corolla, nearly as long as it or far exserted, the tassel either white or distally pink-tinged; ovary linear-ellipsoid, tapering into style, surrounded at base by a variably developed but always perceptible disc. Pods (known only in one species) sessile, in profile oblong or broad-linear, straight or gently decurved but never annular or coiled, sutures not greatly thickened, valves leathery; dehiscence inert, through the seminiferous suture and ultimately both sutures, valves narrowly gaping to discharge seeds.

[20] by James W. Grimes

Distribution: S American genus of 11 species, the majority of them riparian, seasonally flooded forest, 2 of seasonally wet sandy savanna or campirana, most diverse and numerous in the Amazon Basin, extending into the Orinoco valley; 2 species in the Guianas.

KEY TO THE SPECIES

1 No nectary on bracts at base of each unit of inflorescence; inflorescence a depauperate pseudoraceme of capitula arising from defoliate nodes below current foliage. . . . . *1. M. kegelii*
Bracts at the base of each unit of inflorescence with nectaries; inflorescence capitulate . . . . *2. M. pubiramea*

1. **Macrosamanea kegelii** (Meisn.) Kleinhoonte in Pulle, Fl. Suriname 2(2) [= Vereen. Kol. Inst. Amsterdam Meded. 30]: 331. 1940. – *Pithecolobium kegelii* Meisn., Linnaea 21: 249. 1848. Type: Suriname, Cassipoera Cr., Kegel 1173 (holotype NY-herb. Meisner!).

Pithecellobium stipellatum Bernh., nom. nud. in syn. Benth., Trans. Linn. Soc. London 30: 590. 1875, for Kappler 1341E (MO-herb. Bernhardi, U).

Potentially arborescent riparian shrubs attaining 8(-?) m; terete older branches gray, lenticellate, hornotinous ones and all lf-axes densely sordid-pilosulous with short ascending hairs. Lf-formula v-viii/12-16; stipules erect, firm lanceolate, 1.5-3 mm, dorsally glabrate and coarsely 3-6-veined, deciduous; lf-stks 10-19 cm, shallowly or obscurely sulcate ventrally, true petiole 6-12 mm; longer interpinnal segments 23-35 mm; pinna-pairs strongly decrescent proximally and less so distally (expanded lf-outline obovate), rachis of longer ones (6.5)7-12 cm, interfoliolar segments ± as long as greatest width of lft-blades, these subequilong except for a few abruptly decrescent proximally and for the longer, heteromorphic furthest pair; lfts thinly papery, remaining olivaceous when dry, sublustrous above, a little paler beneath, lfts at and near mid-rachis rhombic around an almost straight diagonal midrib, broadly cuneate at base, at apex triangular- or deltate-mucronulate, basal and adaxial sides nearly straight, distal side low-convex, larger blades 21-28 x 5.5-9 mm, 2.3-3.1 times as long as wide; venation of midrib and one strong (sometimes few outer weak) intramarginal vein produced well beyond mid-blade, midrib giving rise on each side to ± 7-10 major and to several intercalated minor secondary veins. Inflorescence little known, primary axis apparently only 5-15 mm and individual peduncles ± 3-4 mm; bracts subtending peduncles and individual fls spatulate, ± 2 mm, some of the former charged internally with a horny nectary; capitula ± 4-6-flowered, stout pedicels (0.5)1-2.5 mm. Calyx cylindric, slightly widened distally, 14-20 x 3.5-6 mm, prominulously many-veined, densely puberulent below middle but more thinly so toward teeth; corolla at least when young more densely silky-puberulent with paler hairs, 24-31 mm; androecium 4-5 cm, staminal sheath included in corolla, filaments creamy-white with pink tips. Fruit unknown.

Distribution: NE Suriname and adjacent French Guiana, on the Suriname, Maroni, and Tapahoni Rs.; on river and creek banks below 200 m; flowering November and December (SU: 4; FG: 2).

Selected specimens: Suriname: Jai Cr., Tapanahoni R., Gonggrijp 4150 (NY, U). French Guiana: commune de Grande Santi, Bassin du Maroni, Capegras 7 (CAY, NY); village Boni de Loca, Bassin du Maroni-Lawa, Fleury 740 (NY).

2. **Macrosamanea pubiramea** (Steud.) Barneby & J.W. Grimes, Mem. New York Bot. Gard. 74(1): 194. 1996. – *Inga pubiramea* Steud., Flora 26: 759. 1843. Type: Suriname, Hostmann & Kappler 171 (holotype P-herb. Steudel.!).

   In the Guianas only: var. **pubiramea** – Fig. 30

   Pithecolobium miquelianum Meisn., Linnaea 21: 250. 1848. Type: Suriname, in humid forest near Jodensavanna, Kegel 1162 (holotype NY-herb. Meisner!).
   Pithecolobium adiantifolium (Kunth) Benth. var. *multipinnum* Benth. in Mart., Fl. Bras. 15(2): 445. 1876. Lectotype (designated by Barneby & Grimes 1996: 197): "In Guiana frequens….", Hostmann & Kappler 1277 (hololectotype K!)

Potentially arborescent and when crowded sarmentose or vinelike shrubs (1.5)2-6(10) m tall; old branches and trunks gray glabrate, hornotinous ones together with axes of lvs and inflorescences sordidly pilosulous with short, incurved-ascending hairs. Lf-formula iii-ix(xiv)/(9)10-26(31); stipules erect-appressed, firm, lance-triangular, 0.8-2.8 mm, at most faintly veined dorsally, caducous; lf-stks including coarsely wrinkled pulvinus (3)5-21 cm, true petiole commonly no longer than first nectary, exceptionally up to 6-15 mm, but always shorter than interpinnal segments, these up to 12-45(55) mm; pinnae strongly decrescent proximally, less or not so distally, short first pair ordinarily reflexed, rachis of longer pairs (6)7-15(18) cm, longer interfoliolar segments (2.5)3-12(17) mm; lfts brown-olivaceous, glabrous or microscopically ciliolate along midrib above, when mature stiffly chartaceous, low-convex upper face

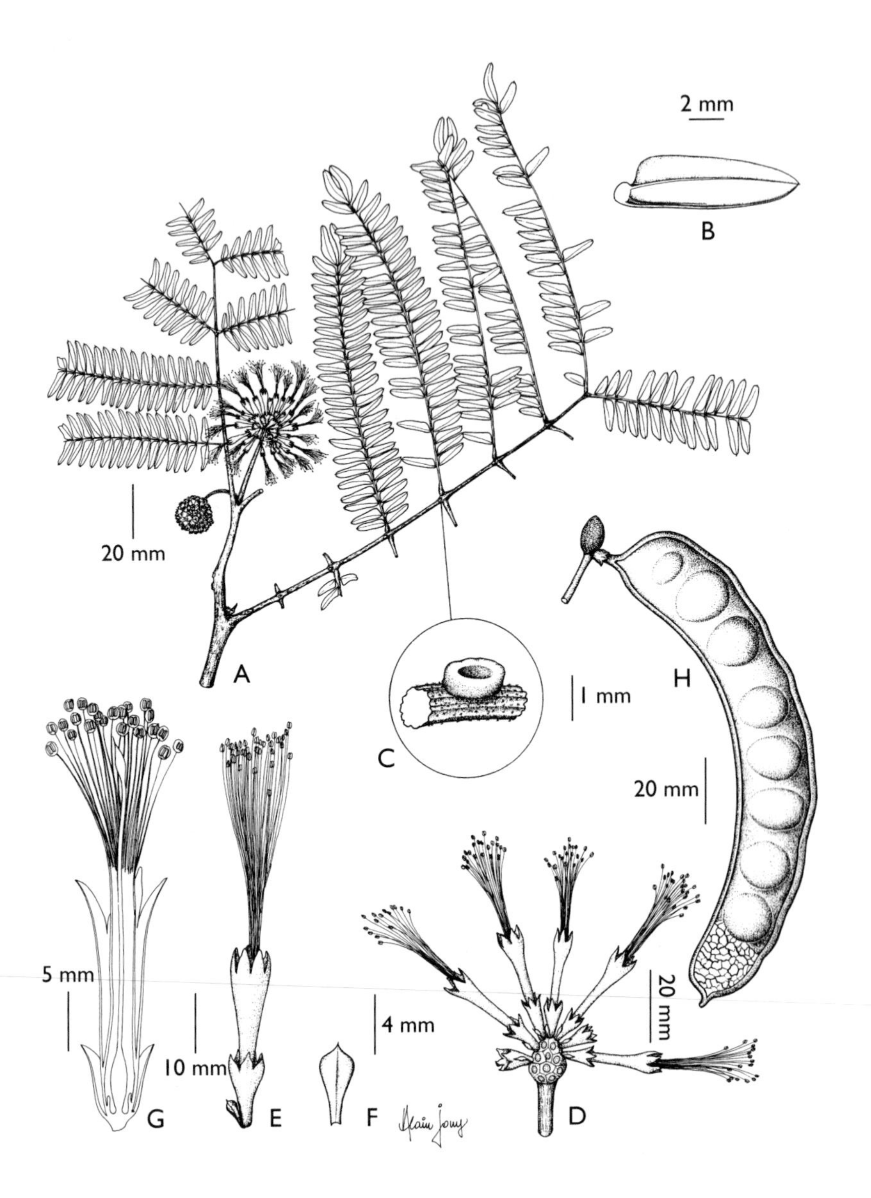

Fig. 30. *Macrosamanea pubiramea* (Steud.) Barneby & J.W. Grimes var. *pubiramea*: A, habit: leaves and inflorescences; B, leaflet; C, foliar nectary; D, flower head; E, flower; F, flower bract; G, longitudinal section of flower; H, fruit (the inferior part of the drawing figures the interior surface of the pod). (A-D, Mélinon 233; E-F, Sagot 169; G, Cowan 38896; H, Mélinon 233). Drawing by Alain Jouy; reproduced with permission from P.

smooth lustrous, veinless or almost so, lower face a little paler, finely venulose, lfts except for few, gently decrescent proximal pairs and for longer narrower distal pair all subequilong, first pair either contiguous to pinna-pulvinus or represented by minute subulate paraphyllidia; blades varying from linear-lanceolate to oblong or rhombic-oblong, mostly obtuse or obtuse and mucronulate, deltately acute when relatively narrow, at base cuneate on proximal and obtusangulate on distal side, largest (10)12-30(38) x (2)2.5-17 mm, 2.1-5.2(6) times as long as wide, midrib of proportionately narrow lfts subcentric, straight or obscurely sigmoid, that of proportionately broad ones diagonal, one weak (and sometimes an outermost very short), narrowly intramarginal, posterior primary vein produced to or beyond mid-blade, costa pinnate, secondary veins giving rise to weak sinuous tertiary venules, all venation usually immersed on upper face, always prominulous on the lower. Peduncle either solitary or 2-3 together, (1)1.5-5.5(9) cm, charged immediately below capitulum with 1-2(3) cycles of reflexed deltate bracts, each of these bearing on its adaxial (exposed) face a sessile, reddish or brownish, plane or crumpled, corneously margined nectary as wide as itself or wider; capitula (12)15-25(40)-flowered, receptacle ellipsoid or globose, alveolate, 3-6.5 mm in greatest diam.; floral bracts fugacious, dry and brown before anthesis or possibly sometimes lacking, when present linear or linear-spatulate, 0.7-2.5 mm. Calyx and corolla faintly many-veined externally, sordid- or pallid-puberulent overall, corolla often a little more densely so, but sometimes both glabrescent in age; calyx either sessile or contracted at often slightly bulbous base into a stout pedicel up to 1.5 mm long, overall 6.5-12 mm long, in lower 2/3 cylindric but expanded distally; corolla (16)18-32(34) mm, tube slender, very gradually dilated upward. Pod 1-4 per capitulum, sessile, in profile oblong or broad-linear, usually a little decurved, (6)8-14 x (2.2)2.5-5.6 mm, broadly rounded at each end, apiculate, planocompressed but early umbonate and finally tumescent over developing seeds, valves brown, stiffly chartaceous, framed by thickened sutures, thinly puberulent overall, transversely venulose; seeds plumply discoid and in broad profile elliptic-obovate, 16-19 x 10-14 mm.

Distribution: The Guianas, S to the lower Amazon valley in SW Venezuela, and Brazil (Amazonas and Pará); along streams and rivers; flowering sporadically throughout the year (GU: 2; SU: 5).

Selected specimens: Guyana: Upper Mazaruni R., de la Cruz 2358 (NY); Upper Demerara-Berbice Region, Linden-Soesdyke Hwy, Pipoly 9728 (NY, US). Suriname: Zanderij I, forest margins, Pulle 63 (NY, U); Para R., Lobin Savanna between Zanderij and Hanover, Lindeman 6595 (NY, U).

Note: The var. *lindsaeifolia* (Benth.) Barneby & J.W. Grimes occurs in the Upper Amazon, Vaupés and Orinoco Rs. in Brazil, Colombia and Venezuela.

17. **MIMOSA**[21] L., Sp. Pl. 516. 1753; emend. Benth., J. Bot. (Hooker) 4: 358. 1841.
Type: M. sensitiva L.

Trees, shrubs, vines, subshrubs, herbs; stems and leaf-stalks either unarmed, or armed near nodes with 1-3 epidermal aculei, or aculeate throughout, in addition nearly always pubescent with random combinations of a) minute incurved hairs, b) short straight hairs, c) short clavate or granular, orange-livid but non-secretory trichomes, d) filiform setae, e) straight and tapering or flagelliform setae dilated at base (these sometimes laterally attached or basally calcarate), f) setae (especially of leaflet-margins) backwardly produced into the epidermis, g) scabrous or arborescently plumose setae (these becoming stellae by shortening of the primary axis), h) gland-tipped setae, and i) either sessile or stalked, peltate or squamiform trichomes. Stipules setiform or variably dilated and sometimes many-veined, commonly persistent. Leaves pari-bipinnate, leaflets (in the Guianas) in opposite pairs, almost always pulvinate and pulvinulate, often sensitive, folding at night or when touched; leaf-stalks often bearing on ventral side between pairs of pinna-pulvinules a spicule or foliaceous bracteole, and in few species one or more nectaries, either near base of petiole or between pinna-pairs; leaflets highly diverse in number, shape and size, number and size reciprocally adjusted, first pair arising immediately beyond pinna-pulvinus, often reduced to stipelliform paraphyllidia. Inflorescences composed of spikes (racemes) or capitula either solitary or fasciculate in leaf-axils, or arranged in a pseudoraceme or panicle, in some xeromorphic species fasciculate on brachyblasts, each flower bracteate, bracts either shorter or longer than flower buds, capitulum in advanced bud in the first case moriform (resembling a *Morus* syncarp), in the second case conelike; bracteoles 0. Flowers either sessile or shortly pedicellate, 3-6-merous, either all haplostemonous, or all diplostemonous, and then all hermaphrodite or some lower ones of inflorescence-unit staminate with rudimentary gynoecium, or rarely distal flowers haplostemonous hermaphrodite and a few proximal ones diplostemonous staminate; corolla sympetalous, lobes either plane or cucullate, 1-several-veined; filaments pink-purple, pink fading whitish, white, or ochroleucous to sulfur-yellow, free or connate at base around

[21] by Rupert C. Barneby

gynoecium, anther-connective in dorsal view either round or broadly ovate, girdled or almost so by thecae, lacking a terminal gland; pollen grains united into simple, double, or triple tetrads. Pods plano-compressed or turgid with continuous replum (margo) corresponding with sutures, valves at maturity separating from replum either in one piece or, more often, both separating and breaking between seeds into 1-seeded, either indehiscent or individually dehiscent articles (fruit then termed a craspedium); seed-testa bearing a pleurogram (linea fissuralis) on each face; endosperm present around embryo, but commonly thin.

Distribution: A genus of ca. 490-510 species, the majority endemic to the American Tropics but some extending, in diminishing number, N into the United States and S into Uruguay and Argentina, few in continental Africa and tropical Asia, ca. 30 endemic to Madagascar, and few circumtropically weedy; in the Guianas 20 species (28 taxa).

Notes: The two Guianan mimosas with prickly stems and amentiform flower-spikes may be distinguished at anthesis from *Piptadenia* by anther-thecae bent around the periphery of a round or ovate connective, not striaght, parallel and contiguous, *M. invisa* further by lack of petiolar nectary, and *M. myriadenia* further by lepidote indument and haplostemonous flowers.
Mimosa viva L. and *M. ceratonia* L. are vaguely attributed in the literature to the Guianas, but there is no material evidence of their occurrence outside the West Indies and eastern Brazil.

LITERATURE

Barneby, R.C. 1991. Sensitivae Censitae. A description of the genus Mimosa Linnaeus (Mimosaceae) in the New World. Mem. New York Bot. Gard. 65: 1-835.

KEY TO THE SPECIES

1 Leaf-petioles charged ventrally with a mounded or urceolate nectary 2
Leaf-petioles without nectary 8

2 Flowers at once spicate and haplostemonous; leaflets of longer pinnae 16-42 pairs, largest less than 1.5 cm long 3
Flowers at once capitulate and diplostemonous; leaflets of longer pinnae 1-11 pairs, largest 1.5-12 cm long 4

3 Leaflets of longer pinnae 16-26 pairs, separated by intervals 1.5-3.5 mm long, larger blades broadly or narrowlyoblong, 5-9.5 x 1.5-3 mm . . . . *11a. M. myriadenia* var. *myriadenia*
Leaflets of longer pinnae 24-42 pairs, separated by intervals 0.4-1.4 mm long, larger blades linear-oblong, 3.5-6.5 x 0.6-1.4 mm . . . . *11b. M. myriadenia* var. *dispersa*

4 Stipules setiform 2-9 mm; paraphyllidia at base of pinna-rachises subulate-acicular 1-3 mm; most leaflets deltate-acute 5
Stipules either wanting or less than 2 mm and deciduous; paraphyllidia less than 1 mm or obsolete; leaflets mostly broadly obtuse 6

5 Pinnae of larger leaves 5-7 pairs; leaflets of longer pinnae 8-11 pairs, and longer distal ones ± 1.5-2.5 cm long . . . . *1a. M. annularis* var. *odora*
Pinnae of larger leaves 2 pairs; leaflets of longer pinnae 2-3 pairs, and longer distal ones ± 4-9 cm long. . . . . *1b. M. annularis* var. *xinguënsis*

6 Pinnae 4-10 pairs and leaflets of distal pinnae (3)4-11 pairs . . . . *16. M. rufescens*
Pinnae 2-4 pairs and leaflets of distal pinnae 1-3 pairs. 7

7 Pinnae 2-3(4) pairs and leaflets of distal pinnae 2(3) pairs, blades thinly lepidote dorsally with sessile scales, otherwise glabrous or almost so, largest ones 3.5-6.5 cm; flowers 5-merous glabrous . . . . *8a. M. guilandinae* var. *guilandinae*
Pinnae 2 pairs and leaflets of distal pinnae 1-2 pairs, blades densely scaberulous with shortly stalked scales, in addition pilosulous along veins dorsally, largest ones 4-12 cm; flowers mostly 4-merous, gray-puberulous at least on corolla-lobes . . . . *8b. M. guilandinae* var. *duckei*

8 Flowers spicate, the spike without filaments 2-many times as long as its diam. 9
Flowers capitulate, the globose or plumply ovoid capitula without filaments less than twice as long as their diam. 10

9 Indument of leaflets composed of simple trichomes; leaflets eglandular; flowers haplostemonous (stamens 5); pod 10-22 mm wide, glabrous or puberulent with simple hairs; diffuse or sarmentose shrubs and vines . . . . *9. M. invisa*
Indument of leaflets composed of stellate trichomes; leaflets glandular dorsally; flowers diplostemonous (stamens 8 or 10); pod 6-8 mm wide, densely stellate; stiffly branched shrubs and trees . . . *17. M. schomburgkii*

10 Flowers diplostemonous (stamens 8 or 10) 11
Flowers haplostemonous (stamens 4). 21

11 Lobes of corolla prominently several-veined 12
Lobes of corolla weakly 1-3-veined 14

12 Trichomes of stem, leaf-stalks and fruits arborescently plumulose .... *20. M. surumuënsis*
Trichomes of stem, leaf-stalks and fruits simple and smooth, or sometimes lacking 13

13 Stems aculeate........................ *19a. M. somnians* var. *somnians*
Stems unarmed .... *19b. M. somnians* var. *viscida*

14 Setae of stem and leaf-stalks attached laterally near or below middle, tapering at each end 15
Setae of stem and leaf-stalks basifixed, sometimes lacking 16

15 Rachis of longer pinnae 3-5.5 cm; prickles of stem sharply recurved .... *10a. M. microcephala* var. *cataractae*
Rachis of longer pinnae 2-3 cm; prickles of stem ascending, straight horizontal, or obscurely recurved..... *10b. M. microcephala* var. *lumaria*

16 Trees and bushy treelets 3-10 m at maturity, cultivated near the coast for hedging, perhaps locally naturalized; stipules obscurely veined; inflorescence an efoliate panicle of white capitula; pod glabrous, 5-9-seeded .... *2. M. bimucronata*
Either a) herbs, shrubs and weak scandent vines, native, and if ruderal less than 3 m tall, or b) if habitally similar to the preceding then panicle early foliate, pod either strigose or hispid and 12-20-seeded, and filaments pink 17

17 Stipules dilated and 5-25-veined dorsally, larger ones 1-3 mm wide; leaf-stalks either unarmed or armed with straight aculei................. 18
Stipules linear-attenuate or setiform 1-veined, mostly less than 1 mm wide; leaf-stalks armed with cat's-claw aculei.......................... 20

18 Arborescent shrubs usually 2-5 m; pod 4-12 cm long, 12-22-seeded .... *12. M. pellita*
Monocarpic herbs sometimes basally lignescent in age, seldom over 1 m; pod less than 2 cm long, (1)2-3(4)-seeded 19

19 Interpinnal segments of leaf-stalk unarmed or almost so, sometimes stiffly setose; interpinnal spicules less than 2.5 mm; stipules mostly 12-25-veined; pod in profile narrowly oblong-elliptic 3-4 mm wide, not widened upward; plants of disturbed woodland, savanna, and waste places .... *3. M. camporum*
Interpinnal segments of leaf-stalk armed laterally near middle with a pair of subopposite straight aculei; interpinnal spicules 3-17 mm; stipules ± 5-8-veined; pod in profile obliquely obovate-cuneate, 6-8 mm wide near apex; plants of seasonally inundated savanna and muddy shores .... *7. M. dormiens*

20 Pinnae of primary leaves 4-9 pairs, inserted at intervals of less than 1 cm; pod a laterally compressed craspedium in profile ± 1-2.5 x 0.3-0.45 cm, 3-8-seeded, replum less than 1 mm wide, valves breaking up when ripe into free-falling articles . . . . *6. M. diplotricha*
Pinnae of primary leaves 2-4 pairs, inserted at intervals of up to 1-2 cm; pod an obtusely tetragonal legume 8-17 x 0.3-0.45 cm, 20-30-seeded, replum 2-3 mm wide, valves as wide or narrower, when ripe separating in one piece from replum or only erratically fractured between seeds. . . . . *15. M. quadrivalvis* var. *leptocarpa*

21 Pinnae of leaves subtending the peduncles 3-5 pairs, subpalmately crowded near tip of leaf-stalk; leaflets of longer pinnae 26-65 pairs . . . . *13. M. polydactyla*
Pinnae of leaves subtending the peduncles 1-2 pairs; leaflets 2-30 pairs . . . 22

22 Pinnae of all leaves exactly 1 pair and leaflets of each pinna exactly 2 pairs . . . . *5. M. debilis*
Pinnae 1-2 pairs and leaflets of longer pinnae (3)4-30 pairs 23

23 Pinnae of all or most leaves 2 pairs 24
Pinnae of all leaves 1 pair. 27

24 Corolla-lobes glabrous externally; pinnae either 1 or 2 pairs. . . . . *14d. M. pudica* var. *unijuga*
Corolla-lobes gray-puberulent dorsally 25

25 Floral bracts at and beyond mid-capitulum 1.7-2.5 mm, bent inward near or below middle, distally ciliate with setae up to 1-1.6 mm, just prior to anthesis as long as or longer than subtended flower bud . . . . *14a. M. pudica* var. *hispida*
Floral bracts at and beyond mid-capitulum 0.7-2 mm, bent inward beyond middle, distally ciliolate with setules less than 0.9 mm, just prior to anthesis shorter than subtended flower bud 26

26 Stems aculeate; faces of pod usually glabrous, rarely puberulent or both puberulent and weakly setulose . . . . *14c. M. pudica* var. *tetrandra*
Stems unarmed; faces of pod at once densely puberulent overall and densely setose over each seed . . . . *14b. M. pudica* var. *pastoris*

27 Herbs and low subshrubs armed at nodes with a pair of aculei and sometimes on internodes with one infra-petiolar aculeus, but petioles unarmed; pod 3-4.5 mm wide; conjugately pinnate form of. . . . . *14. M. pudica*
Prickly vines, armed alike on stems, leaf-stalks and peduncles with files of cat's-claw aculei; pod 6-14 mm wide. 28

28 Leaflets of longer pinnae (8)10-17 pairs. . . . . *18. M. schrankioides*
Leaflets of longer pinnae (3)4-6 pairs. . . . . *4. M. casta*

1. **Mimosa annularis** Benth., Trans. Linn. Soc. London 30: 419. 1875. Type: Brazil, Amazonas, on Rio Vaupés near Panuré, Spruce 2893 (holotype K!, isotypes M!, W!).

Bush-ropes potentially 30 m long but often diffusely sarmentose, 3-10 m; armed on ribs of stem, leaf-stalks, and randomly on inflorescence-branches with files of recurved brown-tipped aculei 0.5-2 mm long and pilosulous with short plain hairs sometimes mixed with minute reddish scales, chartaceous leaflets either glabrous or puberulent facially, ciliolate; pyramidal, largely efoliate panicle of small globose capitula exserted 1.5-6 dm above foliage. Stipules stiffly subulate or setiform 2-11 x 0.3-0.8 mm, persistent; leaf-stalks 3.5-15 cm, petiole 2-6 cm, interpinnal segments 1-3.5 cm; nectary ovoid or flask-shaped, or rarely broad-based and depressed, small-pored, 1-2.6 x 0.8-2.5 mm near base of petiole, smaller ones on rachis of most pinnae below furthest pairs of leaflets; pinnae 2-7 pairs, rachis of furthest and longest pair 3.5-10 cm, interfoliolar segments mostly 6-25 mm; leaflets of distal pinna-pair 3-11 pairs, first pair (next to pulvinus) reduced to deflexed, subulate-setiform paraphyllidia 0.7-2.8 mm, blades of rest obliquely rhombic acute, diagonally costate, those of furthest pair very obliquely elliptic, ovate- or rhombic-elliptic from semicordate base, 15-90 x 9-38 mm, all acute or shortly acuminate and mucronulate, forwardly curved midrib giving rise on each side to ± 5-6 pairs of major secondary veins and a 3-4-nary reticulum of venules prominulous dorsally or on both faces. Peduncles mostly fasciculate by 2-4(5), each fascicle subtended by a rudimentary leaf-stalk bearing a flask-shaped nectary, longer peduncles 4-11 mm, all 2-3-bracteolate; capitula without filaments 4-5 mm diam., prior to anthesis moriform; flower buds obovoid, either glabrous or thinly puberulent at tip. Flowers whitish, either 4- or 5-merous and diplostemonous, in some capitula all staminate, in others bisexual distally; calyx campanulate, 0.6-0.8 mm, glabrous, rim subtruncate; corolla turbinate 1.7-2.3 mm; filaments white, free or shortly connate, longer ones exserted 2-4 mm, connective of anthers commonly produced beyond thecae as a minute point. Pod 1-3 per capitulum, sessile, in profile linear, either straight or variably, sometimes strongly recurved, 60-100 x ± 9 mm, plano-compressed, replum either smooth or randomly retro-aculeate, forming a raised rim, valves firmly papery, either glabrous or minutely reddish-lepidote, when ripe breaking up into transversely oblong or rectangular, free-falling, individually indehiscent articles of ± 4-8 mm long.

Distribution: Widespread in Amazonia and the Guianas; segregated into 3 varieties, 2 of which occur in the Guianas.

N o t e : *Mimosa annularis* differs from other climbing species in the Guianas which have nectaries on the leaf-stalks: in the acutely rhombic outline of the leaflets, the subulate paraphyllidia at base of each pinna, and the ordinarily narrow, flask-shaped nectaries (not obtusely mounded). It varies greatly in number and size of leaflets.

1a. **Mimosa annularis** Benth. var. **odora** Barneby, Mem. New York Bot. Gard. 65: 35. 1991. – *Mimosa paniculata* Benth., J. Bot. (Hooker) 2: 131. 1840 (non *M. paniculata* J.C. Wendl., 1798, nec Poir., 1810). Type: Guyana, Ro. Schomburgk ser. II, 254 (holotype K!).

Leaflets relatively small and 2(3)-veined from pulvinule; scale-like or claviform reddish trichomes sparse or almost lacking.

D i s t r i b u t i o n : The Guianas, NE Brazil (Amapá and Pará); climbing in riparian forest, near or below 100 m; the large fragrant panicles arising obliquely or pendulous from branch-tips; local or seldom collected (GU: 2; SU: 2; FG: 1).

S e l e c t e d s p e c i m e n s : Guyana: s. loc., Ro. Schomburgk 64.S (K). Suriname: Lawa R., affluent of Marowijne R., Kappler 2176 (U); Courantijn R., Hulk 28 (U). French Guiana: Haut Oyapock, Trois Sauts, Grenand 495 (P).

N o t e : See note to *M. guilandinae* and *M. rufescens*.

1b. **Mimosa annularis** Benth. var. **xinguënsis** (Ducke) Barneby, Mem. New York Bot. Gard. 65: 36. 1991. – *Mimosa xinguënsis* Ducke, Arch. Jard. Bot. Rio de Janeiro 4: 32. 1925. Lectotype (designated by Barneby 1991: 36): Brazil, Pará, on Rio Xingú near Altamira, Ducke RB 16818 (hololectotype RB!).

Leaflets relatively ample and mostly 3-4-veined from pulvinule; reddish peltate or claviform trichomes either lacking or scattered over dorsal face of leaflets, usually plentiful on ovary and pod-valves.

D i s t r i b u t i o n : Western Amazonia in Peru, Colombia, and Brazil, but extending downstream to Pará, and apparently disjunct on the Oyapock and Approuague rivers in French Guiana; climbing into and over trees, especially along river and lake shores and in forest clearings or tree-falls, below 250 m (FG: 2).

Selected specimens: French Guiana: Station Nouragues, Prévost 2504 (NY); Camopi R., entre Roche Habillée des Dames et Saut Chien, Oldeman 2541 (CAY, NY).

2. **Mimosa bimucronata** (DC.) Kuntze, Rev. Gen. Pl. 1: 198. 1891. – *Acacia bimucronata* DC., Prod. 2: 469. 1825. Type: Brazil, Raddi s.n. (holotype G!).

Slender trees 2-10 m, armed on some sterile branches and some leaf-stalks with stout internodal aculei 1-8 mm long, but most flowering branchlets unarmed, new stems and all axes of efoliate terminal panicle of small white capitula thinly puberulent and granular, sublustrous leaflets often ciliolate. Stipules lanceolate, 2-5 mm, persistent; leaf-stalks 4-9 cm, petiole 7-24 mm, longer interpinnal segments 7-16 mm; pinnae mostly 5-9 pairs, rachis of longer ones 3.5-7 cm, longer interfoliolar segments 1.4-2.5 mm; leaflets of longer pinnae 24-32 pairs, linear or linear-oblong from obtusangulate base, acute or abruptly apiculate, straight or slightly arched forward, longer ones 6.5-10 x 1-2.3 mm, midrib displaced to divide blade ± 1:2, weakly 1-3-branched, prominulous only on dorsal face. Panicle exserted from foliage 1-4 dm, peduncles 1-4 per node, 6-16 mm; capitula without filaments 5-6 mm diam., prior to anthesis moriform; flower buds glabrous or almost so; bracts less than 1 mm. Flowers 4-merous diplostemonous, fragrant, some lower ones staminate; calyx campanulate, 0.5-0.9 mm, teeth less than 0.2 mm; corolla 2-2.8 mm, nearly plane lobes ± 1 x 0.7 mm; filaments white, free, exserted 4.5-6 mm. Pod 1-5 per capitulum, in profile broad-linear, straight or gently decurved, cuneately contracted at base into a stipe 1.5-5 mm, body plano-compressed, 35-65 x 5.5-7.5 mm, slender replum constricted only where an ovule aborts, valves papery, reddish-brown but finally nigrescent, glabrous but sometimes remotely granular, breaking up into 5-8(9) transversely oblong or almost square segments of 5-7 mm.

Distribution: Native on the coastal plain and Atlantic foothills of Brazil (Bahia), S-ward to Paraguay and NE Argentina; along streams, around ponds, and along ditches, at low elevations; cultivated as a hedge and thereby randomly dispersed and naturalized in Venezuela, on Jamaica, in tropical Africa and SE Asia; introduced and semi-spontaneous around Georgetown, Guyana, and possibly elsewhere in the Guianas (GU: 1).

Representative specimen examined: Guyana: near Georgetown, Jenman 5254 (NY).

3. **Mimosa camporum** Benth., J. Bot. (Hooker) 2: 130. 1840. Type: Guyana, Ro. Schomburgk ser. I, 725 (holotype K!, isotypes F!, M!, NY!, US!, W!).

Mimosa flavescens Splitg., Tijdschr. Natuurl. Gesch. Physiol. 9(2): 110. 1840. Type: Suriname, Jodensavanne, Splitgerber s.n. (isotypes BM!, W!).
Mimosa flaviseta Benth., London J. Bot. 5: 90. 1846. Type: Suriname, Hostmann & Kappler 813 (holotype K!, isotypes NY!, U).

Monocarpic herbs, commonly 3-12(15) dm at maturity, sometimes ephemeral and fruiting when only 2-10 cm, or in age lignescent at base; stems usually erect, sometimes diffuse-assurgent, either simple or branched, together with leaf-stalks and peduncles at once puberulent and hispid with erect yellowish, finely tapering setae variable in length and caliber, longest 1-4.5(6) mm and some often dilated basally, small, subvertically imbricate leaflets mostly glabrous facially but setose-ciliate; small ovoid or subglobose capitula shortly pedunculate in leaf-axils. Stipules broadly lanceolate or ovate-acuminate, mostly 3-9 x 1-2.5 mm, striately (7)12-25-veined, ciliate, persistent; leaf-stalks of larger leaves 1.5-7(8.5) cm, petiole hardly longer than interpinnal segments; an acicular or linear spicule 0.3-2.4 mm between each pair of pinnae; pinnae of longer leaves in robust plants 2-6(9) and in depauperate ones only 1-2 pairs, rachis of longer ones mostly 3-7 cm; leaflets of longer pinnae 16-38 (in depauperates only 5-12) pairs, blades linear or linear-oblong from obtusangulate base, either obtuse or apiculate, longer ones 4-10(11) x 0.8-2.2 mm, all veinless above, 4-6-veined beneath. Peduncles commonly solitary, sometimes paired or paired with a branchlet, 2-20 mm; capitula without filaments 3-7 x 2-7 mm, prior to anthesis cone-like and hispid; bracts resembling stipules in texture and venation, lance-elliptic 1.7-4.5 x 0.5-1 mm. Flowers 4-merous diplostemonous; calyx membranous 0.15-0.3 mm, subtruncate rim glabrous or minutely setulose; corolla 1.4-2.3 mm, lobes thinly setulose dorsally or rarely glabrous; filaments lilac-pink fading whitish, free, exserted 2-4.2 mm. Pods ascending in dense clusters, sessile, in profile oblong or oblong-elliptic, (5)7-15 x 3-4 mm, (1)2-3(4)-seeded, slender replum and papery valves both puberulent and hispid, replum at maturity splitting at apex to release free-falling, individually dehiscent articles of 3-6 mm long.

Distribution: Widespread in tropical lowland America, from S Mexico to NE Bolivia, E to NE Brazil, the Guianas, and Lesser Antilles; colonially weedy in forest clearings, in cultivated ground, on roadsides, and in disturbed brush-savanna, now mostly ruderal; in the Guianas locally common on the coastal plain of Guyana and Suriname, not seen from French Guiana but known from adjoining Amapá (GU: 1; SU: 4).

Selected specimens: Suriname: along the road from Bergendal to Afobaka, S of Blauwe Berg, Distr. Brokopondo, Wessels Boer 423 (NY); experimental farm Coebiti, Distr. Saramacca, Everaarts 628 (NY).

Vernacular name: Suriname: singsing-tapikoto.

4. **Mimosa casta** L., Sp. Pl. 518. 1753. Type: "India", a mistake for West Indies. Lectotype (designated by Wijnands, The Botany of the Commelins 1983: 151): Hb.-Cliffort. 208.2 (a cultivated specimen) (hololectotype BM!).

Prickly vines, procumbent when unsupported, attaining 2-4 m, armed on ribs of stem and leaf-stalks with files of recurved, stramineous-tipped aculei 1-3.5 mm, otherwise either glabrous or puberulent throughout, thin-textured leaflets sometimes appressed-setose dorsally, finely setose-ciliate; small globose capitula solitary or 2-3 together in leaf-axils. Stipules lance-acuminate 2.5-4.5 mm, 2-3-veined, persistent; leaf-stalks 4-10.5 cm; pinnae 1 pair, rachis (2)2.5-5 cm, interfoliolar segments 5-12 mm; leaflets (3)4-6 pairs, varying from subfalcately lanceolate to obliquely elliptic or ovate-elliptic from deeply semicordate base, deltately acute or acuminulate, distal pair 20-36 x 5-16 mm, sharply 5-7-veined dorsally, midrib gently incurved, dividing blade ± 1:2, ascending cilia free through 0.3-1 mm. Peduncle 6-20 mm; capitula without filaments 5-6 mm diam., prior to anthesis moriform; flower buds glabrous; bracts 1.5-2.5 mm, glabrous or minutely ciliolate. Flowers 4-merous, haplostemonous, either all bisexual or some lower ones staminate; calyx membranous 0.15-0.25 mm, obscurely toothed; corolla 1.7-2.6 mm, lobes shallowly concave, 1-veined; filaments pink or whitish, free, exserted 4-7 mm. Pod 1-12 per capitulum, subsessile, in profile oblong, obtuse mucronulate, straight or almost so, 20-40 x 10-14 mm, 3-5(6)-seeded, slender replum armed on back and sides with straight tapering setae 2.5-4 mm, papery valves glabrous but sometimes setulose over each seed, when ripe breaking up into free-falling indehisced articles of 5-8 mm long.

Distribution: Windward Islands, Trinidad, Panama, Venezuela, Colombia, NE Brazil; in thickets, climbing over hedges, and on rock-outcrops, below 200 m; collected near Cayenne, French Guiana (FG: 1).

Representative specimen examined: French Guiana: s. loc., Leblond 148 (G, P).

N o t e : *Mimosa casta* closely resembles *M. schrankioides* in habit and armature, but differs in leaflets about one half as many and twice larger. Its native status in the Guianas requires confirmation, although probable, as the species is well established on Trinidad and Tobago and known from NE Brazil by several collections.

5. **Mimosa debilis** Humb. & Bonpl. ex Willd., Sp. Pl. 4: 1029. 1806. Type: Venezuela, Sucre or Monagas, near Caripe, Humboldt & Bonpland 299 (holotype B-Willd., seen in microfiche!, isotype P-HBK!).

Mimosa adhaerens H.B.K., Nov. Gen. Sp. 6(qu): 249. 1824. Type: Venezuela, on the Orinoco, Humboldt & Bonpland s.n. (holotype P-HBK!).
Mimosa notata Steud., Flora 26: 758. 1843. Type: Suriname interior, Hostmann & Kappler 1205 (holotype P!, isotypes MO!, NY!, W!).
Mimosa hostmannii Benth., London J. Bot. 5: 84. 1846. Type: Suriname, Hostmann 1233 (holotype K!, isotype BM!).

Herbs and weakly frutescent subshrubs, at first or permanently erect, or as often diffuse, scrambling, or sarmentose, potentially flowering precociously and appearing monocarpic, at maturity commonly 8-20 dm, armed on stems and leaf-stalks with scattered recurved aculei 1-3.5 mm and usually hispid with fine sordid setae up to 1-2 mm mixed with fine puberulence (but setae and hairs and aculei sometimes independently lacking), leaflets subequally strigose on both faces or on dorsal face only; capitula solitary or geminate, rarely 3-4-nate, globose, forming an efoliate or only basally leafy bracteate pseudoraceme exserted ± 1-4 dm from foliage. Stipules lanceolate or ligulate 3-8 x 0.7-2 mm, 3-5(9)-veined, persistent; leaf-stalks 1-6.5 cm; pinnae 1 pair, rachis including pulvinus 4-20 mm; leaflets 2 pairs, anterior of first pair much smaller than rest or reduced to a lanceolate paraphyllidium, distal pair longest, blades obliquely obovate or oblanceolate from semi-cordate base 2-6.5 x 0.6-3 cm, obtuse mucronulate or deltately apiculate, all palmately 3-5-veined from pulvinule, margin cilate with forwardly appressed setae free for ± 0.6-1.5 mm, venation immersed on upper face, usually prominulous and pallid beneath. Peduncle 5-22 mm; capitula without filaments 4.5-7.5 mm diam., prior to anthesis either cone-like or moriform; bracts 0.6-3.5 mm, either densely or residually setulose-ciliolate, persistent. Flowers 4-merous haplostemonous, most or all bisexual; calyx 0.15-0.4 mm, rim glabrous or minutely ciliolate; corolla narrowly funnelform or subtubular 2-2.6 mm, lobes concave, 1-veined; filaments lilac-, rose- or magenta-pink, free, exserted 4-7.5 mm. Pods radiating in a dense cluster, sessile, in profile oblong-elliptic or broad-linear 10-18 x 3-5 mm, 2-4-seeded, almost straight replum armed on back and sides with erect, brown or purple-tipped setae up to ± 1.5-4 mm, valves thin, greenish or later stramineous to purplish-brown variably setose or both puberulent and setose, when ripe breaking up into free-falling articles of 3-5 mm long.

Distribution: Interruptedly widespread over the Orinoco and Amazon basins and extending sporadically to Costa Rica, Bolivia, Paraguay, and SE Brazil; in savanna and brush-woodland, becoming weedy in disturbed places, along fencerows, and in pasture thickets, from below 50 m up to 1600 m; Suriname, to be expected in lowland Guyana (SU: 2).

Notes: *Mimosa debilis* is distinguished in the Guianas by exactly 8-foliolate leaves, some of which may appear 6-foliolate by reduction of the anterior leaflet of the first pair to a minute blade or paraphyllidium. Possibly to be expected in the Guianas is the related *M. sensitiva* L., erratically widespread in eastern S America both north and south of the Guianas. This resembles *M. debilis* in leaf-pattern, but has stems and leaf-stalks armed with dense files of cat's-claw prickles, leaflets obliquely lanceolate or half-ovate and usually acute (thus broadest near or below, not beyond mid-blade), and a paleaceous-pappiform calyx. In both these species the leaves are highly sensitive, the leaf-stalk reflexing and the leaflets folding together when touched.

6. **Mimosa diplotricha** C. Wright ex Sauvalle, Anales Acad. Ci. Méd. Habana 5: 405. 1868. Type: Cuba, Santa Cruz de los Pinos, Wright 3541 (holotype GH!, isotypes K!, NY!, US!).

   Mimosa invisa Mart., Flora 20(2), Beibl. 8: 121. 1837, a posterior homonym of *M. invisa* Mart. ex Colla, 1834. Lectotype (designated by Barneby, Brittonia 39: 50. 1987): Brazil, Guanabara, Luschnath in Mart., Herb. Fl. Bras. 172 (hololectotype BR!, isolectotypes BM!, G!, K!).
   Schrankia brachycarpa Benth., J. Bot. (Hooker) 2: 130. 1840. Type: Brazil, on the Rio Negro, Ro. Schomburgk ser. I, 903 (holotype K!), not *Mimosa brachycarpa* Benth., 1842.

Potentially suffrutescent, diffuse or scrambling herbs (0.5)1-2.5 m, usually densely armed on ribs both of stem and leaf-axes with recurved stramineous aculei 1-3.5 mm and often also villosulous (exceptionally glabrous), thin-textured leaflets either glabrous or villosulous on both faces, nearly always minutely ciliolate; short-pedunculate globose capitula all axillary or latest ones pseudoracemose. Stipules lance-attenuate or setiform, 2-7 x 0.25-0.6 mm, 1-veined, persistent; leaf-stalks 4-10.5(12) cm, petiole mostly 2.5-7 cm and ± half as long as or longer than rachis, this 2-5.5 cm, its interpinnal segments 3-8(10) mm; pinnae of larger leaves 4-9 pairs, very sensitive and in most dried specimens deflected back-to-back behind leaf-stalk, rachis of longer ones 1.5-4 cm, its longer interfoliolar segments 0.5-2 mm; leaflets of longer pinnae 17-28 pairs, blades in outline linear-oblong or linear, subacute or apiculate, those near mid-rachis 2.5-6(7) x 0.6-1.3 mm, all veinless above, faintly 1-veined dorsally by very slender subcentric midrib. Peduncle solitary or 2-3 together, 5-16(20) mm; capitula without filaments 4-6 mm diam., prior to anthesis moriform; bracts 1 mm or less. Flowers normally 4-merous, diplostemonous, either all bisexual or few lowest ones staminate; calyx campanulate, 0.2-0.4 mm, glabrous, rim entire or obscurely denticulate; corolla narrowly turbinate, 1.7-2.8 mm, lobes concave 1-veined, reddish or greenish, either glabrous or thinly hispidulous externally; filaments pink, obscurely monadelphous at very base, exserted 2-5 mm. Pods usually 10 or more per capitulum, crowded into a dense umbelliform cluster close against stem, sessile, in profile oblong or linear, straight or gently decurved, (8)10-25 x 3-4.5 mm, (2)3-8-seeded, slender replum aculeate along its 3 ribs with erect or decurved prickles 0.4-1.3 mm, valves papery either villosulous, or stiffly aculeate, or both, when ripe breaking up into free-falling, individually dehiscent, rhombic or square articles of 3-3.5 mm long.

Distribution: Discontinuously dispersed from NE Argentina and SE Brazil to SW tropical Mexico and the Greater Antilles, and pestilently weedy in tropical Africa, Pacifica and N Australia; in pastures and disturbed woodland, fallow or presently cultivated fields, at forest margins, and becoming weedy on roadsides and in waste places, mostly below 300 m; apparently not common in the Guianas (SU:1; FG: 1).

Representative specimens examined: Suriname: Boven Sipaliwini, kamp 17, Rombouts 418 (NY). French Guiana: Fleuve Lawa, Sastre 8160 (P).

Notes: At anthesis *Mimosa diplotricha* closely resembles *M. quadrivalvis* var. *leptocarpa* in habit, armament, and small, short-pedunculate capitula, but differs in more numerous and more crowded pinnae; the small, laterally compressed pods borne in dense clusters are instantly recognizable. The species varies in density of armament, and a completely smooth variant (var. *inermis* (Adelb.) Verdc.) is cultivated for cover and fodder in Africa and the East Indies.
A confusion of long standing between *M. invisa* Mart. (1837), the name under which *M. diplotricha* has passed in the literature since 1875, and *M. invisa* Mart. ex Colla (1834), heterotypic homonyms based on examples of different species, was recently dispelled by Barneby (Brittonia 39: 50. 1987).

7. **Mimosa dormiens** Humb. & Bonpl. ex Willd., Sp. Pl. 4: 1035. 1806. Type: Venezuela, near mouth of Río Apure, Humboldt & Bonpland 819 (holotype B-Willd., seen in microfiche!, isotype P-HBK!).

Either simple erect or diffusely branched, monocarpic herbs; armed on internodes with horizontal stramineous aculei 2-6 mm, near middle of each interpinnal segment of leaf-stalks with a pair of similar ones, and between each pair of pinnae with an ascending vulnerant spicule up to 3-17 mm, leaf-stalks and peduncles at once puberulent and variably hispid-pilose or strigose with setae up to 1-3.5 mm, leaflets either glabrous or dorsally puberulent; plumply ellipsoid capitula axillary, much surpassed by leaves. Stipules herbaceous, ovate-acuminate 2-4.5 mm, prominulously 5-8-veined, setose-ciliate, persistent; leaf-stalks 2-9 cm, longer interpinnal segments (4)6-20 mm; pinnae 3-7 pairs, interfoliolar segments 0.6-1.7 mm; leaflets of longer pinnae 10-17 pairs, blades linear from obtusangulate base, obtuse or minutely apiculate, longer ones 4-8.5 x 0.8-1.5 mm, on dorsal face 4-6-veined from pulvinule, veins on either side of moderately displaced midrib produced almost to blade's apex. Peduncle 0.8-4 cm; capitula without filaments 4-7.5 x 3.5-4.5 mm, prior to anthesis cone-like; bracts finely setose-ciliate. Flowers 4(5)-merous, diplostemonous; calyx obsolete; corolla campanulate ± 1.5 mm, lobes thinly scabrous-setulose; filaments free, pale lilac fading whitish, exserted 2-3 mm. Pods usually several from near mid-receptacle, in profile asymmetrically obovate from broad-cuneate base, obliquely rounded or subtruncate at apex, 7-11 x 6-8 mm, 2(3)- or by abortion 1-seeded, slender replum shallowly constricted near middle, produced into an erect terminal cusp, papery valves weakly hispid, separating at maturity from apically ruptured replum and breaking into 2(3) free-falling, individually indehiscent articles sealed along line of fracture by a septum 0.2-0.45 mm wide.

Distribution: Interruptedly widespread in the Orinoco and Amazon basins, E in Brazil to Maranhão, N to Mexico; on muddy shores, in seasonally flooded savannas, and in openings of varzea forest mostly below 150 m; reported from Guyana by Bentham (1875) but no specimens seen, nevertheless to be expected on the Guianan coastal plains.

Note: This species is recognized with ease by the combination of annual root, low stature, sharp spicules arising from between each pair of pinnae, and ordinarily 2-seeded pod.

8. **Mimosa guilandinae** (DC.) Barneby, Brittonia 37: 85. 1985. – *Acacia guilandinae* DC., Prod. 2: 465. 1825. Type: French Guiana, Cayenne, Patris 26 (holotype G-DC!).

Vines and sarmentose shrubs, potentially attaining 16 m or more but sometimes precociously flowering at 1 m upward; armed on young stems and some leaf-stalks or inflorescence-axes with recurved aculei 0.2-2 mm and pilosulous or not with sordid hairs up to 0.1-0.4 mm, at least lower face of leaflets scabrellous with red or brownish stelliform trichomes, broad thin-textured leaflets dark brown or purplish on upper face, paler brown or golden-olivaceous beneath; pyramidal, efoliate or proximally few-leaved terminal panicle of small globose, white or greenish-yellow capitula exserted 2-5 dm from foliage. Stipules triangular-subulate up to 2 mm, or very small, or obsolete; leaf-stalks 5.5-19 cm, petiole 3.5-10 cm, longer (or only) interpinnal segment 1.5-8 cm; nectary near base of leaf-stalk obese verruciform or pyramidal, broad-based and narrow-pored, 0.7-1.5 x 2-3.5 mm, similar but smaller nectaries near tip of pinna-rachises; pinnae 2-3(4) pairs, rachises of furthest pair 2-9.5 cm, longer interfoliolar segment (when present) 1.5-6.5 cm; paraphyllidia at base of pinna-rachis either conic and less than 0.3 mm or more often lacking; leaflets 1-2(3) pairs, blades (in the Guianas) obovate-suborbicular or obovate-elliptic obtuse, largest 2.5-12 x 2-7.5 cm, all 3-4-veined from pulvinule, forwardly incurved midrib 3-7-branched on each side, venation discolored but immersed on upper face of blade, prominulous beneath. Peduncles fasciculate by 2-7, rarely solitary, longer ones 3-10 mm, randomly 1-3-bracteolate, fascicles subtended either by a rudimentary leaf-stalk bearing a nectary or by a bract; capitula without filaments 2.5-5 mm diam., prior to anthesis moriform, green or yellowish flower buds either glabrous or ashy-puberulent; bracts less than 1 mm, persistent. Flowers white, ochroleucous, or greenish, fragrant, either 4- or 5-merous, diplostemonous, mostly bisexual; calyx campanulate 0.4-1 mm, either glabrous or puberulent externally, rim obscurely deltate-denticulate; corolla turbinate 1.3-2.3 mm, 1-veined lobes at anthesis either ascending or recurved; filaments white, free or rarely united through nearly 1 mm, exserted 1-2.5 mm. Pod 1-3 per capitulum, subsessile, in profile oblong or broad-linear, abruptly contracted at both ends, 7-11 x 2.5-3 cm, plano-compressed, 8-16-seeded, replum sometimes shallowly undulate either smooth or randomly retro-aculeolate, valves papery lustrous, finely brownish-stramineous, colliculate over each seed, thinly or densely red-lepidote, when ripe breaking up into free-falling, individually dehiscent articles of 5-7 mm long and 3-5 times as wide.

Distribution: Widespread over Amazonia and extending N to Panama, in the Guianas in the form of 2, mutually distinct varieties that are neverthless intimately connected by intermediate forms outside the Guianas.

Note: *Mimosa guilandinae* is a racially complex species. It is related to *M. annularis* var. *odora* and *M. rufescens*, see note to *M. rufescens* for commentary.

8a. **Mimosa guilandinae** (DC.) Barneby var. **guilandinae** – Fig. 31

Pinnae 2-3(4) pairs and leaflets of distal pinnae 2(3) pairs, blades thinly lepidote dorsally with sessile scales, otherwise glabrous or almost so, largest ones 3.5-6.5 cm. Flowers 5-merous glabrous.

Distribution: French Guiana and Brazil (Amapá); in disturbed lowland forest and at forest or creek margins, below 300 m, locally plentiful on and near the coastal plain and along lower reaches of rivers (FG: 3).

Selected specimens: French Guiana: Camp Petit-Approuague, route de Bélizon, Bassin de l'Orapu, Hoff *et al.* 6294 (NY); 40 km northeast of Cayenne on the Piste de Risque-tout near Montsinery, Harley *et al.* 24785 (NY).

Note: *Mimosa guilandinae* was mistaken by Bentham for *M. obovata* Benth., which has grossly similar foliage but is a free standing tree with 3-merous flowers and no petiolar nectary, moreover endemic to the cerrado of SE Brazil, and now properly known as *M. laticifera* Rizz. (*M. obovata* Benth., 1842, non Roxb., 1832).

8b. **Mimosa guilandinae** (DC.) Barneby var. **duckei** (Huber) Barneby, Mem. New York Bot. Gard. 65: 46. 1991. – *Mimosa duckei* Huber, Bol. Mus. Goeldi 5: 381. 1909. Type: Brazil, Pará, near Almeirim, Ducke 3447 (holotype MG, seen in F Neg. 39878!, isotypes K!, P!, RB!, S!, US!).

Pinnae 2 pairs and leaflets of distal pinnae 1-2 pairs, blades densely scaberulous with shortly stalked scales, in addition pilosulous along veins dorsally, largest ones 4-12 cm. Flowers mostly 4-merous, gray-puberulous at least on corolla-lobes.

Distribution: Widely scattered in Brazilian Amazonia and apparently isolated on middle Suriname R. in Suriname; climbing into and over trees of regenerating woodland and along margins of virgin forest, mostly below 150 m (SU: 1).

Representative specimen examined: Suriname: Brokopondo Distr., village Brokopondo, van Donselaar 3258 (NY).

Vernacular name: Suriname: akamaka.

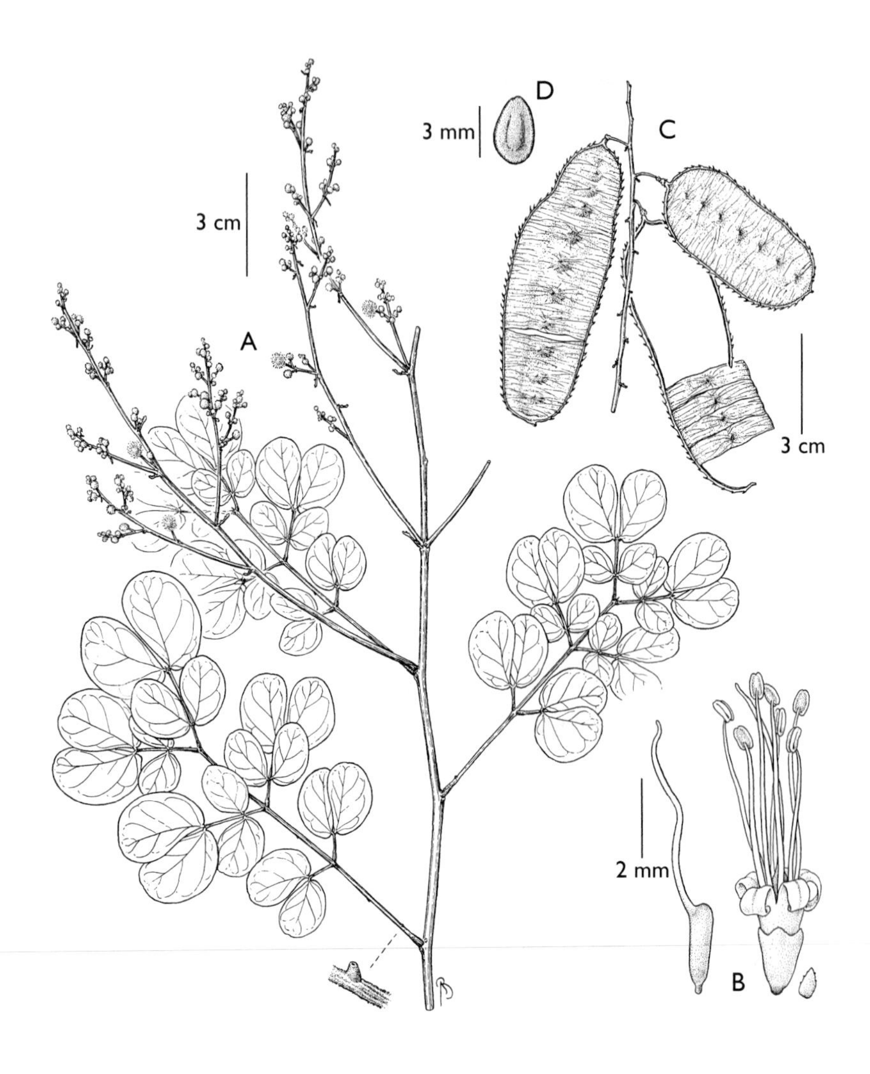

Fig. 31. *Mimosa guilandinae* (DC.) Barneby var. *guilandinae*: A, part of stem showing bipinnately compound leaves and inflorescences; B, gynoecium (left) showing subapical attachment of style and flower (righ) with bract (far right); C, fruits showing partial disperal of articles; D, seed (A-B, Feuillet 1554; C-D, Harley 24785). Drawing by Bobby Angell; reproduced with permission from Mori *et al.*, 2002: 502.

9. **Mimosa invisa** Mart. ex Colla, Herb. Pedem. 2: 255. 1834. Type: Brazil, Martius s.n. (holotype TO, seen in photograph!), – not *M. invisa* Mart., 1837; nor Benth. & later authors, which = *M. diplotricha* C. Wright ex Sauvalle.

Distribution: Discontinuously widespread from Colombia and Venezuela to SE Brazil and Paraguay.

Note: A racially complex species consisting of 4 varieties.

In the Guianas only: var. **spiciflora** (H. Karst.) Barneby, Mem. New York Bot. Gard. 65: 307. 1991. – *Mimosa spiciflora* H. Karst., Fl. Columb. 2: 61, t. 131. 1863. Type: Venezuela, Distr. Federal, near Caracas, Karsten s.n. (holotype W!).

Mimosa brevispica Harms, Notizbl. Königl. Bot. Gart. Berlin 6: 303. 1915. Syntypes: Brazil, Rio Branco, on Serra de Carauma and near Boa Vista, Ule 7726, 8478 (destroyed at B, but fragment of the first survives at F, and a duplicate of the same at K).

Vines and scrambling shrubs attaining 2-5 m or trailing when lacking support; stems and leaf-stalks armed with files of recurved aculei up to 0.5-2 mm and also minutely reddish-granular, otherwise varying from glabrous to densely pilosulous or tomentulose with plain hairs up to 0.2-0.5 mm, small and numerous, thin-textured leaflets ciliolate but varying from glabrous to puberulent on both faces; capitula narrowly oblong or cylindric, mostly fasciculate by 2-4 either in axils of contemporary leaves or forming an efoliate pseudoraceme. Stipules lance-subulate 1.5-5 mm, 1-veined; leaf-stalks (4)7-16 cm, petiole 1-5 cm, interpinnal segments 3-16(20) mm; no leaf-nectaries; pinnae 6-16 pairs, rachis of longer ones 2.5-5.5 cm, interfoliolar segments 0.5-2 mm; leaflets of longer pinnae 26-37 pairs, blades linear or linear-oblanceolate from rectangulate or deeply obtusangulate base, apiculate or deltately acute, longest ± 2-8(9) x 0.4-1.7 mm, all veinless above, beneath 1-2(3)-veined from pulvinule, slender midrib subcentric. Peduncle 3-16 mm; spikes without filaments 4-5.5 mm diam., axis becoming 5-30 mm. Flowers (4)5(6)-merous haplostemonous; calyx campanulate 0.4-0.6 mm, very short teeth often granular-ciliolate; corolla turbinate-funnelform 1.8-2.7 mm; filaments free, pink, exserted 4.5-7.5 mm. Pod in profile broad-linear tapering at base into a stipe 1-5 mm, abruptly cuspidate at broadly rounded apex, when well fertilized 7-14.5 x 1-2.2 cm, (7)10-18-seeded, replum either aculeolate or not, papery valves plano-compressed, either glabrous and granular or thinly to densely puberulent overall, when ripe breaking up into free-falling, individually indehiscent, oblong articles 6-9(10) mm long.

Distribution: Interruptedly dispersed from NE Colombia to Monagas (Venezuela), and disjunct in Upper Rio Branco (Brazil) and Rupununi valley in adjacent Guyana; climbing in thickets or hedges, and trailing in open rocky places, 100-1200 m (GU: 1).

Representative specimen examined: Guyana: Sand Creek, FD s.n. (= WB 84) (NY).

Note: This is the only *Mimosa* species in the Guianas in which spicate capitula with pink filaments coincide with glandless leaf-stalks and broad-linear craspedia.

10. **Mimosa microcephala** Humb. & Bonpl. ex Willd., Sp. Pl. 4: 1039. 1806. Type: Venezuela, on the Orinoco, Humboldt & Bonpland 1163 (holotype B-Willd., seen in microfiche!, isotype, P-HBK!).

Erect and diffusely sarmentose shrubs 1-4(5) m; erratically armed on some or all internodes and on dorsal rib of leaf-stalks with recurved or straight, then perpendicular or antrorse aculei 1-2.5(4) mm, stems and all leaf- and inflorescence-axes strigose with appressed or narrowly ascending, yellowish-brown setae up to 0.5-2 mm attached laterally near or below middle and tapering at each end, and often in addition gray-puberulent, narrow crowded leaflets glabrous facially, ciliolate or not; capitula small, globose or ellipsoid, fasciculate by 2-5 at each node of an efoliate or proximally leafy pseudoraceme or panicle of pseudoracemes exserted 1-2.5 dm from foliage. Stipules lance-triangular 1-5 x 0.6-1.4 mm, either finely 3-5-veined or externally veinless; leaf-stalks of cauline leaves 7-14 cm, petiole and longer interpinnal segments 3-21 mm; pinnae of larger leaves 10-18 pairs, rachis of longer ones 2-5.5 cm, longer interfoliolar segments 0.3-0.9 mm; leaflets of longer pinnae 30-60 pairs, blades linear or linear-elliptic, either acute or obtuse, longer ones 2-6.5 x 0.4-0.9 mm, all finely 3-veined dorsally, midrib subcentric. Peduncles 3-11 mm; capitula without filaments 2.5-4.5 mm diam., prior to anthesis moriform; flower buds either glabrous or puberulent; bracts less than 1 mm, persistent. Flowers 4-merous diplostemonous, fragrant; calyx 0.3-0.5 mm, rim truncate or obscurely denticulate, either glabrous or ciliolate; corolla turbinate 1.6-2.5 mm, lobes plane, either 1- or 3-veined, ± half as long; filaments white (in Guianan varieties), free or almost so, exserted 3.5-5 mm. Pod 1-4 per capitulum, sessile or basally attenuate, in profile linear, linear-elliptic or undulately linear, in Guianan varieties ± 10-12 x 3-5 mm, (2)3-5(6)-seeded, replum and stiffly papery valves strigulose or hispidulous with setae up to 0.5-1.4 mm, when ripe breaking up into rhombic-oblong or elliptic, biconvex, individually dehiscent, free-falling segments 4-7 mm long.

Distribution: Guayana Highland and northern Amazonia, barely penetrating the Guianas; populations isolated on rock outcrops, patches of savanna, and river banks within climax forests; 5 varieties have been recognized, 3 lacking aculei, and 2 aculeate, these both marginally present in the Guianas.

Note: A racially complex species; it may be recognized by the structure of the setae on stems and leaf-stalks, which are laterally attached near or below middle and attenuate at each end, by the small pseudo-racemose-paniculate capitula, and by the narrow fruits.

10a. **Mimosa microcephala** Humb. & Bonpl. ex Willd. var. **cataractae** (Ducke) Barneby, Mem. New York Bot. Gard. 65: 468. 1991. – *Mimosa cataractae* Ducke, Arch. Jard. Bot. Rio de Janeiro 3: 75. 1922. Type: Brazil, Pará, banks of Río Tapajóz near the cataract of Maranhão Grande, Ducke in MG 17070 (holotype MG!, isotypes G!, P!, US!).

Mimosa plumaeifolia Kleinhoonte in Pulle, Receuil Trav. Bot. Néerl. 30: 167. 1933. Type: Suriname, on upper Tapahoni R. near Toeboeberg, Versteeg 796 (holotype U!).

Rachis of longer pinnae 3-5.5 cm; leaflets of longer pinnae (33)40-60 pairs; prickles of stem sharply recurved.

Distribution: Suriname, French Guiana and Brazil (Roraima, Pará, Amapá); on rocky shores, islands in river rapids, outcrops in forest, and becoming sarmentose at forest-margins, 200-500 m, locally plentiful along streams flowing N from the Wilhelmina Mts. in Suriname and along the upper Oyapock R. in French Guiana and adjoining Amapá, discontinuously in Brazil to Pará and Roraima (SU: 3; FG: 1).

Selected specimens: Suriname: Coppename R., Pulle 266 (NY, U); Brokopondo Rapids 3 hours above Pakka Pakka, Maguire 23985 (NY). French Guiana: Oyapock R., "Canari Zozo", Oldeman B2488 (NY).

Vernacular name: French Guiana: queue-lézard, applied also to some prickly *Piptadenia* etc.

Note: The var. *cataractae* was mistaken by Pulle (1906, p. 58) for *Mimosa adversa* Benth., a species of the Brazilian Planalto distinguished by continuous pod-valves.

10b. **Mimosa microcephala** Humb. & Bonpl. ex Willd. var. **lumaria** Barneby, Mem. New York Bot. Gard. 65: 469. 1991. Type: Venezuela, Bolívar, E of Cano Azul, Wurdack & Guppy 178 (holotype NY!).

Rachis of longer pinnae 2-3 cm; leaflets of longer pinnae less than 50 pairs; prickles of stem ascending, straight horizontal, or obscurely recurved.

Distribution: Scattered over Venezuelan Guayana and adjoining SE Colombia, and seemingly isolated on the upper Rupununi R. in Guyana, and on the Tumuc Humac divide in Amapá, Brazil and probably adjacent French Guiana; in habitats of var. *cataractae*, 200-470 m (GU: 1).

Representative specimen examined: Guyana: Upper Takutu - Upper Essequibo Region, Rupununi savannas, Towatawan Mt., Gillespie 1980 (NY).

11. **Mimosa myriadenia** (Benth.) Benth., Trans. Linn. Soc. London 30: 408. 1875. – *Entada myriadenia* Benth., J. Bot. (Hooker) 2: 133. 1840. Type: Brazil, Amazonas, Rio Negro, Ro. Schomburgk ser. I, 917 (holotype K!).

Acacia paniculiflora Steud., Flora 26: 760. 1843. Type: Suriname, interior, Hostmann & Kappler 151 (holotype P, not seen, isotypes K!, MO!).

Potentially high-climbing vines, forming tangled thickets in forest-glades and tree-falls; armed on stems and most leaf-stalks with retrorse aculei, new stems and inflorescence pilosulous with sordid hairs mixed with minute reddish scales, bicolored leaflets either glabrous or pilosulous, dorsally lepidote. Stipules subulate ± 1-3.5 mm, 1-veined; leaf-stalks 6-28 cm, petiole 1.3-4.5 cm, interpinnal segments 5-17 mm; nectary sessile, shallowly patelliform, elliptic, near base of each petiole and smaller ones on some distal segments of both leaf-stalk and pinna-rachises; pinnae 5-33 pairs, rachis of longer ones 2.5-7.5 cm, longer interfoliolar segments 0.4-3.5 mm; leaflets of longer pinnae 16-42 pairs, varying from linear- to rhombic-oblong, either acute or obtuse, longest 3-9.5 x 0.6-3 mm, 1-5-veined from pulvinule. Inflorescence an effuse, efoliate or only basally foliate panicle of loose spike-like racemes of small white fragrant flowers, racemes 2-5 per node, including peduncle 3-7 cm long; pedicels not over 0.5 mm; calyx campanulate 0.5-0.7 mm, either glabrous or puberulent, teeth 0.1-0.2 mm; corolla (nigrescent when dry) 1.3-1.8 mm, haplostemonous; white filaments exserted ± 1.5-2.5 mm. Pod pendulous, shortly stipitate, in profile linear-oblong 5-10 x 1-1.8 cm, 8-14-seeded, valves thinly papery, olivaceous or fuscous dotted with minute red scales, glabrous or rarely pilosulous, breaking up into free-falling, individually dehiscent articles 4-8 mm long.

Distribution: Widely dispersed through the Amazon Basin and the Guianas, extending interruptedly to Pacific Colombia and Costa Rica; represented in the Guianas by 2 (of 5 known) varieties.

Note: This is the only *Mimosa* in which leaf-nectaries coincide with racemose, haplostemonous flowers and lepidote fruits.

11a. **Mimosa myriadenia** (Benth.) Benth. var. **myriadenia** – Fig. 32

Leaflets of longer pinnae 16-26 pairs, separated by intervals 1.5-3.5 mm long, larger blades broadly or narrowly oblong, 5-9.5 x 1.5-3 mm.

Distribution: Locally abundant in the Guianas, adjacent Venezuela, and disjunctly to E Colombia and Costa Rica; in riparian forest, forest glades, surviving in pasture thickets and along roadsides, mostly in lowland non-inundated woodlands but attaining 490 m in Guyana (GU: 2; SU: 3; FG: 2).

Selected specimens: Guyana: Mazaruni Station, FD 3665 (= Fanshawe 929) (NY); Cuyuni-Mazaruni Region, Chi-Chi Mts., Pipoly 10317 (NY). Suriname: Saramacca R. near Mamadam, Florschütz 1150 (NY); Saramacca R. above Kwatta hede, Maguire 23934 (NY). French Guiana: Piste-de-St-Elie, Sauvain 42 (NY); Mt. Galbao, de Granville 8632 (NY).

Vernacular names: Guyana: hiri-hiri-balli. Suriname: kifini-maka, kifoundi-maka.

11b. **Mimosa myriadenia** (Benth.) Benth. var. **dispersa** Barneby, Mem. New York Bot. Gard. 65: 30. 1991. Type: Venezuela, T. F. Amazonas, on the Casiquiari R., Spruce 3183/2058 (holotype NY!).

Leaflets of longer pinnae 24-42 pairs, separated by intervals 0.4-1.4 mm long, larger blades linear-oblong, 3.5-6.5 x 0.6-1.4 mm.

Distribution: Discontinuously dispersed through Amazonia, in the Guianas on the Wilhelmina Mts. in Suriname, and approaching French Guiana on Serra Tumuc Humac in Pará, Brazil; in habitats of var. *myriadenia*.

Selected specimens: Suriname: upper Tapanahoni R., mouth of Paloemeu Cr., Rombouts 670 (NY); Wilhelmina Mts., Lucie R., 2-10 kms below confluence of Oost R., Irwin *et al.* 55377 (NY). French Guiana: Roche ruine Roche Koutou, Bassin du Haut-Marouini, de Granville 9497 (NY).

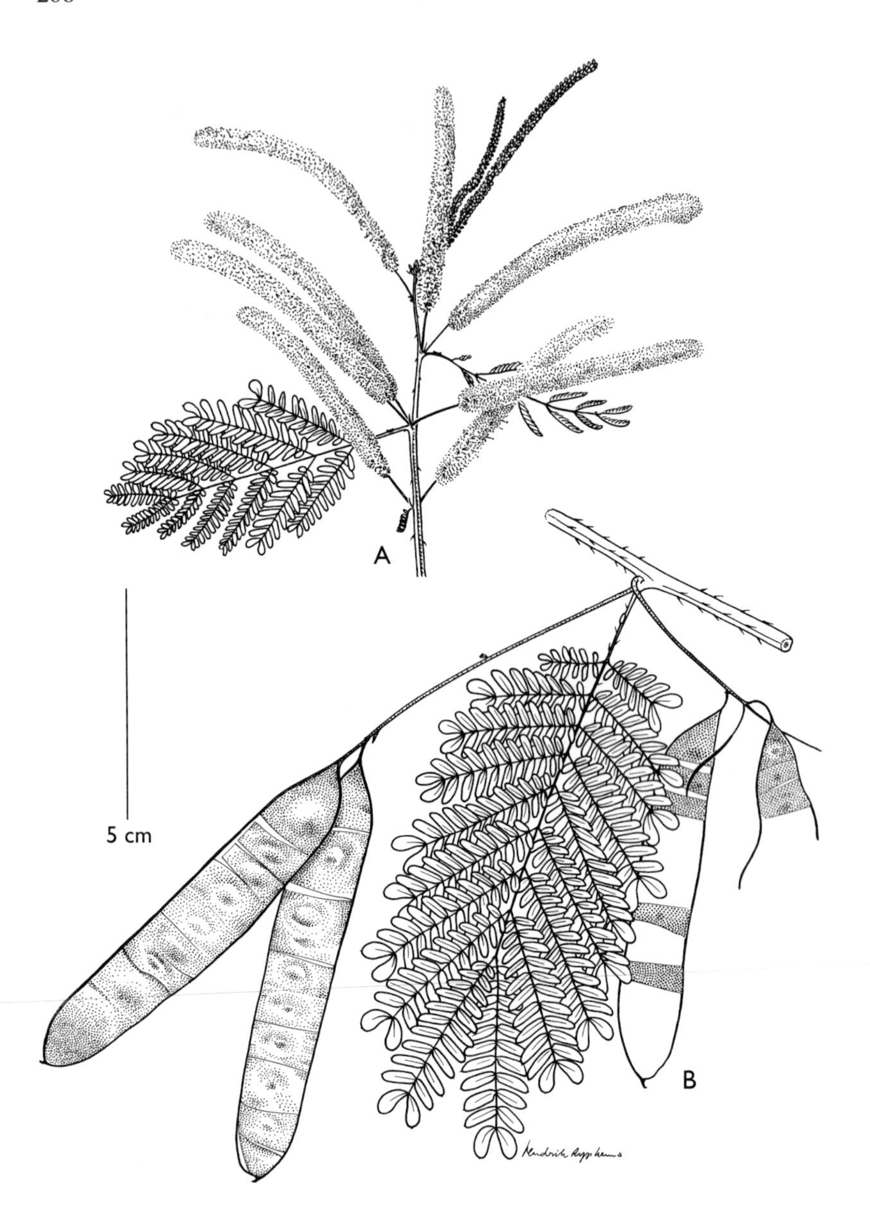

Fig. 32. *Mimosa myriadena* (Benth.) Benth. var. *myriadena:* A, flowering branch; B, fruiting branch (A, Polak 444; B, Hahn 5535). Drawing by Hendrik Rypkema; reproduced with permission from U.

12. **Mimosa pellita** Humb. & Bonpl. ex Willd., Sp. Pl. 4: 1037. 1805. Type: Venezuela, Sucre, Cumaná, Humboldt & Bonpland 90 (holotype B-Willd., seen in microfiche!, isotype P-HBK!).

Mimosa ciliata Willd., Enum. Pl. Hort. Berol. 1048. 1809. Type: Brazil, Sieber s.n. (holotype B-Willd., seen in microfiche!).
Mimosa asperata L. var. *scandens* Ducke, Bol. Técn. Inst. Agron. N. 2: 8. 1944. Lectotype: Brazil, Pará, near Obidos, Ducke 1609 (hololectotype IAN, isolectotypes NY, US.
Mimosa asperata of Bentham, 1875, for the most part, & of most authors since 1900; not L.
Mimosa pigra of most recent authors; not L.

Shrubs of bushy outline or when crowded incipiently sarmentose, maturing at 1-3 m but attaining 5 m; variable in armature and pubescence, stems and all leaf- and inflorescence-axes either strigose or hispid with forwardly appressed, ascending or spreading, scaberulous setae up to 1-3.5 mm often mixed with minute puberulence, stems and commonly also the dorsal or lateral ribs of leaf-stalks erratically armed with straight declined, or recurved, hornlike or broad-based aculei up to 1.5-12 mm, those of leaf-stalks either solitary or opposite on interpinnal segments (randomly wanting), leaflets crowded small, either glabrous or puberulent facially, at least remotely, often densely ciliolate; capitula globose or ellipsoid, solitary or fasciculate by 2-3 in axil of coevally expanding or hysteranthous leaves, becoming immersed in foliage, fruits or empty repla long persistent. Stipules ovate or lanceolate 2-6 x 1-2 mm, silky-strigose dorsally, persistent; leaf-stalks 6-20 cm, petiole and interpinnal segments (3)6-20 mm; a stiffly ascending, glabrous or puberulent spicule 1.5-14 mm between each pair of pinnae; pinnae mostly 8-14 pairs, rachis of longer ones 3-7.5 cm, interfoliolar segments 0.6-1.2 mm; leaflets of longer pinnae 25-44 pairs, blades linear from obtusangulate base, acute or apiculate, larger ones 5-12 x 0.6-2 mm, all veinless above, 4-5-veined beneath, subcentric midrib and parallel lateral veins all finely prominulous. Peduncles 1.5-7.5 cm; capitula without filaments 8-12 x 6-9 mm, prior to anthesis usually moriform but bracts sometimes as long as obtuse, scabrous-setulose flower buds; bracts 1-3.5 mm, setulose-ciliolate; calyx paleaceous, consisting of a shallow brown campanulate cup 0.3-0.6 mm tall surmounted by 4 lobes setose-decompound to variable depth, whole calyx 1-1.8 mm; corolla 4(5)-merous, diplostemonous, narrowly turbinate 3-4.5 mm; filaments pale pink-lilac fading whitish or whitish from the first, obscurely monadelphous at very base, exserted 1.5-5 mm. Pods usually 2 or more and up to 18 per capitulum, in profile linear or broad-linear, abruptly contracted into a stipe 1-5 mm, broadly rounded or abruptly cuspidate at apex, straight or slightly decurved, 4-12 x 0.8-1.5 cm, compressed but turgid, valves densely hispid with straight vertical or proximally erect and distally incurved setae up to

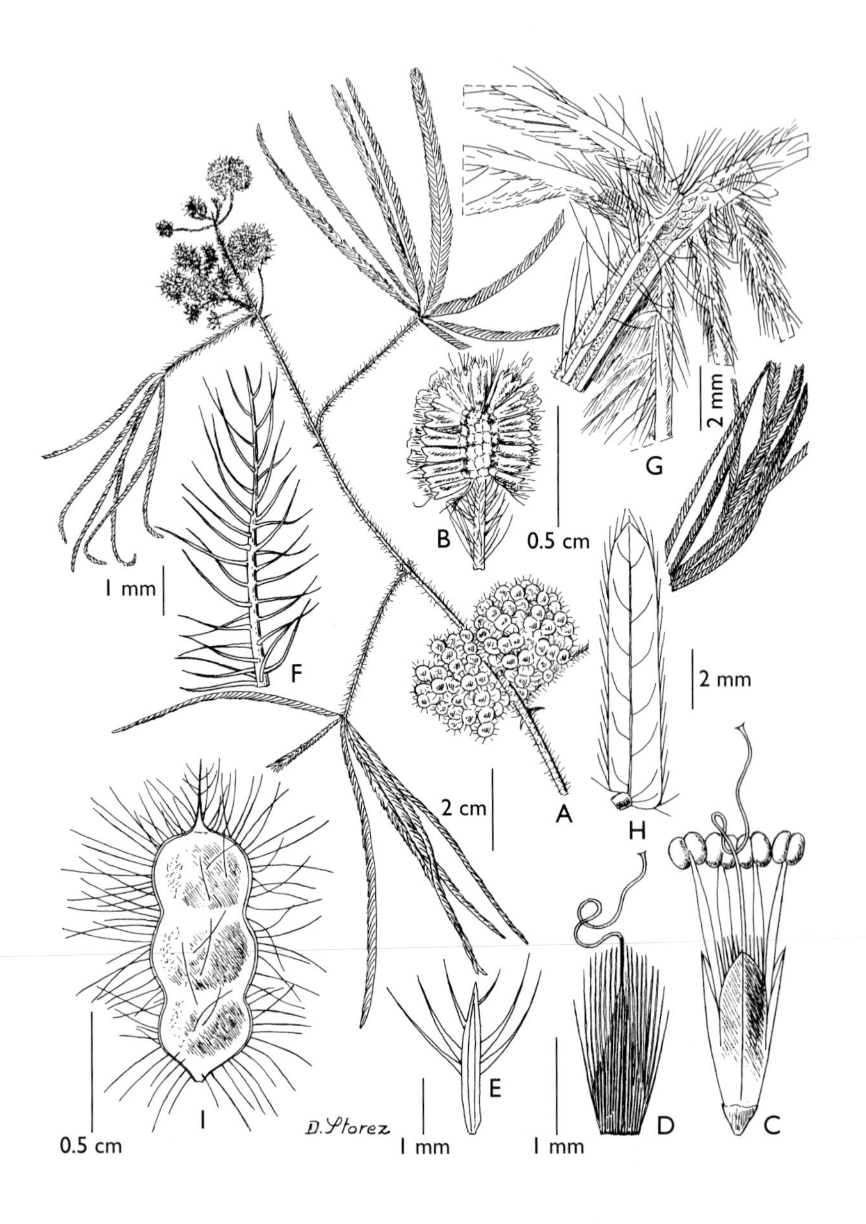

Fig. 33. *Mimosa polydactyla* Humb. & Bonpl. ex Willd.: A, habit: leaves, inflorescences (terminal), fruits (at base of the twig); B, flower head; C, flower; D, very young fruit; E, flower bract; F, stipule; G, leaf: detail of the rachis, petiolules, base of pinnae; H, leaflet; I, fruit (A, van Andel 767; B-I, Sastre 4044). Drawing by Dominique Storez; reproduced with permission from P.

2-4.5 mm, or less often strigose with forwardly subappressed setae only 2 mm, when ripe breaking up into 12-22 free-falling articles, these sealed at each end by a membrane 0.2-2 mm wide and individually indehiscent, readily water-borne.

Distribution: Common and locally abundant nearly throughout the lowland American tropics and warm-temperate S America, the most abundant *Mimosa* of the Guianas; also widespread in tropical Africa and naturalized in SE Asia; on stream banks and seasonally flooded shores, along ditches, in wet savanna, becoming a pest of irrigated land, bushy when solitary or in pure stands but in gallery forest becoming sarmentose, from sea level up to ± 1000 m, exceptionally 1600 m.

Selected specimens: Guyana: Parika, 18 miles W of Georgetown, Hitchcock 16802 (NY); Essequibo Ils.-W Demerara Region, Naamryck Canal, Pipoly 11283 (NY). Suriname: Corantyne R., near Matapie, Florschütz 2267 (NY); Turco Tabbetje, Gonggrijp BW 5331 (NY). French Guiana: vicinity of Cayenne, Broadway 115 (NY); Route de Baduel, Ile de Cayenne, Hoff 5084 (NY).

Note: *Mimosa pellita* is equivalent to the *M. asperata* of older authors who followed Bentham and to the *M. pigra* of modern floras, but has been shown (Barneby, The Davis & Hedge Festschrift 137-142. 1989) to be neither of the species described under these names by Linnaeus. With the aim of avoiding a vexatious change of epithet for this widely dispersed and weedy species, Verdcourt proposed (Taxon 38: 522. 1989) to neotypify *M. pigra* by a specimen of *M. pellita* from Mozambique.

Taxonomic note: Two years after publication of Barneby`s (1991) monograph on *Mimosa*, Verdcourt`s proposal to conserve *Mimosa pigra* L. was accepted and is listed in Appendix 3b to the Tokyo Code (Greuter 1994). For further discussion on this nomenclatural issue refer to Glazier & Mackinder in Kew Bull. 52(2): 459-463. Following Verdcourt, *Mimosa pellita* in the Guianas should be referred to as *Mimosa pigra*.

13. **Mimosa polydactyla** Humb. & Bonpl. ex Willd., Sp. Pl. 4: 1033. 1806. Type: Venezuela, T. F. Amazonas, San Carlos de Río Negro, Humboldt & Bonpland s.n. (holotype B-Willd. = F Neg. 1358!, isotype P-HBK!). – Fig. 33

Diffuse, ascending or sometimes scrambling, in age suffrutescent herbs (2)5-20 dm; armed on stems, at or shortly below nodes, with a pair of recurved aculei 1.5-5 mm and often on internodes with 1 downwardly displaced one, stems and leaf-stalks thinly hispid or strigose with spreading or subappressed setae up to 1-2.3 mm mixed or not with fine puberulence, leaves sensitive, subpalmately pinnate, bicolored, leaflets paler beneath, glabrous on upper or on both faces, setose-ciliate; capitula small ovoid-ellipsoid, solitary or fasiculate in a long succession of leaf-axils on peduncle shorter than subtending leaf-stalk. Stipules linear-attenuate or subsetiform 4-8 x 0.25-0.6 mm, 1-veined, setose-ciliate, persistent; leaf-stalks 2-6(7) cm, petiole 1.5-6.5 cm, 2-4 interpinnal segments commonly 1-2 mm, or almost 0; pinnae of well-developed leaves 3-5 pairs, rachis of longer ones 4-11 cm, interfoliolar segments 1-2 mm; leaflets of longer pinnae 28-55(65) pairs, in outline linear from semicordate base, acute, straight or nearly so, larger ones 4-9 x 0.7-1.8 mm, all veinless above, beneath 2-3-veined from pulvinule, midrib subcentric, weakly 2-4-branched above middle. Peduncle 7-15 mm; capitula without filaments 3-4 mm diam., prior to anthesis moriform, but hispid with emergent bracteal setae; bracts not over 1 mm, 1-veined. Flowers 4-merous, haplostemonous; calyx submembranous 0.2-0.4 mm; corolla 1.1-1.7 mm, lobes either glabrous or minutely hispidulous; filaments pale pink or whitish, almost free, exserted 2-2.5 mm. Pods radiating from receptacle in dense spherical clusters, subsessile, in profile narrowly oblong 9-15 x 3.5-4.5 mm, (2)3-4-seeded, replum slender, undulately constricted, hispid with straight tapering setae up to 2-4 mm, valves papery nigrescent, also thinly or randomly setose and at times minutely puberulent, when ripe breaking up into free-falling biconvex, individually indehiscent articles 2.5-3.5 mm long.

Distribution: Widespread in continental S and C America and Lesser Antilles, common on the coastal plain of the Guianas, less frequent inland, up to 300 m; disturbed forest, brush-woodland, pasture thickets, hedgerows and on ditch-banks.

Selected specimens: Guyana: Amakuna R., NW Distr., de la Cruz 3511 (NY); Mapuera R., FD 7514 (= Guppy 499) (NY). Suriname: above village Kwatta hede, Maguire 23944 (NY). French Guiana: Assici, Bassin du Maroni, Fleury 503 (NY); vicinity of Cayenne, Broadway 274 (NY).

Note: Leaves composed of 6 to 10 pinnae crowded toward the tip of a relalatively long leaf-stalk distinguish this weedy *Mimosa* from sympatric *M. pudica*, which it resembles in flower and fruit.

14. **Mimosa pudica** L., Sp. Pl. 518. 1753. Lectotype (designated by Brenan, Kew Bull. 1955: 185): Brazil, collector unknown (hololectotype BM (hb. Cliffort.)), a sterile specimen not identifiable below level of species and hence (1955) constituting of itself the var. *pudica*.

Herbaceous or weakly suffrutescent, either diffuse, scrambling, or erect-assurgent subshrubs precociously flowering as prostrate or ascending herbs, stems attaining 12 dm but commonly shorter, with rare exceptions (var. *pastoris*); armed at or close to each node with a pair of straight or declined aculei 1-6.5 mm and sometimes also with a few internodal or petiolar ones, stems and leaf-stalks subglabrous, puberulent or hirsute with fine tapering setae up to 1-3.5 mm, leaflets thin-textured, either glabrous or puberulent facially, setose-ciliate; capitula ellipsoid, solitary in some early leaf-axils but verticillate by 2-7 in most distal ones. Stipules lanceolate or linear-attenuate 3-11 x 0.7-2.4 mm, striately (5)7-13-veined, setose-ciliate, persistent; leaf-stalks (1)2-5.5 cm, petiole nearly or quite as long; pinnae of all leaves commonly 2 pairs or of some upper leaves 1 pair, rarely 1 pair in all leaves and exceptionally 3 pairs in some, interpinnal segments, if any, not more than 2 mm; rachis of longer pinnae 3-6.5(8, on some distal branchlets 1.5-3) cm, interfoliolar segments 1-2.5 mm; leaflets of longer pinnae 11-27(36) pairs, in outline linear-oblong from obtusangulate base, straight or subfalcate, abruptly acute-apiculate, longer ones (4)5-13(14) x 1-2.5 mm, all veinless above, beneath (4)5-veined from pulvinule, distally subcentric midrib remotely 3-5-branched on each side. Peduncle 1-3(3.5) cm; capitula without filaments either moriform or conelike and hispid, linear receptacle 5-10 mm; bracts linear-oblanceolate 1-2.5 mm, their margin either glabrous or setose. Flowers 4-merous, haplostemonous, some lower ones staminate; calyx membranous campanulate 0.1-0.2 mm, rim subtruncate or asymmetrically denticulate; corolla narrowly vase-shaped or tubular, or in some bisexual flowers turbinate, bisexual ones 1.5-2.1 mm, concave lobes 0.45-0.7 mm, either glabrous or puberulent externally; filaments pink, free or almost so, exserted 3.5-6 mm. Pods several or up to ± 30 per capitulum, subsessile, in profile undulately narrow-oblong 8-15 x (2.7)3-4.5(5) mm, (2-3)4-seeded, replum slender, hispid dorsally and laterally with tapering setae up to 2-4(5) mm, valves papery, usually glabrous, sometimes densely pilosulous, when ripe breaking up into free-falling biconvex, indehisced articles (2)2.5-4 mm long.

Distribution: A weedy *Mimosa* of American origin now established throughout the tropics of both hemispheres, *M. pudica* has been cultivated since pre-Linnaean times for the curiosity of its irritable leaves, that shrink in response to shock or touch; apart from var. *pudica*, reserved for the sterile type-specimen, it consists of 4 varieties, all present in the Guianas, 1 or 2 of which are probably anthropochorous weeds.

Note: *Mimosa pudica* is close only to *M. polydactyla*, from which it differs in fewer pairs of pinnae and on the average more numerous leaflets per pinna.

14a. **Mimosa pudica** L. var. **hispida** Brenan, Kew Bull. 1955(2): 186. 1955. Type: Java, Junghuhn 719 (holotype K!).

Mimosa striato-stipula Steud., Flora 26: 758. 1843. Type: not found at P; probable isotype, Kappler s.n. from Marowyne R. (NY).

Stems characteristically long-hispid. Floral bracts at and beyond mid-capitulum 1.7-2.5 mm, bent inward near or below middle, distally ciliate with setae up to 1-1.6 mm, just prior to anthesis as long as or longer than subtended flower bud.

Distribution: Nativity uncertain, erratically widespread in the Americas from S Mexico to SE Brazil, in the Old World extensively naturalized and invasively weedy from tropical Africa to Oceania; known mostly from secondary habitats and in cultivation; in the Guianas known only from old collections from Suriname.

14b. **Mimosa pudica** L. var. **pastoris** Barneby, Mem. New York Bot. Gard. 65: 628. 1991. Type: Guyana, Sandcreek on the Rupununi R., Wilson-Browne 23 (holotype NY!).

Stems unarmed. Faces of pod at once densely puberulent overall; seeds densely setose.

Distribution: Known only from SW Guyana, Rupununi R. near 3° S, to be expected in adjacent state of Roraima, Brazil; in damp places on savanna, near 200 m.

14c. **Mimosa pudica** L. var. **tetrandra** (Willd.) DC., Prod. 2: 426. 1825. – *Mimosa tetrandra* Humb. & Bonpl. ex Willd., Sp. Pl. 4: 1032. 1806. Type: Colombia, Magdalena valley, Humboldt & Bonpland 1631 (holotype B-Willd., seen in microfiche!, isotype P-HBK!).

Mimosa hispidula Humb. & Bonpl., Nov. Gen. Sp. 6(qu): 252. 1824. Type: Venezuela, banks of the Orinoco R. near Santa Bárbara, Humboldt & Bonpland 1160 (holotype P-HBK!).

Stems variably hispidulous or loosely strigose. Faces of pod usually glabrous, rarely puberulent or both puberulent and weakly setulose.

Distribution: Widespread in continental N and S America from S Mexico to SE Brazil, uncommon on some West Indian Islands, sparingly naturalized in the Old World; along wayside ditches, in fields and pastures, in disturbed brush-woodland, in vernally wet savanna, and in urban wasteland, mostly in the lowlands but as a weed at elevations up to 1200 m; in the Guianas the commonest variety of *M. pudica*, frequent on the coastal plain, occasional inland.

Selected specimens: Guyana: Marudi Mts., Mazoa Hill near Norman mines camp, Stoffers *et al.* 293 (NY); E Kanuku Mts., 5 miles north of Tionek wau, Cook 117 (NY). French Guiana: Ile de Cayenne, cité Bonhomme, Oldeman B3900 (NY).

14d. **Mimosa pudica** L. var. **unijuga** (Walp. & Duchass.) Griseb., Abh. Königl. Ges. Wiss. Göttingen 7: 211. 1857. – *Mimosa unijuga* Walp. & Duchass., Linnaea 23: 744. 1850. Lectotype (designated by Brenan 1955: 189): Guadeloupe, Duchassaing s.n. (hololectotype GOET, not seen, isolectotype P!).

Stems varying from hispidulous to subglabrous. Pinnae either 1 or 2 pairs. Corolla lobes glabrous externally.

Distribution: Abundant and seemingly native around and within the Caribbean basin, randomly Mexico and Pacific Ecuador, to Amazonian Bolivia and E Brazil, and locally established in SE Asia and Oceania; in habitats of var. *tetrandra*; in the Guianas sporadic on the coastal plain and occasional, presumably native, on savannas of interior Guyana.

Selected specimens: Guyana: N Rupununi Savanna, 2 miles S of St. Ignatius, Goodland & Persaud 747 (NY). Suriname: Paramaribo, Wullschlägel 122 (NY), 1414 (W). French Guiana: Cr. Passoura, Région de Kourou, Cremers *et al.* 10741 (CAY, NY); vicinity of Cayenne, Broadway 684 (NY).

15. **Mimosa quadrivalvis** L.,Sp. Pl. 522. 1753.

In the Guianas only: var. **leptocarpa** (DC.) Barneby, Mem. New York Bot. Gard. 65: 298. 1991. – *Schrankia leptocarpa* DC., Prod. 2: 443. 1825. Type: "Santo Domingo", but perhaps actually from French Guiana, Poiteau s.n. (holotype G-DC!).

Herbaceous or weakly frutescent prickly vines with sensitive leaves, attaining 1-2 m, prostrate in open places or clambering through shrubs; quadrangular stems and ribs of leaf-axes armed with slender recurved aculei up to 1-2.5 mm, otherwise varying from glabrous to finely villosulous and in addition minutely reddish-granular; capitula shortly pedunculate, solitary or paired in a succession of leaf-axils. Stipules linear-setiform 1-8 mm, 1-veined, persistent; leaf-stalks of cauline leaves 4-16 cm, petiole 2.5-6.5 cm, longer of 1-3 interpinnal segments 0.8-3.5 cm; pinnae 2-4 pairs, rachis of longer ones 2-6 cm, interpinnal segments 0.8-2.5 mm; leaflets 12-22(24) pairs, linear or linear-oblanceolate from obtusangulate base, abruptly apiculate, longer ones 5-13 x 1-3 mm, all veinless above, slenderly 2-4-veined beneath, midrib subcentric. Peduncle 4-13 mm; capitula without filaments 4-5 mm diam., prior to anthesis moriform; bracts linear-oblanceolate 0.8-1.4 mm. Flowers (sub)sessile, 5-merous or lower ones 4-merous, all diplostemonous; calyx membranous, shallowly campanulate 0.3-0.6 mm, rim truncate or obscurely denticulate; corolla 1.8-2.3 mm; filaments pink fading whitish, obscurely united at very base, exserted usually 2.5-3 mm, rarely to 5 mm; ovules 20-30. Pods 1-10 per capitulum, ascending sessile, in profile linear, attenuate into a beak ± 6-20 mm, when well fertilized 8-17 cm x 3-4.5 mm, in section obtusely 4-angular, replum 4-5-ribbed, 2-4 mm wide, valves papery, 1-3 mm wide and 1-ribbed lengthwise, ribs of both valves and replum armed with stiff perpendicular, stout setae (1.5)2-4 mm, in addition sometimes minutely granular, valves at maturity separating in one piece; seeds basipetal, contiguous in one series, plumply oblong, obliquely truncate at each end.

Distribution: Widespread over S America E of the Andes, to Bolivia, Paraguay, adjoining Argentina, and in Brazil to Santa Catarina, feebly into SE Mexico and into the Lesser Antilles; in campo, savanna, light woodland, and becoming abundant in disturbed places, along highways or ditches and in abandoned pasture, mostly below 300 m; in the Guianas mostly on the coastal plain where locally common.

Representative specimen examined: French Guiana: vicinity of Cayenne, Broadway 120 (NY).

Note: The var. *leptocarpa* is the common S American representative of the racially complex *Mimosa quadrivalvis* that is most highly differentiated in Mexico and central United States. It is notable in the Guianan flora for the narrow, bluntly quadrangular, many-seeded fruit, the replum of which is as broad as or broader than the valves, which eventually fall away in one piece. The configuration of the fruit, and the seeds truncate by mutual pressure, have been thought to provide the species with generic status as *Schrankia* or *Leptoglottis*, but similar structures are found in other groups of *Mimosa* with clear ties to the center of the genus. The mimosas in the Guianan flora most nearly resembling *M. quadrivalvis* when the fruit is lacking are *M. invisa* and *M. diplotricha*. The former differs by more finely divided leaves, oblong or subcylindric (not globose) capitula, and haplostemonous flowers; the latter by more numerous pinnae and leaflets, and by fewer (4-8, not 20-30) ovules per ovary.

16. **Mimosa rufescens** Benth., Trans. Linn. Soc. London 30: 418. 1875. Lectotype (designated by Barneby 1991: 41): Brazil, Amazonas, São Gabriel da Cachoeira, Spruce 3096 (lectotype K!, isolectotypes C!, E!, F!, G!, NY!, RB!).

Mimosa micracantha Benth. var. *plurijuga* Ducke, Notizbl. Bot. Gart. Berlin-Dahlem 11: 583. 1932. Lectotype (designated by Barneby 1991: 41): Brazil, Amazonas, São Gabriel da Cachoeira, Ducke s.n. in RB 23245 (hololectotype RB!).

Lianas potentially 12-35 m long but precociously flowering as diffuse or sarmentose shrubs 3 m upward; armed on trunk but young branchlets, leaf-stalks and axes of amply effuse panicles of small white capitula only sparsely so with recurved prickles 1 mm or less, young growth densely minutely rufous-lepidote thoughout and commonly also puberulent or pilosulous, leaves conspicuously bicolored, leaflets brown above, paler brown or olivaceous and reddish-lepidote beneath, there proximally barbellate along primary veins; terminal panicles exserted 1.5-4 dm from foliage. Stipules subulate 0.5-2 mm, early deciduous (perhaps sometimes 0); leaf-stalks (5)7-20 cm, petiole 2.5-8 cm, longer interpinnal segments 1-3.5(4.5) cm; nectary pyramidal, low-conical or low-mounded, small-pored, ± 1-1.5 mm diam. near base of petiole and similar smaller ones near tip of all leaf-stalks and of most pinna-rachises; pinnae (3)4-10 pairs, rachises of longer distal ones 4-8 cm, longer interfoliolar segments 7-20 mm; leaflets of longer pinnae (3)4-9(11) pairs, strongly accrescent distally, blades passing upward from small rhombic to much larger and broadly obliquely rhombic-obovate or almost half-obovate, inequilaterally cuneate at base, broadly obtuse, terminal pair (14)16-45 x 8-35 mm, all ± 3-veined from pulvinule, midrib slightly displaced, straight or incurved, 3-5-branched on each side, venation prominulous only dorsally. Peduncles fasciculate by 3-7, longer ones 4-9 mm, minutely bracteolate; capitula without filaments 3.5-5 mm diam., prior to anthesis moriform; flower buds puberulent or rarely glabrous; bracts less than 0.7 mm, persistent. Flowers (4)5-merous diplostemonous, all or nearly all bisexual, sweetly fragrant; calyx submembranous campanulate 0.4-0.8 mm, rim minutely deltate-denticulate; corolla white or ochroleucous, turbinate-campanulate 1.4-2 mm, lobes faintly 1-veined, recurved at anthesis; filaments white, longest exserted 1-2 mm. Pods usually solitary, sessile or abruptly contracted into a stipe less than 2 mm, in profile oblong or broad-linear, straight or subdecurved, plano-compressed, when well fertilized 6-12 x 1.7-3.3 cm, 10-16-seeded, replum shallowly crenate forming an elevated rim around papery, lustrous brown-olivaceous valves, these raised as a papilla over each seed, minutely red-lepidote overall but otherwise glabrous, when ripe breaking up into free-falling, individually indehiscent, narrowly oblong articles ± 3-5 times wider than long.

Distribution: Widespread and common almost throughout the Amazon Basin and Venezuelan Guayana; in primary and disturbed forest, in forest clearings, and becoming colonial in logged or burned forest and on roadsides, mostly below 500 m but ascending to 900 m in Andean foothills of Peru; in Guyana along the Essequibo R., to be expected elsewhere in the Guianas.

Representative specimen examined: Guyana: Waramadong-Kamarang, Essequibo, Persaud 197 (NY).

Note: *Mimosa rufescens* is closely related to *M. guilandinae*, yet distinguished at a glance by the more numerous and smaller leaflets. In size and number of leaflets it more nearly resembles *M. annularis* var. *odora*, but differs decisively in broadly obtuse leaflets and in details of the leaf-nectaries and paraphyllidia mentioned in the key to species.

17. **Mimosa schomburgkii** Benth., J. Bot. (Hooker) 2: 133. 1840. Type: Guyana, Pirara, Ro. Schomburgk ser. I, 715 (holotype K!, isotypes BM!, E!, F!, G!, US!, W!).

Broad-crowned trees (3)5-12(15) m with trunk 0.7-4 dm DBH; flowering branches unarmed but sterile ones said (by Ducke, 1958) to be prickly, defoliate annotinous branchlets atrocastaneous glabrescent but hornotinous ones, leaf-stalks and axes of inflorescence rusty-scabrous-tomentulose with short-stalked or sessile, medusoid setae, resinously aromatic foliage bicolored, leaflets brown-olivaceous sublustrous above, finely remotely stellate and either puberulent or not, charged beneath with scattered, pale yellow or whitish, globose glands ± 0.1 mm diam. and, especially along midrib, with medusoid setae; panicle terminal, distally or wholly efoliate, of narrow amentiform flower-spikes. Stipules linear or subulate 2-8 mm, involute blade stellate dorsally; leaf-stalks of larger leaves 8-15 cm, interpinnal segments 6-12 mm; pinnae 7-14 pairs, rachis of longer ones 5-8.5 cm, interfoliolar segments 1.5-3.5 mm; leaflets of longer pinnae 20-26 pairs, ventrally convex blades narrowly oblong or lance-oblong from obtusely auriculate base, larger ones 5-10.5 x 1.6-3.2 mm, all 3-5-veined from pulvinule, midrib slender subcentric, dorsally prominulous, simple or few-branched. Spikes mostly fasciculate by 2-4, without filaments 4-5 mm diam., axis including short peduncle 4-10 cm. Flowers all 4-merous or both 4- and 5-merous in one spike, diplostemonous, perianth densely silky-tomentulose with pallid intermeshed stellae; calyx campanulate 0.9-1.2 mm; corolla 2-2.5 mm; filaments white or cream, longer ones exserted 3-5 mm. Pods in profile oblong or broad-linear contracted at base into a stipe 2-6 mm, (2)2.5-6.5 x 0.6-0.85(1) cm, (3)4-12-seeded, replum shallowly constricted, ± 0.5 mm wide, valves papery, densely minutely stellulate-lepidote, breaking up into free-falling indehisced articles 5-7 mm long.

Distribution: Discontinuously dispersed from middle and lower Orinoco valley to state of Roraima in Brazil and shortly into adjacent Guyana, disjunct on coastal plain of Honduras; in riverine forest overhanging the water, at gallery-margins, and in bush- or tree-islands in seasonally wet savanna, below 300 m; widely cultivated in botanical gardens and to be expected as an ornamental in the Guianas.

Representative specimen examined: Guyana: Wenamu R., Cuyuni R., FD 5621 (= Fanshawe 2822) (NY).

Note: The only *Mimosa* in the Guianas with catkin-like flower-spikes and stellate indument of leaflets, perianth, and pod.

18. **Mimosa schrankioides** Benth., London J. Bot. 5: 86. 1846. Type: Guyana, Ro. Schomburgk ser. II, 470 (= Ri. Schomburgk 765) (holotype K!, isotypes G!, W!).

Prickly, sarmentose or when unsupported diffuse, weakly frutescent shrubs; armed on stems and leaf-stalks with files of recurved aculei and commonly hispidulous with tawny setulae up to 0.5-1 mm but sometimes glabrate, leaflets either puberulent on both faces, or strigose on one or both faces with appressed setae, or glabrous except for slenderly appressed-setose margin; capitula solitary and 2-3 in leaf-axils. Stipules lanceolate 2-6.5 x 0.3-0.7 mm; leaf-stalks 2.5-10 cm; pinnae 1 pair, rachis of longer ones 4-6.5 cm; leaflets of longer pinnae (8)10-17 pairs, blades broad-linear or linear-oblong from obtusangulate or shallowly semicordate base, deltately acute or obtuse mucronulate, straight or slightly incurved, larger ones 11-26 x 2-6.5 mm, all 3-5-veined from pulvinule, midrib nearly centric. Peduncle ± 1-2 cm; capitula without filaments 3.5-5.5 mm diam., prior to anthesis either moriform with scarcely emergent bract-tips or 4-angulate flower buds covered by enmeshed bracteal cilia; bracts 1-2.2 mm, distally ciliate with setulae 0.4-1.2 mm. Flowers 4-merous haplostemonous, some lower ones often staminate; calyx either minute submembranous or paleaceous-pappiform 0.1-1.2 mm; corolla of *M. casta*, lobes either glabrous or puberulent; filaments white or pink, free, exserted 6-7.5 mm. Pods commonly 3-10 per capitulum, sessile, in profile narrowly or broadly oblong 18-40 x (6)7-12 mm, 2-4-seeded, replum scarcely constricted, contracted into an apical cusp 1.5-5 mm, hispid along back and sides with erect setae up to 2-5 mm, valves papery, setose overall but between setae either glabrous or puberulent, articles free-falling, 4-8 mm long.

D i s t r i b u t i o n: Discontinuously scattered in S America from N Colombia and N Venezuela to the Amazon Basin and the Guianas; on river banks, at edge of seasonally flooded forest, in savanna, and becoming weedy in disturbed places, below 500 m; in the Guianas represented by 2 varieties.

18a. **Mimosa schrankioides** Benth. var. **sagotiana** (Benth.) Barneby, Mem. New York Bot. Gard. 65: 529. 1991. – *M. sagotiana* Benth., Trans. Linn. Soc. London 30: 394. 1875. Type: French Guiana, Cayenne, Sagot 1063 (holotype K!, isotypes BM!, G!, M!, P!, S!, W!).

Floral bracts shortly ciliolate, prior to anthesis not concealing flower buds; calyx membranous, less than 0.5 mm, rim irregularly lobulate or nearly entire, eciliate; corolla glabrous.

D i s t r i b u t i o n: In habitats of the species, and nearly coextensive in range, but known in the Guianas only from the type-collection at Cayenne.

18b. **Mimosa schrankioides** Benth. var. **schrankioides**

Floral bracts strongly ciliate, flower buds prior to anthesis covered with a mesh of setae; calyx paleaceous-pappiform ± 1 mm, lobes setiform-decompound; corolla-lobes puberulent.

D i s t r i b u t i o n: In habitats of the species, less widespread than var. *sagotiana*, best known from the Orinoco valley in Venezuela, around periphery of the Guayana Highland to state of Roraima in Brazil and Guyana.

19. **Mimosa somnians** Humb. & Bonpl. ex Willd., Sp. Pl. 4: 1036. 1806. Type: Colombia, Tolima, Magdalena valley, Humboldt & Bonpland 1845 (holotype B-Willd., seen in microfiche!, isotypes P!, P-HBK!).

In the Guianas erect or assurgent subshrubs 4-25 dm; either unarmed or armed with scattered, straight or gently curved aculei up to 1-4.5 mm, stems, leaf- and inflorescence-axes either glabrous or variably puberulent, strigose, or hispid with sometimes gland-tipped setae, leaflets firm plane, glabrous facially, ciliolate or not; inflorescence simply pseudoracemose or paniculate, of subglobose capitula either proximally foliate or wholly exserted from foliage. Stipules ovate or lance-attenuate 2-7 mm, prominently several-veined, glabrous or puberulent dorsally, ciliolate or not, persistent; leaf-stalks 3-18 cm, petiole 0.4-4.5 cm, longer interpinnal segments 6-35 mm; between each pinna-pair an ovate, entire or denticulate, or triangular-subulate spicule 0.4-1.5 mm; pinnae 3-8(10) pairs, rachis of longer ones 2-6.5 cm, longer interfoliolar segments 0.4-1.3 mm; leaflets of longer pinnae 20-50 pairs, subvertically imbricated in sleep, blades narrowly oblong, obtuse or apiculate, longest 2.5-7.6 x 0.6-1.5 mm, all dorsally 2-4-veined from pulvinule, midrib subcentric, 1-3-branched on each side. Peduncles solitary and 2-3 together, 8-35 mm; capitula without filaments 4.5-7.5 mm diam., prior to anthesis moriform; flower buds obovoid or pyriform; bracts elliptic-oblanceolate or ovate 0.5-4 mm, 1-several veined. Flowers 4(5)-merous diplostemonous, lowest of each capitulum often smaller and staminate; calyx membranous ± 0.3-0.6 mm, rim entire or obscurely toothed, glabrous or minutely ciliolate or glandular-fimbriolate; corolla turbinate 2-3.5 mm, lobes firm, stramineous or reddish, shallowly concave, 1.2-2 mm, striately 7-9-veined; filaments pink, monadelphous through less than 1 mm, exserted 4-7 mm. Pods often more than 1 per capitulum, in profile linear, attenuate at both ends, straight or slightly curved, stipe 3-30 mm, body when well fertilized 30-90 x 3-6 mm, replum shallowly undulate, valves brown or reddish, becoming papery, either glabrous, or setulose, or both strigose and villosulous, or stipitate-glandular, when ripe breaking up into square or oblong-elliptic, bullately biconvex, free-falling, individually indehiscent articles 3-8 mm long.

Distribution: Consists of numerous races, most highly differentiated on the Brazilian Planalto, but extending N of the Amazon valley, with diminished variety, to Trinidad and S Mexico; only feebly represented in the Guianas by 2 varieties.

Note: Collectively distinguished from the other Guianan species, except *Mimosa surumuënsis*, by firm striate corolla-lobes, which generally persist in fruit at base of the pod's stipe. It differs from *M. surumuënsis* in lack of plumose setae.

19a. **Mimosa somnians** Humb. & Bonpl. ex Willd. subsp. **somnians** var. **somnians**

Mimosa tobagensis Urb., Repert. Spec. Nov. Regni Veg. 15: 307. 1918. Type: Tobago, near Scarborough, Broadway 4776 (holotype †B, isotype NY!).

Herbaceous when young but potentially fruticose and attaining 3 m in favorable conditions; aculei either crowded on stems and leaf-stalks or remotely scattered; indument randomly variable, of fine villi, gland-tipped setulae, and plain setae, or some combination of these.

Distribution: Interruptedly widespread from NE Argentina to tropical Mexico, Brazil, lower Orinoco in Venezuela, and Tobago, uncommon in the Guianas in N Guyana and French Guiana; in savanna, on open hillsides, in disturbed brush-woodland, and at edge of gallery forest, becoming weedy in plantations and hedgerows, and forming thickets in moist pastures, mostly below 600 m but locally higher.

Selected specimens: Guyana: N Rupununi Savanna, 2 miles S of St. Ignatius, Goodland & Persaud 769 (NY). French Guiana: Savane Roche du Quatorze Juillet, Bassin du Bas-Oyapock, Cremers 12135 (NY); Fleuve Oyapock, Savanes-Roches "Sikini", Oldeman B2561 (NY).

19b. **Mimosa somnians** Humb. & Bonpl. ex Willd. subsp. **viscida** (Willd.) Barneby var. **viscida**, Brittonia 37: 144. 1985. – *Mimosa viscida* Willd., Enum. Pl. Hort. Berol. 1048. 1809. Type: Brazil, Pará, Sieber s.n. (holotype B-Willd., seen in microfiche!).

Mostly herbaceous from woody rootstock, 6-20 dm; aculei 0, and indument variable, stems, leaf-axes and fruits subglabrous or strigose-setose, or glandular-setose.

Distribution: Common on the Brazilian Planalto and periphery at 400-1500 m, N interruptedly, at lower elevations, of the lower Amazon basin in Pará and Amapá, Brazil, barely entering French Guiana, where known by modern collections only from the Oyapock R.; in campo, about rock outcrops, and on rocky stream banks, white sand campirana.

Selected specimens: French Guiana: Haut Oyapock, Aubert de la Rüe s.n. (P); Maripa, Oldeman 1272 (CAY).

Note: The earliest known collection of var. *viscida* is in the Richard herbarium (P), collected at an unrecorded locality in Guiana (presumably French Guiana) prior to 1800, perhaps by Patris.

20. **Mimosa surumuënsis** Harms, Notizbl. Königl. Bot. Gart. Berlin 6: 304. 1915. Type: Brazil, Roraima, Serra do Mel in district of Surumú, Ule 8131 (holotype †B, isotypes G!, NY!, US!).

Stiffly branched shrubs 0.5-4 m; either unarmed or randomly armed with straight or decurved aculei 1.5-4.5 mm, young branchlets and leaf-axes shaggy-tomentulose with rusty (when old gray) plumulose setae up to 0.5-1.4 mm, small crowded leaflets finely puberulent dorsally; capitula small ovoid or globose, solitary and fasciculate in axils of coevally expanding leaves, forming a narrowly pyramidal panicle leafy throughout in fruit.

Stipules lance- or triangular-acuminate 1.5-4 mm, striately veined, persistent; leaf-stalks of cauline leaves (2)3-7 cm, petiole and interpinnal segments 1.5-7 cm; pinnae 5-12 pairs (in inflorescence fewer), rachis of longer ones 13-32 mm, interfoliolar segments not over 0.5 mm; leaflets of longer pinnae 25-36 pairs, blades linear from deltately auriculate base, obtuse or minutely mucronulate, longer ones 2-4.6 x 0.35-0.6 mm, veinless on upper face, midrib subcentric, tenuously prominent beneath. Peduncle 6-14 mm; capitula without filaments 3.5-8 x 3.5-5 mm, prior to anthesis moriform, obovoid flower buds gray-puberulent; bracts lance-elliptic 0.8-1.6 mm, 3-6-veined, persistent. Flowers 4-merous diplostemonous; calyx turbinate-campanulate 0.6-1.4 mm, 12-veined tube glabrous externally, triangular teeth less than 1 mm, ciliolate; corolla turbinate ± 1.7 mm, shallowly concave lobes 7-10-veined; filaments pale pink or whitish, free, exserted 4.5-6 mm. Pods 1-10 per capitulum subsessile, in profile undulately linear 12-24 x 2.5-3 mm, 3-6-seeded, replum shallowly constricted, valves alike densely hispid-pilose with plumulose setae up to 0.6-1.5 mm, valves breaking up when ripe into free-falling, individually dehiscent articles 3.5-5 mm long.

Distribution: Range bicentric, about the NE and SE periphery of the Guayana Highland: in Bolívar, Venezuela and on headwaters of Rio Branco in Roraima, Brazil and adjoining Guyana (Ireng R.); on rock outcrops in savanna and on cliff ledges, ± 300-600 m.

Selected specimens: Guyana: Orinduik, Ireng R., FD 7812 (= Fanshawe 3582) (NY); Orinduik Falls, Irwin 460 (US).

Note: *Mimosa surumuënsis* is closely related to *M. microcephala*, but readily distinguished by the indument of plumose setae, not duplicated elsewhere in Guianan *Mimosa*.

18. **NEPTUNIA**[22] Lour., Fl. Cochinch. 653. 1790.
Type: N. oleracea Lour. [N. natans (L.f.) Druce]

Erect or diffuse, taprooted herbs, one free-floating in quiet waters; indument of minute white hairs or 0. Stipules membranous, ovate-acuminate from cordate base, several-veined, persistent or deciduous. Leaves bipinnate, sensitive, pinnae and leaflets in opposite pairs; petiolar nectary present or not; venation of leaflets palmate and pinnate, faint. Inflorescences composed of dense capituliform spikes axillary to contemporary leaves; peduncles 2-bracteolate at and below middle;

[22] by Rupert C. Barneby

capitula oblong-ellipsoid in bud, receptacular axis bent through 40°-90° so as to bring lower, barren flowers uppermost and further, bisexual flowers ± geotropic to one side. Flowers 5-merous, diplostemonous (in the Guianas), lower ones of each spike neuter, their stamens modified into linear-oblanceolate petaloid yellow staminodes, upper flowers perfect, their terete white filaments ± twice as long as petals; calyx campanulate short-toothed, open in bud; corolla either yellow or greenish, erect oblanceolate petals valvate in bud, separating to base at anthesis; stamens free; anther-connective either gland-tipped or not; ovary shortly stipitate, glabrous; style distally dilated, truncate; pollen simple. Pods spreading in umbelliform cluster, stipitate, body in outline oblong or linear-oblong from oblique base, straight or gently decurved, abruptly apiculate, plano-compressed, valves firmly papery, dehiscence inert, follicular, through ventral suture; seeds obliquely horizontal, compressed-obovoid, testa areolate, endosperm present.

Distribution: Circumtropical and warm temperate genus of 12 species, 4 American species; 2 species occur in the Guianas.

Note: *Neptunia* in the Guianas is readily recognized at anthesis by the dimorphic flowers borne in capitula recurved so as to orient the proximal, neuter and showy yellow ones more or less erect while the distal perfect ones hang to one side below them. The bright yellow color of the staminodes in neuter flowers distinguishes the genus from *Desmanthus*, which differs in fruit in narrowly linear pods and in the basipetal, not horizontal posture of the seeds.

KEY TO THE SPECIES

1 Stems floating, simple or almost so, early detached from primary root; all internodes of stem ± inflated and parenchymatous, each node adventitiously rooting and bearing 1 leaf and an axillary peduncle; petiolar nectary 0; anther-connective without gland; pod 4-8-seeded . . . . *1. N. natans*

Stems erect, ascending, or diffuse, openly branched; lower internodes of stem inflated and parenchymatous only when standing in shallow waters, lower nodes then adventiously rooting but not floriferous, peduncles arising only from aerial nodes; petiolar nectary present; anther-connective gland-tipped; pod (8)10-20-seeded. . . . . *2. N. plena*

1. **Neptunia natans** (L.f.) Druce, Rep. Bot. Soc. Exch. Club Brit. Isles 4: 637. 1917. – *Mimosa natans* L.f., Suppl. Pl. 439. 1782. – *Neptunia prostrata* Baill., Bull. Mens. Soc. Linn. Paris 1: 356. 1883. Type: Peninsular India, Tranquebar, Koenig, LINN 1228/4.

Neptunia oleracea Lour., Fl. Cochinch. 654. 1790. Type: Cochinchina, Loureiro s.n.(holotype BM).

Free-floating aquatic herb with long simple stems early detached from primary root, internodes swollen, nodes giving rise on one side to a fascicle of adventitious fibrous roots and on other side to a sensitive leaf and an erect axillary peduncle. Stipules 5-14 mm; leaf-stalks 2.5-11 cm; pinnae (2)3 pairs, rachis of longer ones 3-7.5 cm; leaflets of longer pinnae 12-20 pairs, blades linear-oblong from inequilateral base, obtuse or apiculate, larger ones 7-15(18) x 1.5-3 mm. Peduncle 5-26 cm, bracteoles 2 resembling stipules in shape and texture, 5-12 mm; axis of capitula 5-12 mm; calyx 2-3 mm; petals erect 3-4.5 cm; filaments of neuter flowers yellow 7-14 mm, or perfect flowers whitish; anther-connective lacking gland. Pods spreading in an umbelliform cluster, stipe 4-8 mm, body in profile oblong, abruptly apiculate, gently decurved, ± 20-26 x 8-10 mm, valves lustrous brown glabrous; seeds 4-8.

Distribution: Widespread in tropical America, Africa, and SE Asia, frequent on the coastal plain of Guyana and Suriname, to be expected in French Guiana; floating in canals, ditches, ponds, and backwaters, at low elevations; often cultivated for ornament in tanks and fountain pools; flowering throughout the year.

Selected specimens: Guyana: Coastline 10 km E of Georgetown, Boom *et al.* 8034 (NY); Demerara-Mahaica Region, near Mahaica village, Kvist 389 (NY). Suriname: lower Nickerie R. near Paradise, Lanjouw 597 (NY).

Note: This well marked species is well illustrated by Brenan, Fl. Trop. E. Afr., Legum. subfam. Mimosoideae fig. 12. 1959; Windler, Austral. J. Bot. 14: 402, fig. 10. 1966; McVaugh, Fl. Novogalic. 5: fig. 30. 1987.

2. **Neptunia plena** (L.) Benth., J. Bot. (Hooker) 4: 355. 1841. – *Mimosa plena* L., Sp. Pl. 519. 1753. Type: Mexico, Veracruz, Houston s.n., BM. – Fig. 34

Neptunia surinamensis Steud., Flora 26: 759. 1843. Type: Suriname, Hostmann & Kappler 587 (holotype P, isotype K, U).

Taprooted herbs, either erect or diffuse, 4-20 dm, branched distally; lower internodes, when standing in shallow water, sometimes edematous, rest slenderly cylindric; glabrous overall except for minutely ciliolate leaf-axes or leaflets. Stipules submembranous 4-12 mm; leaf-stalks (1.5)3-7.5 cm; nectary sessile, shallowly cupular or patelliform, round or elliptic, 0.5-2.7 mm diam., at insertion of first pinna-pair, thick-rimmed;

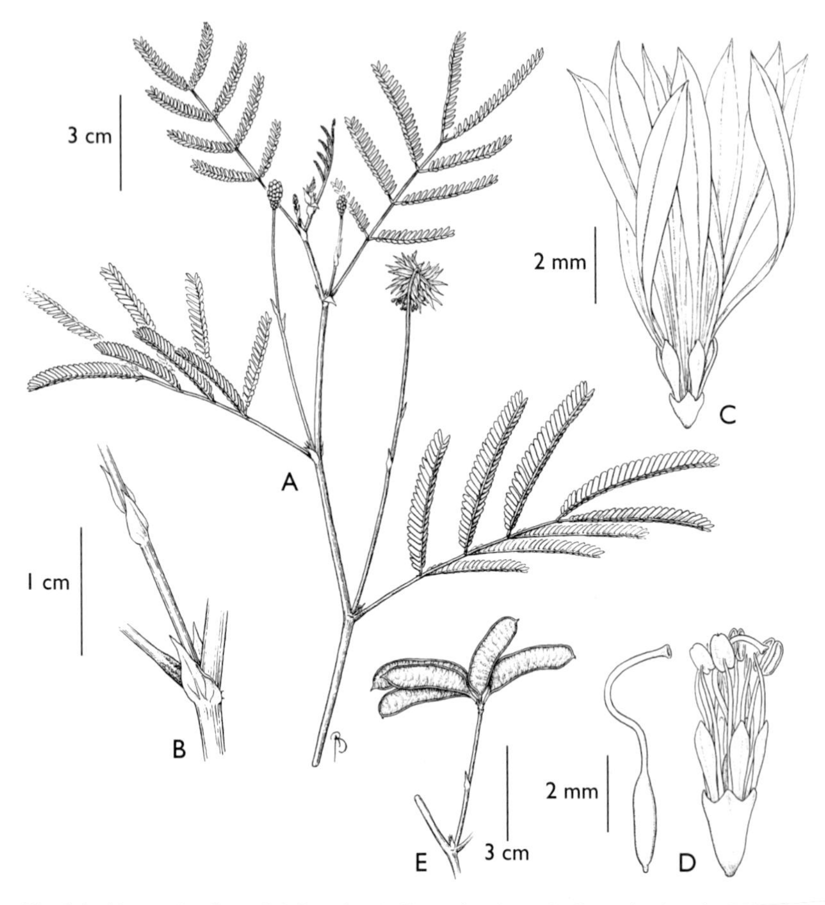

Fig.34. *Neptunia plena* (L.) Benth.: A, flowering branch; B, node showing stipules and peduncular bracts; C, staminodal flower; D, fertile flower and gynoecium; E, pods (A-B, Pennell 3951, from Colombia; C-D, Pittier 12515, from Venezuela; E, Smith 298, from Colombia). Drawing by Bobby Angell; reproduced with permission from NY.

pinnae 2-4(5) pairs, rachis of longer ones 1.5-5.5 cm; leaflets of longer pinnae (9)12-28("38") pairs, linear-oblong, obtuse or apiculate, longer ones (2.5)3-14 x 0.7-2.5 mm. Peduncles (3)5-16 cm, 2 bracteoles near and below middle resembling stipules in form and texture; capitula obovoid-ellipsoid, receptacle 10-18 mm, recurved at full anthesis; flowers dimorphic, essentially like those of *N. natans* except for yellow perianth and gland-tipped anthers. Pods shortly stipitate, spreading in umbelliform cluster, body in profile linear-oblong 2.5-4.5 x 0.8-1.1 cm, straight or slightly decurved, obtuse and mucronulate but not beaked, 8-20-seeded, valves papery green or fuscous-castaneous, glabrous.

Distribution: Tropical Mexico, around the Caribbean Basin, widespread over S America from Pacific Ecuador to NE Brazil, common in the lowlands of the Guianas; introduced in India; on ditch banks, at edge of ponds, in wet savannas, and becoming weedy in moist pastures; flowering almost throughout the year.

Selected specimens: Guyana: Along coastline about 10 kms E of Georgetown, Boom *et al.* 8041 (NY); Vreed-en-Hoop, W bank of Demerara R., opposite Georgetown, Hitchcock 16729 (NY). Suriname: From Wia wia-bank to Grote Zwiebelzwamp, Lanjouw & Lindeman 1122 (NY); Nickerie, in triangle W of van Waw Canal, Lanjouw & Lindeman 3133 (NY). French Guiana: Dégrad Canard, Bassin de la Basse-Mana, Cremers & Hoff 10575 (NY); Haute rivière de Kaw, de Granville 6862 (NY).

Vernacular name: Suriname: sammasamari.

Note: *Neptunia plena* is notably plastic in amplitude of the foliage and in number and size of leaflets, responding readily to a fluctuating depth of water.

19. **PARKIA**[23] R. Br. in Denham & Clapperton, Narr. Travels Africa, App. 22: 289. 1826.
Type: P. africana R. Br., nom. illeg. = P. biglobosa (Jacq.) R. Br. ex G. Don (Mimosa biglobosa Jacq.)

Tall trees to 30 m high or more; trunk regular and straight, generally buttressed in big individuals. Leaves bipinnate, mostly alternate or opposite (3 species); stipules absent; petiole and rachis woody; one foliar nectary present on petiole, most often longitudinally elliptic, sometimes orbicular, often distally below first pair of pinnae, plus nectaries on rachis between distal pairs of pinnae; pinnae 10-20 pairs; leaflets sessile, to (15)20-100 pairs per pinna. Inflorescences erect or pendent; axes branched or not, peduncles woody, thick, to 1 m long; flowers densely gathered on flowerheads (capitula); capitula spherical or biglobose or clavate, homomorphic (all flowers similar, fertile) to heteromorphic with sterile and fertile flowers, in some species with conspicuous staminodial flowers; bracts linear-spatulate, generally longer than calyx, apex thickened and pubescent. Sterile flowers of one or two kinds: nectar-secreting ones smaller than fertile flowers and staminodial ones, with conspicuous, long staminodes forming proximal "globe" of biglobose flowerheads. Fertile flowers hermaphrodite (or male): calyx tubular or infundibuliform, lobes

[23] by Odile Poncy

5, slightly to markedly unequal, imbricate; corolla tubular, slightly to markedly longer than calyx; stamens 10, generally shortly exserted from corolla, filaments often connate at base, anthers sometimes with apical gland; ovary shortly stipitate, style as long as stamens or longer, stigma tubular. Infructescences pendent or erect, swollen rachis bearing one to several pods; pods coriaceous or ligneous, dehiscent or not, glabrous or velutinous, endocarp fleshy or gummy; seeds numerous, flattened, ellipsoid, testa black, hard, with a pleurogram.

Distribution: Pantropical genus with ca. 35 species, half of them native to the Neotropics; in the Guianas 8 species.

Note: Citations of types and specimens combine personal observations completed by citations by Hopkins (1986) when available.

LITERATURE

Hopkins, H.C.F. 1986. Parkia (Leguminosae: Mimosoideae). Flora Neotropica 43:1-129.

KEY TO THE SPECIES

1 Leaves opposite. 2
  Leaves alternate 4

2 Capitulum in flower biglobose 3
  Capitulum in flower pyriform. . . . . *1. P. decussata*

3 Leaflets ≥ 30 mm; inflorescence axis pendent, capitulum > 15 cm long; pod stipe > 6 cm long . . . . *2. P. gigantocarpa*
  Leaflets ≤ 20 mm; inflorescence axis erect, capitulum < 15 cm long; pod stipe < 6 cm long . . . . *4. P. nitida*

4 Capitulum longer than wide: biglobose or clavate, with a ring of staminodial flowers at base 5
  Capitulum not longer than wide: spherical or wider than long. 6

5 Leaflets > 1 cm long; capitulum > 8 cm long . . . . *3. P. igneiflora*
  Leaflets < 1 cm long; capitulum < 8 cm long . . . . *6. P. reticulata*

6 Capitulum somewhat flattened distally, with smaller flowers at apex, peduncle very long and pendent; pod black, glabrous . . . . . . *5. P. pendula*
Capitulum spherical, homomorphous (all flowers similar), peduncle short (<10 cm) and erect; pod brownish to rubiginous, velvety to velutinous 7

7 Capitula < 1 cm diam., yellow . . . . *7. P. ulei*
Capitula > 1 cm diam., red . . . . *8. P. velutina*

1. **Parkia decussata** Ducke, Notizbl. Bot. Gart. Berlin-Dahlem 11: 472. 1932. Lectotype (designated by Hopkins 1986: 83): Brazil, Amazonas, Ducke RB 23262 (hololectotype RB, isolectotypes K, P, NY, S, U, US).

Tall tree to 40 m high; bark reddish brown. Leaves opposite, decussate; petiole and rachis ca. 20 cm long, pinnae 4-6 pairs, opposite; leaflets ca. 20-30 pairs per pinna, opposite, oblong, 20-25 x 4-6 mm, apex rounded. Inflorescence compound, rachis not branched, erect, to 1 m long, bearing several (to 8) pairs of capitula on erect opposite peduncles. Capitula 6-7 cm long, clavate, ca. 2 cm diam. at base, bearing yellow sterile and fertile flowers; sterile flowers (staminodial and nectar secreting) similar in size, smaller than apical fertile flowers; bulb-like apical part of capitula, 3-3.5 cm diam., bears fertile flowers: calyx 12-14 mm long, lobes to 2 mm long; corolla as long as calyx or slightly longer; stamens exserted. Pod to 50 x 6 cm, sutures somewhat thickened, valves coriaceous to subligneous, indehiscent, brown, velutinous with transverse cracks; seeds to 25 per pod, ca. 20 x 10 mm.

Distribution: Brazil (Amazonas and Pará), French Guiana; dense terra firme forest (FG: 3).

Selected specimens: French Guiana: Saül, Mori *et al.* 15482 (CAY, NY, P); Nouragues Station, Mori *et al.* 25564 (CAY, NY, P); Haute Wanapi, de Granville *et al.* 16002 (CAY).

Note: The description of the pod follows Hopkins (1986).

2. **Parkia gigantocarpa** Ducke, Arch. Jard. Bot. Rio de Janeiro 1: 19. 1915. Lectotype (designated by Hopkins 1986: 80): Brazil, Pará, Oriximina, lower Trombetas R., Ducke MG11482 (hololectotype RB, isolectotype MG)

Tall tree to 40 m high; trunk cylindrical, to 1 m diam. Leaves opposite, petiole and rachis 18-22 cm long, pinnae 10-15 pairs, opposite; twig and leaf-rachis velvety tomentose; leaflets 16-20 pairs per pinna, opposite, oblong, 30-40 x 7-10 mm, base broadened, slightly auriculate, apex acute, main vein straight, secondary veins (at least one) visible up to apex. Inflorescence compound, rachis not branched, pendent, 30-60 cm long, bearing 1-3 pairs of capitula on pendent opposite peduncles to 40 cm long. Capitula ca. 16 x 7 cm, biglobose; basal part is a ring of long staminodial flowers (staminodes to 5 cm), white at anthesis; apical part, bearing fertile flowers, broadly elliptic, as wide as staminodial ring, with a constricted ring of small nectar-secreting flowers (ca. 15 mm long) in between; fertile flowers white to pale yellow: calyx 15-20 mm long (including lobes 4-5 mm); corolla 20-25 mm, lobes ca. 2 mm; stamens to 30 mm long, sligthtly exserted. Pod flat, coriaceous, rigid, indehiscent, stipitate, 50-60 (including stipe up to 10 cm) x 5-6 cm, valves brown, glabrous, corrugated; seeds to 30 per pod, 1-2 x 1 cm, black.

Distribution: Guyana, French Guiana, Brazil (Amapá, Amazonas, Pará); high forest, infrequent (GU: 1; FG: 2).

Selected specimens: Guyana: Shodikar Cr., A.C. Smith 3021 (A, F, G, K, MO, NY, P, S, US). French Guiana: Haute-Camopi, Mt. Belvédère, Sabatier & Prévost 1671 (CAY, NY); Mt. St-Marcel, Grenand 1064 (CAY, P).

Note: Grenand 1064 is included here with doubt, given the smaller capitulum and pod.

3. **Parkia igneiflora** Ducke, Notizbl. Bot. Gart. Berlin-Dahlem 11: 472. 1932. Lectotype (designated by Hopkins 1986: 69): Brazil, Amazonas, Rio Madeira, Ducke RB 23261 (hololectotype RB, isolectotype K, P, U, US).

Tree to 20 m high; trunk cylindrical, straigtht, narrow; bark dark reddish brown. Leaves alternate; petiole and rachis to 55 cm long; pinnae 10-12 pairs; rachises and main vein of leaflets puberulous at least on young leaves; leaflets ca. 30 pairs per pinna, opposite, ± perpendicular to secondary rachis, oblong, ca. 30 x 8 mm, base auriculate, apex rounded or obtuse, base straight or slightly curved, main vein straight, secondary vein often visible up to apex. Inflorescence compound, up to 3 m long, sparsely branched, branches woody, erect, each bearing several pedunculate capitula, peduncles pendent, alternate or opposite, up to 40 cm long. Capitula ca. 10 cm long, biglobose; basal part ca. 4 cm diam., bearing a ring of staminodial flowers, staminodes to 30 mm long; apical part elliptic, bearing a ring of small nectar secreting flowers at base, rest of clavate receptacle bears pinkish to purplish fertile flowers: calyx ca. 10 mm long, tubular with short unequal lobes; corolla 2 mm longer than calyx, tubular with short lobes. Pod flat, straight or slightly curved or coriaceous, rigid, indehiscent, stipitate, 20-30 x 3-4 cm, valves dark brown, glabrescent, weakly corrugated; "the cavity when fresh containing amber-colored sticky or crystalline gum" (Hopkins, 1986); seeds black, hard, up to 25 per pod, ca. 16 x 10 mm.

Distribution: Widespread in Amazonia (Colombia, Venezuela, Peru, Brazil Amazonas); known in the Guianas from a single record (GU: 1).

Representative specimen examined: Guyana: Rupununi, Kuyuwini R., Jansen-Jacobs *et al.* 2537 (CAY, NY, P, U).

4. **Parkia nitida** Miq., Stirp. Surinam. Select. 7. 1851. Type: Suriname, Suriname R., Hostmann 1012 ("Kappler 1012") (holotype U, isotypes BM, G, K, OXF, P, S, US).

   Parkia oppositifolia Spruce ex Benth., Trans. Linn. Soc. London 30: 363. 1875. Type: Brazil (Amazonas), Spruce 1473 (holotype K, isotypes BM, G, GH, NY, OXF, P).

Tall tree to 40 m high or more; trunk cylindrical, with regular, straight buttresses up to 2 m high in largest individuals; bark smooth, brownish to grayish. Leaves opposite, very variable in size; petiole and rachis 11-19(22) cm long; pinnae 4-8(10) pairs; leaflets 20-40 pairs per pinna, opposite, perpendicular to secondary rachis, oblong, 11-20 x 3-7 mm, base auriculate, apex rounded, only main vein clearly visible, straight distally; lower surface of blade sometimes coated by a waxy whitish layer. Inflorescence compound, up to 1 m long or more, often branched, branches woody, erect, each bearing 1 to 4 pairs of pendent, pedunculate capitula, peduncle 3-12(20) cm long, robust. Capitula 5-10 cm long, biglobose; basal part 5-7 cm diam., a ring of staminodial flowers 25-35 mm long, pale to bright yellow; apical part elliptic, bearing a ring of small nectar secreting flowers at base, rest of clavate receptacle bears yellow fertile flowers: calyx to 12 mm long, tubular with short unequal lobes; corolla 3-5 mm longer than calyx, tubular with short lobes to lobes almost free to base. Pod flat, coriaceous, rigid, indehiscent, stipitate, 15-20(40) (including stipe 2-5 cm) x 3.5-6 cm, valves brown to black, glabrous, weakly corrugated; "the cavity when fresh containing amber-colored sticky or crystalline gum" (Hopkins, 1986); seeds black, hard, ca. 20 per pod, ca. 2 x 1 cm.

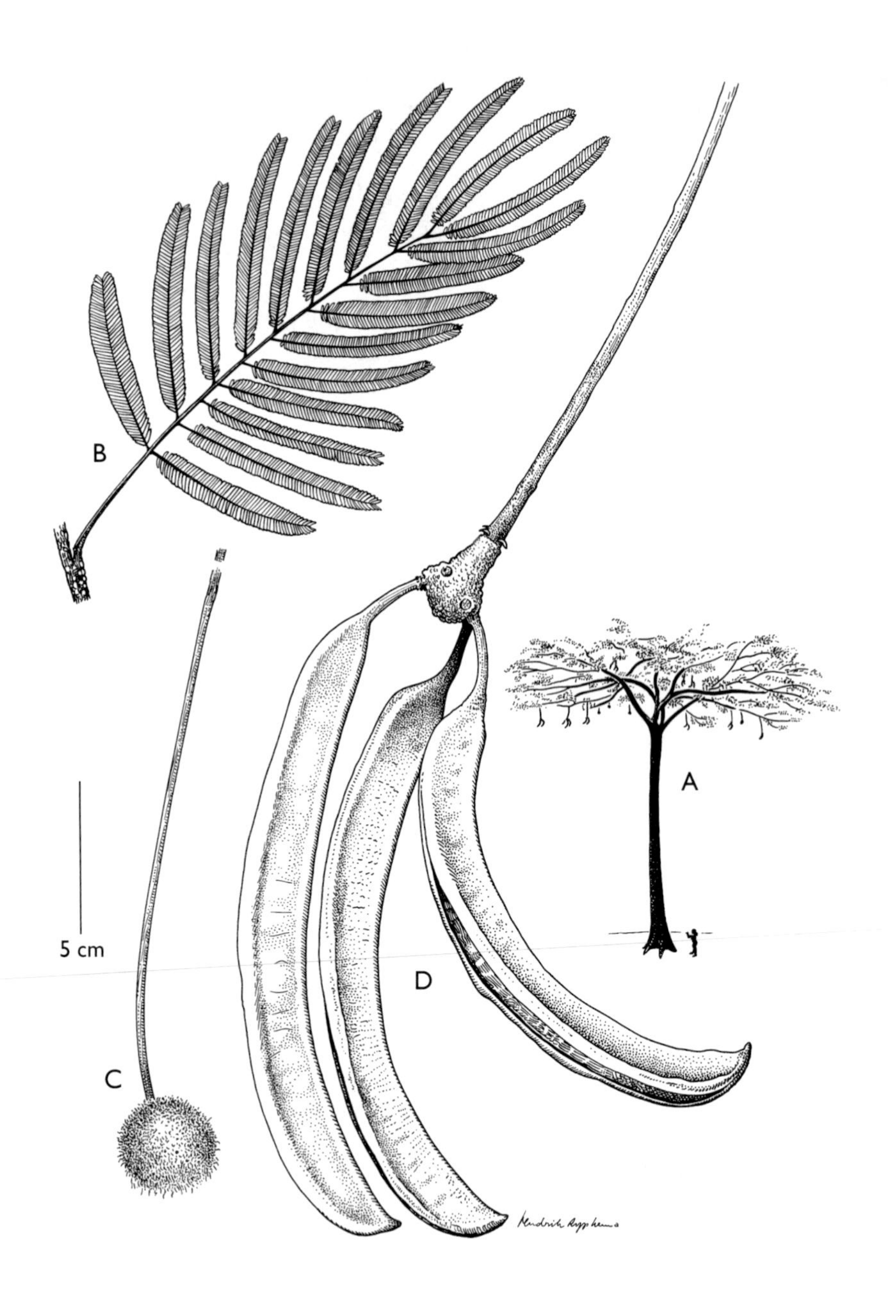

Fig.35. *Parkia pendula* (Willd.) Benth. ex Walp.: A, habitus of tree; B, Leaf; C, inflorescence; D, infructescence (B-D, van Andel 4814). Drawing by Hendrik Rypkema; reproduced with permission from U.

Distribution: Widely distributed throughout Amazonia, the Guianas, Venezuela, up to Panama; high forest; ca. 60 collections studied from the Guianas.

Selected specimens: Guyana: Essequibo R., Siba Cr., FD 3447 (= Fanshawe 711) (K, NY); Upper-Essequibo, Kamoa Mts., Clarke 4905 (NY, P, US). Suriname: Wilhelmina Gebergte, Lucie R., Irwin *et al.* 54459 (F, MO, NY, P, S, SP, U, US); Brownsberg, BW 6736 (K, MO, NY, U). French Guiana: Fourgassié, Oldeman B661 (CAY, P); Saül, Mori & Gracie 18603 (CAY, NY, P); Bassin de l'Approuague, Station des Nouragues, Poncy *et al.* 1843 (CAY, K, NY, P).

Vernacular names: Guyana: black manariballi (Arawak). Suriname: agrobigi, dodomissinga (Paramaca); ajoewa (NE); tontoe awa (Saramaca); tamoene oeloeloe, oeroeroeöe (K); oja, oeja (A); grobigie, tamarindebosch tamarinde, soeloe (Car). French Guiana: acacia mâle, bois macaque (Creole).

5. **Parkia pendula** (Willd.) Benth. ex Walp., Repert. Bot. Syst. 5: 577. 1846. – *Inga pendula* Willd., Sp. Pl. ed. 4, 4:1025. 1806. Type: Brazil, Pará, Sieber s.n. (comm. Hoffmannsegg) (holotype B-Willd. 19055). – Fig. 35

Tall tree up to 40 m high; trunk cylindrical, 1 m diam., buttressed or with buttresses to 1 m in big trees; crown remarkably flat, like a platform spreading to 30 m diam.; bark orangish brown, exsuding a brown gum, exfoliating in longitudinal flakes; twigs thick, angular. Leaves alternate, grouped in terminal whorls; petiole and rachis 12-22(32) cm long; petiolar gland orbicular to elliptical, midway along petiole; pinnae 12-21 pairs, subopposite or alternate; leaflets 50-100 pairs per pinna, opposite, very small, 3-7 x 0.5-1.5 mm, linear. Inflorescence compound, axillary just below leaves, rachis not branched, ca. 50 cm long, bearing 3-5 capitula, peduncle alternate, pendent, flexible, 39-70(100) cm. Capitula wider than long, apex somewhat flattened or depressed, ca. 4 x 3 cm, golden-brown when young, deep-pink to bright red at anthesis, sweet- scented; receptacle discoid or pyriform or clavate, white after flowers have dropped; fertile flowers at base: calyx 8-10 mm long, including lobes 2 mm; corolla 1-2 mm longer, lobes 5-6 mm; stamens exserted 5-6 mm from corolla; nectar-secreting flowers at apex, smaller than fertile ones. Pod flat, straight or curved, 14-24 x 2-3 cm, coriaceous, rigid, dehiscent, valves thick, glabrous, smooth, shiny, reddish when young, turning green, then black, exsuding sticky resin (in which seeds are embedded and released (Hopkins 1986)), dehiscent along adaxial thickened suture; seeds 15-28 per pod, ca. 6-9 x 4-5 mm.

Distribution: Widespread from C America to southern Amazonia; high rainforest, mixed forest, brown sand forest; briefly deciduous; 32 collections studied from the Guianas (GU: 6; SU: 3; FG: 23).

Selected specimens: Guyana: Rupununi, Kuyuwini R., Jansen-Jacobs *et al.* 2864 (CAY, P, U); Cuyuni-Mazaruni, Waramadan, McDowell & Gopaul 3197 (CAY, U, US). Suriname: Tafelberg, Maguire 24659 (BR, F, G, MO, NY, P, RB, US, VEN), as no. "23659" (K); Brownsberg, BW 6669 (IAN, K, NY). French Guiana: Saül, Mori *et al* 14868 (CAY, NY, P); Sinnamary, Piste-de-St-Elie, Sabatier & Prévost 3678 (CAY, K, NY, P, U).

Vernacular names: Guyana: hipanai. Suriname: kwatta kama, kouatakaman (Saramaca). French Guiana: bois macaque.

6. **Parkia reticulata** Ducke, Arch. Jard. Bot. Rio de Janeiro 5: 126. 1930. Type: Brazil, Pará, near Bragança, Ducke RB 16859 (holotype RB, isotypes K, P, U).

Tree to 25 m high or more; trunk cylindrical, to 70 cm diam., buttresses up to 1.5 m; bark with scattered lenticels, exsuding clear sap. Leaves alternate; petiole and rachis 20-25 cm long; petiolar gland close to first pinnae, ca. 4 x 3 mm; pinnae 10-15 pairs, opposite or subopposite; leaflets ca. 40 pairs per pinna, opposite, linear, oblong to sigmoid, 5-9 x 1-1.5 mm, base auriculate, apex acute (pointed), main vein slightly sigmoid. Inflorescence compound, rachis not branched, ca. 40 cm long, bearing 5-8 capitula on pendent alternate peduncles 10-20 cm long. Capitula 5-6 cm long, biglobose; basal part ca. 5 cm diam., a crown of staminodial flowers 20-25 mm long, sulphur yellow; apical part elliptic, bearing a ring of small nectar-secreting flowers at base, rest of clavate receptacle bears yellow fertile flowers: calyx and corolla tubular with inconspicuous short unequal lobes; calyx ca. 5 mm long; corolla ca. 2 mm longer than calyx; stamen filaments free, slightly exserted. Pod flat, coriaceous, rigid, indehiscent, stipitate, 26-35 (including stipe up to 10 cm) x 3-4 cm, valves brown, glabrous, with conspicuous reticulate venation, depressed between seeds; seeds to 15 per pod.

Distribution: Brazil (Pará), French Guiana (FG: 5).

Selected specimens: French Guiana: Cayenne, BAFOG 1243 (BBS, CAY, P); Gourdonville, Benoist 1553 (P); Saül, Mori *et al.* 15440 (CAY, NY), 22998 (CAY, NY), 22304 (NY).

Vernacular name: French Guiana: bois macaque blanc (mâle).

7. **Parkia ulei** (Harms) Kuhlmann, Arch. Jard. Bot. Rio de Janeiro 4: 356. 1925. – *Leucaena ulei* Harms in Ule, Verh. Bot. Vereins Prov. Brandenburg 47:162. 1907. Type: Brazil, Amazonas, R. Madeira, Ule 6085 (holotype B, destroyed, isotypes F, HBG, K, MG, NY, S).

In the Guianas only: var. **surinamensis** Kleinhoonte in Pulle, Recueil Trav. Bot. Néerl. 30: 169. 1933. – *Parkia microcephala* Kleinhoonte in Pulle, Recueil Trav. Bot. Néerl. 22: 411. 1925. Lectotype (designated by Hopkins, 1986: 91): Suriname, Para, Sectie O, BW 2031 (lectotype NY, isolectotypes K, MO, U).

Tall tree up to 40 m; trunk cylindrical, 120 cm diam., buttresses up to 2 m high, narrow, divided bark cream-yellowish, exsuding a small amount of yellowish to orangish resin; twigs thick, angular; twigs, leaf rachises, inflorescences and pods velutinous, reddish brown. Leaves alternate, sometimes whorled at end of twigs; petiole and rachis 20-30 cm long; pinnae 9-20 pairs, opposite or subopposite; leaflets 30-50 pairs per pinna, opposite, narrowly oblong, 3-10 x 0.5-2 mm, apex rounded. Inflorescence compound, axillary to leaves, erect, axis much branched, 30-45 cm long, bearing 12-15 alternate capitula, peduncle 5-10 mm long. Capitula globose, 1-2 cm diam., receptacle clavate ca.1 cm long, brownish in bud; all flowers fertile, creamy to yellow at anthesis, fragrant; calyx funnel-shaped, 3-4 mm long, including lobes 0.5 mm; corolla ca. 1 mm longer than calyx, lobes ca. 1 mm; stamens exserted up to 5 mm from calyx. Pod flat, straight or sometimes curved, shortly stipitate (stipe 1-2 cm), 10-25 (including stipe) x 3-4 cm, dehiscent, valves somewhat coriaceous, velutinous to subglabrous, light brown to rubiginous; seeds 16-22 per pod, 10-16 x 7-12 mm.

Distribution: NE Venezuela, the Guianas, Brazil (Amapá and Pará); high dense rainforest, mixed forest on brown sand, transition high-forest-savanna forest and savanna forest; leafless while flowering (April); 25 collections studied from the Guianas (GU: 6; SU: 3; FG: 9).

Selected specimens: Guyana: Mazaruni, Rupununi R., Simuni Cr., Davis FD 2113 (K); Upper-Takutu Upper Essequibo, Mt. Makarapan, Clarke *et al.* 6966 (NY, US). Suriname: Saramacca R., Maguire 23876 (NY); Wilhelmina Gebergte, Irwin *et al.* 55539 (F, FHO, MO, NY, US). French Guiana: Approuague R., Cr. Tortue, Oldeman 2375 (CAY, P); Route St-Laurent-du-Maroni à Mana, SF 7723 (NY, P, U); Saül, Mori *et al.* 18184 (CAY, NY).

Vernacular names: Suriname: dodomissinga (Saramaca); kojalidan, kleinbloemige agrobigi, tamarin-prokoni (Sranang). French Guiana: acacia mâle, dodomissinga.

Note: The typical variety *Parkia ulei* var. *ulei* is distributed in western Amazonia.

8. **Parkia velutina** Benoist, Notul. Syst. (Paris) 3: 271. 1916. Type: French Guiana, St-Jean-du-Maroni, Benoist 1005 (holotype P).

Tall tree up to 40 m; trunk cylindrical, 70 cm diam., buttresses up to 2 m; twigs thick, angular; twigs, leaf rachises, inflorescences and pods velutinous, red brown. Leaves alternate, grouped in terminal whorls; petiole and rachis 20-25 cm long; pinnae 20-30 pairs, opposite or subopposite; leaflets 40-100 pairs per pinna, opposite, linear to slightly sigmoid, 5-10 x 1-2 mm, apex acute, main vein slightly sigmoid. Inflorescence compound, axillary below leaves, rachis not branched, ca. 20 cm long, bearing 8-10 capitula, peduncle 2-4 cm long. Capitula globose, ca. 2.5 cm diam., receptacle clavate, ca. 1 cm long, reddish brown in bud; all flowers fertile, red at anthesis; calyx 7-8 mm, including lobes 1.5 mm; corolla 3 mm longer than calyx, lobes ca. 2 mm, stamens shortly exserted. Pod flat, coriaceous, rigid, indehiscent, stipitate, 20-30 (including stipe up to 10 cm) x 3-4 cm, valves velutinous, brown to rubiginous; seeds 20-30 per pod, ca. 2 x 1 cm.

Distribution: Widely distributed throughout Amazonia and French Guiana, 1 record from Guyana; 15 collections studied from the Guianas (GU: 1; FG: 14).

Selected specimens: Guyana: Rewa R., Clarke 3631 (NY, P, US). French Guiana: Bassin de l'Approuague, Station des Nouragues, Sabatier & Prévost 1800 (CAY, NY, P, U); Charvein, Benoist 256 (P); Piste-de-St-Elie, Sabatier & Prévost 2345 (CAY, G, INPA, K, MO, NY, P, U); Trois-Sauts, Grenand 631 (CAY, NY, P).

Vernacular name: French Guiana: dodomissinga (Saramaca, Paramaca).

20. **PENTACLETHRA**[24] Benth., J. Bot. (Hooker) 2: 128. 1840.
Type: P. filamentosa Benth. [P. macroloba (Willd.) Kuntze]

Distribution: A genus of 3 species, 1 (described below) widely dispersed from equatorial S America to the lesser Antilles and Nicaragua, 2 (with up to 15 staminodes) in tropical W Africa; 1 species in the Guianas.

1. **Pentaclethra macroloba** (Willd.) Kuntze, Rev. Gen. Pl. 1: 201. 1891. – *Acacia macroloba* Willd., Sp. Pl. 4: 1060. 1805. Type: Brazil, Pará, Sieber in Hoffmannsegg s.n. (holotype B-Willd. 19135 = F Neg. 1498!). – Fig. 36

Acacia aspidioides G. Mey., Prim. Fl. Esseq. 165. 1818. Type: Guyana, Arowabisch Island, Meyer s.n. (holotype GOET).
Pentaclethra filamentosa Benth., J. Bot. (Hooker) 2: 128. 1840. Type: Guyana, Ro. Schomburgk ser. I, 498 (holotype K) = Ri. Schomburgk 408.
Pentaclethra brevifila Benth., J. Bot. (Hooker) 2: 128. 1840. Type: Brazil, Amazonas, Borba, Riedel 1321 (holotype K = NY Neg. 1721!, isotype NY!).
Cailliea macrostachya Steud., Flora 26: 759. 1843. Type: interior Suriname, Hostmann & Kappler 1033 (holotype P, isotype NY).

Unarmed trees, fertile when 5-30 m tall, with narrowly buttressed trunk 0.5-3.5(4) dm DBH and bipinnate, amply multifoliolate leaves; indument of short ascending, sordid or rusty hairs confined on leaves mostly to ventral face of primary and secondary axes, leaflets glabrous facially but sometimes ciliolate, axis of flower-spikes densely silky-tomentulose. Stipules narrowly lanceolate 2-5.5 mm, flaccid, deciduous; leaf-stalks 12-34(40) cm; petiolar nectaries 0; pinnae opposite or almost so, (1)12-20 pairs, decrescent proximally, rachis of longer distal ones 7-13 cm; leaflets opposite, sessile, of longer pinnae 40-56 pairs, blades falcately linear-lanceolate from auriculate base, acute and mucronate at incurved apex, those near mid-rachis 7-12 x 1.2-2 mm, all lustrous dark green and veinless above, paler and weakly penniveined beneath, midrib only a little excentric. Inflorescence a short terminal panicle of narrowly cylindric, many-flowered spikes, axis of each, including short peduncle, 15-27 cm; bracts ovate-acuminate, less than 1.5 mm, caducous before anthesis; bracteoles 0. Flowers sessile, 5-merous, fragrant, in some spikes staminate but in most bisexual, calyx-lobes imbricate in early vernation, petals valvate, whole perianth usually glabrous but calyx sometimes puberulent; hypanthium 0; calyx campanulate, red, red-brown or purple, 2.4-3 mm, short teeth either

[24] by Rupert C. Barneby

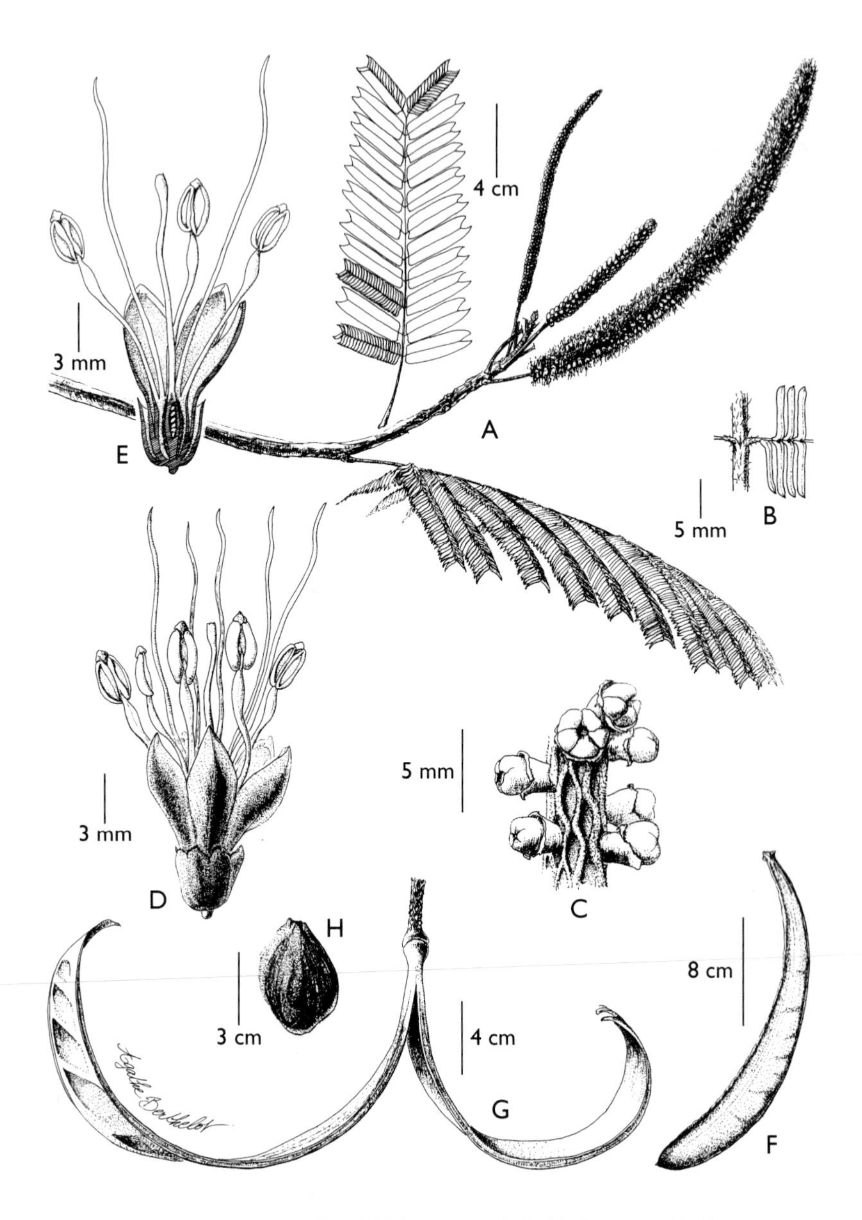

Fig. 36. *Pentaclethra macroloba* (Willd.) Kuntze: A, habit: leaves and inflorescences; B, detail of leaf-stalk and leaflets; C, inflorescence: detail of the rachis with flower-buds; D, flower; E, longitudinal section of flower; F, fruit, view of one valve ; G, fruit, profile view of the mature fruit after dehiscence; H; seed (A-E, de Granville B2646; F-H, Prévost 1354). Drawing by Agathe Berthelot; reproduced with permission from P.

deltate or semicircular, obtuse; corolla green or greenish-white, tube and ascending ovate lobes of ± equal length; androecium 10-merous, united with base of corolla into a stemonozone 1.2-2 mm, stamens dimorphic, 5 opposite sepals fertile 5-9.5 mm, 5 opposite petals sterile 17-31 mm, all filaments dilated distally, white, fertile anthers erect, dorsifixed below middle, 1.3-1.6 mm, thecae parallel, connective tipped with a caducous scutiform appendage; pollen shed in monads; ovary sessile pilosulous, style ± as long as fertile stamens, glabrous, stigma terminal poriform. Pod stiffly erect, in profile narrowly falcate-oblanceolate ± 25-35 x 3-5 cm, planocompressed, valves rigidly woody, glabrous, dark brown, framed by sutures 8-10 mm wide, vertically striate-venulose, dehiscence explosive, through both sutures, valves recurved but not or scarcely twisted; seeds 7-10, obliquely basipetal on short dilated funicle, compressed-obovoid, in broad view ± 4-4.5 x 2.3-3 cm.

Distribution: Interruptedly widespread in S America, from the middle and lower Amazon basin in Brazil to Trinidad and N Colombia, and to Nicaragua, introduced and naturalized on Martinique; in rainforest, on river banks where sometimes annually inundated, on wooded shores, and at edge of mangrove swamp, mostly below 250 m but ascending in Venezuelan Guayana and Guyana to 500-800 m; common nearly throughout Guyana, apparently less frequent in Suriname, known from French Guiana along the Oyapock R.; flowering in the Guianas throughout the year, most prolifically June to December.

Selected specimens: Guyana: ca. 5 km SW of Mabura Hill towards Essequibo R., Stoffers *et al.* 46 (NY); Winipero, Persaud 17 (NY). Suriname: Carolina and vicinity, Archer 2896 (NY). French Guiana: Cr. Cabaret, bassin de l'Oyapock, de Granville 10252 (NY); banks of the Oyapock R., Prévost 1354 (CAY).

Vernacular names: Guyana: trisel [trysel]; kroebara (Arawak); abarkasa-dek (Arelina).

Uses: The bark is used as a fish-poison; the plant in some form is said to provide medicine for wounds and in treatment of leishmaniasis; the seeds yield an industrial oil.

21. **PIPTADENIA**[25] Benth., J. Bot. (Hooker) 2: 135. 1840. – *Pityrocarpa* sect. *Orthocraspedon* Brenan, Kew Bull. 1955: 177. 1955.
Type: P. latifolia Benth., nom. illegit. = P. fruticosa (Mart.) Macbr. (Acacia fruticosa Mart.)

Piptadenia sect. *Pityrocarpa* Benth., J. Bot. (Hooker) 4: 339. 1841. – *Pityrocarpa* (Benth.) Britton & Rose, N. Amer. Fl. 23: 190 (pro gen.). 1928.
Type: P. moniliformis Benth.

Trees, shrubs and vines; either macro- or microphyllidious; either unarmed or aculeate; indument of short plain, often yellowish hairs, eglandular. Stipules mostly small, either persistent or not. Leaves bipinnate, pinnae and leaflets in opposite pairs; nectary cupular, mounded or clavate, on leaf-stalk and often also between some distal leaflet-pairs; pinnae 1-many (in the Guianas only 1 or 2 pairs); leaflets varying in number and size (in the Guianas only 3-8 pairs) and relatively ample, blade of distal pair 3-7 cm; venation of leaflets pinnate. Inflorescences composed of slender spikes either axillary to coeval leaves or collected into an exserted efoliate pseudoraceme or panicle. Flowers sessile, horizontal, 5-merous, diplostemonous; calyx campanulate, open in early bud; petals valvate in aestivation, either erect or recurved at anthesis, sometimes becoming free to base; no intrastaminal disc; filaments either free or shortly monadelphous, anther-sacs linear, parallel, connective (at least in bud) gland-tipped; pollen shed in 8-, 12- or 16-grained polyads; ovary either subsessile or raised on a stipe longer than itself. Pods either stipitate or subsessile, in profile broad-linear or oblong-elliptic, straight or almost so, sutures either straight or undulately constricted, valves papery or subcoriaceous, planocompressed, dehiscence through length of both sutures, valves inertly separating or narrowly gaping to release seeds; seeds plumply discoid, testa hard, areolate.

Distribution: American genus of ± 24 species, widespread from tropical Mexico to Bolivia and N Argentina, most numerous and diverse S from the Equator, absent from the West Indies; forest and scrub communities in climates subject to an annual dry season; in the Guianas 2 species.

Notes: *Piptadenia* was ultimately developed by Bentham (1876) into a comprehensive circumtropical genus characterized by spicate, diplostemonous flowers and gland-tipped anthers, but diverse in structure and dehiscence of the fruit and in form of the seeds. The generic concept has lately been reduced to a core of ± 20 tropical American species

[25] by Rupert C. Barneby

defined principally by an inertly dehiscent, broad-linear or oblong pod resembling that of *Acacia tenuifolia* (L.) Willd. and relatives. This more exactly defined genus comprises 2 sections:
sect. *Piptadenia*, with aculeate stems, and smooth, straight-edged pods
sect. *Pityrocarpa*, with unarmed stems, and scurfy, sinuately constricted pods.
Piptadenia floribunda represents the typical section; a syndrome of epidermal prickles, clavate petiolar nectaries, efoliate panicles, and tiny yellow flowers distinguishes it from all other spicate Guianan Mimosoideae. *Piptadenia leucoxylon* represents sect. *Pityrocarpa*.

KEY TO THE SPECIES

1 Prickly bush-ropes, stems and most leaf-stalks aculeate; pinnae of each leaf 2 pairs and leaflets 3-4 pairs, obovate; perianth yellow-silky-strigose overall; pod in profile oblong-elliptic, ± 3 cm wide, sutures not pinched between seeds, valves not scurfy .... *1. P. floribunda*
Unarmed trees; pinnae in most leaves 1 pair and leaflets 4-7(8) pairs, elliptic of rhombic-elliptic; perianth glabrous or only calyx thinly hispidulous; pod in profile undulately linear, ± 9-11 mm wide, sutures pinched between seeds, valves scurfy at maturity .... *2. P. leucoxylon*

1. **Piptadenia floribunda** Kleinhoonte in Pulle, Receuil Trav. Bot. Néerl. 30: 168. 1933. Type: Suriname, Brownsberg, BW 3186 (holotype U, isotype NY). – Fig. 37

Prickly bush-ropes potentially climbing into forest canopy; armed on stems and most leaf-stalks with declined or subhorizontal aculei; indument of young growth and of all axes of leaf and inflorescence dense, finely silky-strigulose, hairs subappressed, discolored or yellow. Stipules not seen. Leaves bipinnate, pinnae and leaflets in opposite pairs; leaf-stalks 7-12 cm, petiole ± 5-7 cm; nectary near base of petiole ascending, narrowly clavate 1.5-2 mm, small-pored at apex, similar ones at insertion of 2-3 furthest pairs of leaflets; pinnae 2 pairs, rachis of further pair 4-7 cm; leaflets 3-4 pairs, accrescent upward, obovate from subequilateral base, shortly acuminate, blades of furthest pair ± 4-6 x 2.3-3.7 cm; venation pinnate, midrib straight subcentric, giving rise on each side to 6-7 secondary veins, whole venation impressed on upper face, prominent beneath. Inflorescence composed of axillary and terminal, efoliate panicles of narrow amentiform spikes, primary axis of each panicle ± 2-4 dm, that of individual spikes, including short peduncle, 2-4.5 cm. Flowers horizontal, sessile, each subtended by a small bract, both perfect and staminate but all 5-merous, 10-androus, yellow-silky-

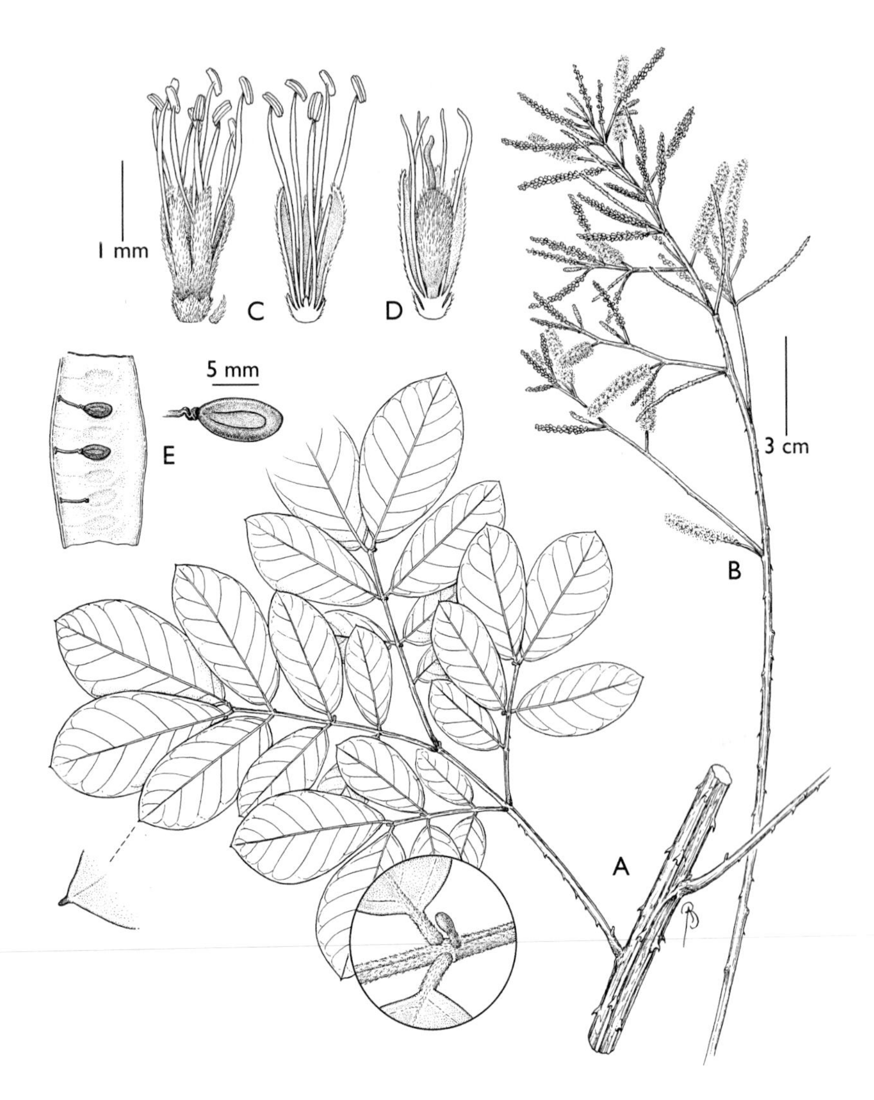

Fig. 37. *Piptadenia floribunda* Kleinhoonte: A, part of stem showing armed ridges and bipinnately compound leaf with prickles and extrafloral nectaries near petiole base and below insertion points of leaflets; detail of stalked nectary (in circle); B, inflorescences; C, lateral view (far left) and medial section (near left) of staminate flower with detail of bract (left); D, pistillate flower with perianth and androecium partially removed; E, part of dehisced fruit with seeds (left) and seed (right) (A-B, Mori & Bolten 8463l; C-E, Rabelo 3097, from Brazil). Drawing by Bobby Angell; reproduced with permission from Mori *et al.*, 2002: 505.

strigulose externally; calyx campanulate, open in bud, 0.6-0.8 mm, triangular teeth 0.1-0.25 mm; corolla 1.7-2 mm, narrowly ellipsoid prior to anthesis, valvate petals separating to middle but remaining erect; filaments 3-3.5 mm, united at base into a stemonozone 0.5 mm or less, anther-sacs linear-oblong, parallel, connective tipped with a caducous gland. Pod oblong-elliptic in profile, ± 10 x 3 cm, plano-compressed, 10-12-seeded, valves papery glabrate, framed by straight slender sutures, dull brown exocarp veinlesss, endocarp lustrous stramineous, dehiscence inert, through length of both sutures; seeds ellipsoid biconvex, testa hard, areolate.

Distribution: Apparently local in Suriname, French Guiana, and Brazil (Amapá); climbing into the canopy of non-inundated forest and draped over riparian bush-forest, ± 150-600 m; flowering Sept.-Nov.

Selected specimens: Suriname: Lely Mts., 175 km SSE of Paramaribo, Mori & Bolten 8436 (NY). French Guiana: on the road from Roura to Kaw Mt., Cremers *et al.* 9223 (NY).

Vernacular name: Suriname: boesi-branti.

Notes: *Piptadenia floribunda* is closely akin to Amazonian *P. uaupensis* Benth., but differs in clavate, not hemispherical, petiolar nectaries and in somewhat more numerous and smaller leaflets. The also related *P. imatacae* Barneby, described from upland forest (± 500 m) on the NE periphery of the Guayana Highland in Venezuela, is to be sought in Guyana. It has obtusely conical petiolar nectaries and yet more numerous and smaller leaflets (less than 3.5 cm), and the pod, similar in form and dehiscence, is puberulent and minutely glandular.

2. **Piptadenia leucoxylon** Barneby & J.W. Grimes, Brittonia 36: 236, fig. 1. 1984. Type: Venezuela, NE of Upata, Bolívar, de Bruijn 1750 (holotype NY!, isotype VEN!).

Unarmed trees (8)20-31 m; young branches and foliage pilosulous; leaflets drying dark brown above, cinnamon-brown beneath; narrow flower-spikes axillary to coeval leaves, immersed in foliage. Stipules deltate ± 1 mm, indurate, persistent; leaf-stalks 7-28 mm; nectary hemispherical, crumpled but obscurely pored, at tip of leaf-stalk and smaller ones between 1-2 furthest pairs of leaflets; pinnae of most leaves 1 pair, in exceptional random leaves 2 pairs, rachis of longer ones 3-9 cm; leaflets 4-7(8) pairs, strongly accrescent distally, blades elliptic or narrowly rhombic-ovate or ovate-elliptic, shortly acuminate and often subfalcate, distal pair 3-7 x 1-1.8 cm, midrib subcentric. Flower-spike including very short peduncle 5-9 cm. Flowers horizontal, sessile, 5-merous, glabrous except for thinly hispidulous calyx, all or most bisexual, perianth white brunnescent; calyx campanulate ± 1 x 1 mm, teeth scarcely 0.2 mm; corolla ± 2.5 mm, lobes recurved; filaments white, 3-4.5 mm; ovary glabrous, stipe ± 2-3.5 mm. Pod undulately linear in profile, straight or gently decurved, 9-14 x 0.9-1.1 cm, contracted at base into a stipe 8-11 mm, valves plane, dull brown, stiffly chartaceous, becoming scurfy when ripe, slender sutures undulately constricted between 8-12 seeds, dehiscence through both sutures, valves narrowly gaping; seeds plumply discoid 5.5-7 mm diam., testa dirty-white, smooth, areolate.

Distribution: Best known from Venezuelan Guayana (Bolívar, Delta-Amacuro), extending NE to Trinidad and S into Brazil (Roraima) and Guyana (Rupununi Savanna); in semideciduous forest, flowering in the canopy, locally dominant at 120-430 m.

Representative specimen examined: Guyana: Rupununi Savanna, Gillespie *et al.* 1688 (NY, US).

Vernacular name: near Venezuela-Guyana boundary: palo blanco (whence the epithet).

Note: The related *Piptadenia moniliformis* Benth., a smaller treelet or bush colonial by means of suckers and different in more numerous and smaller leaflets, is locally abundant in savanna country of the lower Orinoco valley and reappears in comparable habitats in NE Brazil. It should be sought in brush-woodland or grasslands in the interior of the Guianas.

---

22. **PITHECELLOBIUM**[26] Mart., Flora 20(2), Beibl. 8: 114. 1837, nom. et orth. conserv. (Camp *et al.*, Brittonia 6: 65. 1947).
Type: P. unguis-cati (L.) Benth. (Mimosa unguis-cati L.)

Macrophyllidious shrubs and trees, seldom less than 1 m or more than 10 m (but *P. dulce* potentially 20 m) tall; armed at nodes of long-shoots with lignescent stipules but these randomly lacking from flowering branchlets; indument of pallid hairs less than 0.5 mm, often confined to inflorescence or to a dorsal patch on leaflets. Leaves all exactly 4-foliolate (in Guianan species), consisting of 1 pair of pinnae, each 2-foliolate, axes shallowly

[26] by Rupert C. Barneby

openly sulcate, each bearing a sessile cupular nectary at tip, leaflets pinnately and reticulately venulose. Flowers capitulate (in Guianan species), sessile, homomorphic, pentamerous; calyx campanulate; corolla tubular-infundibuliform; androecium 22-42(57)-merous, filaments united into a tube; disc 0; ovary stipitate 8-14-ovulate. Pods sinuously linear in profile (in Guianan species), recurved or coiled and sometimes also twisted, sutures usually undulate, broad but not prominent, valves leathery, red-brown or fuscous, biconvex over seeds, endocarp orange-red, cavity continuous, dehiscence through both sutures, valves erratically recurved and twisted; seeds following dehiscence dangling on and invested through 1/3-2/3 their length by a red, pink, or whitish, spongy arilloid funicle, plumply biconvex, testa hard, lustrous black, with pleurogram; endosperm 0.

Distribution: A genus of 18 species, native of lowland tropical N and S America, 3 extending into subtropical Mexico, Bahamas, and Florida, 1 of these circumtropically cultivated and naturalized; in the Guianas 3 species, 1 native on savannas of S Guyana, 2 probably planted and dubiously naturalized on and near the coast of Guyana, to be expected elsewhere.

Notes: The genus *Pithecellobium*, as now generally received, is distinguished within tribe *Ingeae* by the syndrome of spinescent stipules (inconveniently not shown by many herbarium specimens) and a spongy funicular aril which nidulates the seed, this feature is unique in the tribe. The foregoing description applies to the typical sect. *Pithecellobium*, which has mostly 4-foliolate leaves, capitulate or only shortly spiciform flowers, fruits dehiscent through the length of both sutures, and black, pleurogrammate seeds.
The generic name *Pithecellobium* was derived from Greek *pithecos* (monkey)+*ellobion* (earring). It was modified by Bentham to *Pithecolobium*, as though the second element was from lobos (fruit). Although not strictly orthographic variants, *Pithecellobium*, *Pithecolobium*, and the briefly current compromise *Pithecollobium* are now treated as such.

KEY TO THE SPECIES

1 Peduncles at most nodes of the pseudoracemes solitary; corolla glabrous or finely silky-strigulose only on lobes; coastal plain of Guyana, probably only cultivated or escaped . . . . *3. P. unguis-cati*
Peduncles at most nodes of pseudoracemes fasciculate by 2-4; corolla silky-strigulose or finely pilosulous overall 2

2 Corolla 6-10 mm, androecium 17-25 mm, free filaments red-purple; native on savannas of SW Guyana . . . . *2. P. roseum*
Corolla 2.5-4.5 mm, androecium 7-11 mm, free filaments whitish ochroleucous; planted and perhaps naturalized locally in coastal Guyana, to be expected elsewhere . . . . *1. P. dulce*

1. **Pithecellobium dulce** (Roxb.) Benth., London J. Bot. 3: 199. 1844. – *Mimosa dulcis* Roxb., Pl. Coromandel 1(4): 67, pl. 99. 1798. Type: introduced in E coastal India from the Philippines, Roxburgh s.n. (holotype K!).

Trees fertile when 2-15(20) m, with 1-several trunks, a rounded crown, and pliantly pendulous flowering branches; randomly armed at nodes of long-shoots with pairs of stout ascending spinescent stipules up to 17 mm, but stipules of many young branches scarcely indurated, subulate and less than 2 mm; leaves olivaceous, either glabrous or finely puberulent; capitula small subspherical, of whitish, ochroleucous or faintly pink-tinged flowers borne in axillary and terminal pseudoracemes, mostly fasciculate by 2-4 and each fascicle subtended by a rudimentary leaf with nectary. Leaf-stalk of primary leaves 6-45 mm, dilated upward; a sessile or stoutly short-stipitate cupular nectary at tip of each leaf-stalk and one at tip of each of 2 pinna-rachises; leaflets obliquely elliptic, oblong-elliptic or obovate from inequilateral base, obtuse or emarginate, larger ones ± 2-5.5 x 1-1 cm; midrib subcentric, giving rise to 4-7 pairs of major secondary veins and an open mesh of sinuous tertiary venules, all usually prominulous on both faces of blade. Longer peduncles 4-20 mm; floral receptacle 1-2.5 mm; flowers bisexual or in some capitula mostly staminate; perianth finely gray-silky-puberulent overall; calyx 1.2-2 mm, teeth 0.1-0.4 mm; corolla 2.5-4.5 mm; androecium 25-42-merous, 7-10.5 mm, tube (1)1.4-3.2 mm; ovary minutely pilosulous, raised on a stipe ± 1-2 mm. Pod in profile undulately linear and decurved through ± 1-2 circles to form an open ring or equivalently contorted, body when well fertilized 8-12-seeded, either shallowly or deeply constricted between seeds, 9-17.5 cm long and at each seed 8-17 mm wide, at maturity subcompressed but biconvex over each seed, valves thinly fleshy, reddish-brown, bright red, or green and red-cheeked becoming stiffly papery and coarsely venulose, pilosulous when young but often glabrate at maturity, endocarp reddish-brown, dehiscence through both sutures, valves recurving and coiling; seeds nidulated by scalloped, white or reddish-pink aril, plumply lentiform 7-13 x 6-11 mm.

Distribution: Native in Mexico, C America, inter-Andean Colombia, and W Venezuela; seasonally dry brush and woodland communities in lowland, from sea level up to ± 1550 m; dispersed in early colonial times to West Indies, Philippines, and tropical Asia, where extensively naturalized, and later planted and escaping to waste lots in subtropical Florida, Hawaiian Islands, and elsewhere; a common ornamental and shade tree in Latin America; known from Georgetown and vicinity (Demarara) in Guyana, planted and apparently naturalized locally (GU: 2).

Selected specimens: Guyana: Georgetown, Hitchcock 16838 (NY); Kitty, E coast Demerara, Omawale & Persaud 156 (NY).

Vernacular name: bread-and-cheese.

Use: The astringent bark is employed in folk medicine and furnishes a yellow dye. The acidulous aril is edible, sought by children and by monkeys, and seen often in Mexican markets.

2. **Pithecellobium roseum** (Vahl) Barneby & J.W. Grimes, Mem. New York Bot. Gard. 74(2): 21. 1997. – *Mimosa rosea* Vahl, Eclog. Amer. 3: 33, t. 25. 1807. Type: "Cayenne", von Rohr s.n. (holotype C!), but not since seen in French Guiana, probably collected in northern Colombia. – Fig. 38

Arborescent shrubs 2.5-10 m; randomly armed at nodes of sterile branches with straight, ascending or horizontal, spinescent stipules 3-27 mm (but most fertile branchlets unarmed), resembling *P. dulce* in 4-foliolate leaves but differing in longer flowers and showy purple filaments; leaflets ample, varying from glabrous to finely pilosulous on one or both faces; capitula mostly paired or fasciculate at nodes of axillary and terminal, either efoliate or depauperately foliate pseudoracemes, subtended by a bract, or by a rudimentary leaf-stalk with nectary, or by a small leaf. Leaf-stalks of primary leaves 1.5-5.5 cm; gland at tip of leaf-stalk and of each pinna-rachis subsessile, cupular 0.5-1.3 mm diam.; leaflets obliquely elliptic to ovate- or lance-elliptic or incipiently rhombic-ovate, larger ones 4.5-10 x 2-4.5 cm. Axis of pseudoracemes 6-20 cm; longer peduncles 1-4(6) cm; capitula (without filaments) hemispherical or shortly oblong; perianth greenish, often red-tinged or -tipped, finely silky-strigulose overall; calyx 1.8-3.5 mm, teeth 0.15-0.5 mm; corolla 6-10 mm, lobes 1.2-2.5 mm; androecium 17-25 mm, tube 5.5-10.5 mm, either a little shorter than corolla or shortly exserted; ovary either micropapillate, or minutely puberulent, or glabrous, stipe 1.6-3.5 mm. Pod like in *P. dulce*, 7-12-seeded, 9-13 mm wide at each seed, valves either minutely puberulent or glabrate.

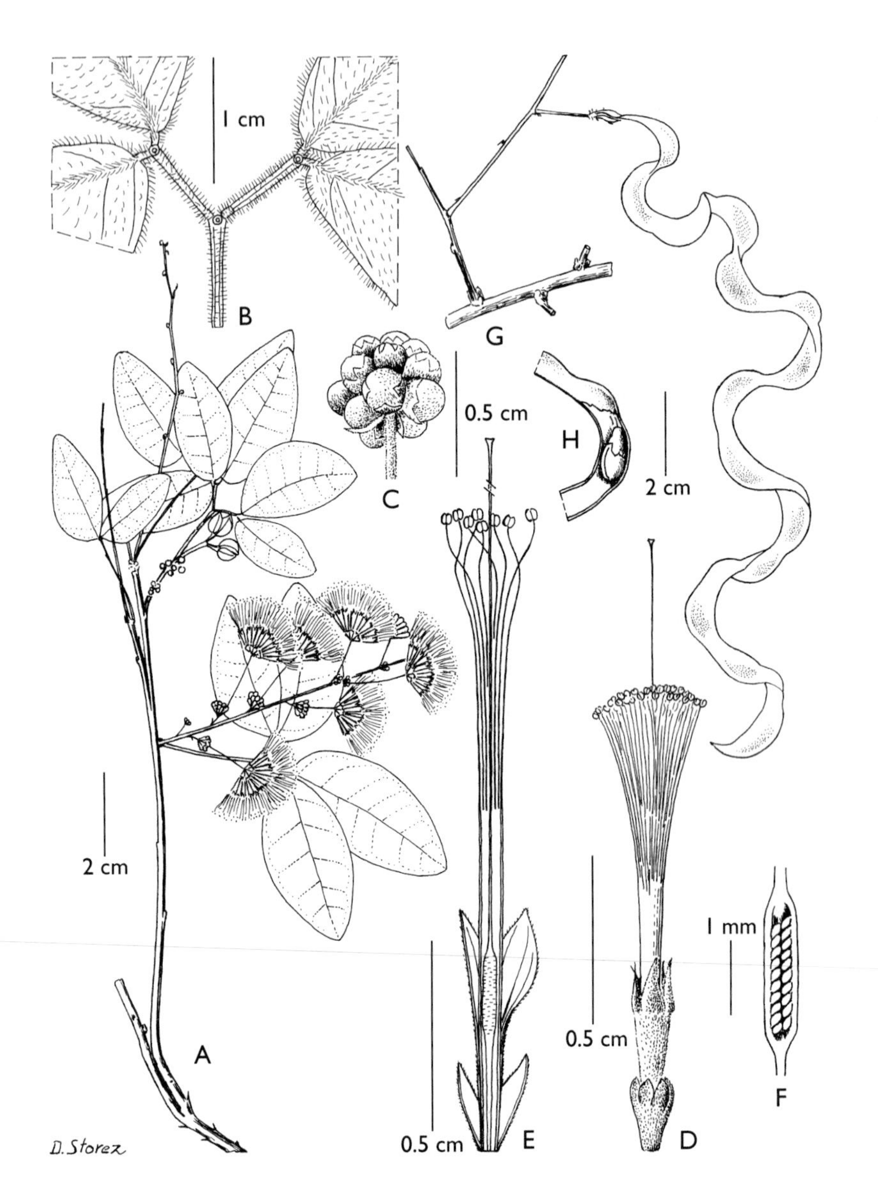

Fig.38. *Pithecellobium roseum* (Vahl) Barneby & J.W. Grimes: A, habit: twig, leaves and inflorescences; B, leaf: detail of leaf-stalk, petiole, petiolules, nectaries; C, young inflorescence with buds; D, flower; E, longitudinal section of flower; F, ovary, longitudinal section; G fruit; H; arillate seed (A-H, Jansen-Jacobs *et al.* 471). Drawing by Dominique Storez; reproduced with permission from P.

Distribution: Interruptedly dispersed from N Colombia to Trinidad, to NE Venezuelan Guayana and disjunctly to S Guyana and adjoining state of Roraima, Brazil; in seasonally dry, semideciduous thickets, in xeromorphic brush-woodland, and in savanna tree-islands, between sea level and ± 350 m (GU: 2).

Selected specimens: Guyana: Rupununi Savanna, Mountain Point, Jansen-Jacobs *et al.* 460 (NY, U): Rupununi Savanna, Chaakoitou, near Mountain Point, Maas & Westra 4122 (NY, U).

3. **Pithecellobium unguis-cati** (L.) Benth., London J. Bot. 3: 200. 1844. – *Mimosa unguis-cati* L., Sp. Pl. 517. 1753. Type: attributed to Jamaica and the Caribbean islands, and based on a) a barren plant grown in Clifford's garden at Hartekamp, Netherlands, and b) prior descriptions by Sloane and Plukenet, these all presumed conspecific and in accord with established usage, but explicit lectotypification awaited from the Linnaean typification project.

Closely resembling *P. dulce*, but smaller, mostly less than 10 m; subglabrous (in Guyana), 4-foliate leaves and flowers completely so. Leafstalks 6-30 mm; pinna-rachises 3-16 mm; petiolar nectaries 0.3-0.8 mm diam.; leaflets obovate-suborbicular, ± 2-5 x 1-4.5 cm (in Guyana). Primary axis of pseudoracemes 3-13 cm, peduncles all or almost all solitary, very slender, 1-3 cm; flowers capitate (in Guyana); calyx ± 1.5 mm; corolla ± 5 mm; androecium ± 12 mm, tube 4 mm; ovary glabrous, stipe 2-3 mm. Pod like in *P. dulce*, when well fertilized 7-17 x 0.7-1.4 cm, 7-12-seeded.

Distribution: Tropical Mexico, C America, West Indies, and along Carribean coast of Colombia and Venezuela, planted and naturalized elsewhere, known from coastal Guyana by one collection; native in semideciduous woodland, scrub, and coppice, from sea level to ± 550 m (GU: 1).

Representative specimen examined: Guyana: Jenman 2113 (K, NY).

Note: The native status of the Guyana plant is doubtful. The species is highly polymorphic.

23. **PLATHYMENIA**[27] Benth., J. Bot. (Hooker) 4: 333. 1842.
Type: P. reticulata Benth.

For description see under species.

Distribution: A monospecific genus dispersed from Paraguay, southern Brazil and Bolivia N to the Guianas.

Taxonomic note: The genus has been revised by Warwick & Lewis (2003) since submission of this manuscript and updates have been made accordingly.

LITERATURE

Warwick, M. & G.P. Lewis. 2003. Revision of Plathymenia (Leguminosae-Mimosoideae). Edinburgh J. Bot. 60: 111-119.

1. **Plathymenia reticulata** Benth., J. Bot. (Hooker) 4: 334. 1842. Type: Brazil, Minas Gerais, Clausen & Delessert 110 (lectotype K!).

Unarmed tree flowering when 3-7 m tall; young growth sparingly to densely hispidulose-villose, becoming glabrate with age. Leaves with 10-22 alternate to subopposite pinnae, and 26-38 alternate leaflets; stipules lacking, replaced by small ± reniform to round nectaries flush with or immersed in bark; lf-rachis 13.6-26 cm, petiole proper, including wrinkled, darkened pulvinus, 1.7-5.0 cm, petiolar nectaries wanting; longer pinnae 5.1-9.5 cm, petiolulate leaflets ovate-elliptic, sometimes falcately so, 7-15 x 3-7 mm, dark above, paler beneath, apiculate at apex, rounded at base, sometimes obliquely so. Inflorescence of axillary spikes usually 1 per node; peduncle 12-21 mm, rachis 5-9 cm; pedicel ± 0.5-0.75 mm; calyx jointed to pedicel, cupular-campanulate, ± 1-1.5 mm, teeth deltate; petals 5, linear-lanceolate to linear-elliptic, 3-5 x 0.5-1 mm, free above stemonozone; stamens 10, stemonozone ± 0.25-0.3 mm; ovary pilose, elevated on a gynophore 1-1.25 mm, this glabrous or distally pilose. Pods 1-5 per spike, elliptic to oblanceolate in outline, 9.5-18.5 x 1.4-2.2 cm, rounded above, at base gently or abruptly attenuate to a stipe 1.5-3 cm, sutural ribs only slightly enlarged, valves smooth, endocarp coherent between seeds, becoming free from mesocarp and deciduous as small tesserae enclosing seed; seeds (not seen mature) round-ovate, pleurogram complete.

[27] by James W. Grimes

Distribution: From Paraguay, Bolivia and Brazil (Goiás, Minas Gerais, Maranhão, Pará and Amapá), barely entering the Guianas in S Suriname; primarily in cerrado, but entering savanna, terra firme and gallery forest, and ascending to rocky ridge tops, flowering mostly September to November (SU: 2).

Selected specimens: Suriname: Sipaliwini savanna area on Brazilian frontier, Oldenburger *et al.* 330 & 454 (NY, U).

24. **PSEUDOPIPTADENIA**[28] Rauschert, Taxon 31: 559. 1982.

Type: P. leptostachya (Benth.) Rauschert (Piptadenia leptostachya Benth.)

Newtonia sect. *Neonewtonia* Burkart in Reitz, Fl. Illustr. Catarinense LEGU 1: 285. 1979. – "American Newtonia" Lewis & Elias in Polhill & Raven (eds.), Advances Leg. Syst. 1: 156, 166. 1981.
Type: N. nitida (Benth.) Brenan (Piptadenia nitida Benth.)
Piptadenia sect. *Eupiptadenia* Benth., Trans. Linn. Soc. London 30: 366. 1875, in small part, excluding the type.
Monoschisma Brenan, Kew Bull. 10: 179. 1955, later homonym.
Pseudopiptadenia sensu G.P. Lewis & M.P. Lima, Arch. Jard. Bot. Rio de Janeiro 30: 43-67. 1991.

Unarmed trees, with buttressed trunk; indument of fine gray hairs. Stipules small, fugacious, or lacking. Leaves bipinnate, pinnae in (sub) opposite pairs, leaflets exactly opposite or sometimes alternate, sessile, small and multijugate (Guianan species). Petiolar nectary sessile concave. Inflorescences composed of narrow, shortly pedunculate spikes solitary or geminate in axil of coeval leaves or terminally paniculate. Flowers 5-merous, diplostemonous, horizontal, silky-puberulent externally; calyx campanulate short-toothed, open in early bud; petals valvate, at full anthesis erect, separating 1/3 their length or finally to base; disc 0; filaments whitish, 2-3 times as long as corolla, anthers parallel, connective tipped with a caducous gland; pollen shed in polyads; ovary pilose, stipitate, style filiform glabrous, 1-2 times as long as ovary, stigma porose. Pods stipitate, in profile elongately linear, either straight, or decurved, or randomly twisted, planocompressed, valves leathery fuscous, glabrous, exocarp and endocarp permanently coherent, dehiscence follicular, through ventral suture, valves narrowly gaping to release anemochorous seeds; seeds attached shortly above base to horizontal funicle, basipetal, planocompressed, in broad view oblong-elliptic or broad-linear, 17-55 mm, testa membranous lustrous translucent, broadly winged; endosperm 0.

[28] by Rupert C. Barneby

Distribution: A genus of 11 species dispersed from N Colombia, Venezuela, and the Guianas through Peruvian and Brazilian Amazonia to NE Bolivia and warm-temperate SE Brazil; in the Guianas 2 species.

Note: The pod and seeds of the Guianan species of *Pseudopiptadenia* closely resemble those of *Anadenanthera*, which differs most obviously in capitulate, not spicate flowers. Lewis & Lima (1991: 56) consider the 2 species recognized below as, at best, only varietally distinct, but the material examined is neatly divisible by the key characters that follow.

LITERATURE

Lewis, G.P. & M.P. Lima. 1991. Pseudopiptadenia no Brasil (Leguminosae: Mimosoideae). Arch. Jard. Bot. Rio de Janeiro 30: 43-67.

KEY TO THE SPECIES

1 Petiolar nectary vertically elongate, mostly 4.5-14.5 mm, its rim as wide as or wider than ventral sulcus of petiole; larger leaflets 7.5-11.5 x 2.5-4.5 mm; pod (14)18-24 mm wide .... *1. P. psilostachya*
Petiolar nectary often obsolete, when present 2-6 mm in long diam. and sunk into ventral sulcus of petiole; longer leaflets 5-9 x 1.4-2.3 mm; pod 9-15 mm wide .... *2. P. suaveolens*

1. **Pseudopiptadenia psilostachya** (DC.) G.P. Lewis & M.P. Lima, Arch. Jard. Bot. Rio de Janeiro 30: 55. 1991, excluding syn. *P. suaveolens*. – *Acacia psilostachya* DC., Prod. 2: 457. 1825. – *Piptadenia psilostachya* (DC.) Benth., J. Bot. (Hooker) 4: 336. 1841. – *Newtonia psilostachya* (DC.) Brenan, Kew Bull. 1955: 182. 1955. Type: French Guiana, Cayenne, Martin s.n. (holotype G-DC!, isotypes K! = NY Neg. 1753, P!). – Fig. 39

Unarmed trees 12-40 m with buttressed trunk 2-10 dm DBH; leaf- and inflorescence-axes minutely puberulent but firm bicolored leaflets glabrous or almost so, lustrous dark green above, paler beneath; long, very narrow flower-spikes either solitary and fasciculate in coeval leaf-axils or collected into a short terminal panicle. Stipules 0. Leaf-stalks 9-22 cm, petiole 2-5.5 cm; nectary vertically elongate, sessile, shallowly patelliform, (2)4.5-14.5 mm, near or above mid-petiole, a smaller one at tip of leaf-stalks and yet smaller ones between furthest 1-2 pairs of leaflets; pinnae 6-10 pairs, rachis of longer ones (6)7-11 cm; leaflets of longer pinnae 21-32 pairs, sessile against rachis, oblong from obliquely

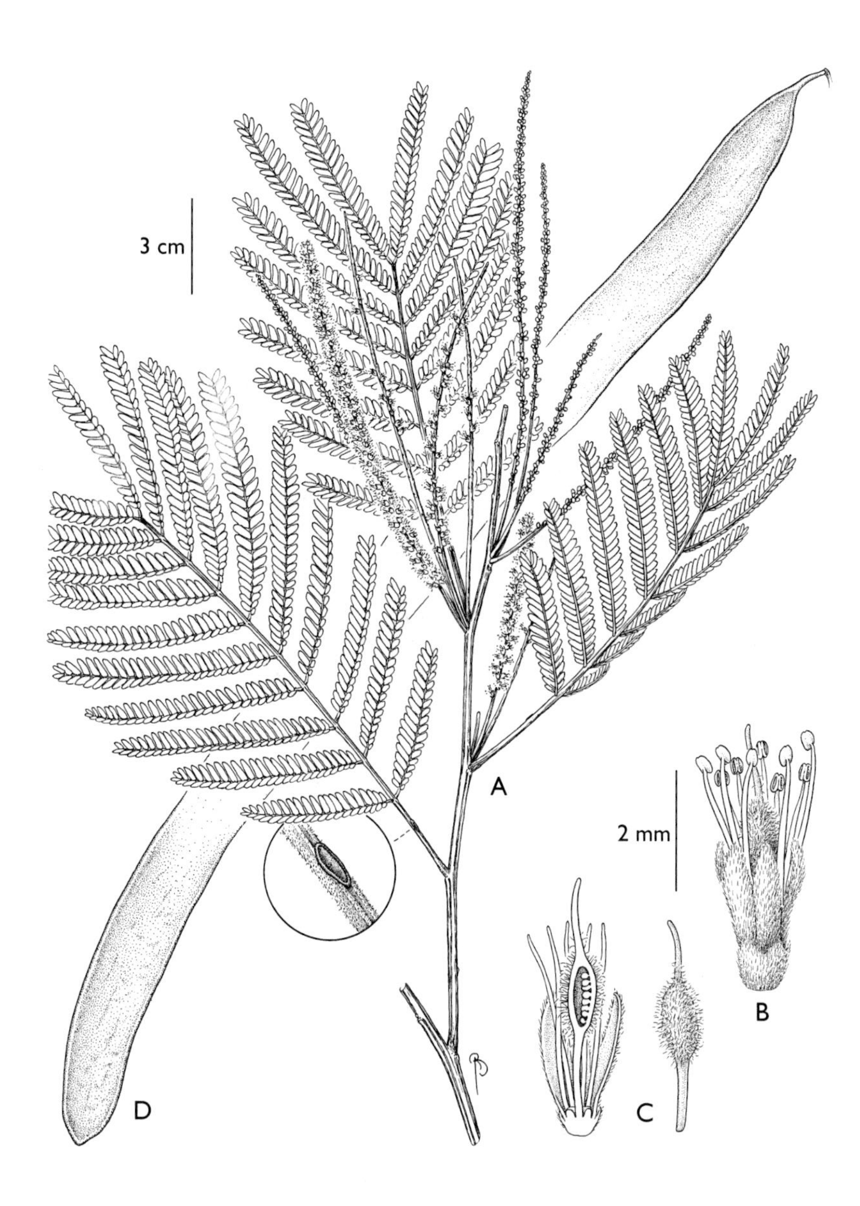

Fig. 39. *Pseudopiptadenia psilostachya* (DC.) G.P. Lewis & M.P. Lima: A, part of stem with bipinnately compound leaves and spicate inflorescences; detail of extrafloral nectary (in circle); B, lateral view of flower; C, medial section (left) of flower showing gynoecium surmounting gynophore and subtended by a nectary disc, and gynoecium (right); D, fruit (A-C, Irwin *et al*. 54449; D, Service Forestier 7226). Drawing by Bobby Angell; reproduced with permission from Mori *et al*., 2002: 506.

truncate base, obtuse or obscurely apiculate, straight or shallowly sigmoid, longer ones (near mid-rachis) 7.5-11.5 x 2.5-4.5 mm; venation palmate, of 3-5 veins from pulvinule, midrib only slightly excentric, simple or few-branched above middle, inner posterior vein produced well beyond mid-blade. Spike including short peduncle 7-13 cm; bracts minute fugacious; perianth silky-puberulent overall, corolla densely so; calyx campanulate 1.2-1.5 mm, deltate teeth ± 0.2 mm; corolla 2.7-3 mm, lobes erect; filaments white, 3.5-4.5 mm; ovary pilosulous ± 1.4 mm, raised on a stipe, 1-1.8 mm, style linear, 0.6-1.5 mm, glabrous. Pod pendulous, in profile linear, straight or nearly so and seldom twisted, stipe 1-2.5 cm, body (15)25-52 x (1.4)1.8-2.4 cm; seeds ± 4.5-5.5 x 1-1.5 cm.

Distribution: Interruptedly widespread in Brazilian Amazonia (Roraima, Amapá, Pará and Maranhão), and the Guianas, related forms in SW Amazonia and coastal Bahia; in non-inundated forest, often in hill-country, mostly 130-600 m, but lower S-ward, flowering August to December.

Selected specimens: Guyana: near Chodikar landing, FD 7459 (= G 444) (NY); 3 miles below Bunawau mouth, Mapuera R., FD 7597 (= G 582). French Guiana: Mt. Galbao, de Granville *et al.*, 8846 (NY); Approuague R., Arataye R., Saut Pararé, Sastre 5921 (NY).

Vernacular names: Guyana: shawaratu, yaifufichó.

2. **Pseudopiptadenia suaveolens** (Miq.) J.W. Grimes, Brittonia 45: 27. 1993. – *Piptadenia suaveolens* Miq., Linnaea 18: 589. 1844. – *Newtonia suaveolens* (Miq.) Brenan, Kew Bull. 1955: 182. 1955. Type: Suriname, Bergendal, Focke 936 (holotype U). – Fig. 40

Closely resembling *P. psilostachya* in habit, indument, and inflorescence, but potentially taller, attaining 35-60 m, with strongly buttressed trunk, and with slightly different leaf-nectaries and leaflets. Leaf-stalks (7)10-22 cm, petiole 3.5-6 cm; petiolar nectary often suppressed, when present sunk in ventral sulcus, linear-elliptic, 2-6 mm; pinnae 7-11(12, in Peru to 17) pairs, rachis of longer ones 4.5-8.5 cm; leaflets of longer pinnae 23-33 pairs, longer ones 5-9 x 1.4-2.3 mm, blades mostly 2-veined from pulvinule (sometimes 1-2 faint outer posterior in addition). Spike including short peduncle 5-13 cm; flowers of *P. psilostachya* but slightly smaller, greenish-white or yellowish, strongly disagreably fragrant, calyx 0.7-1 mm; corolla 2.3-2.7 mm, its ascending lobes ± 1/3 as long as tube; filaments white or ochroleucous 4.3-5.5 mm; ovary pilosulous 1-1.7 mm, glabrous stipe (1)1.6-2.7 mm, style 1.4-2.2 mm. Pod like that of *P. psilostachya* but narrower, when well fertilized 20-42 x 0.9-1.5 cm; seeds including broad wing ± 36-46 x 8-10 mm.

Fig. 40. *Pseudopiptadenia suaveolens* (Miq.) J.W. Grimes: A, flowering branch, and flower (right); B, pod, and seed (upper left) (A, Schulz LBB 10324; B, Sabatier 2121). Drawing by Hendrik Rypkema; reproduced with permission from U.

Distribution: Tricentric range in S & C America: lower Amazonian Brazil (Pará, Amapá), whence N through the Guianas into NE Venezuela (Bolívar, Delta-Amacuro), distantly disjunct near W and SW periphery of the Amazon Basin in Peru, Brazil (Rondônia), and Bolivia, and remotely so on Osa Penisula in Costa Rica; in non-inundated moist forest below 400 m, flowering August to December.

Selected specimens: Guyana: Gunn's, Essequibo R., Jansen-Jacobs *et al.* 1843 (NY, U); Mazaruni Station, FD 2491 (= D 496) (NY). Suriname: Brownsberg, tree nr. 1117, BW 6874 (NY). French Guiana: Mts. la Fumée, Mori & Boom 15309 (NY); Piste-de-St-Elie, between Sinnamary/Counamama Rs., Sabatier & Prévost 3684 (NY).

Vernacular names: Suriname: kousá, koesa. French Guiana: alimiao; pitimissiki (Paramaka).

25. **SAMANEA**[29] (Benth.) Merr., J. Washington Acad. Sci. 6: 46. 1916. – *Pithecolobium* sect. *Samanea* Benth., London J. Bot. 3: 197, 215. 1844.
Type: S. saman (Jacq.) Merr. (Mimosa saman Jacq.)

Trees, some attaining great age and size; branching sympodial, new growth arising yearly from buds below prior season's determinate inflorescence; lfts folding forward slowly at night or after shock. Leaves alternate, bipinnate, lf-formula iii-vi(vii)/3-8(9); stipules herbaceous, lanceolate, early deciduous; lft-venation pinnate. Capitula axillary to coeval or quickly hysteranthous lvs, umbelliform; fls heteromorphic, peripheral ones at least shortly pedicellate, terminal one stouter, sessile. Calyx of peripheral fls vase-shaped, 4.5-11 mm; corolla 7-14.5 mm; androecium (16)20-36-merous, tube included, tassel pink or reddish. Pods subsessile, in profile broad-linear, straight or nearly so, either compressed but plump or biconvex, valves composed of thin, glabrous or densely puberulent exocarp, mesocarp thick pulpy, when ripe pitchlike, endocarp crustaceous-lignescent either adherent or narrowly septiferous between seeds, indehiscent, seeds released by weathering, or by predators, or following excretion by cattle; seeds oblong-ellipsoid, testa hard opaque, closely investing embryo, areolate.

Distribution: Genus of 3 species, native from El Salvador to E Brazil, Paraguay, and NE Bolivia, but 1 species (*S. saman*) planted and subspontaneous into Mexico and the West Indies.

[29] by James W. Grimes

KEY TO THE SPECIES

1 Leaflets of longer pinnae 3-5(6) pairs; 1 petiolar nectary at or close to base of petiole, none between pinna-pairs; interfloral bracts linear or linear-oblanceolate, 0.25-1 mm wide, never concealing the flower buds; corolla of terminal flower 14-27 mm; valves of pod strongly convex, horizontally wrinkled and deeply fissured across its whole width from suture to suture, not wrinkled lengthwise; seeds 20-30 . . . . *1. S. inopinata*

Leaflets of longer pinnae 6-8(9) pairs; petiolar nectaries always between first and furthest pinna-pairs and often between all (none at base of petiole); interfloral bracts spatulate, with linear claw and rhombic blade 1-2.3 mm wide, in early praefloration clasping the young flower buds; corolla of terminal flower 9.5-14.5 mm; face of each valve of pod depressed and nearly plane, but shallowly corrugate lengthwise, exocarp continuous, neither transversely fibrous nor wrinkled; seeds 12-20 . . . . . . *2. S. saman*

1. **Samanea inopinata** (Harms) Barneby & J.W. Grimes, Mem. New York Bot. Gard. 74: 123. 1996. – *Serianthes inopinata* Harms, Notizbl. Bot. Gart. Berlin-Dahlem 11: 55. 1930. Type: Brazil, cultivated in the Bot. Garden at Rio de Janeiro [as] *Pithecellobium saman*, Ducke HJBR 15248 (holotype †B, isotypes K!, RB!, U!).

Pithecellobium nuriensis [sic] H.S. Irwin, Mem. New York Bot. Gard. 15: 107, fig. 4. 1966. Type: Venezuela, Edo. Bolívar, between Rancho Alegre and base of cerro, on trail to Quebrada Cabeza Burro, 5 km E of Chicharras, Steyermark 89317 (holotype NY (2 sheets)!).

Trees potentially attaining 40 m with trunk to 1.5 m diam., but commonly smaller; young stems, all lf-axes and peduncles puberulent or pilosulous with pallid, either straight and spreading-ascending or shorter incumbent hairs. Lf-formula iii-iv/3-5; stipules narrowly lanceolate, ± 4-6.5 x 1-1.3 mm; lf-stk of lvs on flowering branchlets 12-20 cm (of lvs from sterile or juvenile stems, no further mentioned, up to 45-55 cm), petiole 4-8 cm, longer interpinnal segments 3-5.5(6.5) cm; rachis of distal pinnae 4.5-11 cm, longer interfoliolar segments 1.7-3 cm; lfts bicolored, dull olivaceous and minutely or remotely puberulent on upper face, on lower face paler and more densely pilosulous (especially along veins), blade of penultimate pair ± 4-6 x 2.1-2.5 cm. Peduncle 6.5-15 cm; capitula 12-20-flowered; receptacle including terminal pedestal 1.5-4 mm; interfloral bracts linear-oblanceolate, 3-10.5 x 0.45-1 mm. Peripheral fls: pedicel of lower ones 4-9.5 x 0.4-1 mm; calyx 7-11 x 2.7-4 mm; corolla 14-18.5 mm; filaments pink or purple at tips, white at base. Terminal fl: differentiated apparently as that of *S. saman*. Pod 1-2 per peduncle, sessile or almost so, in profile broad-linear, 11-25 x 1.5-3.5 cm, only a little compressed and 1-2 cm thick, straight or randomly bent sidewise but not arcuately recurved, girdled by prominent, dorsally either plane or shallowly sulcate sutures 4-6 mm wide, valves densely puberulent, often subvelutinous, closely and deeply transversely wrinkled from suture to suture, internally composed of crustaceous or ligneous endocarp, 0.2-1 mm thick in section and alveolate pulpy (pitchlike) mesocarp, 2.5-6 mm thick, cavity divided by septa into 1-seeded locules, 5-6 mm long.

Distribution: Venezuelan Guayana, Guyana, S to Brazil (Maranhão, Bahia, and Pernambuco); in forest, or in caatinga forest, at 100-700 m elev.; flowering October through March (GU: 1).

Specimen examined: Guyana: Kanuku Mts., drainage of the Takatu R., A.C. Smith 3316 (NY).

2. **Samanea saman** (Jacq.) Merr., J. Washington Acad. Sci. 6: 47. 1916. – *Mimosa saman* Jacq., Fragm. Bot. 15, t. 9. 1801. – *Pithecolobium saman* (Jacq.) Benth., London J. Bot. 3: 216. 1844. Type: Described from plants grown at Vienna from seeds "ex patria Caracas [Venezuela]" (holoype W, not seen, but the illustrated protologue decisive).

Trees of potentially great age and size, flowering as treelets but becoming 25(30), exceptionally 40-50 m tall with short stout trunk up to 2(3) m DBH; hornotinous branchlets, all leaf-axes and peduncles densely pilosulous with fine whitish, either straight spreading or shorter incumbent hairs. Lf-formula iii-vi(vii)/6-8(9); stipules herbaceous, lanceolate or oblanceolate, 2.5-7 mm, densely pilosulous, early caducous; lf-stks of fully developed lvs (4.5)6-25 cm, petiole 3-7.5(8.5) cm, at middle 1.4-3 mm diam., first interpinnal segment (1.2)1.7-4.7 cm, further ones progressively shorter; pinnae strongly accrescent distally, rachis of furthest and penultimate pairs 6-14.5 cm, longer interfoliolar segments 10-22(25) mm; lfts bicolored, dark-olivaceous, glabrous and lustrous on upper face, beneath pallid dull and pilosulous, conspicuously accrescent distally, all but rhombic-obovate and largest terminal pair obtusely rhombic-oblong or -elliptic, blade of penultimate pair (20)24-62 x (8)10-25(27) mm, (1.9)2.1-2.8 times as long as wide. Peduncle (3.5)4-8.5 cm, woody in fruit; capitula hemispherical, (12)15-22(25)-flowered; bracts 3-7 x 1-2.3 mm, heteromorphic, lowest lance- or oblance-elliptic, interfloral ones spatulate, linear claw abruptly expanded into a rhombic blade, all deciduous at anthesis, when young folded over and concealing young flower buds. Flowers heteromorphic, peripheral ones at least shortly pedicellate, terminal one stout, sessile; perianth of all 5-6-, or that of terminal fl 5-9-merous, densely yellowish-pilosulous-tomentulose overall (corolla tube sometimes glabrescent). Peripheral fls: pedicel compressed, up to 1-3.5(5.5) x 0.3-0.6(0.7) mm; calyx slenderly vase-shaped, (4.4)4.8-7(7.8) x 1.3-2 mm; corolla 9-13(13.5) mm, slenderly trumpet-shaped. Terminal fl: sessile; calyx (5.5)7-10.2 x 2.5-4.3 mm; corolla 9.5-14.5 mm. Pod 1(2) per capitulum, broad-linear abruptly contracted at base into a short neck and as abruptly cuspidate at apex, compressed but plumply fleshy, when well fertilized straight or almost so and 10-22 x (1.4)1.7-2.3 x 0.6-1 cm, (10)12-20-seeded, coarsely bicarinate by plane or shallowly sulcate, woody sutures ± 2.5-3.5 mm wide, low-convex valves consisting of a) thin, continuous, when dry papery, livid-castaneous or blackish, ± lustrous, irregularly crumpled, glabrate or minutely puberulent exocarp; b) thin crustaceous endocarp, adherent between seeds to form discreet seed-cavities but not septiferous; and c) thick alveolate mesocarp forming a black, when dry pitchlike (sweet, nutritious) pulp; seeds plumply oblong-ellipsoid, 8-11.5 x 5-7.5 mm.

Distribution: The Orinoco valley and Caribbean slope in Venezuela, in N Colombia, and in C America north perhaps to El Salvador; extensively cultivated both within and outside its presumed natural range; native in and at margins of seasonally dry, deciduous and semideciduous as well as moister evergreen woodland and savanna (llano), mostly below 450 m, but occasionally to 1400 m or more; flowering sporadically throughout the year, but most abundantly toward the end of the dry season (GU: 2; FG: 2).

Selected specimens: Guyana: W Demerara Region, Georgetown, Guyana Botanical Garden, Pipoly 7316 (NY); NW-Distr., Amakura R., de la Cruz 3480 (NY). French Guiana: cultivated in the Garden of ORSTOM, Cayenne, Leeuwenberg 11691 (CAY, NY); Oldeman 1416 (CAY, U).

26. **STRYPHNODENDRON**[30] Mart., Flora 20(2), Beibl. 8: 117. 1837. Lectotype (Britton & Rose, 1928): S. barbatimam Mart. = S. adstringens (Mart.) Coville (Acacia adstringens Mart.)

Unarmed trees, arborescent shrubs, and functionally herbaceous subshrubs from xylopodium; indument of fine brown or gray, simple hairs and often in part of minute, amorphously coralloid trichomes. Stipules very small and caducous, or obsolete. Leaves bipinnate, pinnae opposite or alternate; nectary sessile either between first pair of pinnae or on petiole well below them; lfts variable in size and number, either opposite or alternate, venation pinnate or simple. Inflorescences composed of solitary or fasciculate, densely many-flowered, amentiform spikes borne either in contemporary leaf-axils or in efoliate pseudoracemes or panicles, each spike shortly pedunculate and peduncle charged at top with 2 bracts, these either caducous from oblique scars or united into a cuplike involucre. Flowers

[30] by Rupert C. Barneby

sessile, horizontal to spike's axis, 5-merous diplostemonous; flower buds prior to anthesis oblong; calyx campanulate, short-toothed, open in bud; petals valvate in aestivation, at anthesis separating 1/3-1/2 their length; filaments free or united at very base, either into a free tube or into a short stemonozone, anther-sacs linear, parallel, connective gland-tipped in bud; pollen shed in polyads. Pods sessile or almost so, in profile linear-oblong, either straight or falcately to circinnately incurved, compressed but biconvex, ± pulpy when ripe but granular-fibrous and thick-walled when dry, endocarp raised between seeds as an incipient partition, dehiscence essentially 0, valves becoming brittle and seeds released by predators or by weathering; seeds obliquely basipetal on sigmoid funicle, hard brown testa areolate; embryo encased in a shell of endosperm.

Distribution: Neotropical genus of ca. 30 species, most diverse in Amazonian and planaltine Brazil but extending interruptedly to Costa Rica, Venezuelan Guayana and the Guianas, and to Bolivia, Paraguay, and warm-temperate SE Brazil; 4 species in the Guianas.

Taxonomic note: *Stryphnodendron* is subject of a doctoral dissertation by V. Scalon in Brazil and the reader can anticipate a number of specimen reidentifications and nomenclatural realignments that may possibly impact on the treatment presented here.

Notes: In the Guianas *Stryphnodendron* is not very easily distinguished, when in flower only, from other genera with amentiform spicate units of inflorescence and diplostemonous flowers. *Adenanthera* differs from it in pedicellate flowers, *Entada* in a massive one-sided efoliate pseudoraceme of spikes, and *Pseudopiptadenia* by leaflets at once small (less than 12 mm) and in opposite pairs along the rachis. The sessile, pulpy, indehiscent fruit of *Stryphnodendron* is, however, perfectly distinctive. In the Guianas the spicate species of *Mimosa* are readily known at anthesis by their prickly stems and bushy or vinelike habit, and by lack of anther-glands. The genus is represented in the Guianas by 2 groups of species; the typical one has relatively many and small leaflets (less than 2.5 cm) alternate along the pinna-rachis on pulvinules less than 1.5 mm long, the other has few and ample leaflets in opposite pairs on well developed pulvinules 1.5-9 mm long. Pachycaul species of *Stryphnodendron* in eastern Brazil furnish an astringent bark used in medicine and for tanning.

LITERATURE

Occhioni Martins, E.L. 1981. Stryphnodendron Mart. (Leguminosae Mimosoideae) com especial referencia aos taxa Amazonicos. Leandra 10-11: 3-100.

KEY TO SPECIES

1 Pinnae of larger leaves 7-18 pairs; leaflets alternate along pinna-rachis, relatively many and small, those of longer pinnae 8 or more along each side of rachis and less than 25 mm long 2
Pinnae of larger leaves 2-6 pairs; leaflets in opposite pairs along pinna-rachis, relatively few and large, those of longer pinnae 2-7 pairs and longer ones 5.5-15 cm long 3

2 Leaflets of longer pinnae 8-13 on each side of rachis and longer ones 10-23 x 4.5-10 mm . . . . *1. S. guianense*
Leaflets of longer pinnae 16-38(40) on each side of rachis and longer ones (4)5-9 x 0.8-3 mm . . . . *4. S. pulcherrimum*

3 Pinnae 3-6 pairs and leaflets of longer ones 4-7 pairs, pulvinule of leaflets 1.5-3 mm, blade ovate- or lance-acuminate; pods several per spike, strongly incurved or almost annular and randomly twisted together into an involved cluster . . . . *3. S. polystachyum*
Pinnae 1-3 pairs and leaflets of longer ones 2-5 pairs, pulvinule of leaflets 7-9 mm, blade ovate- or obovate-elliptic, shortly or obscurely acuminate; pods few, gently falcate, not twisted . . . . *2. S. moricolor*

1. **Stryphnodendron guianense** (Aubl.) Benth., Trans. Linn. Soc. London 30: 374. 1875. – *Mimosa guianensis* Aubl., Hist. Pl. Guiane 2: 938, pl. 357. 1775. Type: French Guiana, Cayenne, Aublet s.n. (holotype BM). – Fig. 41

Stryphnodendron microstachyum Poepp. & Endl., Nov. Gen. Sp. Pl. 3: 82. 1845. Type: Brazil, Amazonas, Ega, Poeppig s.n. (holotype W).
Stryphnodendron purpureum Ducke, Arch. Jard. Bot. Rio de Janeiro 1: 16. 1915. Type: Brazil, Pará, on the Tocatins near Alcobaça, Ducke 15556 (holotype MG).

Trees, in subequatorial forest attaining 35 m but sometimes flowering at 8 m or less, smaller elsewhere; young stems and leaf-axes pilosulous with brownish-yellow or sordid-gray hairs, leaves strongly bicolored, lfts lustrous dark green but either glabrous or thinly strigulose above, beneath pallid, when dry cinnamon-brown, and either minutely silky-strigulose overall, or pilosulous overall, or glabrous except for a tuft of hairs in anterior basal angle of midrib. Stipules 0; leaf-stalk of longer leaves 13-28 cm; petiole 4-10(12) cm; nectary verruciform or misshapen low-conical, 3-5 mm diam., well below mid-petiole, and often (but not always) similar but smaller, ± hemispherical ones on some furthest interpinnal and interfoliolar axes; pinnae either alternate or opposite, 7-13 pairs, rachis of distal ones 4-14 cm; lfts always alternate, in longer pinnae

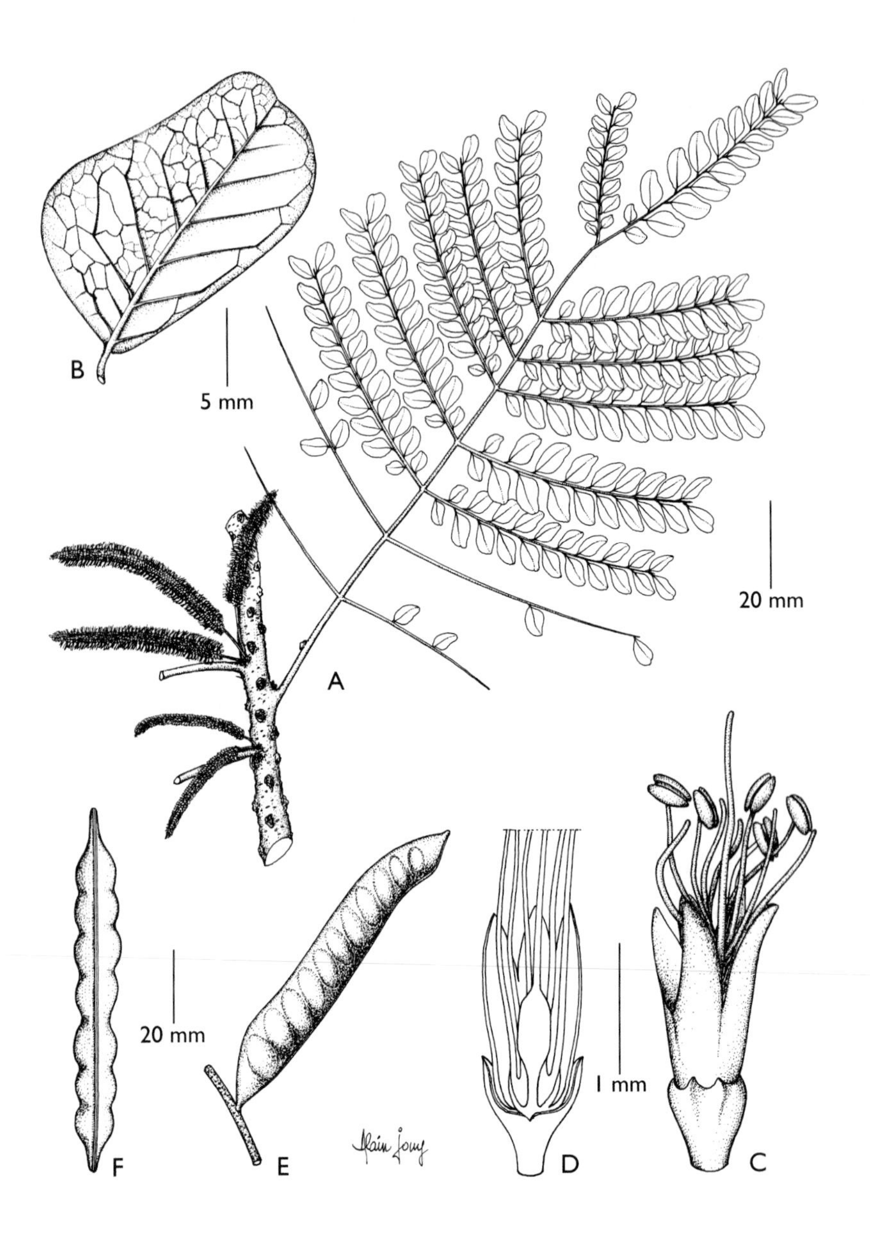

Fig.41. *Stryphnodendron guianense* (Aubl.) Benth.: A, habit, leaf and inflorescences; B, leaflet; C, flower; D, longitudinal section of flower; E, fruit, view of one valve; F fruit, profile view (A-B, Mori & Boom 15361; C-D, Richard s.n.; E-F, Devollet s.n.). Drawing by Alain Jouy; reproduced with permission from P.

(in the Guianas) 8-11(13) on each side of rachis (to 18 elsewhere), blade obtusely rhombic-oblong or oblong-obovate from inequilateral, broadly or narrowly cuneate base, broadly rounded or shallowly emarginate at apex, longest 10-23 x 4.5-10 mm, midrib diagonal, nearly straight, prominulous dorsally, weakly pinnate-branched, margin ± revolute. Inflorescence with flower-spikes solitary or fasciculate by 2-5 in axil of contemporary or immediately hysteranthous leaves; peduncle 8-35 mm; involucre at base of spike early deciduous from an oblique scar; spikes either ascending or recurved, very densely many-flowered, rachis becoming 5.5-13.5 cm; bracts minute caducous. Perianth glabrous except for sometimes minutely puberulent calyx, corolla white, ochroleucous, yellowish, orange, or pink-purple; calyx campanulate 0.5-0.8 mm, minutely toothed; corolla narrowly campanulate 1.7-2.8 mm, erect lobes 0.5-0.8 mm; filaments 3.5-4 cm, united around stipe of ovary into a tube 0.3-0.4 mm, free from corolla; ovary ± 1 mm, either glabrous or pilosulous, raised on a stipe 0.3-1 mm. Pod subsessile, in profile broad-linear, straight or slightly incurved, 6-13 x 1.1-1.7 cm, compressed but plump, 8-13(14)-seeded, sutures not prominent, valves pulpy, fuscous or nigrescent, low-convex over each seed, exocarp densely minutely leprous-puberulent or glabrate, dehiscence very tardy, sutures separating inertly but not gaping, seeds released by weathering; seeds compressed-obovoid, separated by incipient septa less than 1 mm tall, obliquely basipetal on sigmoid funicle ± dilated at apex, ± 8-11 x 5-7 mm, testa dull black, areolate; endosperm copious.

Distribution: Widely dispersed in S America: nearly throughout the Amazon basin, Venezuelan Guayana and the Guianas, in slightly variant forms in Bolivia and SE Brazil (Paraná), in Costa Rica, absent from West Indies; rainforest on terra firme, in second-growth woodland, at margins of campo and campirana, or in gallery forest, mostly below 300 m; flowering in the Guianas mostly October to January, perhaps intermittently later.

Selected specimens: Guyana: Wabuwak, Kanuku Mts., FD 5710 (= WB 236) (NY). Suriname: Boven Suriname R., near Goddo, BW 104 (K, NY). French Guiana: Route de l'Acarouany, km 0.5, SF 7609 (NY); vicinity of Cayenne, Broadway 896 (NY).

Vernacular names: Guyana: mananballi. French Guiana: tamalin.

Note: *Stryphnodendron guianense* sens. lat. varies in stature, in pubescence of leaflets and ovary, in number of pinnae and leaflets, and in flower color. Several syndromes of these independently variable characters are thought by Occhioni Martins to deserve specific or varietal status, but resist definition in substantial or constant terms. The flowers, except for minute differences in indument of calyx, petals and ovary, are indistinguishable, and the fruits offer no known differential characters.

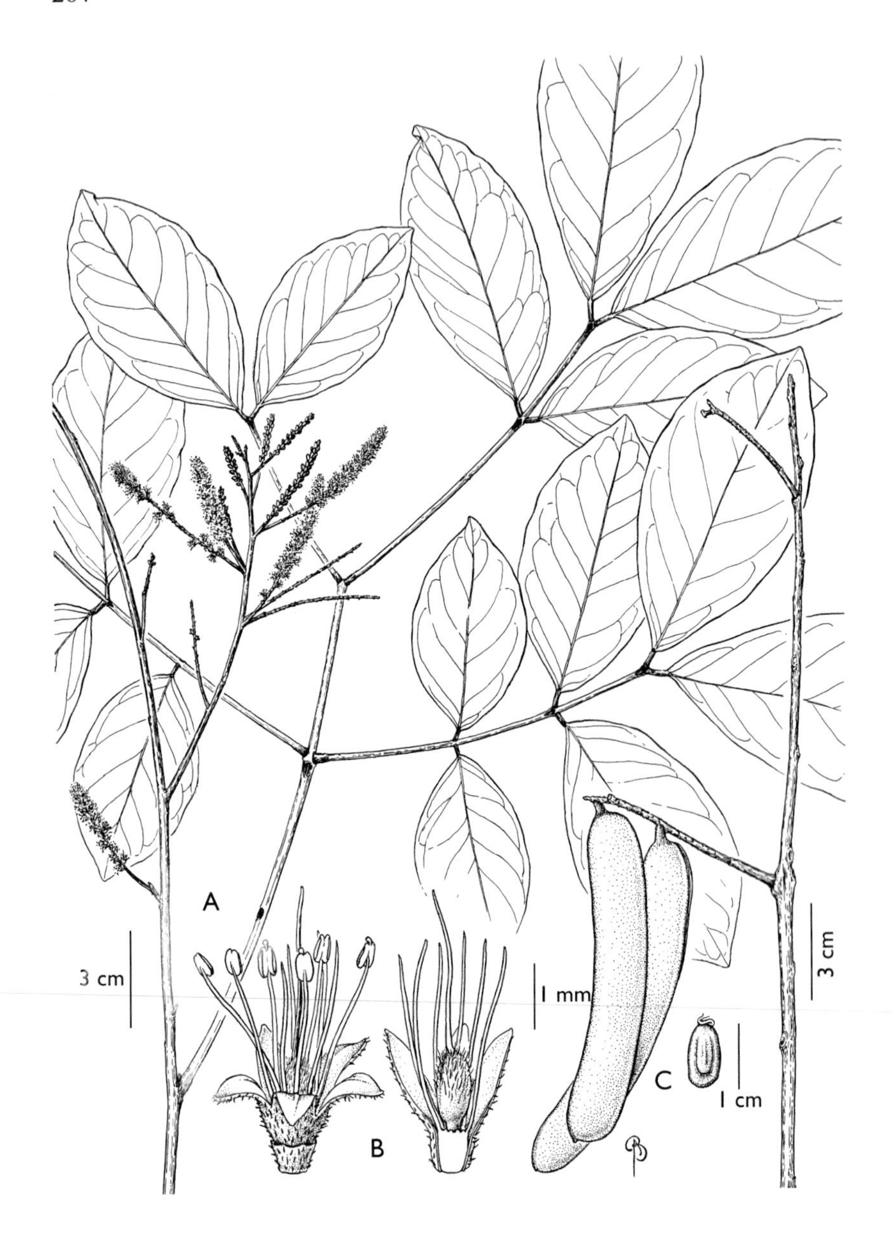

Fig. 42. *Stryphnodendron moricolor* Barneby & J.W. Grimes: A, part of stem with bipinnately compound leaves and inflorescences; B, lateral view of flower (left) and flower with part of perianth and androecium removed to show gynoecium, note the short gynophore; C, stem with pods and seed (below right) (A-B, Mori & Boom 15236; C, Mori & Pipoly 15407). Drawing by Bobby Angell; reproduced with permission from Mori *et al.*, 2002: 508.

Color of the petals appears locally stabilized, but is poorly correlated overall with other variable features.

Taxonomic note: *S. guianense* is divided into a number of infraspecific taxa. Following the specimen annotations of V. Scalon (see note under *Stryphnodendron*), most material from the Guianas represents var. *guianense*. The collection Sagot 1196 (K) from French Guiana represents the less common var. *roseiflorum* Ducke.

2. **Stryphnodendron moricolor** Barneby & J.W. Grimes, Brittonia 36: 45, fig. 1. 1984. Type: French Guiana, Saül, Mori & Boom 15236 (holotype CAY!, isotypes NY!, P!). – Fig. 42

Trees attaining 45 m with trunk to 6.5 dm DBH; young branches and leaf-axes minutely scurfy-puberulent but ample coriaceous leaflets glabrous. Stipules not seen; lf-stks ± 8-17 cm; petiole ± 2-7 cm; nectary plane or lentiform, 2-5 mm diam., immersed in lf-stk between each pinna- pair; pinnae opposite 1-3-jugate, rachis of longer ones ± 10-25 cm; lfts 2-5 pairs, raised on wrinkled pulvinule 7-9 mm, blade ovate- or obovate-elliptic, shortly or obscurely acuminate, furthest pair ± 10-15 x 4-7.5 cm; venation pinnate, costa subcentric, reticulum of venules prominulous on both faces. Inflorescence a terminal efoliate panicle of spikes, primary axis 3-5 dm, secondary axes up to 2 dm, spikes fasciculate by 2-3; peduncle 3-6 mm; axis of flower-spikes 2-4.5 cm; bracts minute. Perianth red-scurfy externally; calyx campanulate ± 0.5 mm; corolla ± 2 mm, red-purple, ascending lobes ± as long as tube; stamens ± 3 mm, filaments red-purple, united at base into a stemonozone ± 0.5-0.8 mm; ovary shortly stipitate, pilosulous, style linear, glabrous, 2-2.5 mm. Pod sessile, in profile broad-linear 10-11 x 1.5-1.7 cm, compressed but plump, 10-12-seeded, valves ripe stiffly papery, scurfy-papillate, wrinkled, produced internally to form incomplete interseminal septa ± 1 mm high; seeds transverse, ± 10-11 x 5-6 mm, testa brown, areolate; endosperm copious.

Distribution: Known only from interior French Guiana (Saül); in forested hill country, 300-400 m; flowering November to December.

Representative specimen examined: French Guiana: Saül, La Fumée, Mori & Pipoly 15407 (K, NY).

Note: This species is closely akin to *S. racemiferum* (Ducke) W.A. Rodrigues of central Amazonia, which differs, however, in almost glabrous flowers and in lack of interseminal septa in the pod. The epithet alludes directly to the mulberry-purple stain imparted to drying-papers by the flowers, and indirectly to Scott Mori, the discoverer.

3. **Stryphnodendron polystachyum** (Miq.) Kleinhoonte, Receuil Trav. Bot. Néerl. 22: 416. 1925, lacking full citation of basionym; in Pulle, Fl. Suriname 2(2): 272. 1940. – *Piptadenia polystachya* Miq., Linnaea 18: 590. 1844. Type: Suriname, Bergendal, Focke s.n. (holotype? U).

Trees 20-36 m with trunk at maturity buttressed and up to nearly 1 m DBH; young growth varying from nearly glabrous to puberulent with minute, amorphously dilated or coralloid trichomes sometimes mixed with short plain hairs, broad leaflets glabrous except for sometimes puberulent costa, on upper face lustrous dark green, paler dull beneath. Stipules triangular-subulate 1 mm or less, caducous, absent from most specimens; pinnae and lfts opposite or almost so; lf-stks 12-40 cm; petiole 5-10.5 cm; sessile hemispherical nectary 1.5-2.5 mm tall on petiole shortly below first pair of pinnae, and progressively smaller ones below each pinna-pair and below some furthest pairs of leaflets; pinnae 3-6 pairs, rachis of longer ones 9-17 cm; lfts 4-7 pairs, accrescent distally, each raised on a pulvinule 1.5-3 mm, blade subequilaterally ovate- or broad-lance-acuminate from rounded base, longer ones 5.5-11 x 1.8-5.2 mm; venation pinnate, dorsally prominent midrib subcentric, straight or nearly so, fine 3-4-nary reticulum of venules more sharply prominulous above than beneath. Inflorescence a terminal efoliate panicle of dense narrow spikes; primary axis of panicle 1-2.5 dm; spikes solitary and fasciculate by 2-3, including short peduncle 3-7 cm; involucre campanulate or randomly 2-lobed, 1.5-3 mm, distant 2-8 mm from base of peduncle, circumscissile in age; bracts 0.2-0.5 mm. Perianth 5(6)-merous; calyx sessile bronze-puberulent, 0.7-0.9 mm, ± as broad, teeth 0.2-0.3 mm; corolla greenish glabrous, 2-2.2 mm, lobes spreading-ascending ± 0.8 mm; androecium crimson; ovary subsessile, densely puberulent. Pod usually several per spike and long persistent on lignescent peduncle, sessile, in profile broad-linear, falcately or circinnately incurved through ± 3/4-circle or further and randomly twisted into an involved cluster, body compressed but biconvex over seeds, 3.5-5 x 1.1-1.7 cm, 8-13-seeded, sutures wiry not dilated, ventral one evenly incurved, dorsal (exterior) one shallowly undulate, valves crustaceous brown transversely sulcate between seeds, densely scaberulous or glabrate, endocarp tan, impressed-venulose, dehiscence essentially 0, seeds released by weathering on the tree; seeds compressed-obovoid 6.5-8 x 4-5.5 mm, testa dull fuscous-brown areolate.

Distribution: Scattered in Brazil (Amapá, Pará, Maranhão), the Guianas, and Venezuelan Guayana (Bolívar, Delta-Amacuro); in non-inundated rain forest, 25-600 m, flowering mostly in August and September, occasional in April and perhaps at other times.

Selected specimens: Guyana: Mazaruni Station, FD 3990 (= Fanshawe 1254) (K, NY); Wabuwak, Kanuku Mts., FD 5711 (= WB 237)

(NY). Suriname: 4 km SW of Juliana Top, 10 km N of Lucie R., Irwin *et al.* 55179 (NY); Brownsberg, tree nr. 1173 (NY, U). French Guiana: Passoura, Sabatier & Prévost 3758 (CAY, K).

Vernacular name: Suriname: hiakantaballi.

Note: According to Ducke and Fróes the pods smell of garlic and are reputed to be poisonous.

4. **Stryphnodendron pulcherrimum** (Willd.) Hochr., Bull. New York Bot. Gard. 6: 274. 1910. – *Acacia pulcherrima* Willd., Sp. Pl. 4: 1061. 1806. Type: Brazil, Pará, Sieber in Hoffmannsegg s.n. (holotype B-Willd. 19136).

Stryphnodendron angustum Benth., Trans. Linn. Soc. London 30: 375. 1875. Type: Brazil, Amazonas, near Manáus, Martius 1920 (holotype M, F Neg. 6202).
Stryphnodendron melinonis Sagot, Ann. Sci. Nat., Bot. sér. 6, 13: 322. 1882. Type: French Guiana, woods on the Maroni, Mélinon 236 (holotype P, isotype US).

Trees attaining 10-35 m, resembling *S. guianense* in habit, pubescence, inflorescence and fruit, but differing in greater number and small size of lfts, these more crowded along pinna-rachises. Lf-stks of longer leaves 11-22 cm; petiole 3-5.5(6.5) cm; nectary sessile verruciform near or often below mid-petiole, 1-2 mm tall, 1.4-4 mm diam. at base; pinnae of longer leaves (8)10-18 pairs, rachis of longer ones 5-8.5 cm; lfts alternate, 18-38(40) on each side of pinna-rachis, narrowly oblong, narrowly lance-oblong, or linear, obtuse, larger ones (4)5-8.5 x 0.8-3 mm, margin revolute, midrib straight, slightly displaced, prominulous only beneath, simple or faintly pinnately branched. Inflorescence and flower of *S. guianense*, perianth puberulent but sometimes thinly so, corolla white, ochroleucous, or pale yellow. Pod of *S. guianense*.

Distribution: Widely scattered through Brazilian Amazonia, adjoining Peru and Bolivia, in Brazil disjunct in restinga and Atlantic forest of Bahia, N into Venezuelan Guayana, and sparingly into the Guianas; in forest on terra firme, especially common locally in second-growth woodland, mostly below 350 m but on Gran Sabana in Venezuela up to 850-1000 m; flowering in the Guianas October to January, elsewhere intermittently through the year.

Selected specimens: Guyana: Waramadong Mission, Kamarang R. - Wenamu Trail, Maguire & Fanshawe 32442-a (NY). Suriname: Zanderij I, BW 3379 (K, NY). French Guiana: Cayenne, Mélinon 236 (P, US).

27. **ZAPOTECA**[31] H.M. Hern., Ann. Missouri Bot. Gard. 73: 757. 1986.
Type: Z. tetragona (Willd.) H.M. Hern. (Acacia tetragona Willd.)

Decumbent, erect or scandent herbs, shrubs or rarely small trees. Leaves alternate, bipinnate; lf-formula i-viii/1-134; stipules striately veined, membranous papery (in 1 species in S America spinescent), persistent; lf-stk (in the Guinanas) lacking nectary; lfts opposite, petiolulate. Inflorescences of capitula solitary or fasciculate in axils of contemporary leaves, or in terminal, efoliate pseudoracemes. Flowers sessile, homomorphic; calyx shortly campanulate; corolla campanulate or infundibuliform; disc present in all flowers; stamens 30-60, staminal tube included, free filaments long-exserted, anthers dorsifixed; ovary sessile or shortly stipitate. Pods straight, or slightly falcate, planocompressed, sutural ribs thickened, dehiscence elastic from apex to base, valves not contorting.

Distribution: Genus of 20 species from SW U.S.A. and Mexico, through C America, the West Indies, and through S America, where most numerous along the Cordillera, though scattered in the Amazon basin, into N Argentina; Hernández (1989) reports 1 species in Guyana, though he cites no collections.

Note: Hernández (1986) segregated *Zapoteca* from *Calliandra* based on cytological and palynological characters. In gross appearance the two genera appear closely related.

1. **Zapoteca portoricensis** (Jacq.) H.M. Hern., Ann. Missouri Bot. Gard. 73: 758. 1986. – *Mimosa portoricensis* Jacq., Collect. 4: 143. 1791. Neotype (designated by Hernández, Ann. Missouri Bot. Gard. 76: 818. 1989): Puerto Rico, Bredemeyer 16 (B-Willd.).

In the Guianas only: subsp. **flavida** (Urb.) H.M. Hern., Ann. Missouri Bot. Gard. 76: 825. 1989. – *Calliandra flavida* Urb., Ark. Bot. 24A: 4. 1931. Lectotype (designated by Hernández, Ann. Missouri Bot. Gard. 76: 825. 1989): Grenada, ad Belvedere, Eggers 6226 (US, not seen).

Scandent shrubs 2-3(7) m tall; young parts commonly villous, becoming glabrate with age. Lf-formula iii-v(vii)/8-18; stipules inequilaterally lanceolate to lanceolate-triangular, 5-13 x 2.5-4 mm, persistent, those at efoliate nodes of inflorescence usually the largest. Inflorescence of solitary or geminate capitula axillary to coeval leaves, or in terminal efoliate pseudoracemes of capitula; peduncle 1.4-7.5 cm long. Calyx ± 1-2 mm; corolla ± 3-4.5 mm; stemonozone ± 0.5 mm; staminal tube ±

[31] by James W. Grimes

1.0 mm, stamens exserted ± 2.0 cm; ovary sessile, glabrous. Pod (fide Hernández, none seen), 16.5 x 1.2 cm, thick membranous; seed ovoid, ± 6 x 4.5 mm, pleurogram "irregular".

Distribution: Mexico (Veracruz and Oaxaca), Grenada, to Venezuela (Delta Amacuro), Peru, C and NE Brazil; from 0-1100 m alt. in Venezuela; along rivers and in inundated forest; reported in Guyana by Hernández (l.c.), but no collections seen (Bentham in J. Bot. (London) cites Ro. Schomburgk ser. II, 820 (= Ri. Schomburgk 1515)), certainly to be expected.

Note: *Zapoteca portoricensis* subsp. *portoricensis* is confined to C America and the Greater Antilles.

28. **ZYGIA**[32] P. Browne, Civ. Nat. Hist. Jamaica 279, t. 22, fig. 3. 1756.
Type: Z. latifolia (L.) Fawc. & Rendle (Mimosa latifolia L.)

Unarmed, macro-and microphyllidious arborescent shrubs and slender, 1-several-stemmed trees, seldom attaining 20 m. Lf-formula variable, in the Guianan species most commonly a) i/(½)1½-5½(6½), with petiole mostly much shorter than rachis of the one pair of pinnae or occasionally as long, or b) ii-x/6-23, lfts then diminished with increasing number; stipules usually small, firm, caducous, veinless or bluntly few-veined, rarely thick-textured and persistent, occasionally papery and striately veined; foliar nectary between or just below first pair of pinnae, plus in most species similar but smaller nectaries between further pairs; lft-venation pinnate, or palmate and pinnate, in some species of sect. *Zygia* the first 1-2 secondary veins on posterior side of midrib stronger and longer than the rest. Inflorescences arising in most species from knots on annotinous and older wood, below current foliage, occasionally axillary to coeval lvs, consisting commonly of capitula or spikes, exceptionally of umbelliform capitula or racemes. Perianth normally 5-merous (random exceptions); calyx varying from shallowly saucer-shaped to campanulate, most commonly ± 0.5-3(4) mm, rarely campanulate to 5.5 mm, rim truncate or denticulate; corolla subtubular or dilated at limb, tube nearly always striately veined, lobes at anthesis erect-ascending; androecium 26-130-merous, stemonozone short or obscure, tube either shorter than corolla or far exserted; ovary sessile or almost so, commonly surrounded at base by a crenately lobed nectarial disc. Pods diverse in size, texture and compression, either straight, or gently decurved (and sometimes twisted), or evenly retrofalcate to decurved through nearly a full circle, in profile

[32] by James W. Grimes

oblong, linear, or undulately linear, exceptionally torulose or tetragonal, valves either leathery and ± planocompressed, or lignescent and then often biconvex or subterete at maturity, either smooth, or rugulose, or less often transversely fissured or elaborately sulcate, sutures forming a frame only around plano-compressed valves, otherwise immersed or almost so, exceptionally dilated and wider than valves, dehiscence inert, follicular or more often through both sutures, but often tardy, valves narrowly gaping or recurving to release seeds; seeds uniseriate on slender crumpled funicle, either discretely spaced along cavity or (especially when large or fat) contiguous or imbricate and then distorted by mutual pressure, thin testa either smooth or wrinkled, often lustrous, becoming papery and fragile, veined or narrowly winged around periphery, lacking pleurogram.

Distribution: American genus of ca. 60 species (9 to 13 of these are recognised as species of *Marmaroxylon* by Lewis & Rico Arce (2005), see note below), widespread from lowland tropical Mexico and Greater Antilles to SE Brazil, NE Argentina, Paraguay, and Amazonian Bolivia, extending W of the Andean cordilleras only between SW Mexico and N Peru; characteristically of riparian forest and shore habitats; 14 species in the Guianas.

Taxonomic notes: This treatment considers *Marmaroxylon* Killip as a section of *Zygia*. Lewis & Rico Arce in Lewis *et al*. (eds.) (2005), recognize *Marmaroxylon* as a genus distinct from *Zygia*. Pending molecular analysis of both taxa, the position of *Marmaroxylon* remains unclear.

Note: Barneby and Grimes (1997) recognize 9 sections within *Zygia* based on leaf-formula and morphology of the inflorescences. The 7 sections to which the species from the Guianas belong are indicated in the key.

KEY TO THE SPECIES

1 Leaves simply paripinnate, nearly all 4-foliolate (sect. *Ingopsis*) . . . . *6. Z. inundata*
  Leaves bipinnate, leaves rarely if ever exactly 4-foliolate 2

2 Pinnae of all leaves exactly 1 pair (sect. *Zygia*) 3
  Pinnae 2-several pairs per leaf 8

3 Leaflets of all pinnae exactly 1 pair . . . . *14. Z. unifoliolata*
  Leaflets of all pinnae 1.5-5.5 pairs 4

4 Peduncles branched. . . . . *7. Z. juruana*
  Peduncles unbranched. 5

5 Leaflets mostly 3-4.5 pairs, rarely only 2.5 pairs per pinna; calyx 2.0-4.25 mm, if as short as 2.0 mm still more than 1/3 the length of the corolla .... *5. Z. inaequalis*
Leaflets (1.5)2-3.5 pairs per pinna; calyx 0.33-2.25 mm, always 1/3 or less the length of the corolla.......................................6

6 Leaflets with all secondary veins ± equal, none of the lower, lateral ones reaching midblade .... *1. Z. cataractae*
Leaflets with 1, 2 or 3 lateral veins longer than the rest and produced from near base to beyond mid-blade.....................................7

7 Tube of corolla glabrous; larger leaflets (9)11-20 x (3)3.5-8 cm; pod 20-33 mm wide .... *8a. Z. latifolia* var. *latifolia*
Tube of corolla pilosulous or subappressed puberulent; larger leaflets (4)5-12 x 2-5 cm; pod 11-19(20) mm wide.......... *8b. Z. latifolia* var. *lasiopus*

8 Leaflets of longer pinnae 1-9 pairs, and the longer at least 3.5 cm 9
Leaflets of longer pinnae 12-22 pairs, the longer ones 1-2(3.2) cm...... 10

9 Pulvinules of leaflets 3.5-5 mm, blade well elevated from pinna-rachis; inflorescence a compact panicle of subsessile, short racemes; calyx 2-2.7 mm and corolla ± 10 mm; androecium ± 5-merous (sect. *Barticaea*)........... .... *4. Z. eperuetorum*
Pulvinule of leaflets in dorsal view 1.5-2.5 mm, but blade in ventral view subsessile against pinna-rachis; inflorescence umbellately capitulate, each capitulum borne on a slender peduncle ± 2.5-4.5 cm; calyx ± 1 mm and corolla 6.5-7 mm; androecium 25-28-merous (sect. *Pseudocojoba*) .... *12. Z. sabatieri*

10 Inflorescences spicate (sect. *Parazygia*)........................... 11
Inflorescences capitulate....................................... 12

11 Leaflets plane, largest 18-30 mm; perianth subglabrous ...... *3. Z. collina*
Leaflets ventrally convex, largest 7-12 mm; perianth densely silky-pilosulous .... *13. Z. tetragona*

12 Peduncles 2-4 at nearly all nodes of raceme; corolla 3.2-4.3 mm (sect. *Marmaroxylon*)................................ *11. Z. racemosa*
Peduncles solitary at each node of inflorescence; corolla 6-19 mm (sect. *Zygiopsis*) 13

13 Pinnae 8-9 pairs, larger leaflets 7-10 mm; pod plurilocular *10. Z. potaroënsis*
Pinnae 4-6 pairs, larger leaflets 11-18 mm; pod unilocular 14

14 Larger leaflets 11-18 x 3-5 mm, obtuse mucronulate; pod plump, ± 7.5 mm diam., valves densely pilosulous .... *2. Z. claviflora*
Larger leaflets 18-32 x 5-9 mm, acute mucronulate; pod planocompressed, ±18 mm wide, valves glabrous....................... *9. Z. palustris*

1. **Zygia cataractae** (Kunth) L. Rico, Kew Bull. 46: 496. 1991. – *Inga cataractae* Kunth in Humboldt, Bonpland & Kunth, Nov. Gen. & Sp. 6: 297. 1823. Type: Brazil, on the falls of the Javari, Humboldt & Bonpland 24 (holotype P, not seen).

   Inga glomerata DC., Prod. 2: 438. 1825. – *Pithecolobium glomeratum* (DC.) Benth., London J. Bot. 3: 213. 1844. – *Zygia glomerata* (DC.) Pittier, Trab. Mus. Comercial Venezuela 2: 70. 1927. Type: French Guiana, Cayenne, no collector given (holotype G, not seen).

Shrubs or most commonly trees 4-8 (rarely to 25) m tall, glabrous throughout or lf-rachises, fls and base of fruit sparingly pubescent with short reddish hairs. Lf-formula i/1½(3½), odd lft inserted just above pinna-pulvinus or up to 1/4 length of pinna-rachis. Stipules deltate or triangular, sometimes slightly falcate, 1.07-1.75 x 0.65-1.1 mm, with obvious midvein, long-persistent; petiole 2-8(19) mm, with sessile, patelliform to shallowly cup-shaped nectary inserted between pinna-pair, and similar but smaller nectaries subtending lfts; pinnae 1.1-5.7 cm; lfts broadly to narrowly elliptic, elliptic-lanceolate or lanceolate, sometimes somewhat falcate, (5.1)-17 x (2.2)2.7-6 mm, inequilaterally attenuate at base, acuminate or acuminate-cuspidate at apex; venation pinnate, secondary veins subequal. Inflorescence of unbranched peduncles 1-4(5)-fasciculate on brachyblasts in lf-axils, or most commonly cauliflorous; peduncles (<1)2-11 mm; capitula ±13-23-flowered; bracts rhombic-spatulate, 0.75-1.25 mm, persistent. Flowers sessile; calyx shallowly cup-shaped to short-tubular-campanulate, 0.33-1.3(2) mm, truncate but with (4)5(6) small, ciliolate teeth; corolla at least slightly asymmetrical, very narrowly tubular- or funnelform campanulate, 3.75-7.5(10) mm, obviously veined when dry; intrastaminal disc 0.31-0.54 mm; stemonozone 0.19-0.85 mm; staminal tube (2.75) 7.5-12 mm, stamens 36-60(71); ovary sessile, glabrous. Pods 1-3 per capitulum, sessile, in profile gently falcate-elliptic, (3)6-22 x 1.2-2.5 cm, valves dull to shiny brown when dry, described as green in the field, dehiscence inert, through ventral suture; seeds broadly elliptic to round, flattened, in broad profile 13-15 x 6-15 mm, seed-coat crumbly-papery when dry, gelatinous when wet.

Distribution: Colombia, Venezuela and the Guianas, through the Amazon Basin of Ecuador, Brazil, Peru and Bolivia; rarely in forest on terra firme, most commonly in swamp-forest or along rivers and streams; flowering February through September.

Selected specimens: Guyana: Puruni R., FD 7770 (= JB 86) (NY); Kanuku Mts., Rupununi R., Bush Mouth near Witaru Falls, Jansen-Jacobs *et al.* 116 (NY, U). Suriname: along the Marowijne R. above base camp,

Lanjouw & Lindeman 2959 (NY); vicinity of Brokolonka, Saramacca R., Maguire 23798 (NY). French Guiana: Village Boni de Loca, Bassin du Maroni, Lawa, Fleury 640 (NY); Bourg de Maripasoula, Bassin du Maroni, Fleury 672 (NY).

2. **Zygia claviflora** (Spruce ex Benth.) Barneby & J.W. Grimes, Mem. New York Bot. Gard. 74(2): 65. 1997. – *Pithecolobium claviflorum* Spruce ex Benth., Trans. Linn. Soc. London 30: 596. 1875. – *Abarema claviflora* (Benth.) Kleinhoonte in Pulle, Fl. Suriname 2(2): 320. 1940. – *Marmaroxylon claviflorum* (Benth.) L. Rico, Kew Bull. 46: 515. 1991. Lectotype (designated by Barneby & J.W. Grimes 1997): Brazil, on the Rio Negro near São Gabriel and the falls of the Panuré on the Rio Vaupés, Spruce 2252 (hololectotype K (hb. Benth.)! = NY Neg. 2030, isotypes K (hb. Hook.)!, NY! P(2 sheets)!).

Slender multifoliolate cauliflorous trees (1.5)3-10 m with smooth terete branches; young stems and all lf-axes densely brown-gold-pilosulous. Lf-formula iv-vi/(10)12-18. Stipules erect or falcately erect, herbaceous but early dry paleaceous, linear-lanceolate or narrowly lance-elliptic 2-9 x 0.3-2 mm, persistent but becoming dry and fragile, hence deciduous; lf-stks of larger lvs (4)5.5-13 cm, petiole including (or consisting of) poorly differentiated pulvinus 2-6 x 1-2 mm, longer interpinnal segments 11-23 mm; nectary sessile, shallowly cupular or almost plane, thick-rimmed, between proximal pinna-pair, smaller ones between most or only between furthest pinna-pairs, and yet smaller, sometimes rudimentary cupular nectaries between all but a few proximal pairs of lfts; pinnae strongly accrescent upward, rachises of furthest pair 4-8 cm, longer interfoliolar segments 3-6.5 mm; pulvinules subobsolete, lfts sessile against rachis, not readily disjointing when dry; lfts decrescent at each end of or only at base of pinna-rachis, first pair very unequal, those near and beyond mid-rachis narrowly oblong from obliquely truncate or very broadly flabellate base, nearly straight or commonly incurved beyond middle, sharply auriculate at posterior basal angle, larger ones 11-18 x (2.8)3-5 mm, 3-4 times as long as wide; venation palmate and pinnate, midrib subcentric, innermost posterior primary vein produced nearly to or well beyond mid-blade, outer 2-3 very short or obscure. Inflorescence of short, pedunculate few-flowered capitula arising for most part in fascicles (but not pseudoracemose) from knots on annotinous and older wood, rarely and randomly solitary and lateral to axil of an older leaf; peduncle 0-5 mm, subtended by a rudimentary, bracteiform lf-stk with small cupular nectary; capitula ± 4-9-flowered, axis less than 2 mm. Flowers sessile; perianth either puberulent overall or almost glabrous, corolla greenish-white, androecium red or pink distally; calyx campanulate 1-2 x 0.8-1.2 mm; corolla 6-10 mm, tube striate subcylindric, abruptly ampliate distally; androecium 24-44-merous, 21-33 mm, stemonozone ± 1 mm, slender tube 12-15 mm; ovary sessile, in profile linear, symmetrically conical at apex, ± 1.8-2 mm, at anthesis glabrous. Pod imprecisely known from few, either immature or over-mature samples, apparently similar to that of *Zygia racemosa* in size and proportions, ± 7.5 mm wide at seed-locules, valves ultimately pithy densely pilosulous overall with yellowish hairs; seeds not seen.

Distribution: Localized on both slopes of the Orinoco-Rio Negro divide in Brazil (NW Amazonas) and Venezuela (SW Amazonas); in rain forest, along non-inundated river banks and in hill country; flowering September to December and March to April; perhaps to be expected in Guyana.

Note: The epithet *claviflora* refers to the clovelike outline of the flower buds, abruptly swollen in the short limb. The species is distinguished in sect. *Zygiopsis* by the combination of striate stipules, relatively small, obtuse but mucronate leaflets, whitish flowers with pink tassel of filaments, and narrow pubescent pods, these, however, known as yet from few examples.

3. **Zygia collina** (Sandwith) Barneby & J.W. Grimes, Mem. New York Bot. Gard. 74(2): 80. 1997. – *Pithecellobium collinum* Sandwith, Kew Bull. 1948: 315. 1948. – *Marmaroxylon collinum* (Sandwith) L. Rico, Kew Bull. 46: 516. 1991. Type: Guyana, Groete Cr., Lower Essequibo R., FD 3976 (= Fanshawe 1240) (holotype K).

Trees 6-35 m tall; new branchlets and all lf-axes puberulent with erect or forwardly incurved, yellowish hairs. Lf-formula v-vii/12-18. Stipules caducous (few seen), triangular or lanceolate 1.5-3 x 1-1.5 mm, externally veinless; lf-stks of longer lvs 7-13 cm, petiole reduced to lf-pulvinus, longer interpinnal segments 15-26 mm; petiolar nectary between first pair of pinnae sessile, shallowly cupular, thick-rimmed or almost plane, similar but smaller and more deeply cupular ones between furthest 2-3 pairs of pinnae and yet smaller ones on pinna-rachis between 2-3 furthest pairs of lfts; pinnae distally accrescent, penultimate pair often longest, rachises of these 6-14.5 cm, longer interfoliolar segments 5-10 mm; lfts glabrous, abruptly decrescent at base of rachis, thence subequilong or somewhat decrescent distally, oblong or lance-oblong from obliquely truncate, obtusangulate base, gently incurved into a deltate, sharply mucronate tip, longer ones (18)20-27(30) x 5-8(11.5) mm; venation palmate and pinnate, slender subcentric midrib distally incurved, finely prominulous on both faces, inner posterior primary vein produced nearly to mid-blade, outer 2-3 very short, slender secondary veins from midrib and open reticulum more sharply defined dorsally. Inflorescence of simple spikes, 7-15-flowered, rachis ± 6-10 mm; bracts broadly ovate ± 1 mm, persistent. Flowers sessile; calyx (4)5-merous, deeply campanulate, 1.7-3 x 1-1.4 mm, depressed-deltate teeth 0.15-0.3 mm; corolla whitish subtubular, very slightly ampliate at limb, 7-10 mm, unequal ovate lobes ± 1-1.2 mm; androecium 26-30-merous, 17-19 mm, stemonozone ± 1 mm, tube 10 mm; ovary in profile linear-elliptic, 1.8-2 mm, contracted at base into a stipe 0.3 mm or less, conical at apex, at anthesis smooth, glabrous; disc up to 0.2 mm tall, sometimes obscure. Fruit unknown.

Distribution: Apparently most common in Venezuela (Bolívar, Delta-Amacuro) and Guyana (Lower Essequibo R.); in upland rain forest, 40-250 m; known to flower in April and July.

Representative specimen examined: Guyana: Groete Cr., Lower Essequibo R., FD 3976 (= F1240) (K).

4. **Zygia eperuetorum** (Sandwith) Barneby & J.W. Grimes, Mem. New York Bot. Gard. 74(2): 83. 1997. – *Pithecellobium eperuetorum* Sandwith, Kew Bull. 1948: 316. 1948. – *Marmaroxylon eperuetorum* (Sandwith) L. Rico, Kew Bull. 46: 516. 1991. Type: Guyana, Bartica-Potaro road, FD 4170 (= Fanshawe 1434) (holotype K).

Trees 3-12 m tall; appearing glabrous but lf-stks minutely puberulent and inflorescence densely gray-puberulent overall. Lf-formula ii/3-5. Stipules unknown; lf-stks 5-8.5 cm, petiole reduced to pulvinus, interpinnal segment 5-8 cm; nectary between each pair of pinnae sessile, plane, ± 1.5-2 mm diam., none seen on pinna-rachises; rachises of further pinna-pair 20-27 cm, longer interpinnal segments ± 4-7 cm; lfts bicolored, little graduated, blade elliptic-acuminate from inequilaterally cuneate base, subfalcately incurved, longer ones ± 10-14 x 3-5.5 cm; venation palmate and then pinnate, midrib subcentric, gently incurved, slenderly prominulous on each face, posterior primary vein ascending to or beyond mid-blade, anterior one weaker or obsolete, first pair of secondary veins narrowly ascending, further 5-8 pairs widely incurved-ascending, weak tertiary venulation raised on both faces. Inflorescence a condensed panicle of racemes, primary axis ± 3-8 mm, axis of subsessile, ± 6-20-flowered racemes ± 8-16 mm, some or all racemes subtended by a rudimentary lf-stk with nectary; floral bracts deltate ± 0.6 mm, persistent; pedicels 1.5-1.8 x 0.35-0.4 mm. Perianth 5-merous; calyx deeply campanulate 2-2.7 x 1-1.4 mm, depressed-deltate teeth 0.2-0.3 mm; corolla cylindric, moderately ampliate at limb, 9.5-10 mm, glaucous-green when fresh, lobes ± 1.5 mm; androecium ± 21 mm, tube ± 12-13 mm; no nectarial disc; ovary sessile, at anthesis glabrous. Fruit unknown.

Distribution: Known only from the type locality in Guyana, see above; locally common in wallaba (*Eperua*) forest on white sand, below 200 m.

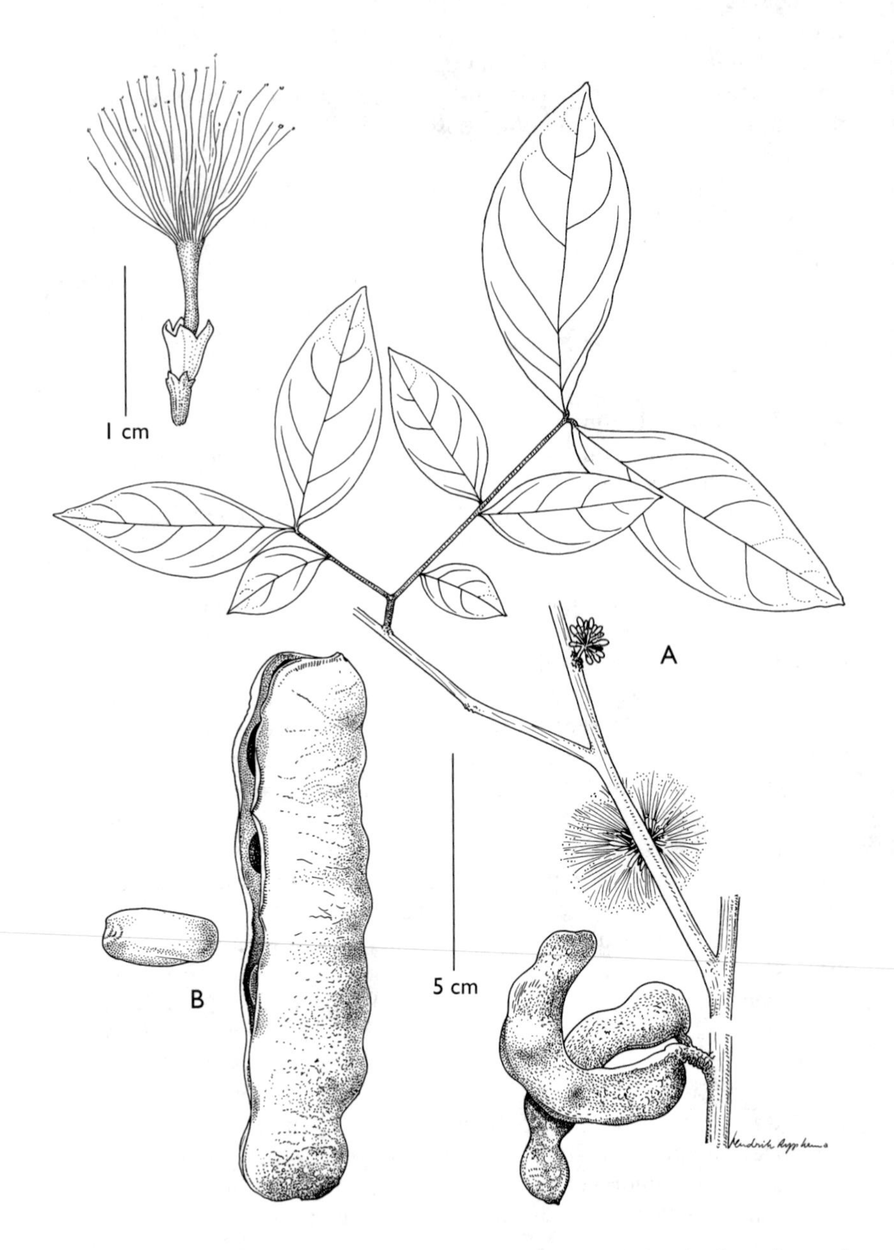

Fig.43. *Zygia inaequalis* (Humb. & Bonpl. ex Willd.) Pittier: A, flowering and fruiting branch, and flower (upper left); B, pod, and seed (left) (A, Wessels Boer 983; B, Clarke 839). Drawing by Hendrik Rypkema; reproduced with permission from U.

5. **Zygia inaequalis** (Humb. & Bonpl. ex Willd.) Pittier, Trab. Mus. Comercial Venezuela 2: 69. 1927. – *Inga inaequalis* Humb. & Bonpl. ex Willd., Sp. Pl. 4: 1019. 1806. Type: Venezuela, on the Orinoco, Humboldt (holotype P, seen in photo NY!). – Fig. 43

Shrub or more commonly tree to 25 m tall; young stems, lf- and infl-axes and flowers glabrate or sparingly to moderately minutely pubescent. Lf-formula i/2½(3)-4½, leaflets commonly 5, 7, or 9. Stipules (few seen) thick, deltate or deltate-triangular ± 1.75 mm, quickly deciduous; petiole reduced to thickened pulvinus, 2-7 mm, with a small patelliform nectary inserted between pinna-pair, smaller but similar nectaries inserted between lfts; pinna-rachis 5.3-22.5 mm; interfoliolar segments 1.7-4.5 cm; lfts bicolored or not, accrescent, elliptic, elliptic-lanceolate or lanceolate, terminal pair somewhat falcately so, 6-18 x 2.4-5.6 cm, gently attenuate at base, apex round-acuminate. Inflorescence cauliflorous, short spikes up to 11-fasciculate at brachyblasts, 11-23-flowered; peduncle 3-10 mm, rachis 2-5 mm. Calyx narrowly to broadly campanulate, sessile though attenuate at base, 2.0-4.25 mm; corolla narrowly to broadly tubular-campanulate, 5.5-8 mm; stemonozone 0.74-1.15 mm; staminal tube 7-9.5 mm, stamens 34-43; intrastaminal disc 0.19-0.43 mm; ovary sessile, glabrous, tapered to subtruncate at apex. Pod 1-2 per peduncle 1-4 per node, on tree green or brownish, sessile, at maturity straight or strongly decurved through about 1/2-circle, in profile 9-27 x 1.5-2.5 cm, valves smooth or more commonly rugulose and cracking between seed chambers, sutural ribs ± winged, dehiscence at first through ventral suture, but dorsal suture eventually dehiscing in part; seeds plump, in broad profile round or round-elliptic, 10-22 mm, as long as broad or 1.4 times longer than broad; seed-coat thin papery.

Distribution: Colombia, Venezuela, the Guianas and Amazonian Brazil, Peru and Bolivia; in moist forest and along edges of water, below 200 m; flowering September to May.

Selected specimens: Guyana: Kanuku Mts, Jansen-Jacobs *et al.* 250 (NY); Sand Cr., Rupununi R., FD s.n. (= WB 108) (NY); Suriname: near airport on the Oelemari R., Wessels Boer 983 (NY). French Guiana: village Boni de Loca, Bassin du Maroni, Lawa, Fleury 742 (CAY, NY).

6. **Zygia inundata** (Ducke) H.C. Lima ex Barneby & J.W. Grimes in Mem. New York Bot. Gard. 74(2): 130. 1997. – *Inga inundata* Ducke, Arch. Jard. Bot. Rio de Janeiro 3: 48. 1922. Syntypes: Brazil: lake Jeretepaua near Obidos, Ducke 16.340; on the island of Cutijuba near Belém, Huber 8.224; affluents of the Amazon near Gurupá, Ducke, HJBR 10.008 (syntypes RB, syntype RB 10008, †B = F Neg. 1153! U!).

Tree attaining 8 m; new branchlets fuscous, sometimes flaking, smooth gray older ones, except for minutely puberulent perianth and nascent foliage glabrous throughout. Leaves simply paripinnate, almost all amply 4-foliolate. Stipules small, caducous; lf-stks ± 5-9.5 cm, petiole and one interfoliolar segment subequilong; between each pair of lfts a low-convex fuscous nectary ± 1 mm diam.; lfts subequilaterally elliptic or ovate-elliptic from cuneately attenuate base, shortly acuminate at apex, blade of distal pairs 6.5-18 x 2-7 cm, 2.2-3.6 times as long as wide, proximal pair somewhat smaller. Inflorescence of capitula fasciculate on old wood, or solitary, or pseudoracemose, primary axis 0-2 cm, peduncle ± 1.5-4.5 cm; subglobose receptacle ± 2 mm diam., up to ± 20-25-flowered; pedicels 0.8-2.5 mm. Calyx 1.6-2 mm, minutely denticulate; corolla 9-12 mm, only slightly dilated distally; staminal tube 12-14 mm; stamens 34-48; intrastaminal disc lacking; ovary glabrous, tapering at apex. Pod sessile, in broad profile oblong ± 12-14 x 3.3-4.3 cm, 7-10-seeded, very slightly decurved, compressed but low-biconvex, bluntly bicarinate by 2-ribbed sutures ± 3 mm wide, valves stiffly coriaceous lignescent becoming dark brown, smooth or almost so, microscopically puberulent or glabrate; dehiscence through narrowly gaping sutures; seeds in broad view ± 2.7-3.5 cm diam., 5-9 mm thick, testa brown, papery fragile, embryo horny nigrescent.

Distribution: Brazil (Amazonas, Pará, Amapá) and French Guiana; in varzea along streams and river banks or lake shores below 10 m, locally plentiful; known to flower June through August.

Selected specimens: French Guiana: Approuague R. between Saut Tourépé and Ipoussin Cr., Poncy 361 (NY); Approuague R. on Matarony Cr., Oldeman B994 (NY).

7. **Zygia juruana** (Harms) L. Rico, Kew Bull. 46: 501. 1991. – *Pithecolobium juruanum* Harms, Verh. Bot. Vereins Prov. Brandenbrug 48: 162. 1907. Type: Brazil, Amazonas, in woods near Marary, Jurua, Ule 5062 (holotype †B, = F Neg. 1202!).

Trees to 10 m tall; the whole, except bark and sometimes upper surface of lfts at least sparingly pubescent with curly or curved, red-brown hairs. Lf-formula i/3½-5½. Stipules (few seen) equi- or inequilaterally deltate-triangular, ± 2.0 x 1.0 mm, pubescent, not striate; petiole reduced to a wrinkled pulvinus, 4-6 mm, with a small mounded or patelliform nectary between insertion of pinna-pair, and similar but smaller nectaries between every pair, most pairs, or two furthest pairs of lfts; pinna 10.2-21.5 cm; interpinnal segments 2-7.6 cm; lfts slightly accrescent, a- or symmetrically lanceolate to broadly elliptic, 7.9-12.9 x 2.6-9 cm, broadly acuminate. Inflorescence arising from brachyblasts on older wood, branched one or rarely two times, main axis 7-21 mm; peduncle subtended by bracts and commonly a stalked nectary resembling those on lvs, 1.5-3.5 mm; rachis of infl 2-15 mm; bracts ± 0.25-0.5 mm, sessile, persistent. Flowers sessile or calyx contracted at base into a pseudostipe ± 0.25 mm; calyx obconic to narrowly campanulate, 1.1-1.3 mm; corolla tubular-funnelform or funnelform, 5.5-6.25 mm; stemonozone 0.33-0.83 mm; staminal tube 6-9 mm; stamens 33-41; disc 0.25-0.5 mm; ovary sessile, tapering at apex, glabrous, becoming pubescent at least on ventral margin after fertilization. Pod not seen mature, when young up to 5 per brachyblast, green, with prominent venation.

Distribution: Brazil (Goiás, Pará, Amazonas, Roraima) and Guyana; forests along streams and rivers; flowering June to September.

Representative specimen examined: Guyana: Kanuku Mts., Rupununi R., Jansen-Jacobs *et al.* 432 (U).

8. **Zygia latifolia** (L.) Fawc. & Rendle, Fl. Jamaica 4: 150. 1920. – *Mimosa latifolia* L., Sp. Pl. ed. 2: 1499. 1763. Type: West Indies (holotype B).

Tree or rarely liane to 15 m; at least lf-axes, young stems and lobes of perianth pubescent with short curved hairs, sometimes densely so, sometimes glabrate. Lf-formula i/(1½)2-3½. Stipules when young reddish, striate, sometimes persistent and becoming thick, horny and striate or not, more often quickly deciduous, this varying on same branch; petiole reduced to pulvinus or not, 1.25-5.5(22!) mm, with small round, mounded nectary between insertion of pinna-pair, and similar but smaller nectaries between insertion of terminal pair of lfts; pinna 2.0-14.5 cm; interfoliolar segments 12-51 mm; lfts accrescent, asymmetrically broadly elliptic to narrowly lanceolate, 5.0-13.1 x 2.1-5.4 cm, inequilaterally attenuate at base, broadly to long-acuminate, venation pinnate-palmate, with 1, 2 or rarely 3 lateral veins larger and produced from near base beyond mid-blade. Inflorescence from brachyblasts axillary to older lvs, or cauliflorous, 1-4 fasciculate per brachyblast, up to 11 peduncles per cauliflower, each 7-13-flowered, peduncle 0-10 mm, rachis much longer than or much shorter than peduncle, 2.5-6 mm; bracts narrowly to broadly rhombic or rhombic-elliptic, 1.07-2.25 x 0.73-1.75 mm, subpersistent. Flowers subsessile; calyx contracted at base but no true pedicel formed, calyx small and cup-shaped, or tubular-campanulate, 1.0-2.25 mm, 5-lobed; corolla tubular, tubular-funnelform, or tubular-campanulate, at least those at base of spike somewhat asymmetrically so, 5-7.5 mm, pubescent at least on lobes; stemonozone 0.4-1.0 mm; staminal tube 7-11.5 mm, stamens 19-39; nectaries surrounding base of ovary 0.23-0.66 mm tall; ovary sessile, glabrous, gently or abruptly tapering at apex. Pod 1-3 per peduncle, up to 12 (immature) per brachyblast, sessile, ± straight or gently falcate, 7.5-42 x 1.75-3 cm, valves smooth or minutely rugulose, described as green on tree, becoming brown when dry, dehiscence from apex ± equally along both sutures; seeds plump, round or mutually distorted in very fertile pods, up to 24 mm on longest axis, seed-coat papery.

8a. **Zygia latifolia** (L.) Fawc. & Rendle var. **latifolia** – Fig. 44

Pithecolobium huberi Ducke, Arch. Jard. Bot. Rio de Janeiro 4: 29. 1925. – *Zygia huberi* (Ducke) L. Rico, Kew Bull. 46: 501. 1991. Lectotype (designated by L. Rico 1991): Brazil, near Belem, Pará, Huber 2085 (hololectotype MG).

Characters as given in key to varieties.

Distribution: Mexico (Tabasco), the Greater and Lesser Antilles, and in S America along and near the coast from the Orinoco Delta through the Guianas to Brazil (Maranhão); moist or flooded woodland and the edges of mangrove swamps, inland to 400 m; flowering throughout the year.

Selected specimens: Guyana: Pomeroon-Supenaam Region, Pomeroon R. between Charity and Araplako R., Gillespie 1182 (NY, US); Demerara, Jenman 6159 (NY).

8b. **Zygia latifolia** (L.) Fawc. & Rendle var. **lasiopus** (Benth.) Barneby & J.W. Grimes, Mem. New York Bot. Gard. 74(2): 120. 1997. – *Pithecolobium lasiopus* Benth., J. Bot. (Hooker) 2: 141. 1840. Type: Guyana, Ro. Schomburgk ser. I, 487 (holotype K!).

Inga ramiflora Steud., Flora 36: 759. 1843. Type: Suriname, Hostmann & Kappler 1173 (holotype P, seen in photo, NY!, isotype U).

Characters as given in key to varieties.

Distribution: Locally abundant in lower Amazonian Brazil and the Guianas, less common in Venezuelan Guayana; seasonally flooded forest, on river banks, and in gallery woodland; flowering throughout the year (GU: 16; SU: 14; FG: 6).

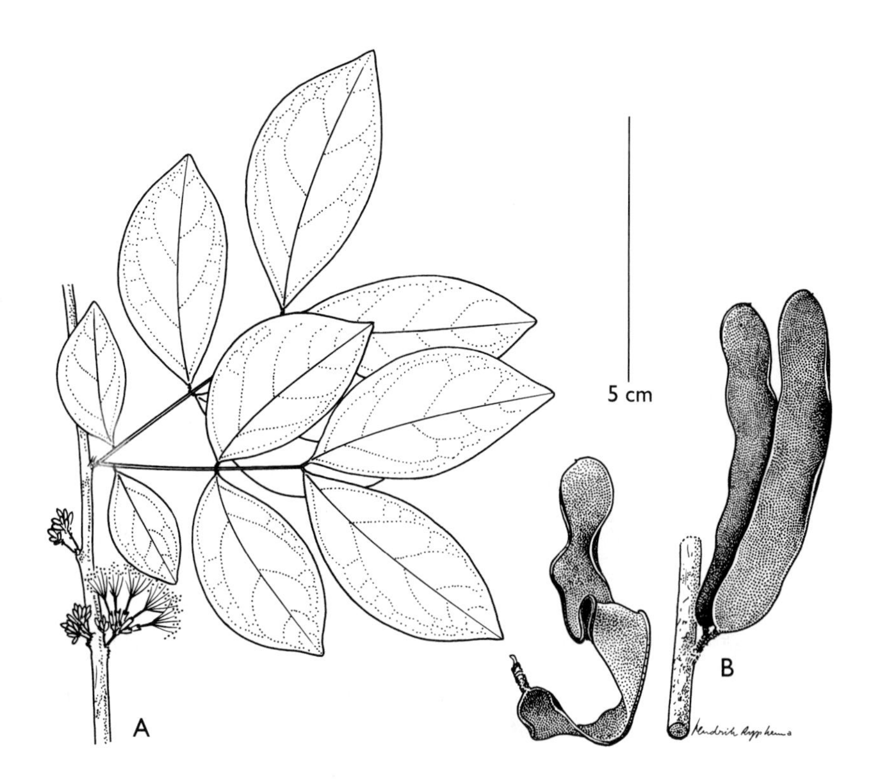

Fig.44. *Zygia latifolia* (L.) Fawc. & Rendle var. *latifolia*: A, flowering branch; B, pods (A, van Andel 5004; B, BBS 2968). Drawing by Hendrik Rypkema; reproduced with permission from U.

Selected specimens: Guyana: Essequibo R., Moraballi Cr., near Bartica, Sandwith 105 (NY); basin of Kuyuwini R. (Essequibo tributary), A.C. Smith 2588 (NY). Suriname: from Moengo tapoe to Grote Zwiebelzwamp, Lanjouw & Lindeman 627 (NY); Zuid R., 2 km above confluence with Lucie R., Irwin *et al.* 55208 (NY). French Guiana: Itany R., upstream from the Indian village Antecum-Pata, Sastre 1812 (NY); Cr. Gabrielle, tributary of the Mahury R., Mori *et al.* 22135 (NY).

Note: Two collections from Suriname (BW 3802 and Lanjouw 728, both NY) probably represent unusually red-hispidulous forms of this species.

9. **Zygia palustris** Barneby & J.W. Grimes, Mem. New York Bot. Gard. 74(2): 66. 1997. Type: Venezuela, Amazonas, uppermost Río Yatua, Maguire *et al.* 41958 (holotype NY!).

Amply multifoliolate cauliflorous small trees, 2-4 m tall; young stems and all lf-axes densely hispidulous with erect straight, yellowish or bronze hairs. Lf-formula iv-v/13-19. Stipules erect, narrowly lance-attenuate, 7-19 x 0.5-1.5 mm, early dry paleaceous, dorsally glabrous or in Guyana thinly pilosulous, persistent unless broken; lf-stk of longer lvs 8.5-15 cm, petiole (consisting largely of lf-pulvinus) 3.5-8(14) x 2-3 mm, longer interpinnal segments ± 25-40 mm; nectary between short first pinna-pair and similar ones between nearly all further pinna-pairs and yet smaller ones on pinna-rachis between all but a few proximal pairs of lfts; pinnae abruptly decrescent proximally, not or scarcely so distally, rachis of longer ones 7-13 cm, longer interfoliolar segments 6-9 mm; pulvinule of lfts obsolete, blade sessile against pinna-rachis; lfts strongly decrescent proximally, first very unequal pair arising next to pinna-pulvinus; blade falcately narrow-oblong, larger ones 19-32 x 5-9 mm; venation palmate and pinnate, midrib subcentric, innermost posterior primary vein produced nearly to or well beyond mid-blade, outer 2-3 very short or obscure. Peduncle 1-6 mm, each subtended by a reflexed bract charged ventrally with a small shallow-cupular nectary; capitula 4-10-flowered, receptacle ± 1.5 mm. Flowers sessile, 5-merous, whitish perianth typically glabrous, but in Guyana sometimes thinly puberulent, everywhere campanulate, including short solid base, 2-3.2 x 1-1.3 mm; corolla narrowly tubular-funnelform, 10.5-13 mm, little ampliate in distal 1/3; stamens 50-56, stemonozone ± 1 mm, slender tube 15-16.5 mm, free filaments pink distally; ovary sessile tapered at apex, at anthesis glabrous. Pod (not seen fully ripe) pendulous, in profile broad-linear straight, 17-20 x 1.8 cm, contracted at base into a stipelike neck 1.5-3 mm long, obtuse at apex, 14-16-seeded, valves thinly leathery, coarsely venulose, glabrous, fuscous-brown, framed by a slightly dilated, almost straight, ciliolate sutures, slightly elevated over each ovule; dehiscence and seed unknown.

Distribution: Venezuela, from the base of C. de la Neblina, and Guyana, on Kaieteur Plateau; in swamp forest and on river-banks subject to flooding; flowering in March, May, October and December, perhaps throughout the year.

10. **Zygia potaroënsis** Barneby & J.W. Grimes, Mem. New York Bot. Gard. 74(2): 68. 1997. Type: Guyana, in forest along trail from Kaieteur Falls to Tukeit, Cowan & Soderstrom 2016 (holotype, NY!).

Slender microphyllidious, cauliflorous small trees 2-6 m tall; young stems and all lf-axes densely pilosulous with erect yellow-brown hairs. Lf-formula viii-ix/15-19. Stipules paleaceous, lanceolate, 6-15 x 1.2-2 mm, persistent at tip of branchlets; lf-stks 8-15 cm, petiole including poorly differentiated pulvinus 5-23 x 1-1.4 mm, longer interpinnal segments 10-16 mm; nectary between or close below proximal pinna-pair and a similar one, only slightly smaller, at tip of lf-stk, those on pinna-rachises minute or 0; pinnae proximally decrescent, rachis of longer distal ones 4-6 cm, longer interfoliolar segments to 2.5-3.5 mm; lft-pulvinules subobsolete, blades sessile against rachis; lfts abruptly decrescent near base of rachis, slightly so distally, blade asymmetrically oblong, longer ones 7.5-10 x 2.4-3.5 mm; venation palmate and pinnate, subcentric midrib incurved beyond middle, inner posterior primary vein produced well beyond mid-blade. Inflorescence not seen intact, and fls unknown. Pod pendulous, in profile broad-linear, nearly straight, ± 17-19 x 1-1.4 cm, plumply biconvex, ± 9-11-seeded, valves thinly leathery, densely brownish-pilosulous overall, weakly venulose, endocarp produced between seeds to form complete papery septa; dehiscence not seen; seeds vertically basipetal, plumply oblong-ellipsoid, ± 18 x 11 x 4.5 mm, testa papery, fragile, dull brown, loosely investing horny embryo, girdled by a fine prominulous vein, brittle when dry.

Distribution: Known only from the Kaieteur Plateau in Guyana and from headwaters of Río Caura, Venezuela (Bolívar); in riparian rain forest.

11. **Zygia racemosa** (Ducke) Barneby & J.W. Grimes, Mem. New York Bot. Gard. 74(2): 71. 1997. – *Pithecolobium racemosum* Ducke, Arch. Jard. Bot. Rio de Janeiro 3: 272a. Corrigenda. 1922. – *Abarema racemosa* (Ducke) Kleinhoonte in Pulle, Fl. Suriname 2(2): 320. 1940. Lectotype (designated by Spichiger, Boissiera 43: 347. 1989): Brazil, Río Jaramacarú, vicinity of Campos do Ariramba, Ducke 14861 (lectotype RB). – Fig. 45

Unarmed microphyllidious trees (6)8-30 m tall; new branchlets, lf-axes, and whole cauliflorous inflorescences pilosulous (strigulose) or tomentulose with sordid or brown-gold, spreading or procumbently curved hairs. Lf-formula (iii)iv-viii(x)/(11)12-19(23). Stipules slenderly lanceolate acute or lance-ovate obtuse, (2)2.5-10(13) x 0.7-2.5 mm, early deciduous; lf-stks of larger lvs (4)5-14 cm, ventrally flattened or shallowly sulcate petiole including pulvinus 3-24(28) mm, longer interpinnal segments 12-28(30) mm, nectary immediately below first pinna-pair and smaller ones at furthest or several further pairs, and yet smaller ones, sometimes distinctly but shortly stipitate, between some distal pairs of lfts; pinnae strongly decrescent proximally, less or not so distally, rachis of penultimate pair 5-9.5(11) cm, longest interfoliolar segments (3)3.5-7.5(8) mm; lft-pulvinules 0.2-0.4 x 0.8-1.2 mm, lfts appearing sessile against pinna-rachis; lfts opposite, decrescent proximally (anterior one of proximal pair often lacking or reduced to a paraphyllidium), scarcely so distally, blade narrowly

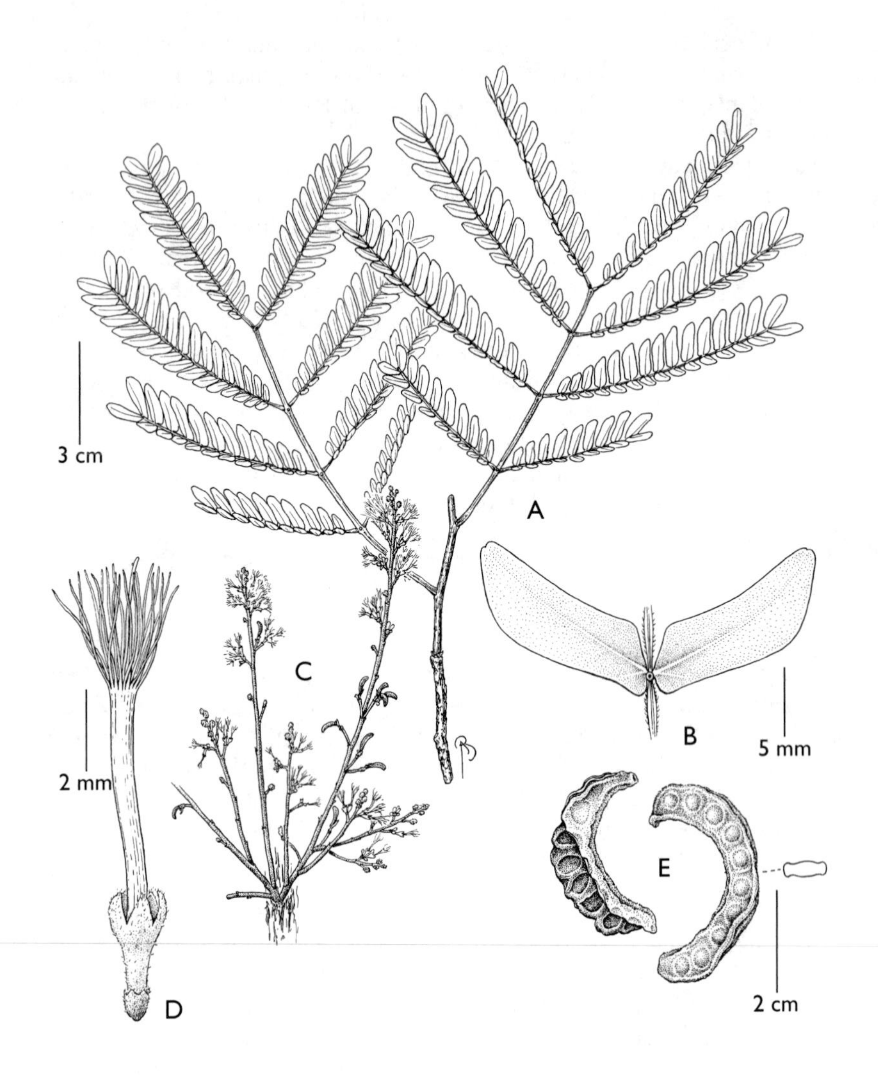

Fig. 45. *Zygia racemosa* (Ducke) Barneby & J.W. Grimes: A, stem and leaves; B, leaflets and nectary; C, cauliflorous inflorescence; D, flower; E, pod (A-B, Silva 1313, from Brazil; C, Daly 1235; D, Daly 1455, both from Brazil; E, closed pod from Lanjouw & Lindeman 2217, open pod from Stahel (Wood Herbarium Surinam) 191). Drawing by Bobby Angell; reproduced with permission from Mori *et al.*, 2002: 509.

oblong, those near mid-rachis (9)10-19(20.5) x (3)3.3-6.5(7) mm; venation palmate and then pinnate, inner posterior vein produced to anastomosis beyond mid-blade. Pseudoracemes either simple or proximally paniculately branched, primary axis (1)2-10(14) cm; peduncles (1)2-4 per node, longer ones (1.5)3-11 mm, each fascicle subtended by a deltate deciduous bract 0.4-1 mm; capitula 12-25-flowered, axis 1-2(2.5) mm; floral bracts ovate or spatulate, 0.25-0.6 mm, persistent. Flowers sessile, 5-merous; calyx campanulate or hemispherical, 0.7-1.1 x 0.7-0.9 mm; corolla narrowly trumpet-shaped, 3.2-4.3(4.5) mm, described as pale brown, salmon-pink, yellow, orange or whitish; stamens 18-22(25), when fresh ochroleucous, whitish, yellow-tipped, or orange, tube (5.2)6.5-9.5 mm, dilated distally, slightly thickened internally at base, stemonozone ± 0.5 mm, disc 0; ovary sessile, tapering at apex. Pods 1-3 per capitulum, falcately recurved or nearly straight, in profile linear, 4-7 x 0.7-1 cm, in cross-section grossly I-shaped, whole fruit densely bronze-brown-tomentulose overall with short crowded (but not "stellate") trichomes, crumpled brown endocarp produced inward as complete or incomplete interseminal septa, dehiscence follicular, through ventral suture; seeds (few seen) in broad view 6-6.5 x 5 mm, testa papery brown.

Distribution: Venezuela, the Guianas, frequent and widespread in the Amazon Basin of Brazil, Colombia, and Peru; in terra firme forest; flowering intermittently throughout the year.

Selected specimens: Guyana: Essequibo R., near mouth of Onoro Cr., A.C. Smith 2721 (NY). Suriname: Nassau Mts., Lanjouw & Lindeman 2217 (NY); Lely Mts., 175 km SSE of Paramaribo, Mori & Bolten 8557 (NY). French Guiana: Piste-de-St-Elie, interfluve Sinnamary/ Counamama, Prévost & Sabatier 2912 (NY).

12. **Zygia sabatieri** Barneby & J.W. Grimes, Mem. New York Bot. Gard. 74(2): 83. 1997. Type: French Guiana: Piste-de-St-Elie, Sabatier & Prévost 3896 (holotype CAY).

Slender macrophyllidious cauliflorous trees attaining 20 m; annotinous and older branchlets pallid gray or cream-color, young ones and all axes of lvs and inflorescences minutely grayish-puberulent. Lf-formula ii/3-5. Stipules erect appressed, ovate-triangular or lanceolate, 2.5-6 x 1.2-4 mm, persistent; lf-stk 5-8 cm, petiole 1.5-3.5 cm, one interpinnal segment somewhat longer; nectary between each pinna-pair; distal pinnae longer than first pair, their rachis 7-11.5 cm, longer interfoliolar segments 2-4 cm; lfts opposite, pulvinules in dorsal view ± 1.5-2.5 mm, much shorter in ventral view; lfts accrescent distally, inequilaterally elliptic or subrhombic-elliptic, longer, furthest two pairs 8-12.5 x 3-4.8 cm; venation pinnate,

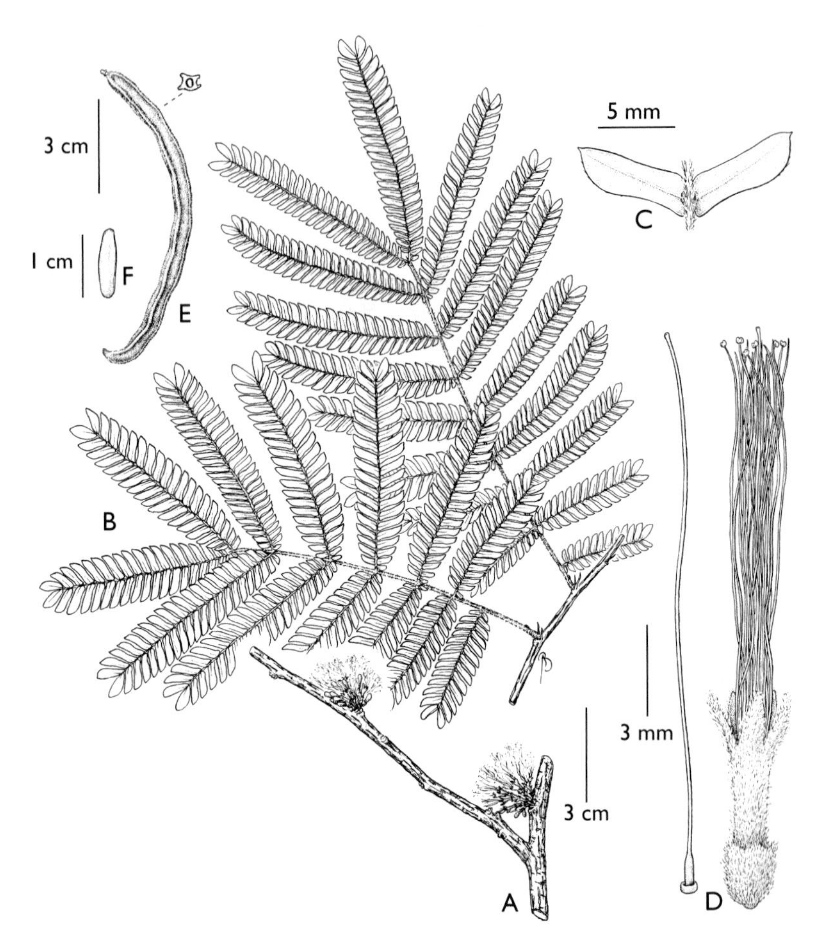

Fig.46. *Zygia tetragona* Barneby & J.W. Grimes: A, inflorescences; B, leaves; C, lower surface of leaflets; D, flower (right), gynoecium (left); E, pod; F, seed (A-D, Sabatier 2561; E-F, Larpin 867). Drawing by Bobby Angell; reproduced with permission from NY.

straight or almost straight, subcentric midrib giving rise on each side to ± 7-9 major secondary veins. Peduncles 1-2 per brachyblast, (1.5)2.5-4 cm; capitula 30-48-flowered; bracts submembranous, early dry deciduous, spatulate, ± 0.5-0.7 mm, blade hispidulous; pedicel 0.8-1.5 x 0.2-0.25 mm. Calyx campanulate, ± 1 x 0.7-0.8 mm; corolla slenderly cylindric, a trifle dilated distally, ± 6.5-7 mm and 1 mm diam.; stamens 25-28, staminal tube 6-6.5 mm, nearly as long as corolla tube, stemonozone ± 0.6 mm; disc 0, but receptacle surrounding base of ovary tumid and slightly discolored; ovary glabrous, tapering at apex. Pod sinuously linear-moniliform and after dehiscence randomly twisted, when well fertilized 14-20 cm, at each seed 6-7 mm and at each interseminal isthmus 2.5-3 mm wide, valves stiffly leathery, when ripe dark brown evenulose, densely minutely puberulent overall, endocarp smooth brown, cavity continuous; funicle and seed unknown, seed described (Sabatier 2092) as green.

Distribution: French Guiana, known only from the Sinnamary-Counamama divide; in lowland virgin forest.

Representative specimen examined: French Guiana: Piste-de-St-Elie, interflue Sinnamary/Counamama, Sabatier 2092 (NY!).

13. **Zygia tetragona** Barneby & J.W. Grimes, Mem. New York Bot. Gard. 74(2): 80. 1997. Type: French Guiana, Bassin de l'Arataye, Station des Nouragues, Sabatier & Prévost 2561 (holotype, CAY!). – Fig. 46

Broad-crowned microphyllidious cauliflorous, understory trees attaining 20 m; young stems and all axes of lvs and inflorescences densely fine pilosulous with spreading, ± sinuous, brown hairs; bark peels in large flakes. Lf-formula vi-x(xi)/(16)18-25. Stipules firm, lanceolate, 2-4 x 0.6-1.2 mm, brown-silky dorsally, glabrous within, deciduous, terminal resting-buds of similar, loosely imbricate scales; lvs subsessile, lf-stk including pulvinus 7-12 cm, very short first pinna-pair inserted close to lf-pulvinus and deflexed amplexicaul, longer distal interpinnal segments 10-18 cm, sessile nectary at base of lf-stk between first pinna-pair and similar but smaller ones between 1-3 furthest pinna-pairs and often a minute one at tip of some pinna-rachises; pinnae strongly decrescent from below mid-lf, thence subequilong, rachis of longer ones 4.5-7 cm, longer interfoliolar segments 2-4.5 mm; lfts abruptly decrescent near base of pinna-rachis (anterior of first pair often reduced to paraphyllidium), slightly or scarcely decrescent distally, all sessile against rachis, pulvinule 0.2-0.4 mm in dorsal view, wider than long, blade narrowly rhombic-oblong, longer ones 7-12 x 2-5 mm; venation of mature lfts mostly reduced to simple midrib, but one weak primary vein sometimes produced, on posterior side of midrib, well beyond mid-blade but not brochidodrome. Spikes sessile, axis ± 1 cm; bracts deltate 1.2-1.5 mm, like whole perianth densely pilosulous externally overall. Calyx campanulate, 1.9-2 mm, nearly as wide; corolla "white", subcylindric, 6.5-7 mm; stamens ± 34, stemonozone 0.6 mm, included tube ± 5 mm, tassel reddish distally; nectarial disc ± 0.3 mm tall, twice as wide; ovary at anthesis glabrous, tapering at apex. Pod pendulous, in profile falcately linear, 8.5-11 x 0.6 cm, bluntly tetragonal (not laterally compressed), thick, rigidly ligneous, transversely dilate, longitudinally shallow-sulcate sutures as wide as remainder of pod, in section ± 1 mm thick, valves as thick but only half as wide, whole fruit densely minutely brownish-pilosulous overall, dehiscence follicular, through seminiferous (convexly arcuate) suture, valves gaping widely; seeds narrowly ellipsoid, 10.5("15" on label) x 3("6") mm, testa when dry papery, castaneous, exareolate, described as "greenish white, peripherally pink" when fresh.

Distribution: Locally plentiful but known only from French Guiana, basins of the Sinnamary and Approuague/Arataye Rs.; in lowland primary forest.

Selected specimens: French Guiana: Mts. des Nouragues, bassin de l'Approuague-Arataye, Larpin 867 (CAY, NY).

14. **Zygia unifoliolata** (Benth.) Pittier, 3rd Conf. Interam. Agric. Caracas 259. 1945. – *Pithecolobium unifoliolatum* Benth., London J. Bot. 3: 212. 1844. Type: Brazil, Río Madeira, Langsdorff (holotype K!).

Shrub or tree to 8 m tall; glabrate throughout. Lf-formula i/1. Stipules (few seen) narrowly triangular, sometimes falcately so, ± 1.75 x 0.75 mm, deciduous or persistent, when young faintly striate, becoming horny-white with age, striations reduced to obvious midvein; petiole reduced to pulvinus, or as long as pulvinus, whole 1.4-7 mm, nectary mounded ± round, between insertion of pinna; pinnae reduced to pulvinules; lfts narrowly to broadly elliptic or elliptic-lanceolate, sometimes slightly asymmetrically so, 3.9-15.5 x 1.9-7 cm. Inflorescence of capitulate to short-spicate, unbranched peduncles 1-7 per cauliflower, 0-11 mm; rachis 0.75-2 cm; bracts broadly lanceolate or spatulate-lanceolate, sometimes somewhat cucullate, persistent through anthesis. Flowers sessile, those at base of rachis usually asymmetric in shape; calyx 0.5-0.95 mm, shallowly cup-shaped, ± truncate; corolla tubular-funnelform, 5.5-6.5 mm, glabrous, with obvious striations; staminal tube 8-12 mm; stamens 23-37; stemonozone 0.25-0.65 mm; intrastaminal disc 0.15-0.4 mm; ovary subsessile, glabrous. Pod 1 per peduncle, up to 4 per cauliflower, falcate to curved through 1/2 circle when mature, 7-16.5 x 1.5-2 cm, valves green on tree, drying shiny, light brown, smooth, though venation obvious, dehiscent initially through ventral suture, dorsal one eventually dehiscing at least slightly at apex; seeds ± round, to 1.7 cm diam., commonly distorted by mutual pressure, seed-coat papery.

Distribution: Colombia, Venezuela (Apuré, Bolívar, Amazonas), Brazil, Peru and Bolivia; to be expected in Guyana; in varzea and gallery forest along rivers, though also reported from terra firme.

WOOD AND TIMBER

by

PIERRE DÉTIENNE[33], JIFKE KOEK-NOORMAN[34], LUBBERT Y.TH. WESTRA[34], IMOGEN POOLE[34, 35, 36] & BEN J.H. TER WELLE[34]

WOOD ANATOMY

The MIMOSOIDEAE are a subfamily of tropical and subtropical taxa with 28 genera known to occur in the Guianas. *Dinizia* has often been included within the MIMOSOIDEAE but recent analyses have placed this genus in a excentric phylogenetic position between MIMOSOIDEAE and CAESALPINIOIDEAE on the base of molecular and morphological characters (Luckow *et al.* 2000, 2003). Even though *Dinizia* is very similar in anatomy to the other mimosoids (Evans *et al.* 2006), it has not been included in this fascicle because of its phylogenetic placement.
This chapter outlines the wood anatomy of mimosoid plants growing in the Guianas. However when a species known to occur in the Guianas was not represented by a sample in the collections from this immediate area, material from adjacent regions was used for completeness. Most samples studied are from the Montpellier (CTFw) and Nationaal Herbarium Nederland, Utrecht (Uw) wood collections (since 2009 housed in NHN-Leiden). If the name of the collector is not documented, only the number of the original wood collection is given. All acronyms used for wood collections follow Stern 1988. All descriptive terms and measurements are in accordance with the IAWA-list of microscopic features for hardwood identification (IAWA-committee 1989).

FAMILY CHARACTERISTICS

Vessels diffuse, solitary and in short radial multiples, rarely in irregular clusters of 4-10, perforations simple, intervessel pits alternate, vestured, vessel-ray pits similar to intervessel pits. Tyloses absent.

[33] C.I.R.A.D., TA-B-40/16, 73 rue Jean-François Breton, 34398 Montpellier, Cedex 5, France.

[34] National Herbarium of the Netherlands, Utrecht University branch, Heidelberglaan 2, 3584 CS Utrecht, The Netherlands (since 2009 housed in NHN-Leiden).

[35] Department of Organic Geochemistry, Faculty of Earth Sciences, Utrecht University, P.O. Box 80021, 3508 TA Utrecht, The Netherlands.

[36] Palaeoecology, Institute of Environmental Biology, Faculty of Science, Utrecht University, Budapestlaan 4, 3584 CD Utrecht, The Netherlands.

Rays homogeneous, composed of procumbent cells, mostly 1-3-seriate.
Parenchyma mostly paratracheal (except in *Calliandra*), vasicentric to aliform-confluent; apotracheal diffuse often present but generally not abundant, marginal bands sometimes present. Strands of (1-)2-4 cells but up to 24 cells in *Cedrelinga* and *Mimosa*.
Fibres in some genera non-septate, in other genera all or partly septate; pits simple, confined to the radial walls.
Rhombic crystals in chambered axial parenchyma cell frequent but absent in the genus *Cedrelinga*.

KEY TO GENERA OR SPECIES

(The anatomical structure of the family is fairly homogeneous, but individual anatomical characters may vary within species.)

DICHOTOMOUS KEY

1 Vessels of two distinct size classes: large vessels solitary, diameter up to 350 µm, 3-6 per sq. mm; narrow vessels in radial multiples or clusters of 2-7 . . . . *Mimosa guilandinae*
Vessel diameter and arrangement otherwise 2

2 Fibres all or partly septate 3
Fibres non-septate 10

3 Rays commonly 3-seriate or more 4
Rays 1-3-seriate 7

4 Rays height < 200 µm, frequency > 6 per mm . . . . *Anadenanthera peregrina*
Rays height > 200 µm, frequency 4-6 per mm 5

5 Fibres partly septate; rays 4-8-seriate, > 300 µm high . . . . *Acacia*
Fibres septate; rays 3-5-seriate, 200-300 µm high 6

6 Vessel diameter < 150 µm . . . . *Leucaena leucocephala*
Vessel diameter > 150 µm . . . . *Albizia*

7 Vessel frequency > 8 per sq. mm; rays height < 200 µm . . . . *Pithecellobium dulce, P. unguis-cati*
Vessel frequency < 8 per sq. mm; rays height > 200 µm 8

8 Fibres all septate . . . . *Inga*
Fibres partly septate 9

9 Intervessel pits diameter 6-7 μm; fibres thick walled .... *Enterolobium oldemannii, E. schomburgkii*
Intervessel pits diameter 8-10 μm; fibres thin walled .... *Enterolobium cyclocarpum*

10 Parenchyma strands with 8-24 cells .... *Cedrelinga cateniformis*
Parenchyma strands with 1-4 cells 11

11 Intervessel pits diameter 8-10 μm 12
Intervessel pits diameter < 8 μm 17

12 Paratracheal parenchyma thick, diffuse parenchyma absent 13
Paratracheal parenchyma thin, diffuse parenchyma present 15

13 Rays 3-6-seriate, > 300 μm high .... *Parkia pendula*
Rays 1-3-seriate, < 300 μm high 14

14 Rays mostly 1-seriate; vessel diameter > 200 μm .... *Stryphnodendron guianense, S. pulcherrimum*
Rays 2-3-seriate; vessel diameter < 200 μm .... *Samanea saman*

15 Parenchyma mostly vasicentric; vessel diameter 130-160 μm .... *Adenanthera pavonina*
Parenchyma lozenge to winged-aliform; vessel diameter > 160 μm 16

16 Parenchyma with oblique confluences; vessel frequency 2-3 per sq. mm .... *Stryphnodendron polystachyum*
Parenchyma with tangential confluences; vessel frequency 1-3 per sq. mm .... *Parkia*

17 Parenchyma apotracheal banded; rays 2- to 4-seriate .... *Calliandra*
Parenchyma mostly paratracheal, including very confluent; rays 1- or 2-seriate 18

18 Parenchyma at least partly confluent, sometimes forming large bands .... *Zygia*
Parenchyma not, or hardly, confluent 19

19 Diffuse parenchyma abundant; 9-12 rays per mm; intervessel pits 4-5 μm .... *Pentaclethra macroloba*
Diffuse parenchyma rare; 5-9 rays per mm; intervessel pits 5-7 μm 20

20 Wood non fluorescent under ultraviolet light 21
Wood fluorescent under ultraviolet light 22

21 Vessel diameter < 150 μm; vessel frequency 5-8 per sq. mm .... *Macrosamanea*
Vessel diameter > 150 μm; vessel frequency 2-4 per sq. mm .... *Balizia pedicellaris*

22 Vessel frequency 6-10 per sq. mm; fibre length < 1000 μm. . . . . *Chloroleucon acacioides*
Vessel frequency 2-6 per sq. mm; fibre length > 1000 μm. . . . . *Abarema* and *Hydrochorea*

SYNOPTICAL KEY

Numbers in the key refer to the number assigned to the genus and thus generic descriptions given below.
Numbers underlined indicate that a character is present only in some species that belong to that particular genus.

1 Abarema
2 Acacia
3 Adenanthera
4 Albizia
5 Anadenanthera
6 Balizia
7 Calliandra
8 Cedrelinga
9 Chloroleucon
10 Enterolobium
11 Hydrochorea
12 Inga
13 Leucaena
14 Macrosamanea
15 Mimosa
16 Parkia
17 Pentaclethra
18 Pithecellobium
19 Pseudopiptadenia
20 Samanea
21 Stryphnodendron
22 Zygia

a Vessels of two distinct size classes, large vessels solitary, small vessels in radial multiples or clusters of 2-7: 15
vs Vessel diameter and arrangement otherwise

b Vessel diameter < 150 μm: 5-9-13-14-18-19
Vessel diameter up to > 200 μm: 1-4-6-8-10-12-15-16- 21
vs Highest vessel diameter between 150-200 μm

c Vessel frequency < 5 per mm: 6-8-11-12-16-20-21
Vessel frequency up to > 10 per mm: 5-7-9-18-19
vs Highest vessel frequency between 5-10 per mm

d Vessels in clusters of up to 6-12 narrow vessels: 3-6-7-8-12-13-15-17
vs Vessels solitary and in short radial multiples

e Rays 1- over 3-seriate: 2-4-5-13-16
vs 1-3-seriate

f Ray frequency over 10 per mm: 8-17-18-22
vs Ray frequency lower

g Apotracheal parenchyma bands present: 7-14-15
Diffuse apotracheal parenchyma frequent: 3-6-8-17
vs Apotracheal parenchyma scanty diffuse, and/or in narrow marginal bands

h Paratracheal parenchyma bands present: 22
vs Parenchyma vasicentric/lozenge to winged-aliform/confluent

i Parenchyma strands up to 24 cells: 8-15
Parenchyma strands up to 2 cells: 5-9-14-18-19-20-22
vs Parenchyma strands up to 4-8 cells

j Crystals in chambered parenchyma cells lacking: 8-21
Crystals up to 8 in chambered parenchyma cells present: 9
vs 16-32 crystals in chambered parenchyma cells or strands

k Fibres septate: 2-4-5-10-12-13-18
vs Fibres non-septate

GENERIC DESCRIPTIONS

ABAREMA Pittier – Figs. 47 A-C; 48 A-D

Vessels diffuse, solitary and in short radial multiples of 2-3, round, 2-6 per sq. mm, diameter 120-210 µm. Vessel member length 430-520 µm. Perforations simple. Intervessel pits vestured, alternate, 5-6 µm. Vessel-ray pits similar to intervessel pits. Red brown deposits present.
Rays 1(-2)-seriate, 5-7 per mm, 150-265 µm high. Homogeneous, composed of procumbent cells.
Parenchyma paratracheal mostly vasicentric to hardly lozenge-aliform, rarely confluent and apotracheal parenchyma diffuse scarce. Strands of 2-4 cells. Prismatic crystals in chambered cells, 8-32 per strand, common.
Fibres thin- to thick-walled, lumen 13-23 µm, walls 4-6 µm. Pits simple, small, confined to radial walls. Length 950-1380 µm. F/V ratio: 2.2-2.8.

Material studied:
Abarema barbouriana (Standl.) Barneby & J.W. Grimes: Guyana: FD 4181 (Uw 971). Brazil: INPAw 1928.
A. jupunba (Willd.) Britton & Killip: Guyana: Polak 185 (CTFw 31900). Suriname: Stahel 20 (Uw 20); Lanjouw & Lindeman 1578 (Uw 1520). French Guiana: Benoist 589 (CTFw 14514), Mélinon 119 (CTFw 16539).
A. laeta (Benth.) Barneby & J.W. Grimes: Guyana: Maas *et al.* 3110 (Uw 23641).
A. mataybifolia (Sandwith) Barneby & J.W. Grimes: Guyana: Cowan 39332 (Uw 4856).

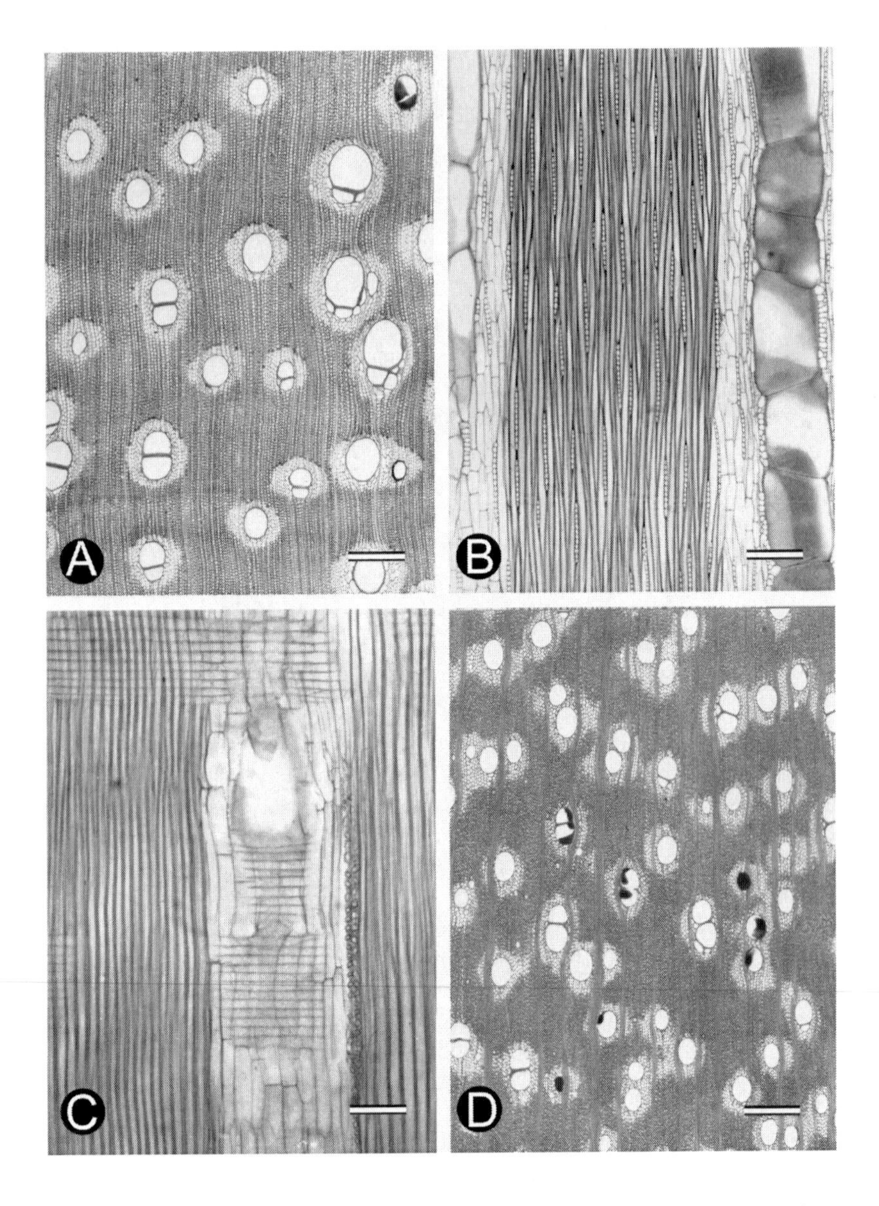

Fig.47. A-C: *Abarema jupunba* (Willd.) Britton & Killip: A, transverse section, scale bar = 400 μm (CTFw 14514); B, tangential section, scale bar = 180 μm (CTFw 14514); C, radial section, scale bar = 400 μm (CTFw 14514). D: *Acacia macracantha* Willd.: D, transverse section, scale bar = 400 μm (CTFw 8255).

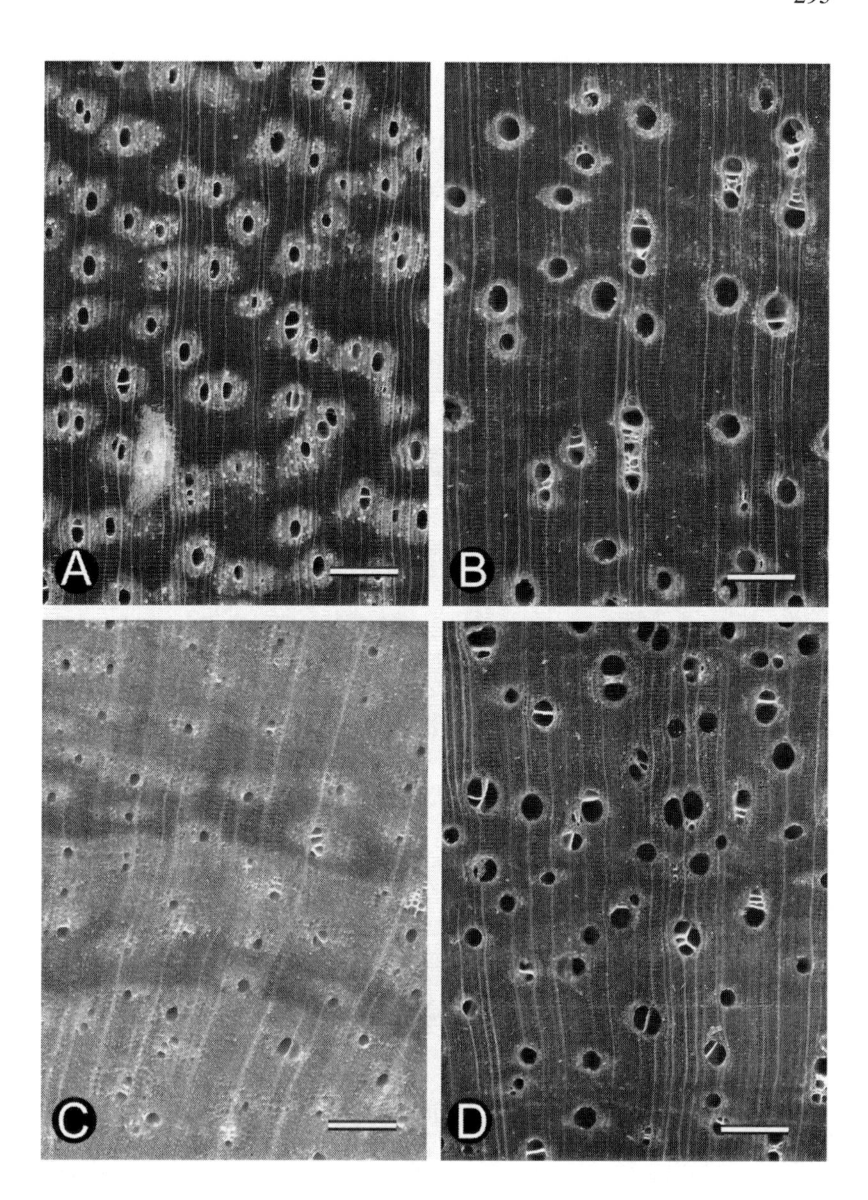

Fig. 48. A-D: transverse sections, scale bar = 500 µm: A, *Abarema barbouriana* (Standl.) Barneby & J.W. Grimes (Uw 971); B, *Abarema jupunba* (Willd.) Britton & Killip (Uw 1520); C, *Abarema laeta* (Benth.) Barneby & J.W. Grimes (Uw 23641); D, *Abarema mataybifolia* (Sandwith) Barneby & J.W. Grimes (Uw 4856).

ACACIA Mill. – Figs. 47 D; 49 A

Vessels diffuse, solitary and in radial multiples of 2-3(-4), round, 3-7 per sq. mm, diameter 110-190 μm. Vessel member length 250-360 μm. Perforations simple. Intervessel pits vestured, alternate, 8-9 μm. Vessel-ray pits similar to intervessel pits. Red brown deposits more or less frequent.
Rays 4-7(-8)-seriate, 4-5 per mm, 340-450 μm high. Homogeneous, composed of procumbent cells.
Parenchyma paratracheal vasicentric to shortly lozenge-aliform not often confluent, apotracheal parenchyma diffuse rare, marginal bands occasionally present. Strands of 2-4 cells. Prismatic crystals, 8-16 per strand in chambered cells in *A. polyphylla*; 4-8, often large, crystals per strand in inflated chambered cells in *A. macracantha*.
Fibres partly septate, thin-to-thick walled, lumen 7-15 μm, walls 4-6 μm. Pits simple, small, confined to radial walls. Length 1000-1300 μm. F/V ratio: 2.8-4.5.
Traumatic canals in tangential lines sometimes present.

Material studied:
Acacia macracantha Willd.: Ecuador: Acosta-Solis 11833 (CTFw 8255); Little 25 (CTFw 23472). Peru: Williams 652 (CTFw 8255).
A. polyphylla DC.: Peru: Gutierrez 46 (MADw 22385). Brazil: Lindeman & de Haas 2338 (Uw 13682). Paraguay: Bernardi 19417 (CTFw 30560).

ADENANTHERA L. – Figs. 49 B; 50 A

Vessels diffuse, solitary and in radial multiples of 2-4, sometimes in small clusters, round, 4-6 per sq. mm, diameter 130-160 μm. Vessel member length 420-700 μm. Perforations simple. Intervessel pits vestured, alternate, 8-9 μm. Vessel-ray pits similar to intervessel pits.
Rays 1(-2) or (1-)2-seriate, 6-8 per mm, 240-320 μm high. Homogeneous, composed of procumbent cells.
Parenchyma paratracheal vasicentric to shortly lozenge-aliform rarely confluent, apotracheal parenchyma diffuse relatively abundant and in fine marginal bands sometimes present. Strands of 2-4 cells. Prismatic crystals in chambered cells, 8-32 per strand.
Fibres thin- to thick-walled, lumen 15-21 μm, walls 5-65 μm. Pits simple, small, confined to radial walls. Length 1050-1280 μm. F/V ratio: 1.7-2.5.

Material studied:
Adenanthera pavonina L.: French Lesser Antilles: Beraud ONF 4 (CTFw 27762 = Uw 27768). China: FRTGw 188. New Caledonia: McKee 16048 (CTFw 20731).

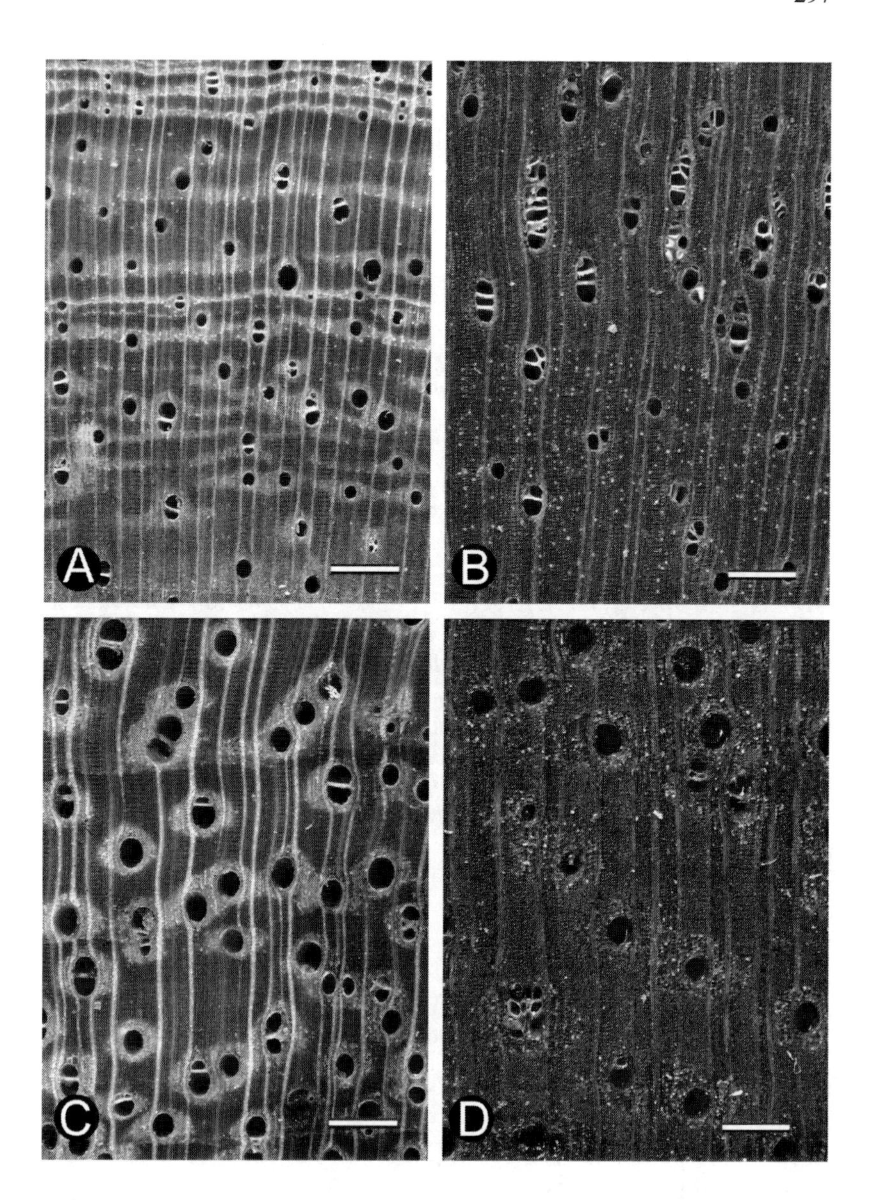

Fig. 49. A-D: transverse sections, scale bar = 500 μm: A, *Acacia polyphylla* DC. (Uw 13682); B, *Adenanthera pavonina* L. (Uw 27768); C, *Albizia glabripetala* (H.S. Irwin) G.P. Lewis & P.E. Owen (Uw 975); D, *Albizia lebbeck* (L.) Benth. (Uw 24769).

ALBIZIA Durazz. – Figs. 49 C, D; 50 B; 51 A

Vessels diffuse, solitary and in radial multiples of 2-4, round, 2-7 per sq. mm, diameter 140-220 µm. Vessel member length 360-450 µm. Perforations simple. Intervessel pits vestured, alternate, 6-9 µm. Vessel-ray pits similar to intervessel pits. Dark brown deposits common.
Rays 2-5-seriate, 4-6 per mm, 230-310 µm high. Homogeneous, composed of procumbent cells.
Parenchyma paratracheal mostly lozenge-aliform, sometimes confluent, apotracheal parenchyma diffuse rare, fine marginal bands occasionally present. Strands of 2-4 cells. Prismatic crystals in chambered cells, 4-16 per strand, common.
Fibres septate, thin- to thick-walled, lumen 10-18 µm, walls 4-7 µm. Pits simple, small, confined to radial wall. Length 950-1200 µm. F/V ratio: 2.3-3.1.

Material studied:
Albizia glabripetala (H.S. Iriwn) G.P. Lewis & P.E. Owen: Guyana: FD 5397 (Uw 975).
A. lebbeck (L.) Benth.: India: Gamble P1193 (FIw 508). Comoro Islands: SF 16609 (CTFw 13038). Madagascar: SF 19284 (CTFw 13054). New Caledonia: SF 96 (CTFw 6165); Sarlin 265 (Uw 24769).
A. niopoides (Spruce ex Benth.) Burkart: Venezuela: MAD-SJRw 45580; MAD-SJRw 45652. Bolivia: Nee 37021.
A. subdimidiata (Splitg.) Barneby & J.W. Grimes: Brazil: Krukoff 6750 (Uw 7914).

ANADENANTHERA Speg. – Figs. 50 C; 51 B

Vessels diffuse, solitary and in radial multiples of 2-4, round, 7-10 per sq. mm, diameter 80-110 µm. Vessel member length 230 µm. Perforations simple. Intervessel pits vestured, alternate, 6-7 µm. Vessel-ray pits similar to intervessel pits.
Rays 2- to 5-seriate, 6-8 per mm, 150-200 µm high. Homogeneous, composed of procumbent cells.
Parenchyma vasicentric, sometimes tangentially or obliquely confluent. Strands of 2-(4) cells. Prismatic crystals in chambered cells, 8-24 per strand, common.
Fibres partly septate, thin-to thick-walled, lumen 10-13 µm, walls 4 µm. Pits simple, small, confined to radial walls. Length 750-850 µm. F/V ratio: 3.4.

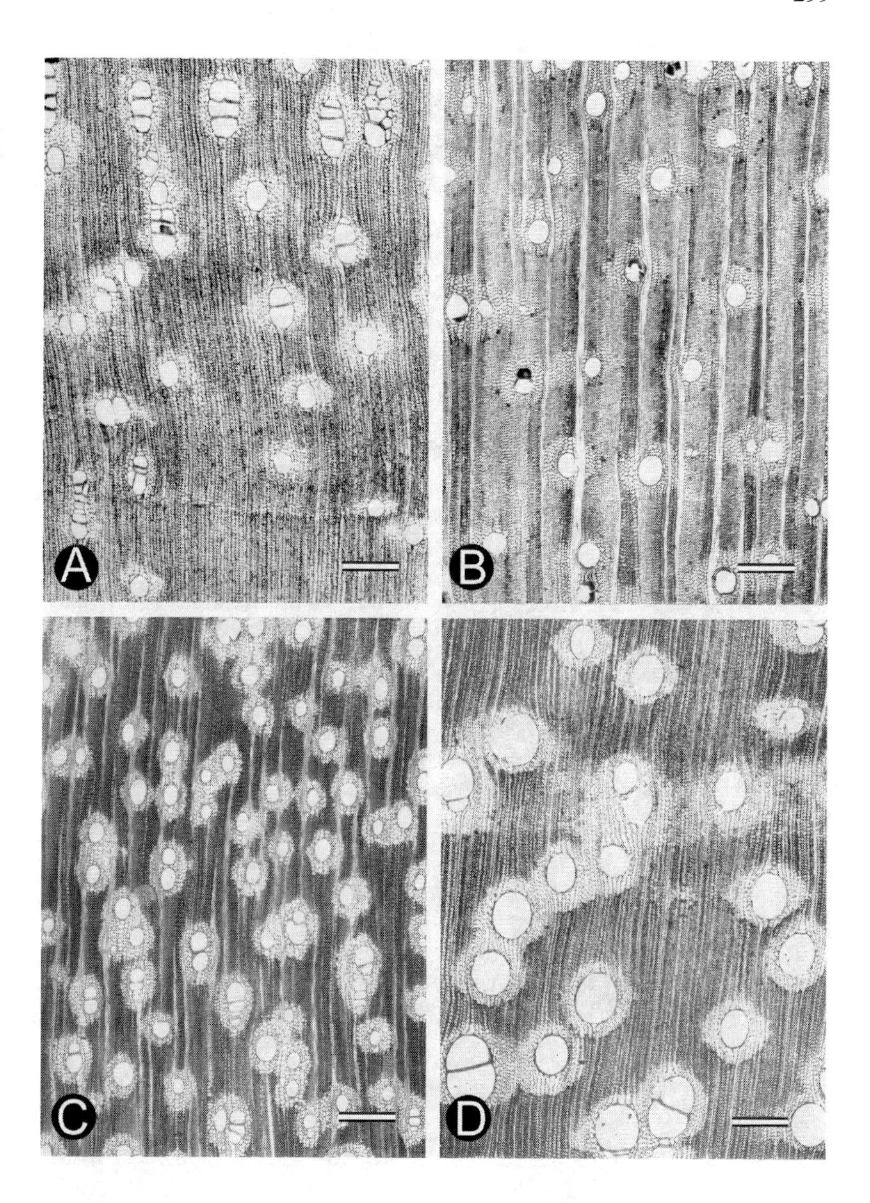

Fig. 50. A-D, transverse sections, scale bar = 400 µm: A, *Adenanthera pavonina* L. (CTFw 27762 = Uw 27768); B, *Albizia lebbeck* (L.) Benth. (CTFw 6165); C, *Anadenanthera peregrina* (L.) Speg. (CTFw 26619); D, *Balizia pedicellaris* (DC.) Barneby & J.W. Grimes (CTFw 16706).

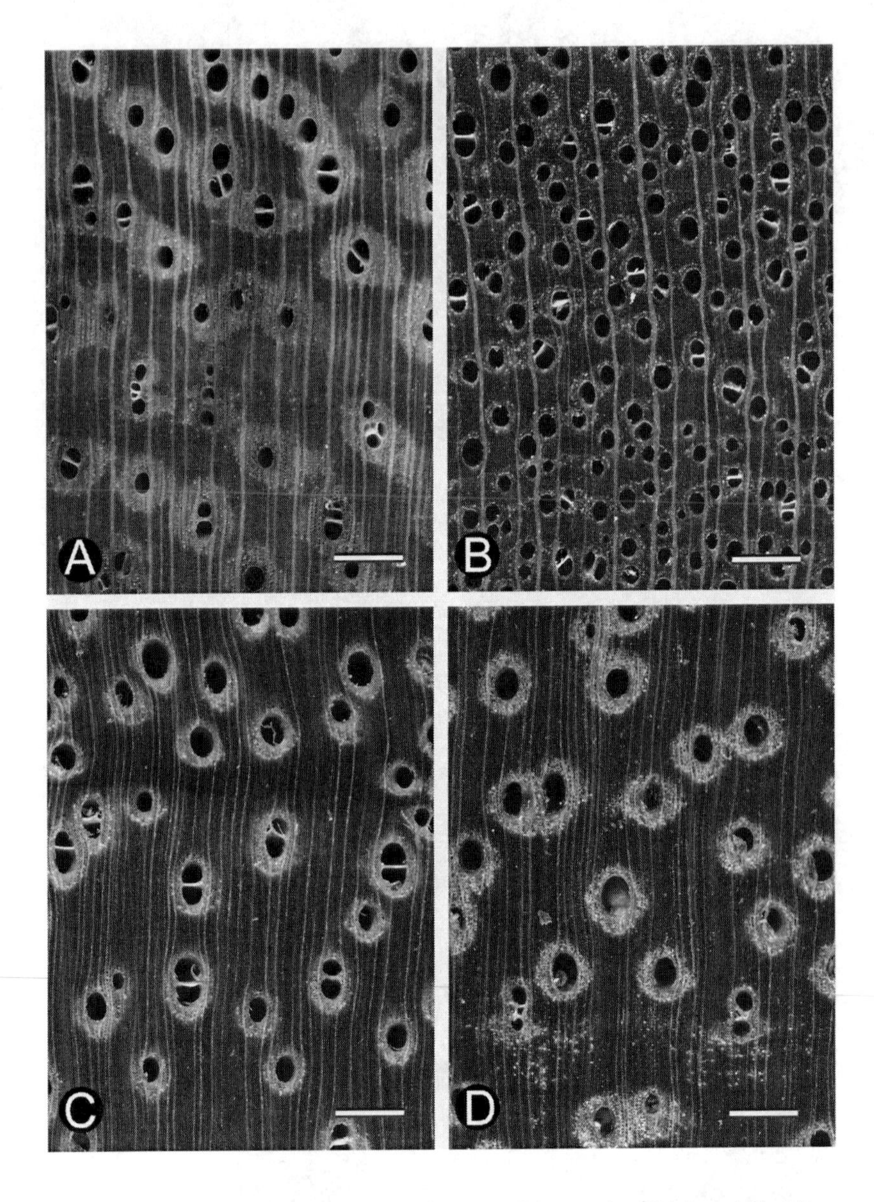

Fig.51. A-D: transverse sections, scale bar = 500 µm: A, *Albizia subdimidiata* (Splitg.) Barneby & J.W. Grimes (Uw 7914); B, *Anadenanthera peregrina* (L.) Speg. (Uw 15360); C-D: *Balizia pedicellaris* (DC.) Barneby & J.W. Grimes: C, (Uw 125); D, (Uw 2100).

Material studied:
Anadenanthera peregrina (L.) Speg.: Suriname: Oldenburger & Norde 543 (Uw 15360). Brazil: Hatschbach 14016 (Uw 14359); Détienne 350 (CTFw 26619).

BALIZIA Barneby & J.W. Grimes – Figs. 50 D; 51 C-D

Vessels diffuse, solitary and in short radial multiples of 2-3(-4), rarely in clusters of 4-10 small vessels, round, 2-4 per sq. mm, diameter 160-210 µm. Vessel member length 400-600 µm. Perforations simple. Intervessel pits vestured, alternate, 5-6 µm. Vessel-ray pits similar to intervessel pits.
Rays in majority 1- or 2-seriate, 7-9 per mm, 100-270 µm high. Homogeneous, composed of procumbent cells.
Parenchyma paratracheal vasicentric to lozenge-aliform rarely confluent, apotracheal parenchyma diffuse rare to frequent. Strands of 2-4 cells. Prismatic crystals in chambered sometimes 2-seriate cells, 8-32 per strand.
Fibres thin-walled, lumen 15-20 µm, walls 4-5 µm. Pits simple, small, confined to radial walls. Length 1250-1400 µm. F/V ratio: 2.1-3.1.

Material studied:
Balizia pedicellaris (DC.) Barneby & J.W. Grimes: Suriname: Stahel 125 (Uw 125); BBS 1059 (Uw 2100). French Guiana: BAFOG 10M (CTFw 7612); Benoist 1554 (CTFw 14586); Petrov 74, 86, 105 (CTFw 16661, 16691, 16706).

CALLIANDRA Benth. – Figs. 52 A; 53 A

Vessels diffuse, solitary and in radial multiples of 2-4, occasionally in clusters of 4-6, round, 4-15 per sq. mm, diameter 110-170 µm. Vessel member length 260-310 µm. Perforations simple. Intervessel pits vestured, alternate, 6-7 µm, often with coalescent apertures. Vessel-ray pits similar to intervessel pits.
Rays (1-)2-3(-4)-seriate, 6-8 per mm, 250-410 µm high. Homogeneous, composed of procumbent cells.
Parenchyma mostly apotracheal in bands 3-7 cells wide, in 1-seriate marginal bands, in rare diffuse cells, and scanty paratracheal. Strands of 2-4 cells. Prismatic crystals in chambered cells, 8 to 16-24 per strand, sometimes 2 crystals in a chamber.
Fibres thick-walled, lumen 4-5 µm, walls 6-7 µm. Pits simple, small, confined to radial walls. Length 1000-1200 µm. F/V ratio: 3.8.
Pith flecks sporadic.

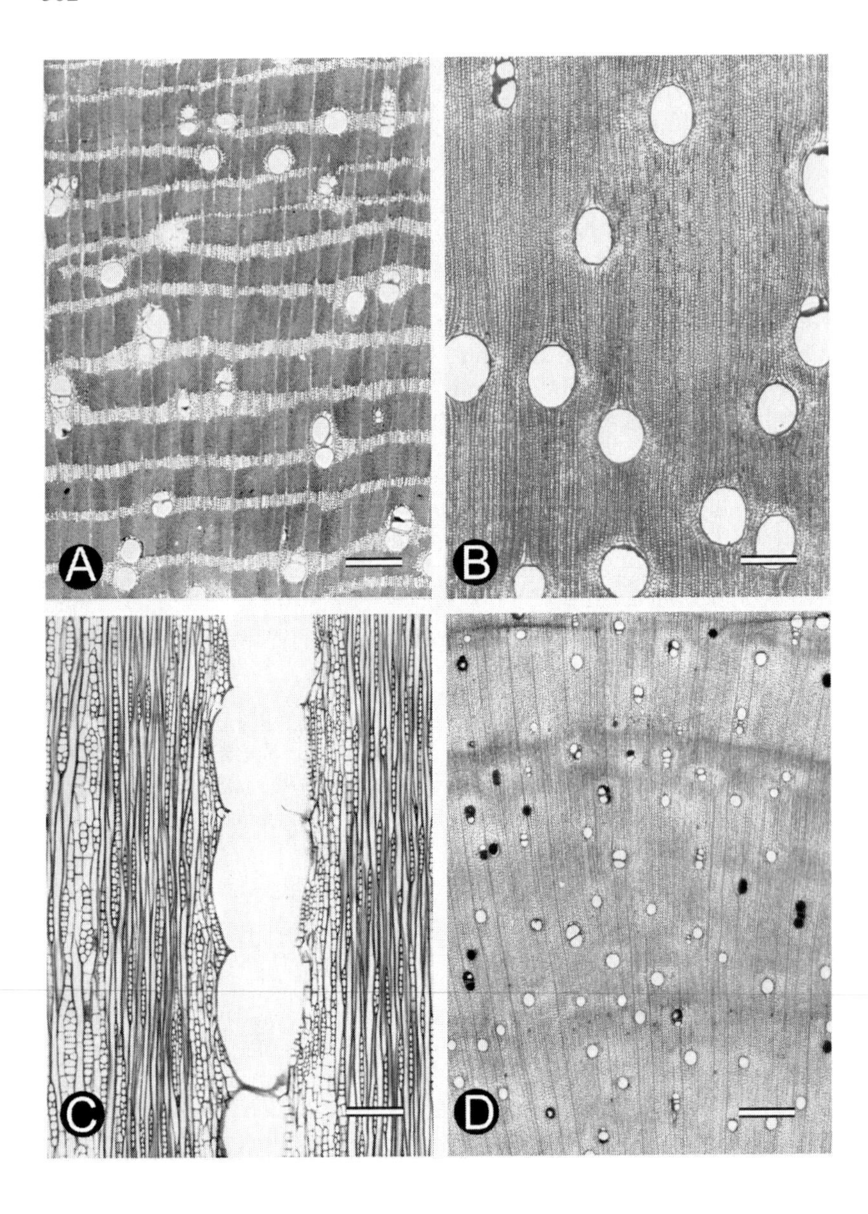

Fig. 52. A, *Calliandra surinamensis* Benth., transverse section, scale bar = 400 μm (CTFw 32365); B-C: *Cedrelinga cateniformis* (Ducke) Ducke (CTFw 14412): B, transverse section, scale bar = 400 μm; C, tangential section, scale bar = 180 μm; D: *Chloroleucon acacioides* (Ducke) Barneby & J.W. Grimes: D, transverse section, scale bar = 400 μm (CTFw 32367).

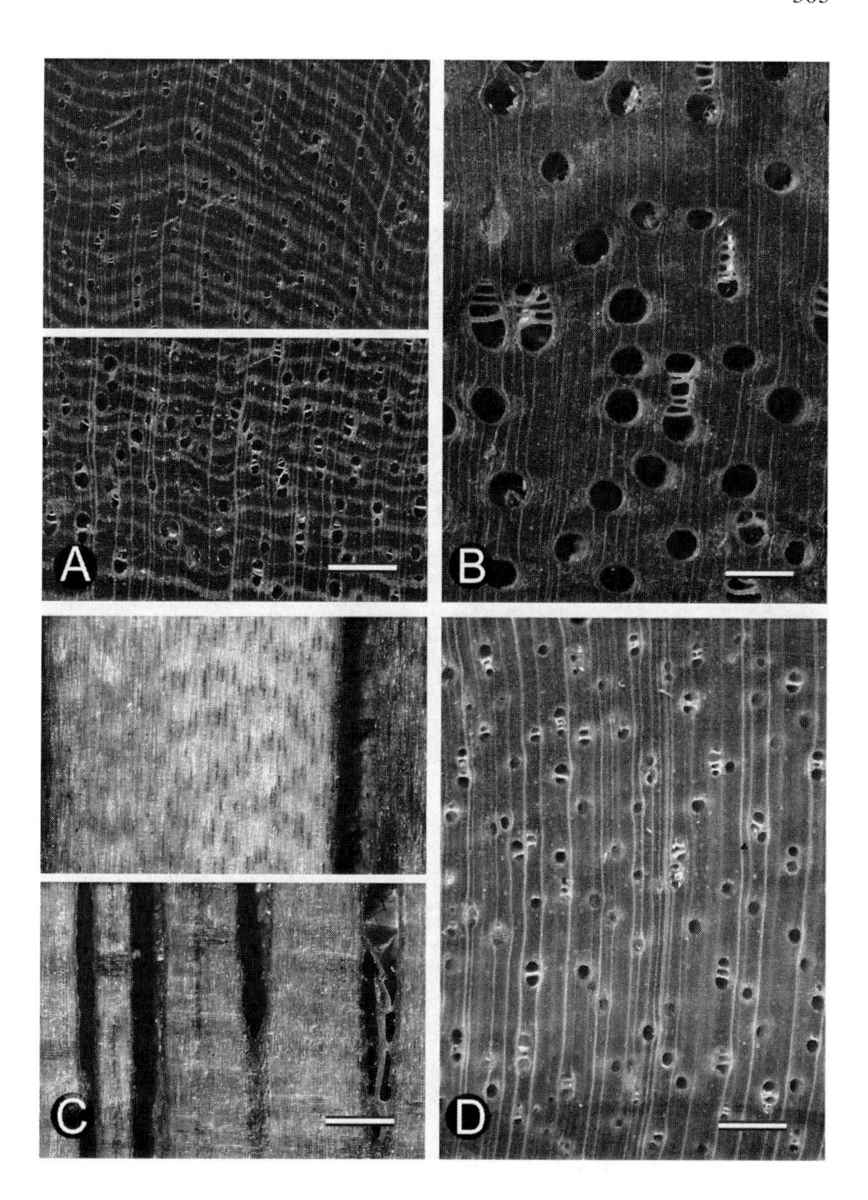

Fig.53. A-D: transverse sections, scale bar = 500 µm: A-above, *Calliandra coriacea* (Willd.) Benth. (Uw 2963); A-below, *Calliandra hymenaeodes* (Pers.) Benth. (Uw 1835); B-C: *Cedrelinga cateniformis* (Ducke) Ducke (Uw 17974): B, transverse section; C-above, tangential section; C-below, radial section; D, *Chloroleucon acacioides* (Ducke) Barneby & J.W. Grimes (Uw 24401).

Material studied:
Calliandra coriacea (Willd.) Benth.: Suriname: Mennega 397 (Uw 2963).
C. hymenaeodes (Pers.) Benth.: Suriname: Lanjouw & Lindeman 2568 (Uw 1835).
C. laxa (Willd.) Benth. var. **stipulacea** (Benth.) Barneby: Guyana: A.C. Smith 2378 (MAD-SJRw 35539).
C. surinamensis Benth.: Brazil: Capucho 499 (CTFw 32365).

CEDRELINGA Ducke – Figs. 52 B-C; 53 B-C

Vessels diffuse, solitary and in short radial multiples of 2-3, occasionally in clusters of 1- or 2-seriate radial multiples of 6-12 small round vessels, 1 or 2 per sq. mm, diameter 230-330 µm. Vessel member length 500-580. Perforations simple. Intervessel pits vestured, alternate, 7-9 µm. Vessel-ray pits similar to intervessel pits.
Rays 1(-2)-seriate non storied but sometimes in horizontal arrangement, 8-10 per mm, 190-240 µm high. Homogeneous, composed of procumbent cells.
Parenchyma paratracheal vasicentric to shortly lozenge-aliform rarely confluent and apotracheal parenchyma diffuse abundant. Strands of 8-12 cells in paratracheal parenchyma, 14 to 24 in apotracheal parenchyma. Crystals absent or very scarce.
Fibres very thin-walled, lumen 25-30 µm, walls 4-5 µm. Pits simple, small, confined to radial walls. Length 1250-1450 µm. F/V ratio: 2.4-2.6. Pith flecks sporadic.

Material studied:
Cedrelinga cateniformis (Ducke) Ducke: Suriname: LBB 12198 (Uw 17974). French Guiana: Thiel 704 (CTFw 26024). Colombia: Mariaux 1061 (CTFw 17342). Peru: Arostegui 146 (MADw 22148); Infantes 13 (CTFw 14412).

CHLOROLEUCON (Benth.) Britton & Rose – Figs. 52 D; 53 D

Vessel diffuse, solitary and in radial multiples of 2-3, round, 6-10 per sq. mm, diameter 100-120 µm. Vessel member length 250-290 µm. Perforations simple. Intervessel pits vestured, alternate, 6-7 µm. Vessel-ray pits similar to intervessel pits. Orange red deposits common.

Rays 1(partially -2)-seriate, 6-7 per mm, 140-180 μm high. Homogeneous, composed of procumbent cells.
Parenchyma paratracheal vasicentric to shortly lozenge-aliform rarely confluent, apotracheal parenchyma in diffuse cells. Strands of 2-(4) cells. Prismatic crystals in chambered cells, 4-8 per strand.
Fibres thin-walled, lumen 17 μm, walls 3 μm. Pits simple, small, confined to radial walls. Length 600-700 μm. F/V ratio: 2.1-2.5.

Material studied:
Chloroleucon acacioides (Ducke) Barneby & J.W. Grimes: Brazil: Capucho 356 (CTFw 32367); RTIw s.n. (Uw 24401).

ENTEROLOBIUM Mart. – Figs. 54 A; 55 A-B

Vessels diffuse, solitary and in short radial multiples of 2-3(-4), round, 1-5 per sq. mm, diameter 160-250 μm. Vessel member length 425-500 μm. Perforations simple. Intervessel pits vestured, alternate, 6-7 μm (*E. oldemannii* and *E. schomburgkii*), 8-10 μm (*E. cyclocarpum*). Vessel-ray pits similar to intervessel pits.
Rays 1(-2)- or (1-)2-seriate in *E. cyclocarpum*, mostly (2-)3-seriate in *E. schomburgkii* and *E. oldemannii*, 5-8 per mm, 150-320 μm high. Homogeneous, composed of procumbent cells.
Parenchyma paratracheal lozenge-aliform sometimes confluent occasionally aliform at end of growth rings, apotracheal parenchyma diffuse rare and in fine marginal bands sporadic. Strands of 2-4 cells. Prismatic crystals in chambered cells, 8-20+ per strand.
Fibres rarely to frequently septate, thin-walled in *E. cyclocarpum* (lumen 20-30 μm, walls 3-3.5 μm), thin-to thick-walled in *E. schomburgkii* and *E. oldemannii* (lumen 10-13 μm, walls 5.5-6 μm). Pits simple, confined to radial walls. Length 1000-1250 μm. F/V ratio: 2.3-2.9.

Material studied:
Enterolobium cyclocarpum (Jacq.) Griseb.: Mexico: MADw 11645. Guatemala: Sosa 51 (CTFw 17567 = MADw 23621). Panama: USw 682. Venezuela: Corothie 57 (CTFw 23196).
E. oldemanii Barneby & J.W. Grimes: French Guiana: BAFOG 302M (CTFw 10034); Petrov 92 (CTFw 16695).
E. schomburgkii (Benth.) Benth.: Suriname: Maguire 24664 (Uw 2461). French Guiana: BAFOG 17M (CTFw 7715); Petrov 3, 66 (CTFw 16621, 16654); Thiel 351 (CTFw 23642). Brazil: Krukoff 6336 (CTFw 22746).

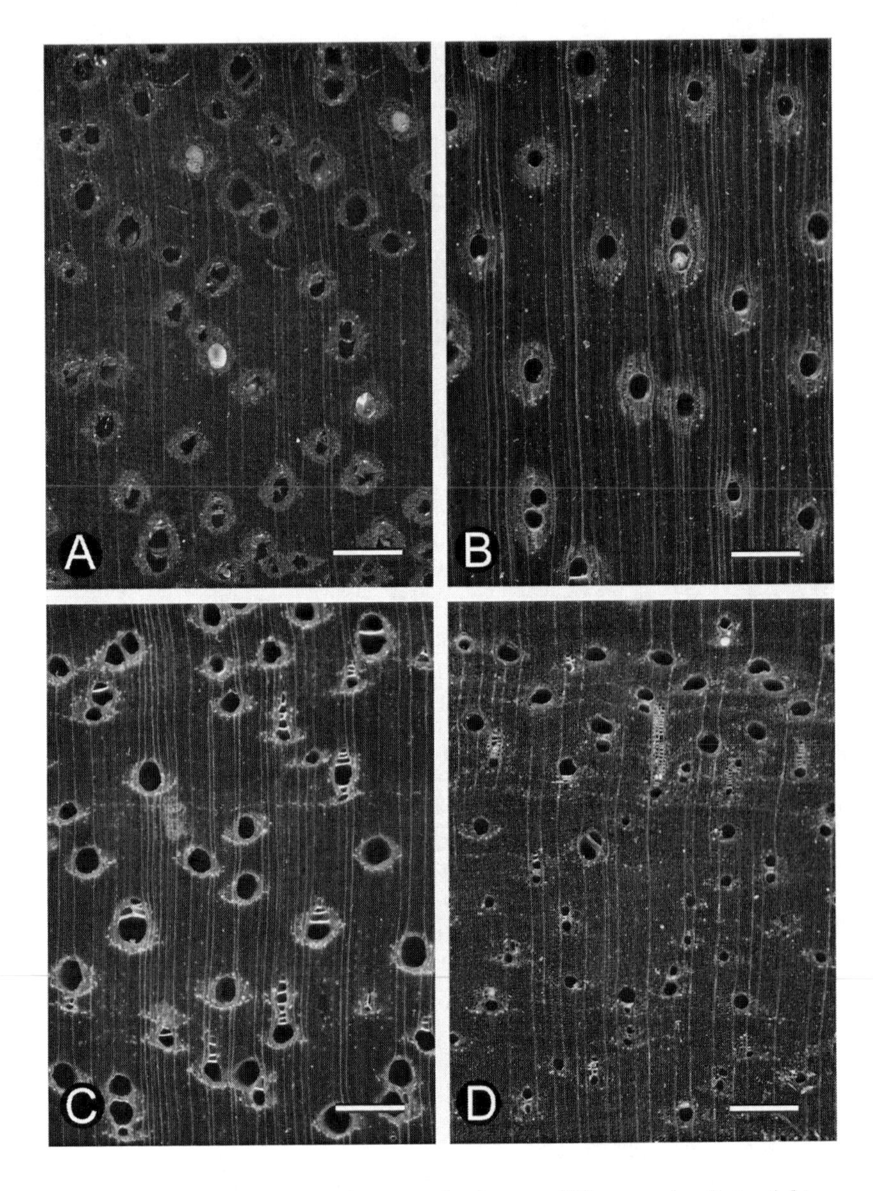

Fig. 54. A-D: transverse sections, scale bar = 500 μm: A, *Enterolobium schomburgkii* (Benth.) Benth. (Uw 2461); B, *Hydrochorea corymbosa* (Rich.) Barneby & J.W. Grimes (Uw 2063); C-D: *Hydrochorea gonggrijpii* (Kleinhoonte) Barneby & J.W. Grimes: C, (Uw 88a); D, (Uw 2460).

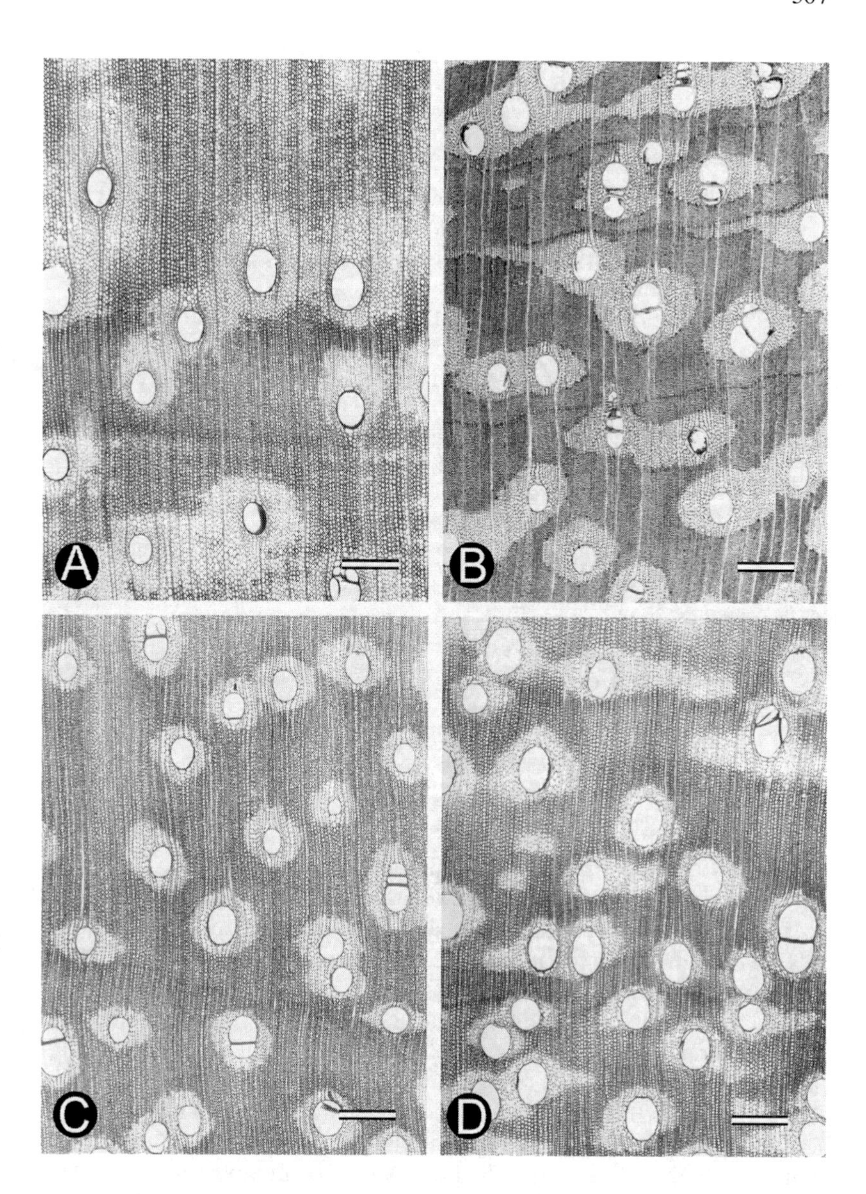

Fig. 55. A-D, transverse sections, scale bar = 400 µm: A, *Enterolobium cyclocarpum* (Jacq.) Griseb. (CTFw 17567); B, *Enterolobium schomburgkii* (Benth.) Benth. (CTFw 7715). C, *Hydrochorea corymbosa* (Rich.) Barneby & J.W. Grimes (CTFw 7607); D, *Inga alba* (Sw.) Willd. (CTFw 9324).

HYDROCHOREA Barneby & J.W. Grimes – Figs. 54 B-D; 55 C

Vessels diffuse, solitary and in short radial multiples of 2-3, round, 2-4 per sq. mm, diameter 150-200 µm. Vessel member length 460-620 µm. Perforations simple. Intervessel pits vestured, alternate, 5-6 µm. Vessel-ray pits similar to intervessel pits. Dark red brown deposits infrequent.
Rays 1(-2)-seriate, 6-8 per mm, 130-180 µm high. Homogeneous, composed of procumbent cells.
Parenchyma vasicentric paratracheal to lozenge-aliform rarely confluent, apotracheal parenchyma diffuse absent or rare and in marginal interrupted bands sporadic. Strands of 2-4 cells. Prismatic crystals in chambered cells, 8-32 per strand, frequent.
Fibres thin-to thick-walled, lumen 20-22 µm, walls 4-6 µm. Pits simple, small, confined to radial walls. Length 1150-1450 µm. F/V ratio: 2.1-2.6.

Material studied:
Hydrochorea corymbosa (Rich.) Barneby & J.W. Grimes: Suriname: BBS 1047 (Uw 2063); Stahel 360 (Uw 360). French Guiana BAFOG 4M (CTFw 7607), 30M (CTFw 7621).
H. gonggrijpii (Kleinhoonte) Barneby & J.W. Grimes: Suriname: Maguire 24273 (Uw 2460); Stahel 88a (Uw 88a). French Guiana: Normand 518 (CTFw 16436).

INGA Mill. – Figs. 55 D; 56 A-65 B; 66 D-left

Vessels diffuse, solitary and in radial multiples of 2-4(-5), round, 2-4(-6) per sq. mm, diameter 130-250 µm. Clusters of small vessels sometimes present. Vessel member length 420-640 µm. Perforations simple. Intervessel pits vestured, alternate, 6-8(-9) µm. Vessel-ray pits similar to intervessel pits. Small whiteish deposits common.
Rays 1(-2)-seriate or up to 3-seriate, 6-9 per mm, 180-330 µm high. Homogeneous or near, composed of procumbent cells, sometimes with one marginal row of shorter cells. Crystals absent or very scarce.
Parenchyma paratracheal variable, thin vasicentric to thick winged-aliform sometimes confluent, apotracheal parenchyma diffuse and sometimes in round to elongated clusters. Marginal bands, continuous or interrupted, irregularly present. Strands of 2-4 cells. Prismatic crystals in chambered cells, 8-32 per strand, mostly common. Some inflated parenchyma cells observed in few species.
Fibres septate, thin- to thick-walled, lumen 7-20(-25) µm, walls 4-10 µm. Pits simple, confined to radial walls. Length 1150-1700(-2000) µm. F/V ratio: 2.2-3.5.
Pith flecks sometimes present.

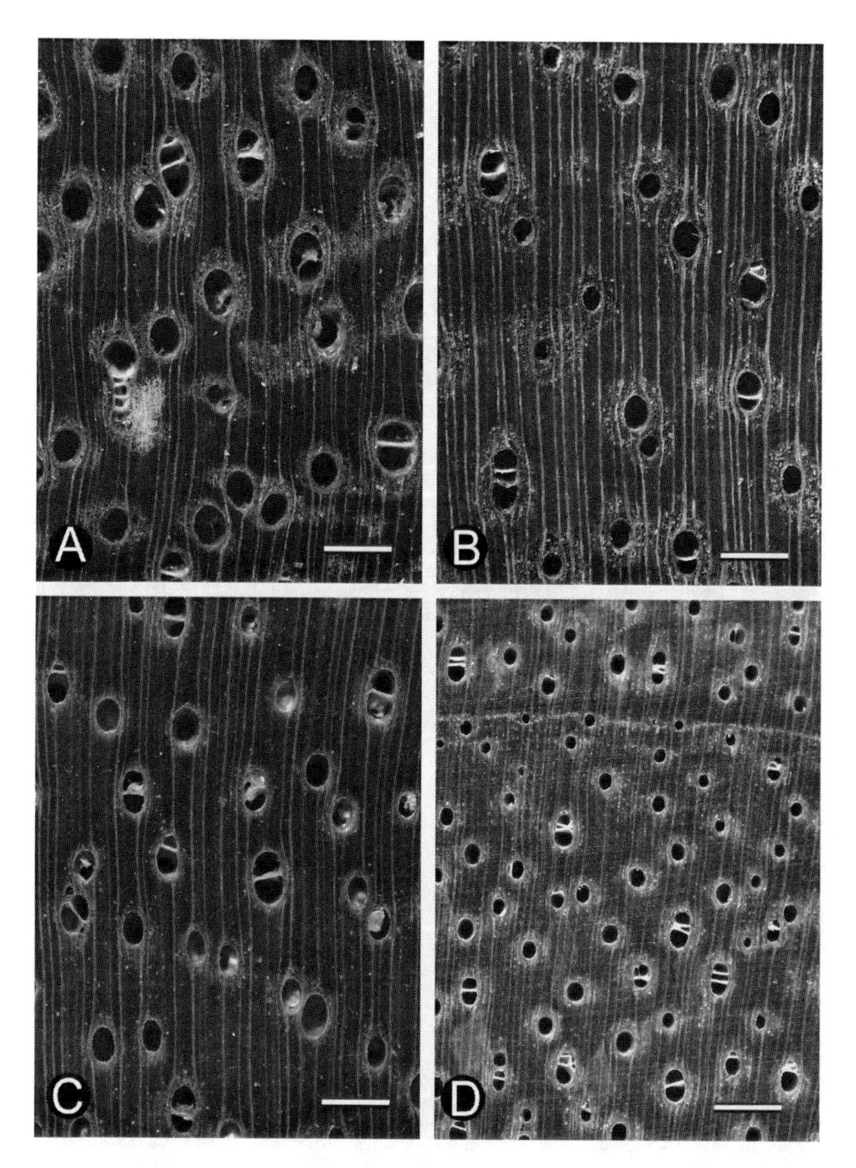

Fig.56. A-D: transverse sections, scale bar = 500 µm: A-B: *Inga alba* (Sw.) Willd.: A, (Uw 2116); B, (Uw 7766); C-D: *Inga bourgonii* (Aubl.) DC.: C, (Uw 624); D, (Uw 8775).

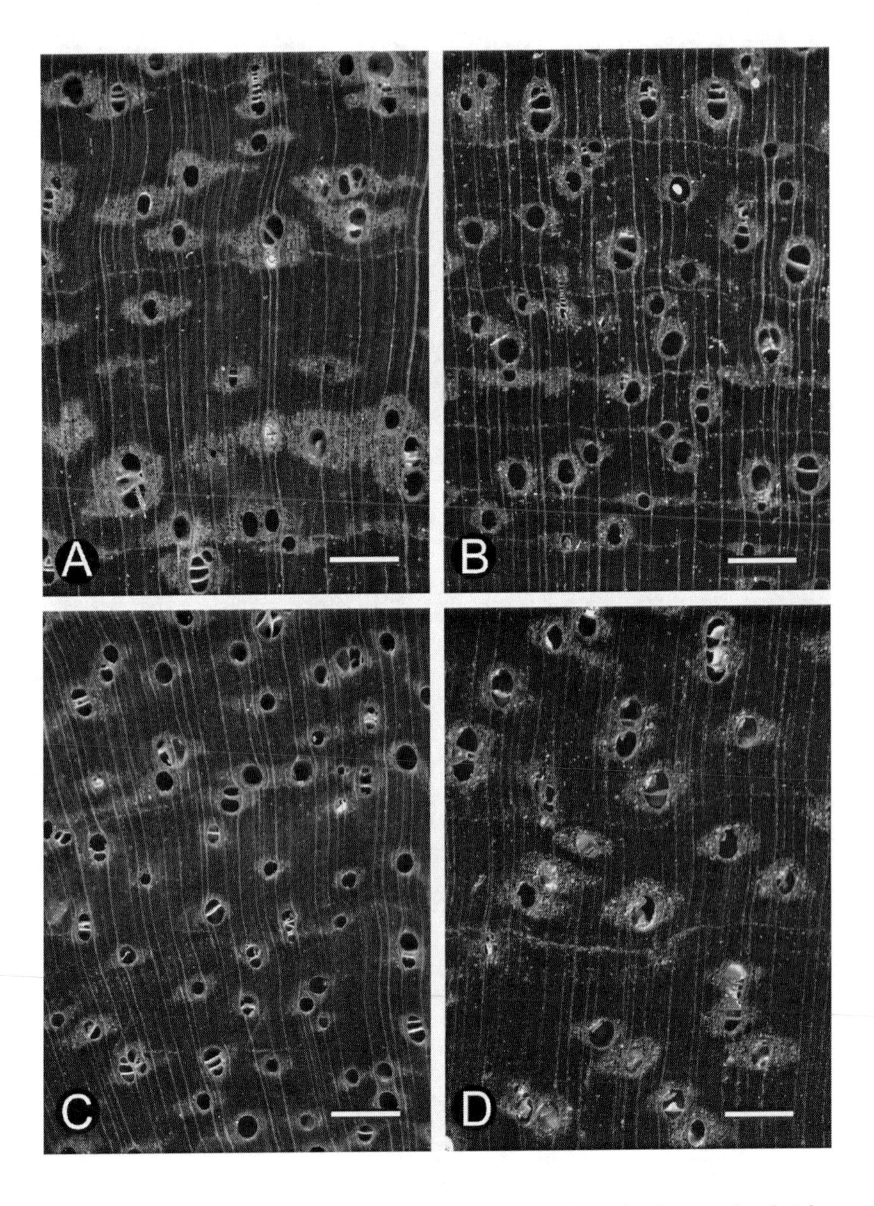

Fig.57. A-D: transverse sections, scale bar = 500 µm: A, *Inga calanthoides* Amshoff (Uw 2456); B, *Inga capitata* Desv. (Uw 3746); C-D: *Inga cayennensis* Sagot ex Benth.: C, (Uw 12151); D, (Uw 15304).

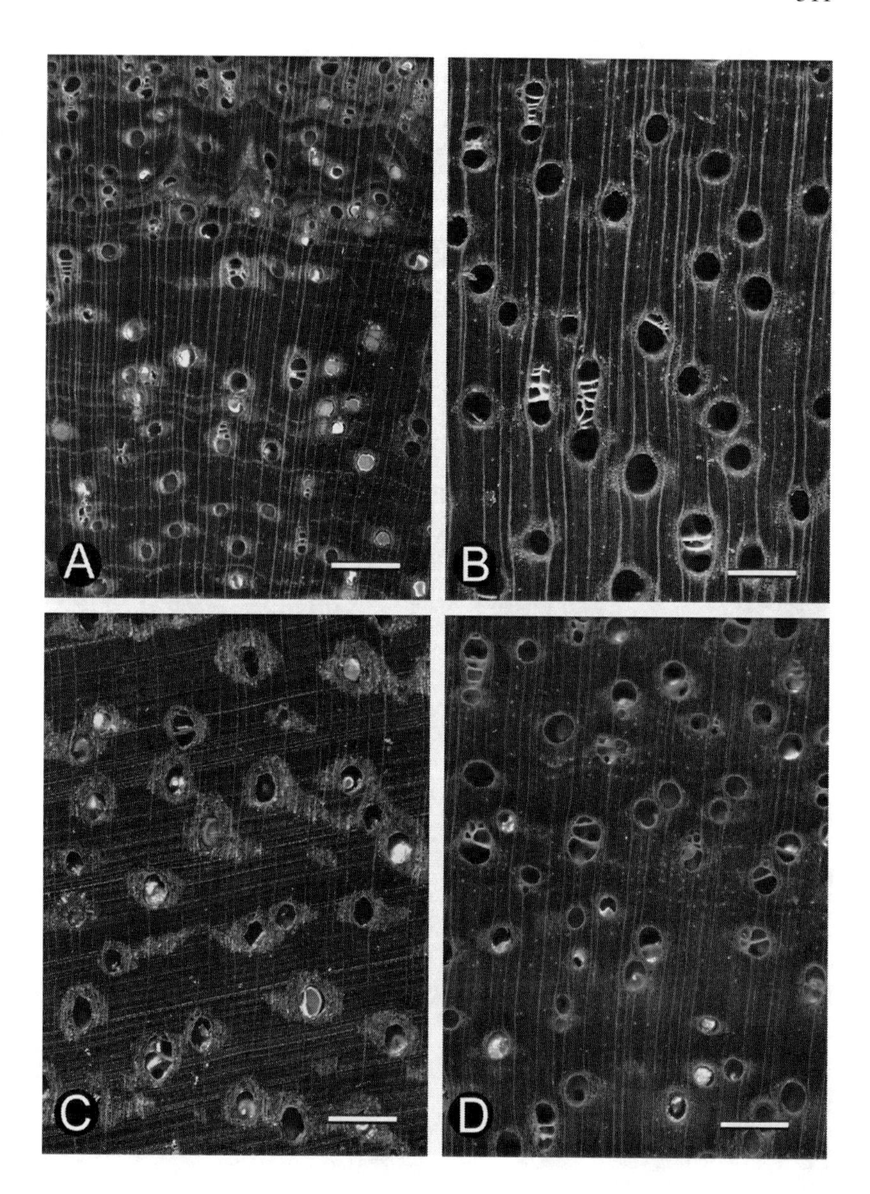

Fig.58. A-D: transverse sections, scale bar = 500 µm: A, *Inga disticha* Benth. (Uw 4139); B, *Inga edulis* Mart. (Uw 5344); C, *Inga gracilifolia* Ducke (Uw 4874); D, *Inga heterophylla* Willd. (Uw 250).

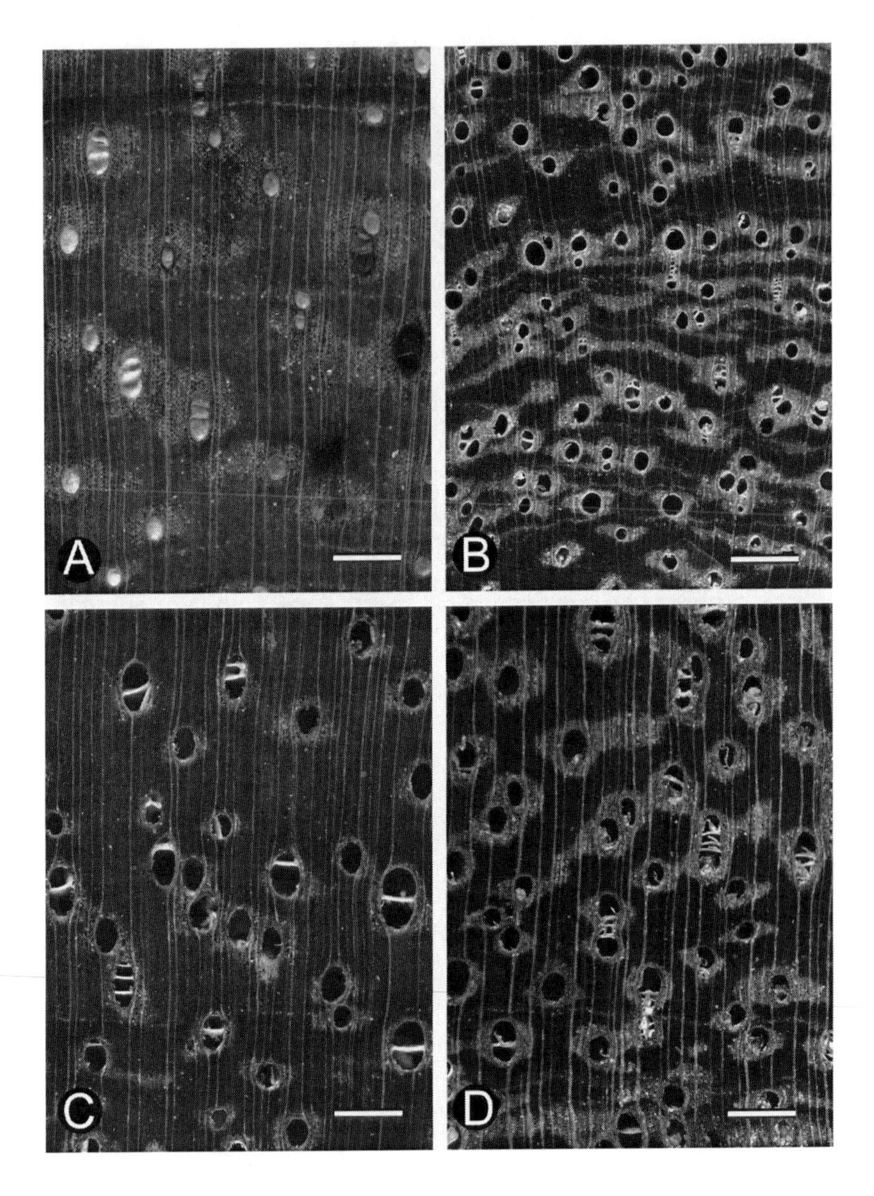

Fig.59. A-D: transverse sections, scale bar = 500 µm: A, *Inga huberi* Ducke (Uw 2415); B, *Inga jenmanii* Sandwith (Uw 17536); C, *Inga lateriflora* Miq. (Uw 149a); D, *Inga leiocalycina* Benth. (Uw 5270).

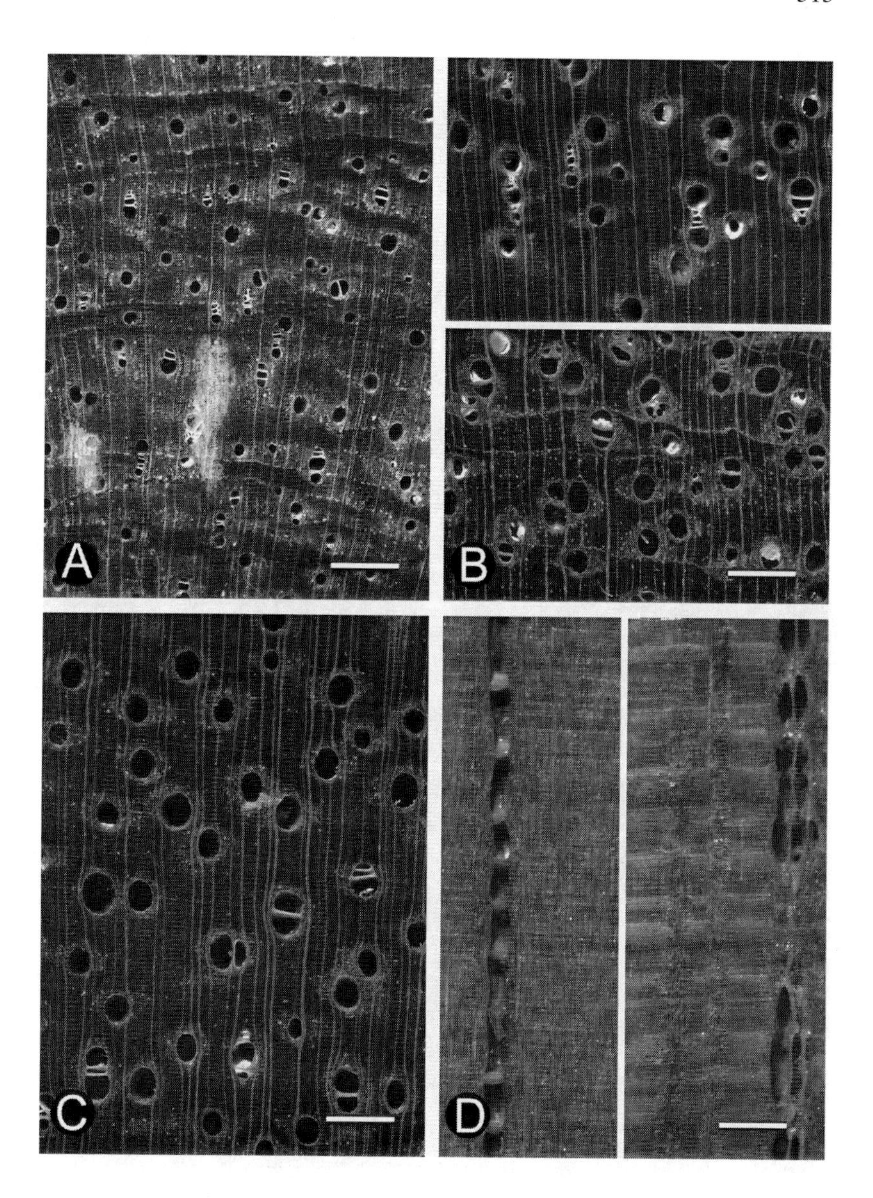

Fig. 60. A-D: transverse sections, scale bar = 500 µm: A, *Inga longiflora* Spruce ex Benth. (Uw 4509a); B: *Inga marginata* Willd.: B-above, (Uw 957); B-below (Uw 4740); C-D: *Inga melinonis* Sagot (Uw 959): C, transverse section; D-left, tangential section; D-below, radial section.

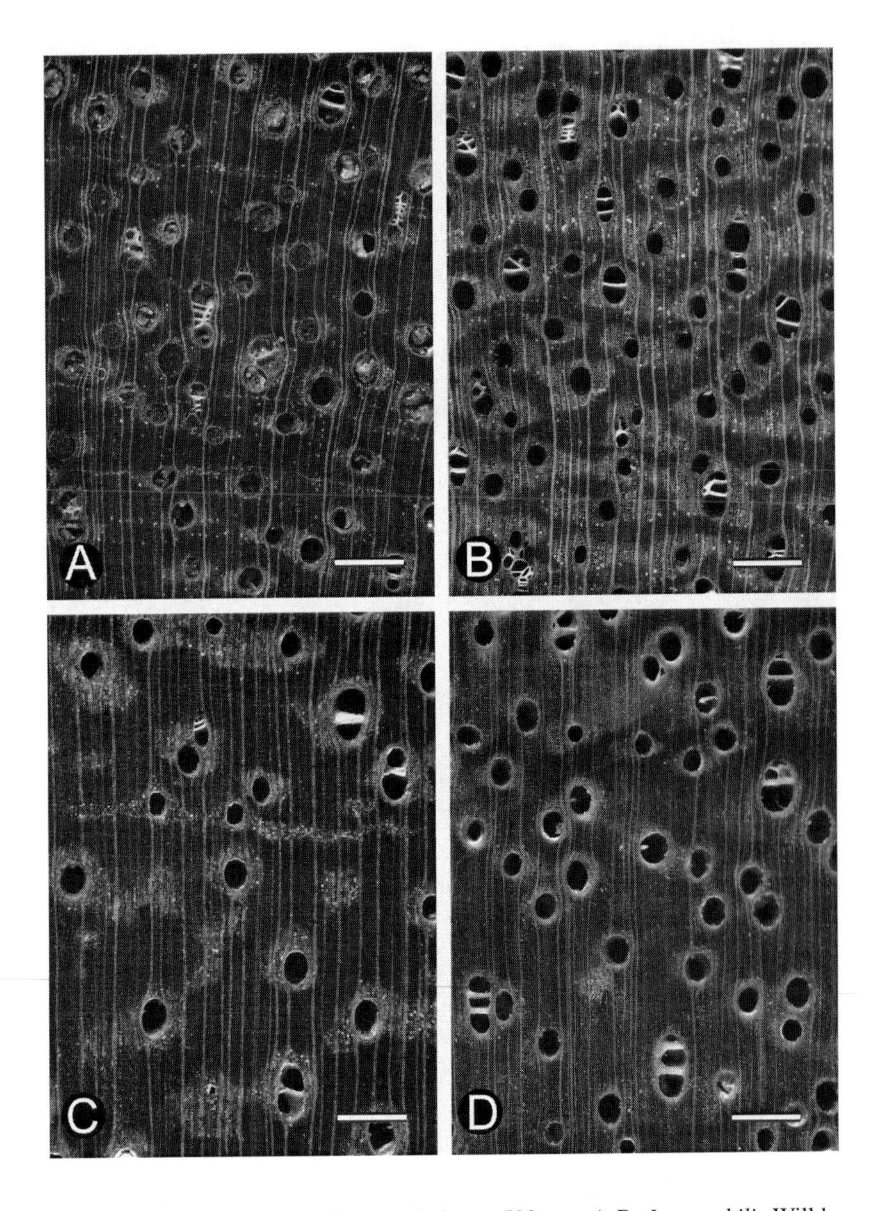

Fig.61. A-D: transverse sections, scale bar = 500 μm: A-B: *Inga nobilis* Willd.: A, (Uw 3046); B, (Uw 17147); C, *Inga paraensis* Ducke (Uw 18951); D, *Inga pezizifera* Benth. (Uw 230).

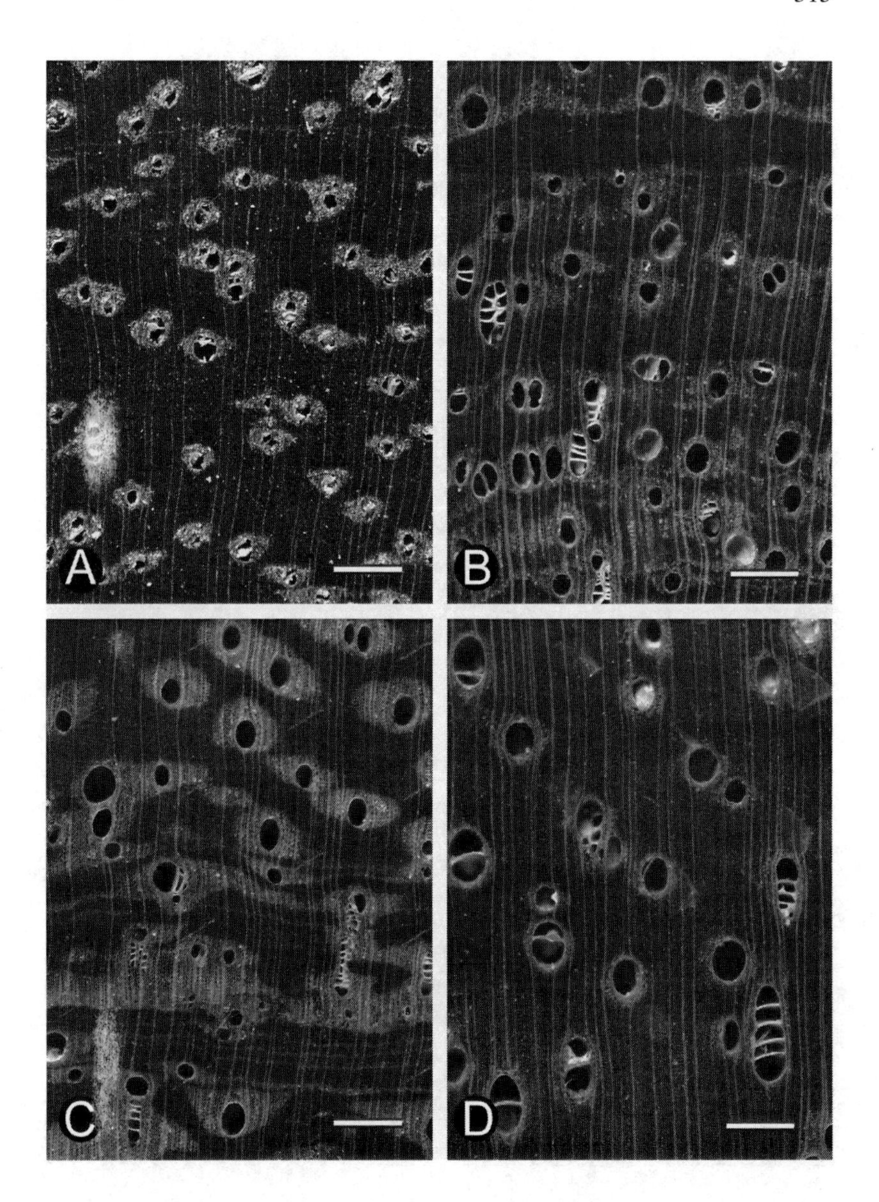

Fig.62. A-D: transverse sections, scale bar = 500 µm: A-B: *Inga pilosula* (Rich.) J.F. Macbr.: A, (Uw 1823); B, (Uw 17472); C, *Inga retinocarpa* Poncy (Uw 4865); D, *Inga rhynchocalyx* Sandwith (Uw 960).

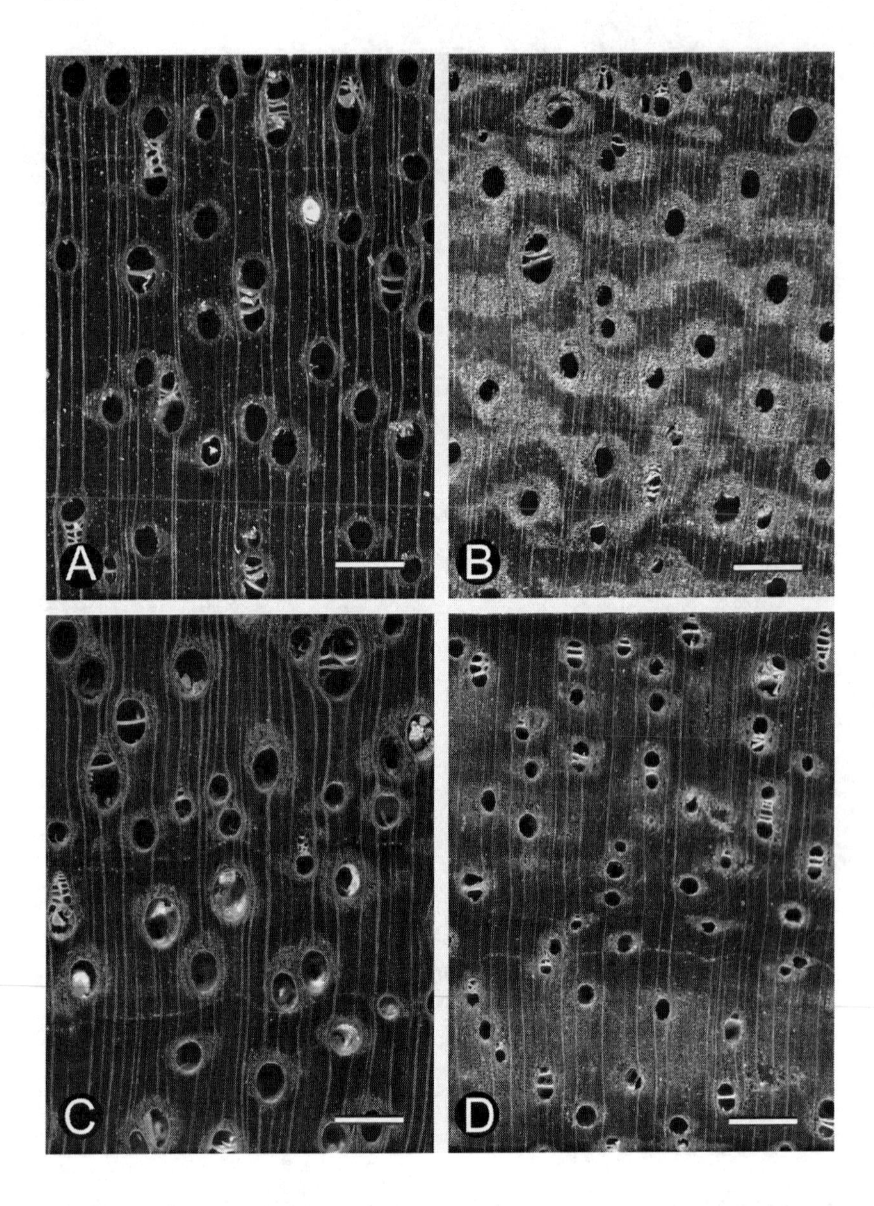

Fig. 63. A-D: transverse sections, scale bar = 500 µm: A, *Inga rubiginosa* (Rich.) DC. (Uw 4945); B, *Inga sertulifera* DC. (Uw 7813); C, *Inga splendens* Willd. (Uw 701); D, *Inga stipularis* DC. (Uw 2451).

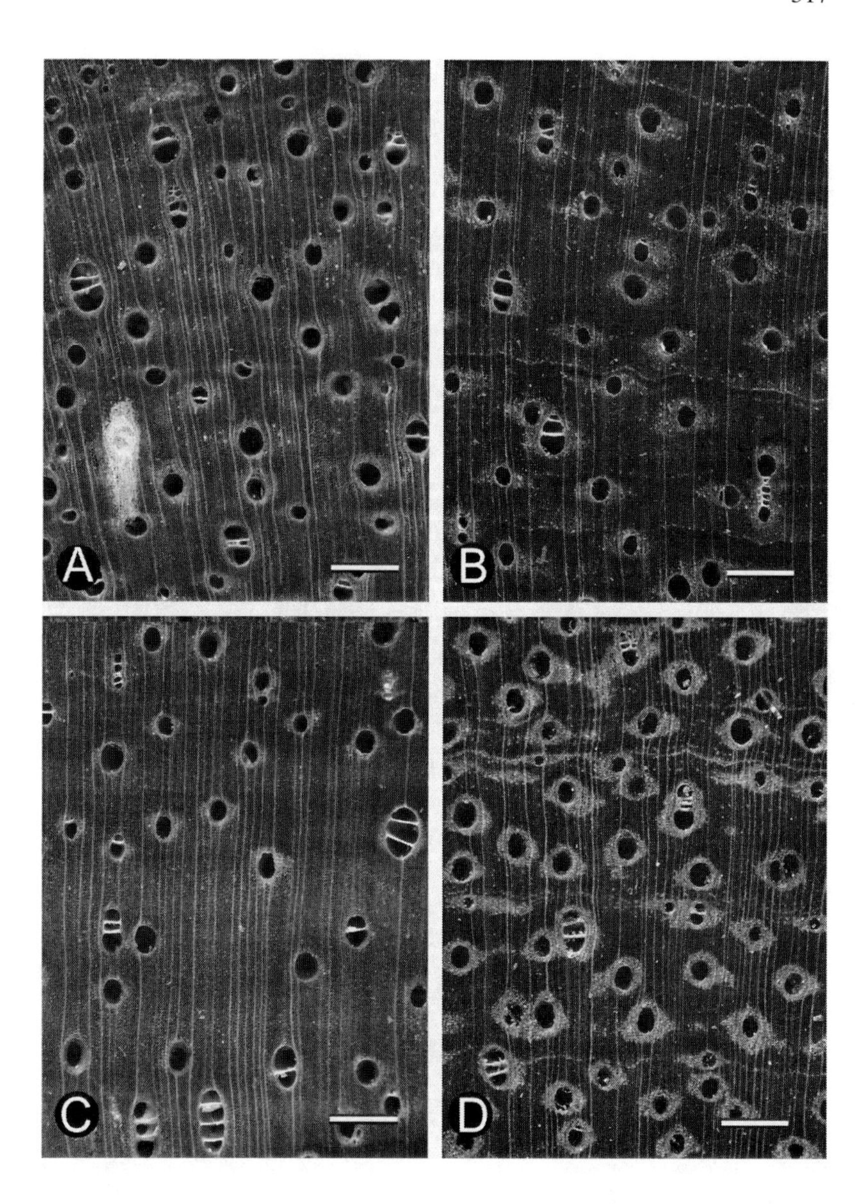

Fig.64. A-D: transverse sections, scale bar = 500 µm: A, *Inga striata* Benth. (Uw 14529); B-C: *Inga thibaudiana* DC.: B, (Uw 2452); C, (Uw 4582); D, *Inga umbellifera* (Vahl) Steud. ex DC. (Uw 1985).

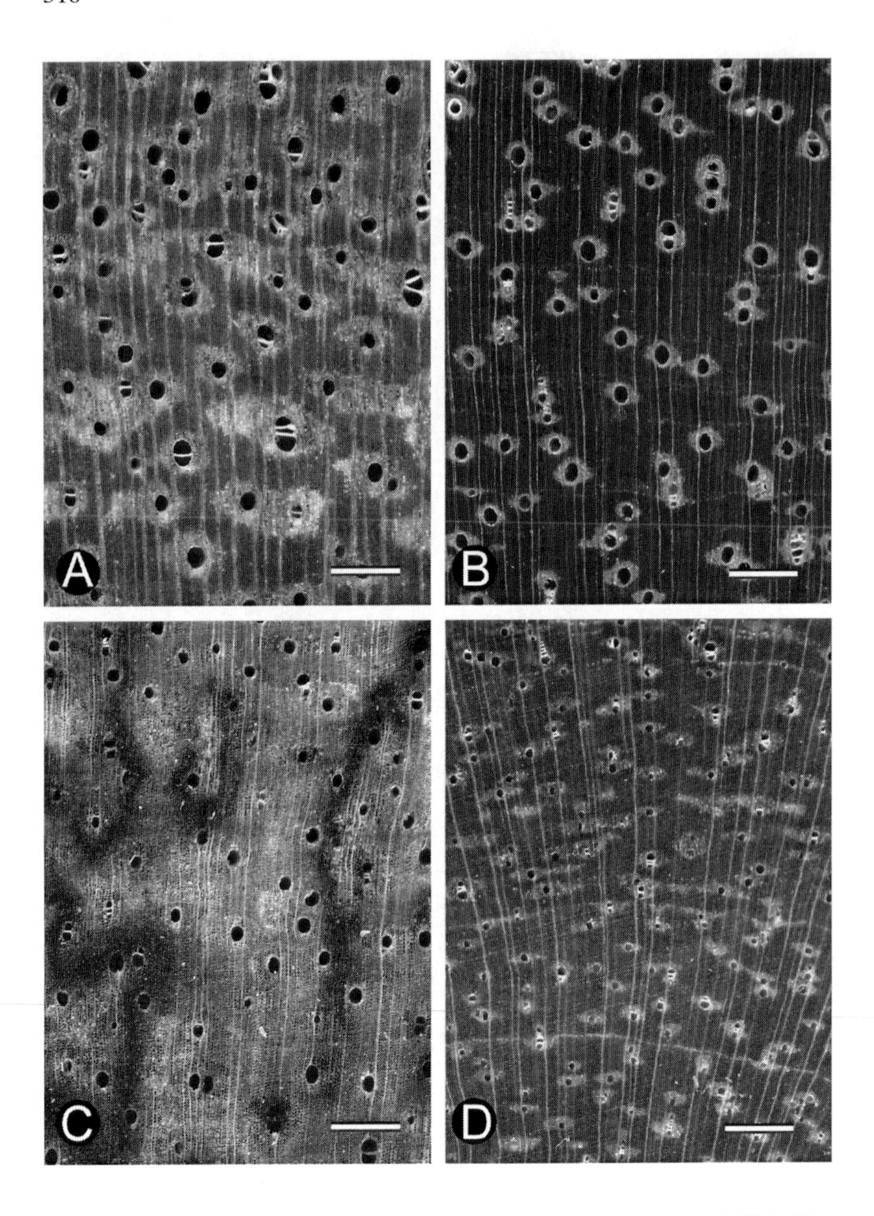

Fig.65. A-D: transverse sections, scale bar = 500 µm: A, *Inga vera* Willd. (Uw 13324); B, *Inga virgultosa* (Vahl) Desv. (Uw 1834); C, *Macrosamanea kegelii* (Meisn.) Kleinhoonte (Uw 4512); D, *Macrosamanea pubiramea* (Steud.) Barneby & J.W. Grimes (Uw 4351).

Material studied:
Inga acrocephala Steud.: French Guiana: BAFOG 31M (CTFw 7877).
I. alba (Sw.) Willd.: Guyana: Polak 92 (CTFw 31874). Suriname: BBS 1075 (Uw 2116). French Guiana: BAFOG 160M, 206M (CTFw 8076, 9324); SF 6128 (CTFw 7515). Brazil: Krukoff 6491 (Uw 7766).
I. bourgonii (Aubl.) DC.: Suriname: BBS 25 (Uw 624). Peru: Ellenberg 2244 (Uw 8655).
I. calanthoides Amshoff: Suriname: Maguire 24527 (Uw 2456).
I. capitata Desv.: Suriname: Lindeman 5401 (Uw 3746); Stahel 341 (Uw 341).
I. cayennensis Sagot ex Benth.: Suriname: van Donselaar 3758 (Uw 12151); Oldenburger & Norde 345 (Uw 15304). French Guiana: BAFOG 297M (CTFw 9646); Sabatier 1518 (CTFw 30895).
I. disticha Benth.: Suriname: Lindeman 6068 (Uw 4139). French Guiana: Loubry 1534 (CTFw 32556).
I. edulis Mart.: Suriname: Stahel 315 (Uw 315). French Guiana: BAFOG 269M (Uw 5344). Peru: Williams 2690 (CTFw 8200).
I. fanchoniana Poncy: French Guiana: Sabatier 1758 (CTFw 31130).
I. gracilifolia Ducke: Suriname: Maguire 40496 (Uw 4874). French Guiana: Loubry 615 (CTFw 32084).
I. heterophylla Willd.: Suriname: Stahel 250 (Uw 250).
I. huberi Ducke: Suriname: Lindeman 3663 (Uw 2415). French Guiana: Loubry 1565 (CTFw 32586).
I. jenmanii Sandwith: Suriname: Maguire *et al.* 55780 (Uw 17536).
I. lateriflora Miq.: Suriname: Stahel 149 (Uw 149a).
I. leiocalycina Benth.: French Guiana: BAFOG 33M (CTFw 7878), 195M (Uw 5270). Brazil: Krukoff 6607 (CTFw 22828).
I. lomatophylla (Benth.) Pittier: French Guiana: Sabatier & Prévost 2174 (CTFw 31325).
I. longiflora Spruce ex Benth.: Suriname: Lindeman 6658 (Uw 4509a).
I. longipedunculata Ducke: French Guiana: Sabatier & Prévost 2185 (CTFw 31340).
I. marginata Willd.: Guyana: FD 3536 (Uw 957). French Guiana: Loubry 571 (CTFw 32048). Brazil: Krukoff 7037 (CTFw 22921).
I. melinonis Sagot: Guyana: FD 3989 (Uw 959). French Guiana: Loubry 1780 (CTFw 32744).
I. mitaraka Poncy: French Guiana: Loubry 541 (CTFw 32021).
I. nobilis Willd.: Suriname: Lindeman 4081 (Uw 3046). French Guiana: BAFOG 269M (CTFw 9407). Brazil: Maguire *et al.* 51741 (Uw 17147).
I. nouragensis Poncy: French Guiana: Loubry 812 (CTFw 32203); Mori 23456 (CTFw 32886); Sabatier 2386 (CTFw 31479).
I. paraensis Ducke: French Guiana: Loubry 604 (CTFw 32075). Brazil: Maas *et al.* P12647 (Uw 18951).

I. pezizifera Benth.: Suriname: Stahel 230 (Uw 230). French Guiana: Benoist 340, 544 (CTFw 14489, 14511); Mélinon 43 (CTFw 16505).
I. pilosula (Rich.) J.F. Macbr.: Suriname: Lanjouw & Lindeman 2552 (Uw 1823). French Guiana: Sabatier 1768 (CTFw 31137). Brazil: Maguire *et al.* 55344 (Uw 17472).
I. retinocarpa Poncy: Guyana: Maguire & Cowan 39348 (Uw 4865).
I. rhynchocalyx Sandwith: Guyana: FD 3622 (Uw 960).
I. rubiginosa (Rich.) DC.: Suriname: Schulz 7280 (Uw 4945); Stahel 162 (Uw 162). French Guiana: BAFOG 52M, 216M (CTFw 7883, 9354).
I. sarmentosa Glaziou ex Harms: French Guiana: BAFOG 58M (CTFw 7886); Sabatier 2223, 2372 (CTFw 31354, 31465).
I. sertulifera DC.: French Guiana: Loubry 1549, 1861 (CTFw 32570, 32812); Mori 23654; Sabatier 2257 (CTFw 31413). Brazil: Krukoff 6608 (Uw 7813).
I. splendens Willd: Suriname: BBS 108 (Uw 701); Stahel 314 (Uw 314); Schulz 7165 (Uw 4915).
I. stipularis DC.: Suriname: Maguire 24450 (Uw 2451).
I. striata Benth.: Brazil: Reitz & Klein 3696 (Uw 14529).
I. thibaudiana DC.: Suriname: Lindeman 2491, 6769 (Uw 4582); Maguire 24549 (Uw 2452); Schulz 7201 (Uw 4936).
I. umbellifera (Vahl) Steud. ex DC.: Suriname: Lanjouw & Lindeman 3019 (Uw 1985); Lindeman 146 (Uw 21764).
I. vera Willd.: Brazil: Lindeman & de Haas 1783 (Uw 13324).
I. virgultosa (Vahl) Desv.: Suriname: Lanjouw & Lindeman 2567 (Uw 1834). French Guiana: Sabatier & Prévost 4349 (CTFw 33126).

LEUCAENA Benth. – Fig. 66 A

Vessels diffuse, solitary and in radial multiples of 2-4, occasionally in clusters of 4-10 small vessels, round, 4-9 per sq. mm, 100-130 µm diameter. Vessel member length 400-430 µm. Perforations simple. Intervessel pits vestured, alternate, 7-9 µm. Vessel-ray pits similar to intervessel pits. Dark brown deposits common.
Rays 2- to 4-seriate, 5-6 per mm, 250-300 µm high. Homogeneous, composed of procumbent cells.
Parenchyma paratracheal vasicentric to lozenge-aliform not often confluent, apotracheal parenchyma diffuse and in fine 1-seriate marginal bands. Strands of 2-4 cells. Prismatic crystals in chambered cells, 8-24 per strands, frequent.
Fibres septate, thin- to thick-walled, lumen 21-24 µm, walls 5-6 µm. Pits simple, small, confined to radial walls. Length 950-1150 µm. F/V ratio: 2.3 – 2.8.

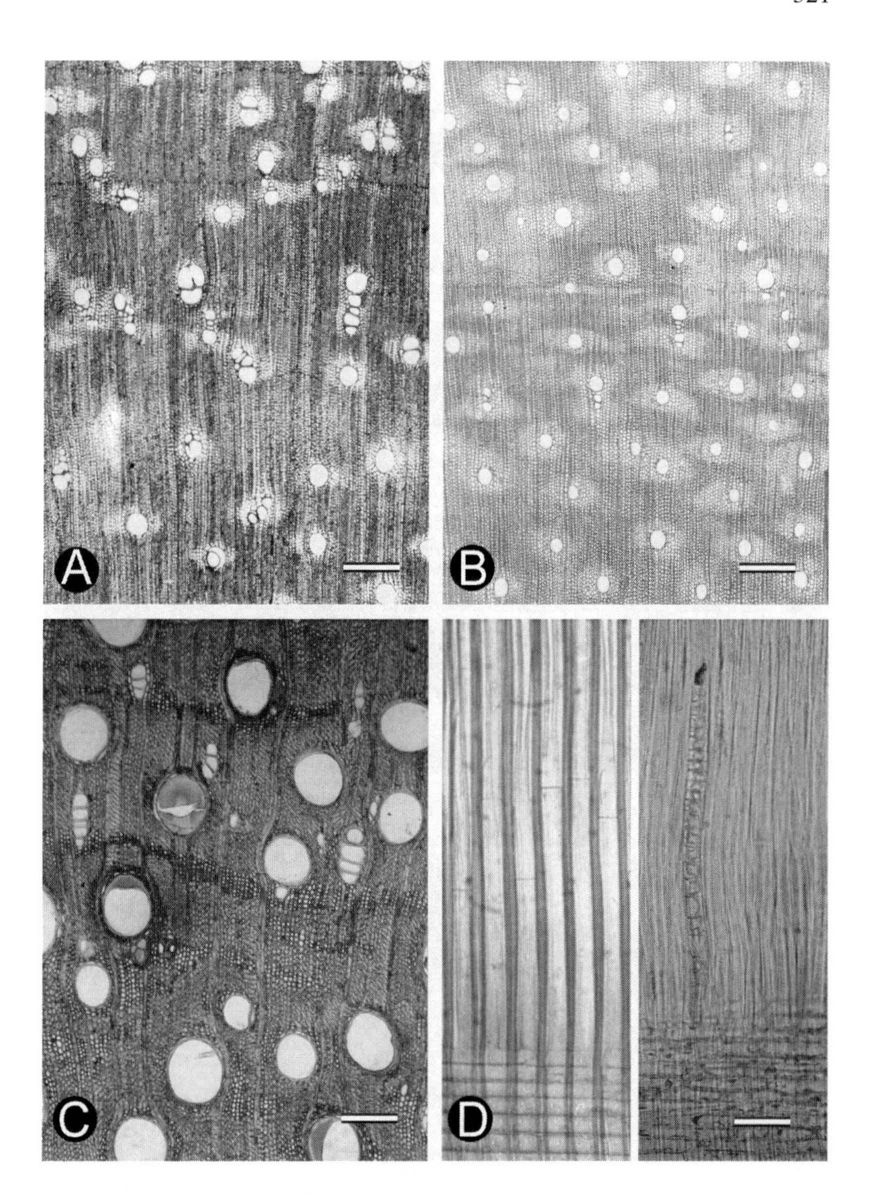

Fig. 66. A, *Leucaena leucocephala* (Lam.) de Wit, transverse section, scale bar = 400 µm (CTFw 26492); B, *Macrosamanea kegelii* (Meisn.) Kleinhoonte, transverse section, scale bar = 400 µm (CTFw 32368); C, *Mimosa guilandinae* (DC.) Barneby, transverse section, scale bar = 200 µm (Uw 11884); D-left, *Inga nobilis* Willd., radial section, scale bar = 50 µm, fibres septate (CTFw 9407); D-right, *Mimosa guilandinae* (DC.) Barneby, radial section, scale bar = 50 µm, crystals in axial parenchyma (Uw 11884).

Material studied:
Leucaena leucocephala (Lam.) de Wit: U.S.A. (Florida): BWCw 80796 (CTFw 26492). French Lesser Antilles: Rollet 1087 (CTFw 28473).

MACROSAMANEA Britton & Rose ex Britton & Killip
– Figs. 65 C-D; 66 B

Vessels diffuse, solitary and in short radial multiples of 2-3, round, 5-8 per sq. mm, diameter 120-130 µm. Vessel member length 460 µm. Perforations simple. Intervessel pits vestured, alternate, 6 µm. Vessel-rays pits similar to intervessel pits.
Rays 1-seriate (1-partially 2-seriate rare), 6-7 per mm, 100-120 µm high. Homogeneous, composed of procumbent cells.
Parenchyma paratracheal lozenge-aliform sometimes confluent and apotracheal parenchyma in 1-seriate continuous or interrupted marginal bands. Strands of (1-)2 cells. Prismatic crystals in chambered cells, 8-32 per strand.
Fibres thin-walled, lumen 18-19 µm, walls 3-3.5. Pits simple, small, confined to radial walls. Length 1000 µm. F/V ratio: 2.2.
Pith flecks sporadic.

Material studied:
Macrosamanea kegelii (Meisn.) Kleinhoonte: Suriname: Lindeman 6662 (CTFw 32368 = Uw 4512).
M. pubiramea (Steud.) Barneby & J.W. Grimes: Suriname: Lindeman 6346 (Uw 4351).

MIMOSA L. – Figs. 66 C, D-right; 67 A

Vessels of two distinct sizes: Large vessels diffuse, solitary, round, rarely in contact with small vessels, 3-6 per sq. mm, diameter 90-343 µm, vessel member length 150-650 µm; small vessels diffuse, in short radial multiples of 2-5 and clusters of 2-7, round, 2-7 per sq. mm, diameter 20-55 µm. Perforations simple. Intervessel pits vestured, alternate, 3.5-10 µm. Vessel ray pits similar to intervessel pits. Red brown deposits present in vessels.
Rays 1-3-seriate, 5-8 per mm, 120-1080 µm, composed of procumbent body cells and up to 4(-5) square-upright marginal rows. Globular and granular deposits frequent, sometimes cell entirely filled with a red deposit.
Parenchyma paratracheal mostly vasicentric, aliform and confluent,

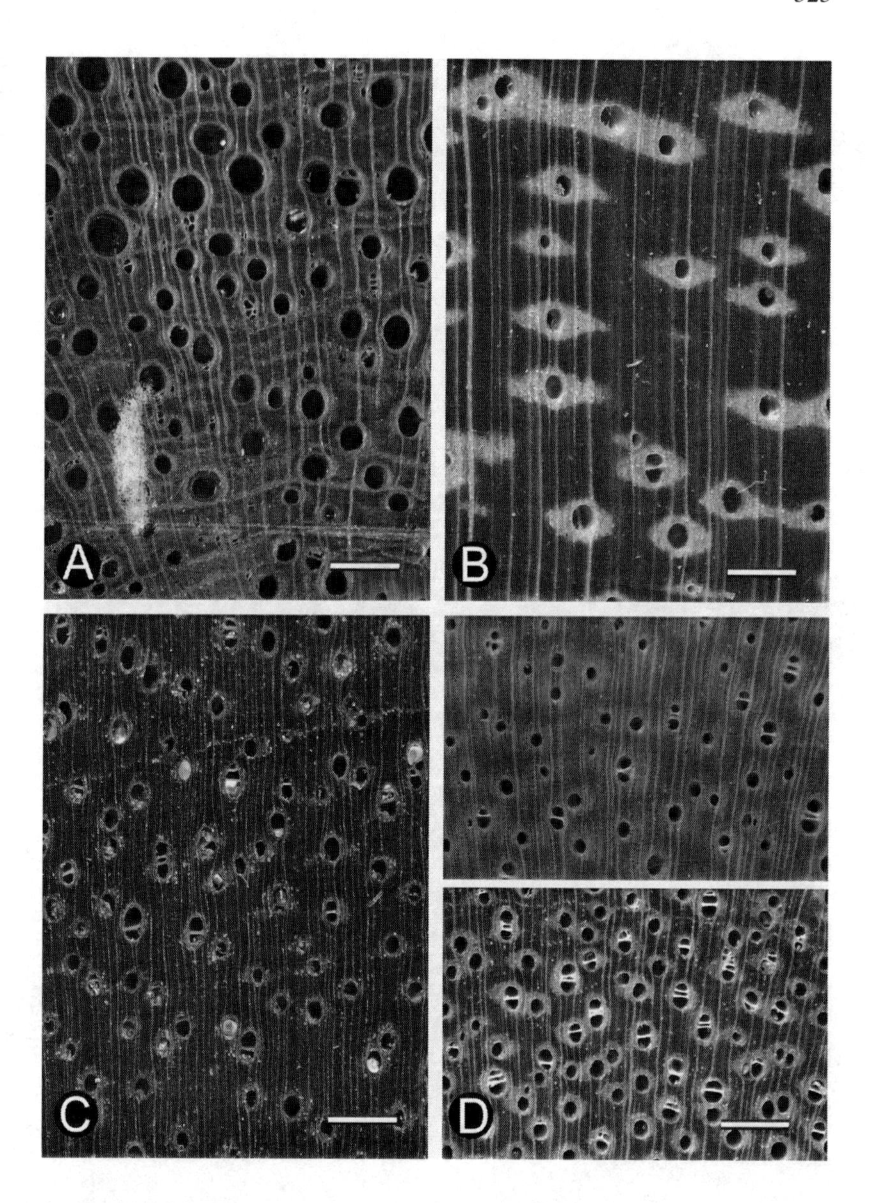

Fig.67. A-D: transverse sections, scale bar = 500 µm: A, *Mimosa guilandinae* (DC.) Barneby (Uw 11884); B, *Parkia pendula* (Willd.) Benth. ex Walp. (Uw 663); C, *Pentaclethra macroloba* (Willd.) Kuntze (Uw 623); D, *Pithecellobium dulce* (Roxb.) Benth.: D-above, (Uw 16681); D-below, (Uw 24653).

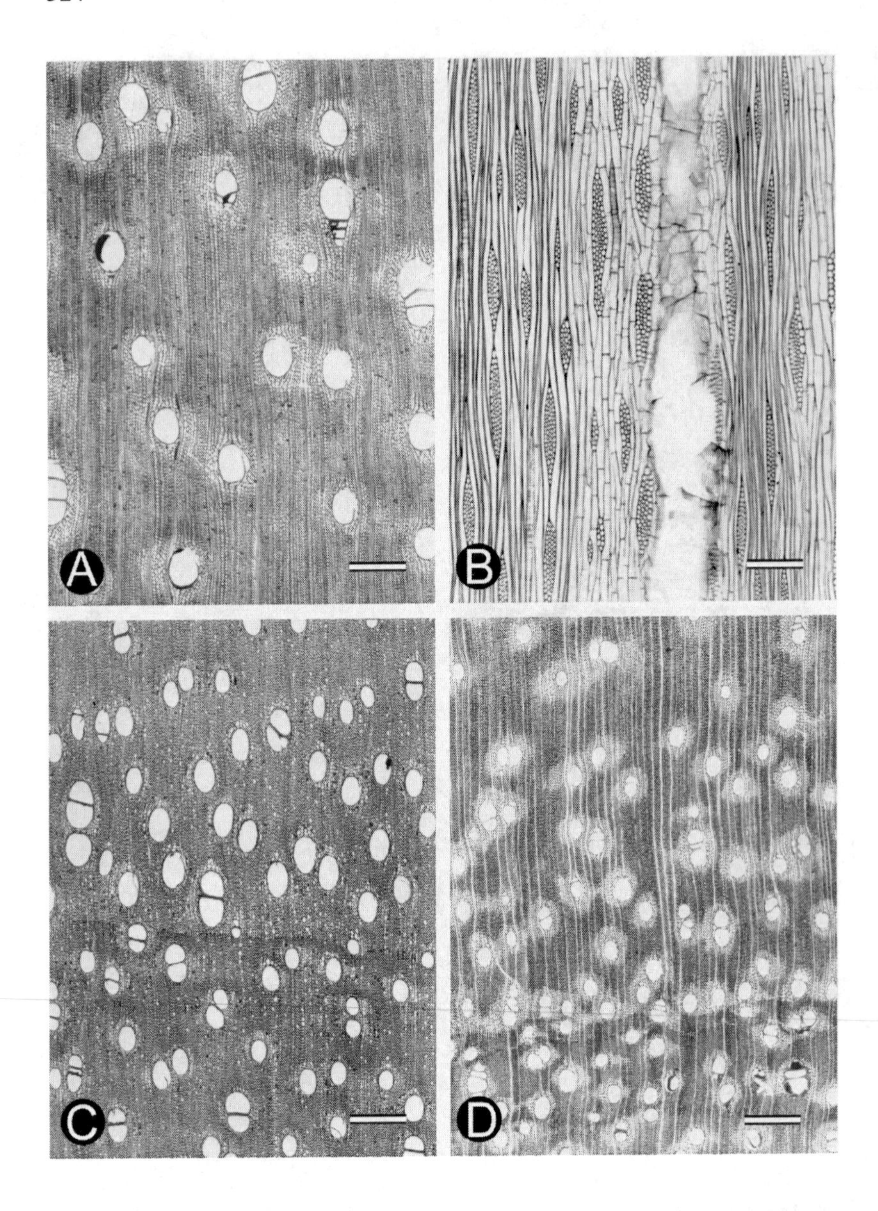

Fig. 68. A, C-D: transverse sections, scale bar = 400 µm: A-B, *Parkia ulei* (Harms) Kuhlmann (CTFw 7606); B, tangential section, scale bar = 180 µm; C, *Pentaclethra macroloba* (Willd.) Kuntze (CTFw 4713); D, *Pithecellobium dulce* (Roxb.) Benth. (CTFw 20841).

apotracheal parenchyma diffuse, scarce, banded parenchyma up to and over 3 cells wide. Strands of many cells. Prismatic crystals in chambered cells 4-30 per strand, relatively frequent especially in radial section.
Fibres thick walled, lumen ~2.5 µm, walls ~2.5 µm. Pits simple to minutely bordered, small, rare, confined to radial walls.

Note: The wood structure deviates from the other mimosoids by the two distinct vessel types. This phenomenon is often found in species with a climbing habit.

Material studied:
Mimosa guilandinae (DC.) Barneby var. **duckei** (Huber) Barneby: Suriname: van Donselaar 2815 (Uw 11884).

PARKIA R. Br. – Figs. 67 B; 68 A-B

Vessels diffuse, solitary and in radial multiples of 2-3, round, 1-3 per sq. mm, diameter (150-)180-300 µm. Vessel member length 450-580(-640). Perforations simple. Intervessel pits vestured, alternate, sometimes in rows, 7.5-9 µm. Vessel-ray pits similar to intervessel pits. Red brown deposits present.
Rays generally 1-2(-3)-seriate, sometimes up to 6-seriate, 4-7 per mm, 270-510 µm high. Homogeneous, composed of procumbent cells, rarely with a terminal row of square cells.
Parenchyma paratracheal lozenge-aliform to short winged-aliform more or less confluent, apotracheal parenchyma diffuse, sporadically in fine marginal, continuous or interrupted bands. Strands of 2-4 cells. Prismatic crystals in chambered cells, 8-32 per strand, common.
Fibres very thin- or thin- to thick-walled, lumen 15-34 µm, walls 3-6 µm. Pits simple, small, confined to radial walls. Length 1000-1650 µm. F/V ratio: 2.0-3.3.

Material studied:
Parkia nitida Miq.: French Guiana: BAFOG 308M (CTFw 10040), Petrov 48, 50, 115 (CTFw 16641, 16643, 16765).
P. pendula (Willd.) Benth. ex Walp.: Suriname: BBS 66 (Uw 663). French Guiana: BAFOG 204M (CTFw 9322); Benoist 435 (CTFw 14496); Petrov 73, 130, 145 (CTFw 16660, 16778, 16787).
P. reticulata Ducke: French Guiana: Benoist 1553 (CTFw 14585).
P. ulei (Harms) Kuhlmann: Suriname: Stahel 316 (Uw 316). French Guiana: BAFOG 3M (CTFw 7606), 119M (CTFw 8630).
P. velutina Benoist: French Guiana: Benoist 440, 706 (CTFw 14530, 14729).

PENTACLETHRA Benth. – Figs. 67 C; 68 C

Vessels diffuse, solitary and in short radial multiples of 2(-3), sometimes in clusters of 4-8, round, 4-7 per mm, diameter 160-180 µm. Vessel member length 410-530 µm. Perforations simple. Intervessel pits vestured, alternate, 4-5 µm. Vessel-ray similar to intervessel pits. White deposits common, dark brown deposits less frequent.
Rays 1(-1-partially 2)-seriate, 9-12 per mm, 180-220 µm high. Homogeneous, composed of procumbent cells.
Parenchyma paratracheal vasicentric to slightly lozenge-aliform sometimes confluent, apotracheal parenchyma diffuse frequent, and in 1-3-seriate marginal bands. Strands of 2-4 cells. Prismatic crystals numerous in chambered cells, 8-24 per strand.
Fibres thin- to thick-walled, lumen 10-16 µm, walls 5-6.5 µm. Pits simple, small, confined to radial walls. Length 1300-1400 µm. F/V ratio: 2.5-3.5.

Material studied:
Pentaclethra macroloba (Willd.) Kuntze: Suriname: BBS 24 (Uw 623); Stahel 344 (Uw 344 = CTFw 4713). Venezuela: Breteler 4973, 4992 (CTFw 17134, 17140).

PITHECELLOBIUM Mart. – Figs. 67 D; 68 D; 69 A

Vessels diffuse, solitary and in radial multiples of 2-4, round, 10-20 per sq. mm, diameter 70-140 µm. Vessel member length 220-310 µm. Perforations simple. Intervessel pits vestured, alternate, 6-8 µm. Vessel-ray pits similar to intervessel pits. Red brown deposits common.
Rays 1-3-seriate, 9-12 per mm, 115-180 µm high. Homogeneous composed of procumbent cells.
Parenchyma paratracheal mostly lozenge-aliform, sometimes confluent, apotracheal parenchyma diffuse and in marginal bands frequent. Strands of 1-2 cells. Prismatic crystals in chambered cells, 4-24 per strand, common.
Fibres generally septate, thin- to thick-walled, lumen 8-13 µm, walls, 4-5 µm. Pits simple, small, confined to radial walls. Length 670-800 µm. F/V ratio: 2.1-3.6.

Material studied:
Pithecellobium dulce (Roxb.) Benth.: Venezuela: Curran & Haman 502. U.S.A. (Florida) BWCw 8755 (CTFw 20841); Larsen s.n. (Uw 24653). Micronesia (Mariana Islands): Dutton 121 (Uw 16681).
P. unguis-cati (L.) Benth.: Florida: BWCw 8763; Stern & Brizicky 257 (Uw 6123).

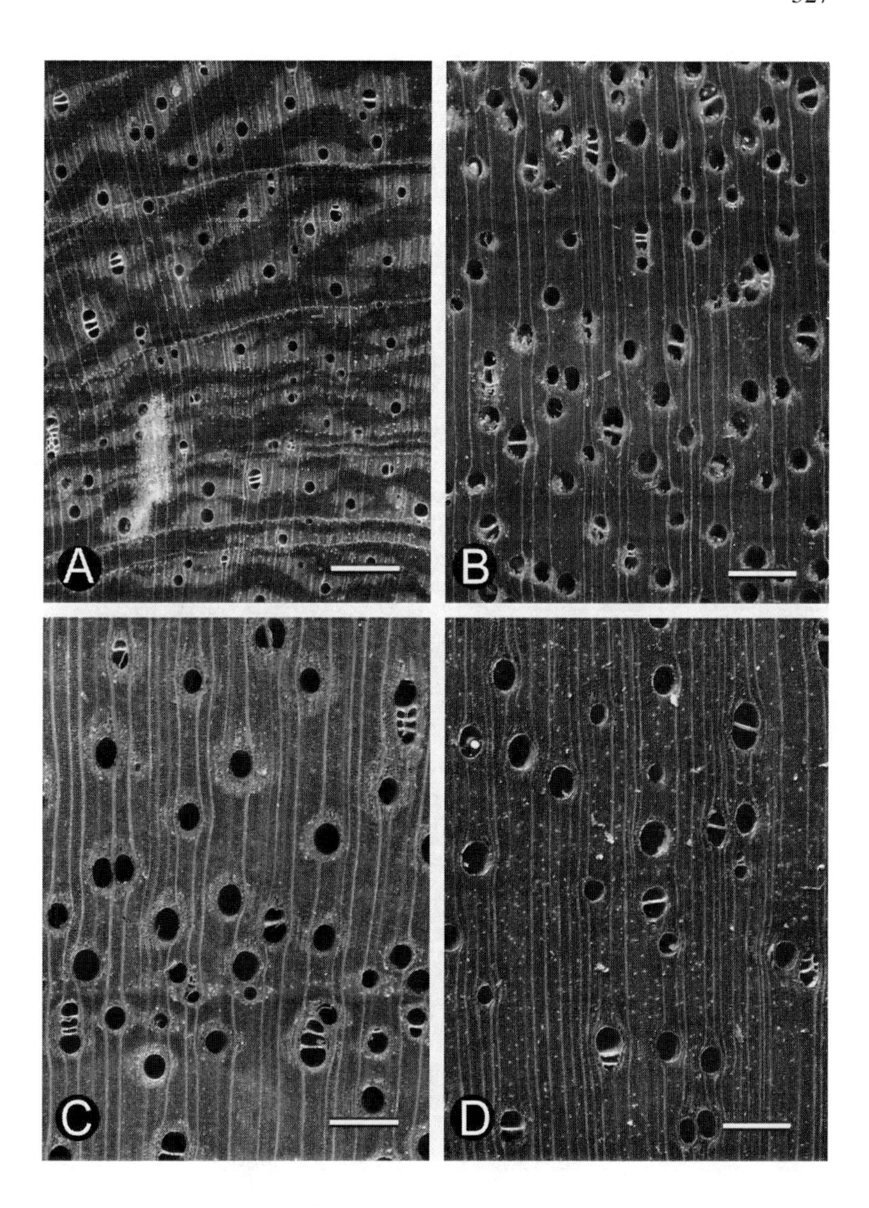

Fig.69. A-D: transverse sections, scale bar = 500 μm: A, *Pithecellobium unguis-cati* (L.) Benth. (Uw 6123); B, *Pseudopiptadenia suaveolens* (Miq.) J.W. Grimes (Uw 228); C, *Samanea saman* (Jacq.) Merr. (Uw 27921); D, *Stryphnodendron pulcherrimum* (Willd.) Hochr. (Uw 19960).

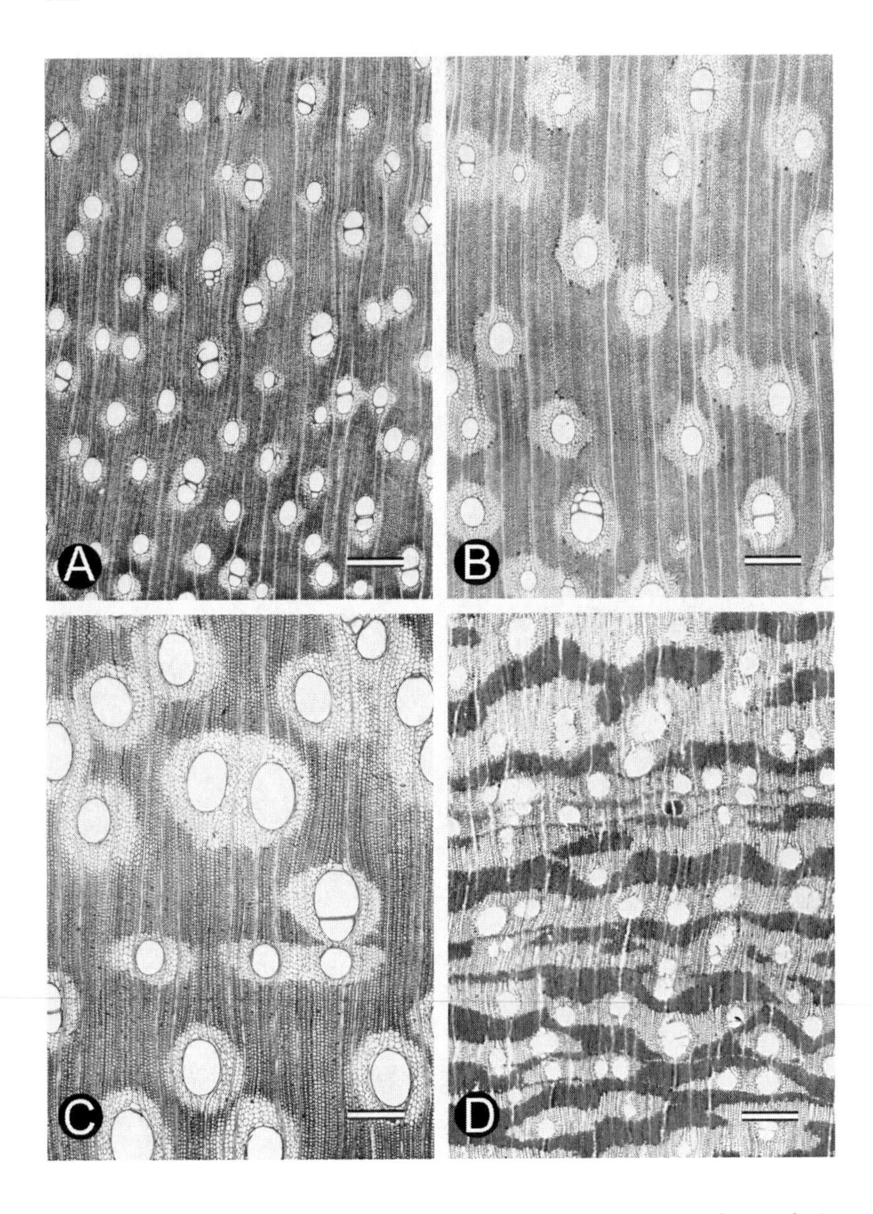

Fig. 70. A-D: transverse sections, scale bar = 400 μm: A, *Pseudopiptadenia suaveolens* (Miq.) J.W. Grimes (CTFw 16640); B, *Samanea saman* (Jacq.) Merr. (CTFw 6135); C, *Stryphnodendron polystachyum* (Miq.) Kleinhoonte (CTFw 17071); D, *Zygia latifolia* (L.) Fawc. & Rendle (CTFw 29419).

PSEUDOPIPTADENIA Rauschert – Figs. 69 B; 70 A

Vessels diffuse, solitary and in short radial multiples of 2-3, round, 6-12 per sq. mm, diameter 120-150 µm. Vessel member length 375-500 µm. Perforations simple. Intervessel pits vestured, alternate, 6-8 µm. Vessel-ray pits similar to intervessel pits.
Rays 1(-partially 2)-seriate, 7-9 per mm, 180-325 µm high. Homogeneous, composed of procumbent cells.
Parenchyma paratracheal vasicentric to shortly lozenge-aliform rarely confluent, apotracheal parenchyma diffuse rare and in fine marginal bands sporadic. Strands of 2(-4) cells. Prismatic crystals in chambered cells, 8-20 per strand.
Fibres thin- to thick-walled, lumen 8-15 µm, walls 3.5-5 µm. Pits simple, small, confined to radial walls. Length 1000-1250 µm. F/V ratio: 2.4-3.0.

Material studied:
Pseudopiptadenia psilostachya (DC.) G.P. Lewis & M.P. Lima: French Guiana: BAFOG 40M (CTFw 7717). Venezuela: Corothie 60 (CTFw 23199). Brazil: Détienne 346 (CTFw 26615).
P. suaveolens (Miq.) J.W. Grimes: Suriname: Stahel 228 (Uw 228). French Guiana: Petrov 11, 47, 175, 176, 177 (CTFw 16627, 16640, 17428, 17429, 17430).

SAMANEA (Benth.) Merr. – Figs. 69 C; 70 B

Vessels diffuse, solitary and in radial multiples of 2-3(-4), round, 3-4 per sq. mm, diameter 130-170 µm. Vessel member length 250-310 µm. Perforations simple. Intervessel pits vestured, alternate, 8-9 µm. Vessel-ray pits similar to intervessel pits. Brown deposits more or less frequent.
Rays 2(-3) or (2-)3-seriate, 5-7 per mm, 170-300 µm high. Homogeneous, composed of procumbent cells. Crystal absent or very rare.
Parenchyma paratracheal thick vasicentric to slightly lozenge-aliform sometimes confluent, apotracheal parenchyma diffuse not abundant and occasionally in fine marginal bands. Strands of (1-)2(-4) cells. Prismatic crystals in chambered cells, 8-24 per strand.
Fibres thin- to thick-walled, lumen 15-17 µm, walls 3-4 µm. Pits simple, small, confined to radial walls. Length 380-1000 µm. F/V ratio: 3.0-3.6.

Material studied:
Samanea saman (Jacq.) Merr.: Costa Rica: s.n. (Uw 27921). Porto-Rico MADw 54. Trinidad: CTFw 10603. Bolivia: Schmidt 99 (CTFw 11183). New Caledonia: Service Forestier 50 (CTFw 6135).

STRYPHNODENDRON Mart. – Figs. 69 D; 70 C

Vessels diffuse, solitary or in radial multiples of 2-4, round, 2-4 per sq. mm, diameter 200-290 μm. Vessel member length 380-560 μm. Perforations simple. Intervessel pits vestured, alternate, 7.5-9 μm. Vessel-rays pits similar to intervessel pits.
Rays 1(-2)- to 2-3-seriate, 6-8 per mm, occasionally in horizontal arrangement. Homogeneous, composed of procumbent cells.
Parenchyma paratracheal vasicentric to lozenge-aliform rarely to frequently confluent, apotracheal parenchyma diffuse and sporadically in large marginal bands. Strands of 1-4 cells. Prismatic crystals in chambered cells, 8-32 per strand frequent in *S. polystachyum*, not observed in *S. guianense* and *S. pulcherrimum*.
Fibres thin-walled, lumen 18-23 μm, walls 7-9 μm. Pits simple, small, confined to radial walls. Length 1100-1650 μm. F/V ratio: 2.5-3.6.

Material studied:
Stryphnodendron guianense (Aubl.) Benth.: French Guiana: Benoist 477 (CTFw 14502).
S. polystachyum (Miq.) Kleinhoonte: French Guiana: Petrov 15, 54, 158 (CTFw 16631, 16616, 17071).
S. pulcherrimum (Willd.) Hochr.: Suriname: Schulz 7272 (Uw 4943). Brazil: INPAw 294; Krukoff 5426 (Uw 19960).

ZYGIA P. Browne – Figs. 70 D; 71 A-D

Vessels diffuse, solitary and in short radial multiples of 2-3, round, 2-11 per sq. mm, diameter 100-180 μm. Vessel member length 310-520 μm. Perforations simple. Intervessel pits vestured, alternate, 5-7 μm. Vessel-ray pits similar to intervessel pits.
Rays 1(-2) or 1-2(-3)-seriate, 8-11 per mm, 130-250 μm high. Homogeneous, composed of procumbent cells.
Parenchyma paratracheal mostly aliform, often confluent forming broad irregular tangential bands, apotracheal diffuse, sometimes in spots, and in fine marginal bands. Strands of 1-2(-4) cells. Prismatic crystals in chambered cells, generally 8-24 per strand, common.
Fibres thick-walled, lumen 4-8 μm, walls 5-9 μm. Pits simple, confined to radial walls. Length 900-1550 μm. F/V ratio: 2.8-3.9.

Material studied:
Zygia cataractae (Kunth) L. Rico: Suriname: Lanjouw & Lindeman 2007 (Uw 1746). French Guiana: BAFOG 172M (CTFw 8654).

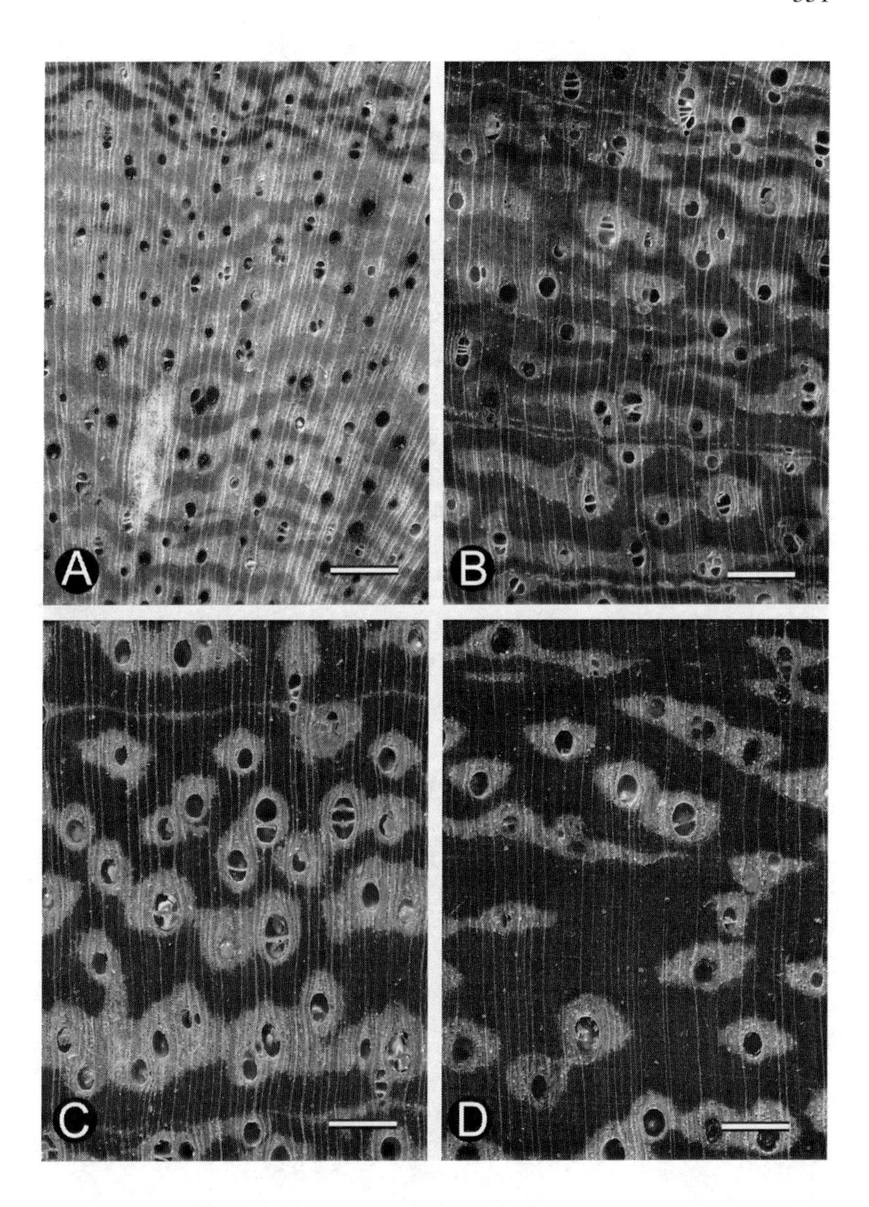

Fig. 71. A-D: transverse sections, scale bar = 500 µm: A, *Zygia cataractae* (Kunth) L. Rico (Uw 1746); B, *Zygia juruana* (Harms) L. Rico (Uw 19597); C, *Zygia latifolia* (L.) Fawc. & Rendle (Uw 284); D, *Zygia racemosa* (Ducke) Barneby & J.W. Grimes (Uw 72).

Z. collina (Sandwith) Barneby & J.W. Grimes: Guyana: FD 3976 (CTFw 32369).
Z. inaequalis (Humb. & Bonpl. ex Willd.) Pittier: Venezuela: Williams 15464 (CTFw 32374).
Z. juruana (Harms) L. Rico: Brazil: Krukoff 4746 (Uw 19597).
Z. latifolia (L.) Fawc. & Rendle: Suriname: Stahel 284 (Uw 284). French Guiana: Normand 527 (CTFw 16442). Colombia: van Rooden, ter Welle & Topper 692 (CTFw 27607). French Lesser Antilles: Rollet 1637 (CTFw 29419).
Z. racemosa (Ducke) Barneby & J.W. Grimes: Suriname: Stahel 72 (Uw 72). French Guiana: BAFOG 81M (CTFw 8033); Benoist 687 (CTFw 14520).
Z. sabatieri Barneby & J.W. Grimes: French Guiana: Loubry 1747 (CTFw 32715); Mori 23397, 23468, 23605 (CTFw 32838, 32897, 33001).
Z. tetragona Barneby & J.W. Grimes: French Guiana: Loubry 1752 (CTFw 32720); Mori 23638 (CTFw 32026); Tellier 359 (CTFw 31168).
Z. unifoliolata (Benth.) Pittier: Brazil: Krukoff 6716.

Calliandra, Chloroleucon, Macrosamanea, and *Mimosa* are of no commercial value.

Abarema

A. jupunba is the only potential timber producing species.

Tree	Canopy tree up to 30 m high, with low irregular buttresses. Bole moderately well formed, often cracked, 10-20 m long, diameter 50-80 cm.
Description of the wood	Heartwood pale pinkish brown (with yellow fluorescence under ultraviolet light) well demarcated from yellowish-white sapwood. Texture medium. Grain straight to moderately interlocked. Lustre medium to high.
Weight	Specific gravity (550-)600-700(-760) kg per cubic meter (12%).
Shrinkage	From green to ovendry: volumetric 15-16% (only 1 test).
Working properties	Works easily, finishes smoothly.
Durability	Reported as moderately resistant to decay.
Supply	Occasional to frequent.
Trade names	Huruasa, Kwatupana, Soapwood (GU). Fijnbladige Sopo-oedoe (SU). Bois Macaque (FG).

Acacia

Tree	Shrub or small tree.

Description of the wood	Heartwood pale brown not sharply demarcated from sapwood (*A. polyphylla*), or pinkish brown (fluorescent under ultraviolet light) and demarcated from yellowish white sapwood. Texture medium.
Weight	Specific gravity variable, 600-900 kg per cubic meter (12%).
Trade names	Parica (GU).

Adenanthera

A. pavonina has been introduced from Asia.

Description of the wood	Heartwood, very little in the trunk, brown (fluorescent yellow under ultraviolet light), sharply demarcated from greyish white sapwood. Texture medium to fine. Grain slightly interlocked.
Weight	Specific gravity 600-800 kg per cubic meter (12%).

Albizia

A. lebbeck, an introduced species, is the only species to have wood suitable for timber production.

Tree	Tree up to 20-27 m high, with trunk diameter of 60-90 cm.
Description of the wood	Heartwood golden brown when fresh turning to dark brown (yellow fluorescence with dark streaks under ultraviolet light), sharply demarcated from whitish sapwood. Texture medium to coarse. Grain moderately to deeply interlocked. Lustre medium.
Weight	Specific gravity 650-820 kg per cubic meter (12%).

Shrinkage	From green to ovendry: radial 2.9-3.2%, tangential 5.5-6%, volumetric 8.5-11%.
Seasoning properties	Moderately difficult to air dry.
Mechanical properties	Crushing strength: 510-560 kg/sq. cm. Static bending: 980-1200 kg/sq. cm. Modulus of elasticity: 62 000 kg/sq. cm.
Working properties	Easy to saw but sometimes difficult to machine because of deep interlocked grain. Sawdust may be irritating to eyes, nose and throat.
Durability	Rated as moderately durable.
Preservation	Heartwood resistant to preservative treatments.
Uses	Furniture and cabinetwork, decorative veneers, flooring, joinery.

Anadenanthera

Description of the wood	Heartwood reddish brown, demarcated from pale pinkish white sapwood. Texture medium, lustre medium to high.
Weight	Specific gravity 900-950 kg per cubic meter (12%).
Trade name	Parika (GU).

Balizia

Tree	Canopy tree up to 40 m high, unbuttressed. Bole poor or moderately well formed, 15-25 m long, diameter 60-100 cm.

Description of the wood	Heartwood pale pinkish to pale brown, sharply demarcated from whitish sapwood. Texture medium. Grain straight to slightly interlocked. Lustrous.
Weight	Specific gravity (500-)550-700(-750) kg per cubic meter (12%).
Shrinkage	From green to ovendry: radial 2.7-3.5%, tangential 6.4-8%, volumetric 10-14%.
Seasoning properties	No data available. Because of the high ratio tangential/radial shrinkage kiln-drying should be slow.
Mechanical properties	Crushing strength: 410-550 kg/sq. cm. Static bending: 1030-1380 kg/sq. cm. Modulus of elasticity: 90 000-135 000 kg/sq. cm.
Working properties	Works well and takes a high natural polish.
Durability	Heartwood resistant to decay by brown-rot fungi, moderately resistant to decay by white-rot fungi and termites.
Uses	To date only for fuel. Suitable for joinery.
Supply	Rare to locally abundant, especially in marsh and riparian forest.
Trade names	Manariballi, Kairaimai, Chiti, Koded (GU). Tamaren-prokoni (SU). Assao (FG).

Cedrelinga

Tree	Canopy tree up to 40-50 m high. Bole well formed, cylindrical, 18-25 m long, diameter 70-150 cm, occasionally more.

Description of the wood	Heartwood pale pinkish brown (high fluorescent yellow under ultraviolet light), gradually demarcated from lighter sapwood. Texture coarse. Grain straight to slightly interlocked. Lustre medium.
Weight	Specific gravity 420-600 kg per cubic meter (12%).
Shrinkage	From green to ovendry: radial 3-4.5%, tangential 6-9%, volumetric 12-15%.
Seasoning properties	Seasons quickly with minor risk of warping.
Mechanical properties	Crushing strength: 300-450 kg/sq. cm. Static bending: 800-1200 kg/sq. cm. Modulus of elasticity: 80 000-100 000 kg/sq. cm.
Working properties	Saws uncleanly, works easily. Fine dust may cause irritation to nose. Peeling easy.
Durability	Moderately resistant to decay fungi, vulnerable to attack by termites.
Preservation	Poorly treatable.
Uses	Interior joinery, panelling, moulding, furniture of medium quality, plywood and particle board.
Supply	Very rare.
Trade names	Don Cede (FG).

Enterolobium

E. schomburgkii and *E. oldemannii* are the most important timber producing species in this genus. *E. cyclocarpum*, a species less frequently encountered and with very different wood, similar in wood characteristics to *Cedrelinga* (except for shrinkage and durability).

Tree	Canopy tree up to 45 m high, with low branched buttresses 1(-2) m high. Bole well formed, cylindrical 15-25 m long, diameter 50-80 cm.
Description of the wood	Heartwood uniformly brown ochre to red brown or red brown with brown ochre streaks (slighly fluorescent yellow under ultraviolet light), demarcated from yellowish sapwood. Texture medium. Grain rather straight, sometimes lightly interlocked or curly. Lustrous.
Weight	Specific gravity 750-900 kg per cubic meter (12%).
Shrinkage	From green to ovendry: radial 3.5-5.5%, tangential 9.0-11.0%, volumetric 14-17% (for *E. cyclocarpum*, radial 2-3%, tangential 4-5%, volumetric 7-9%).
Seasoning properties	Case hardening possible in thickness > 50 mm during air-seasoning and kiln-drying too with risk of subsequent warping.
Mechanical properties	Crushing strength: 550-750 kg/sq. cm. Static bending: 1300-2000 kg/sq. cm. Modulus of elasticity: 120 000-170 000 kg/sq. cm.
Working properties	Easy – moderately difficult to work. Dust may cause skin irritation.
Durability	Durable to decay fungi (*E. cyclocarpum* particularly). All are relatively resistant to termites.
Preservation	Heartwood is resistant to preservative treatments.

Uses	Interior and exterior joinery, carpentry, flooring, panelling and furniture.
Supply	Relatively common in SU and FG (diameter > 40 cm: 0.4 cubic meter per hectare), less frequent in GU.
Trade names	Devil's ear, Aratabana, Suburutin (GU). Kabbes, Kadiouchi, Maka (SU). Acacia franc, Bougoubatibatra (FG).

Hydrochorea

The species *H. corymbosa* and *H. gonggijpii* are easily confused with *Balizia pedicellaris*.

Description of the wood	Heartwood pale pinkish brown (slight to high fluorescence yellow under ultraviolet light), fairly demarcated from yellowish white sapwood. Texture medium. Grain usually straight, occasionally slightly interlocked or curly. Lustre low to medium.
Weight	Specific gravity 550-700 kg per cubic meter (12%).
Trade names	Manariballi, Örükorong, Chiti, Koted (GU). Bos-tamarinde, Boesi-tamaren (SU). Tamarin, Assao (FG).

Inga

Although numerous in species, only *I. alba* is reported as being a timber producing species.

Tree	Tree up to 30 m high, with a large crown and thick buttresses up to 1 m high. Bole moderately well formed, 50-80 cm diameter, up to 15 m high.

Description of the wood	*I. alba*: heartwood pink-brown sometimes with darker streaks (with a poor yellow fluorescence under ultraviolet light) not distinctly demarcated from paler sapwood. Texture medium. Grain straight to slightly interlocked. Lustre medium. Other species: heartwood variable from cream colour to deep dark red (with or without yellow fluorescence under ultraviolet light, or sometimes yellow fluorescence restricted to border of the sapwood).
Weight	*I. alba*: specific gravity 600-750 kg per cubic meter (12%). Other species: variable from 550 to 1100 kg per cubic meter (12%).
Shrinkage	From green to ovendry: radial 3.5%, tangential 7.4%, volumetric 12-15.5%.
Seasoning properties	Seasons rapidly with moderate warping.
Mechanical properties	Crushing strength: 545 kg/sq. cm. Static bending: 1215 kg/sq. cm. Modulus of elasticity: 120 000 kg/sq. cm.
Working properties	Works easily and finishes to a smooth surface.
Durability	Moderately resistant to decay.
Preservation	No data available but certainly poorly treatable.
Uses	Suitable for interior joinery and particle board.
Supply	Relatively frequent in secondary and primary forest.
Trade names	Maporokon, Kurang, Kwari (GU). Prokonie, Aboonkini (SU). Bougouni, Lebi-oueko, Bois Pagode (FG).

Leucaena

L. leucocephala is introduced in the Guianas.

Description of the wood	Heartwood brown, sometimes with darker streaks (without fluorescense under ultraviolet light), sharply demarcated from yellow sapwood. Texture medium. Grain straight to slightly interlocked.
Weight	Specific gravity: 750-900 kg per cubic meter (12%).

Parkia

P. nitida, P. pendula and *P. ulei* are reported as timber producing species.

Tree	Canopy tree of up to 40 m high, with thin, up to 3-4 m high, buttresses. Bole well formed, cylindrical, 20-25 m long, diameter 60-130 cm.
Description of the wood	Heartwood creamy-pink to pale brown, sometimes with irregular red brown streaks (yellow fluorescent or not under ultraviolet light). Texture coarse. Grain generally straight. Lustre medium.
Weight	Specific gravity very variable, from 250 to 600 kg per cubic meter (12%).
Shrinkage	From green to ovendry: radial 2-4.5%, tangential 6-9%, volumetric 9-15%.
Seasoning properties	Air-dries or kiln-dries at a fast rate with slight warping and a little tendency to check.
Mechanical properties	Variable, depending on the species and to the specific gravity. Crushing strength: 200-480 kg/sq. cm. Static bending: 450-1300 kg/sq. cm. Modulus of elasticity: 50 000-110 000 kg/sq. cm.

Working properties	Works easily, finishes generally well, peeling possible.
Durability	Poorly or moderately resistant to decay fungi and to insect attack.
Preservation	Moderately permeable to preservative treatments.
Uses	Plywood, particule board, panelling, interior joinery, ordinary furniture and box manufacturing.
Supply	Occasional (GU) to frequent (FG). Easy regeneration.
Trade names	*P. nitida*: Black Manariballi, Uya, Ululu (GU). Agrobigi, Oeja (SU). Dodomissinga (FG). *P. pendula*: Hipanai, Darina (GU). Kwatakama (SU). Kouatakaman, Acacia mâle (FG). *P. ulei*: Uya (GU). Kwatakama, Kojalidan (SU). Dodomissinga (FG).

Pentaclethra

Tree	Medium to large tree up to 25-30 m high, with low and irregular buttresses. Bole form very poor, 6-9 m long, diameter 40-50(-90) cm.
Description of the wood	Heartwood brown to red brown (yellow fluorescence under ultraviolet light) demarcated from pinkish white sapwood. Texture medium. Grain rather straight. Lustrous.
Weight	Specific gravity: 700-850 kg per cubic meter (12%).
Working properties	Works relatively well, polishes well.
Durability	Reported as moderately resistant to decay.

Uses	Furniture, staircase, house framing.
Supply	Abundant in GU, less in SU, rare in FG.
Trade names	Trysil, Koroballi, Barawakashi, Awaragaik (GU). Kroebara (SU).

Pithecellobium

P. dulce is the only species reported as producing timber.

Tree	Small to medium tree, up to 15 m high, diameter 30-60 cm.
Description of the wood	Heartwood pinkish brown (not fluorescent under ultraviolet light) not sharply demarcated from greyish white sapwood. Texture relatively fine. Grain straight to interlocked. Lustre medium.
Weight	Specific gravity 580-700 kg per cubic meter (12%).

Pseudopiptadenia

P. suaveolens is reported as being a timber producing species.

Tree	Canopy tree, up to 40-45 m high, with very low buttresses. Bole straight but more or less in planks, 15-25 m long, diameter 50-100 cm.
Description of the wood	Heartwood variable from pale brown to red brown (fluorescence yellow ochre under ultraviolet light), gradually to very sharply demarcated from whitish sapwood. Texture fine to medium. Interlocked grain common. Lustrous.

Weight	Specific gravity 700-950 kg per cubic meter (12%). The wood of *P. psilostachya* is lighter, 550-700 kg per cubic meter.
Shrinkage	From green to oven dry: radial 4.0-5.5%, tangential 7-8%, volumetric 13-14%.
Seasoning properties	The wood air-seasons at a moderate rate with little tendency to end splitting. Warps possible during kiln-drying.
Mechanical properties	Crushing strength: 640-760 kg/sq. cm. Static bending: 1500-1800 kg/sq. cm. Modulus of elasticity: 130 000-160 000 kg/sq. cm.
Working properties	Works well except in case of very interlocked grain. Takes a light natural polish.
Durability	Moderately resistant to decay fungi and termites.
Preservation	Heartwood is poorly treatable.
Uses	Carpentry and interior joinery.
Supply	Infrequent occurrence.
Trade names	Shirimai (GU). Pikin-misiki (SU). Alimiao (FG).

Samanea

S. saman is the only timber producing species.

Tree	In its area of origin (Yucatan) and in forest, 30-38 m high, remaining low when grown in open area. Small tree. Bole cylindrical relatively short but thick, diameter 90-120 cm.

Description of the wood	Heartwood uniform light brown to brown with darker streaks (discrete yellow fluorescence under ultraviolet light), clearly demarcated from yellowish white sapwood. Texture medium to coarse. Grain straight to slightly interlocked. Lustre medium to high.
Weight	Specific gravity: 500-650 kg/per cubic meter (12%).
Shrinkage	From green to ovendry: radial 2-4%, tangential 3-6%, volumetric 6-10.5%.
Seasoning properties	Air seasons rather poorly with moderate to severe warp.
Mechanical properties	Crushing strength: 350-510 kg/sq. cm. Static bending: 850-890 kg/sq. cm. Modulus of elasticity: 65 000-110 000 kg/sq. cm.
Working properties	Saws and machines easily. Takes an excellent finish.
Durability	Reported as durable to very durable to decay fungi and resistant to dry wood termites, the durability variable depending on area of origin and rate of growth.
Uses	Fine furniture and cabinet work, joinery.

Stryphnodendron

S. polystachyum is the only potentially timber producing species.

Tree	Large tree up to 30 m high. Bole generaly well formed, cylindrical, diameter 60-90 cm.

Description of the wood	Heartwood red-brown (without fluores-cence under ultraviolet light), not sharply demarcated from paler pink-brown sapwood. Texture medium to coarse. Interlocked grain common. Lustre medium.
Weight	Specific gravity: 500-660 kg per cubic meter (12%).
Shrinkage	From green to ovendry: radial 3.5-4.8%, tangential 6.5-9.7%, volumetric 12-15.5%.
Mechanical properties	Crushing strength: 360-535 kg/sq. cm. Static bending: 700-1000 kg/sq. cm. Modulus of elasticity: 75 000-100 000 kg/sq. cm.
Working properties	Works well, finishes well except when the grain is hard interlocked.
Durability	Very variable. Resistant to vulnerable to decay fungi depending on the individual trees.
Uses	Restricted to local uses, box manufacturing.
Supply	Not frequent to rare.
Trade names	Boschtamalen (SU). Nionoudou (FG).

Zygia

Z. racemosa is the only species reported as timber producing.

Tree	Canopy tree up to 35 m high, unbuttressed, basally swollen. Bole with small bumps, 15-25 m long, diameter 50-70 cm.

Description of the wood	Heartwood yellowish with irregular dark brown streaks (not fluorescent under ultraviolet light), not demarcated from yellowish sapwood except by the presence of streaks. Grain rather straight. Texture medium. Lustre medium.
Weight	Specific gravity: 950-1300 kg per cubic meter (12%).
Shrinkage	From green to ovendry: radial 6-7%, tangential 10-13.5%, volumetric 18-22%.
Seasoning properties	High tendency to end splitting.
Mechanical properties	Crushing strength: 700-900 kg/sq. cm. Static bending: 2000-2300 kg/sq. cm. Modulus of elasticity: 205 000-225 000 kg/sq. cm.
Working properties	Difficult to saw and machine.
Durability	Fairly resistant to termites but probably moderately resistant to decay fungi.
Preservation	Resistant to preservative treatments.
Uses	Cabinet-making, marquetry, decorative panels and turning.
Supply	Rare to occasional.
Trade names	Snakewood, Tureli (GU). Bostamarinde, Boesi Tamaren, Manaliballi Tatara (SU). Bois Serpent, Sinekioudou (FG).

LITERATURE ON WOOD AND TIMBER

Anonymus. 1982. Nomenclature générale des bois tropicaux. A.T.I.B.T. Nogent-sur-Marne.

Anonymus. 1989. Bois des Dom-Tom. Tome 1 Guyane. C.T.F.T. Nogent-sur-Marne.

Baretta-Kuipers, T. 1973. Some aspects of wood-anatomical research in the genus Inga (Mimosaceae) from the Guianas and especially Suriname. Acta Bot. Neerl. (22(3): 193-205.

Baretta-Kuipers, T. 1981. Wood anatomy of Leguminosae: its Relevance to Taxonomy. In R.M. Polhill en P.H. Raven (eds.), Advances in Legume Systematics: 677-705.

Bena, P. 1960. Essences forestières de Guyane. Bureau Agricole et Forestier guyanais. Imprimerie Nationale, Paris.

Benoist, R. 1931. Les bois de la Guyane française. Arch. Bot. Mém. 5(1).

Berni, C.A., E. Bolza & F.J. Christensen. 1979. South American Timbers. The characteristics, properties and uses of 190 species. C.S.I.R.O., Melbourne.

Cassens, D.L. & R.B. Miller. 1981. Wood anatomy of the new world. Pithecellobium (sensu lato). J. Arnold Arbor. 62(1).

Chudnoff, M. 1984. Tropical timbers of the world. Agriculture handbook number 607. U.S. Forest Service, Madison.

Détienne, P., P. Jacquet & A. Mariaux. 1982. Manuel d'identification des bois tropicaux. Tome 3, Guyane française. C.T.F.T., Nogent-sur-Marne.

Détienne, P. & P. Jacquet. 1983. Atlas d'identification des bois de l'Amazonie et des régions voisines. C.T.F.T., Nogent-sur-Marne.

Ducke, A. 1949. Notas sobre a Flora Neotropica-II. As Leguminosas da Amazônia brasileira. Bol. Técn. Inst. Agron. N. 18.

Evans, J.E., P. Gasson & G.P. Lewis. 2006. Wood anatomy of the Mimosoideae (Leguminosae). IAWA J., Suppl. 5.

Fanshawe, D.B. 1954. Forest products of British Guiana. Part 1. Principal Timbers. Forest. Bull. Forest Dept. British Guiana 1 (new series) 2nd ed.

IAWA Committee 1989. The IAWA list of microscopic features for hardwood identification. IAWA Bull. n.s. 10: 219-332.

Japing, C.H. & H.W. Japing. 1960. Houthandboek Surinaamse houtsoorten. 's Lands Bosbeheer, Paramaribo.

Lindeman, J.C. & A.M.W. Mennega. 1963. Bomenboek voor Suriname. 's Lands Bosbeheer, Paramaribo.

Luckow, M., J.T. Miller, D.J. Murphy & T. Livschultz. 2003. A phylogenetic analysis of the Mimosoideae (Leguminosae) bases on chloroplast DNA sequence data. In B.B. Klitgaard & A. Bruneau (eds.), Advances in Legume Systematics 10: 197-220.

Luckow, M., P.J. White & A. Bruneau. 2000. Relationships among the

basal genera of mimosoid legumes. In P.S. Herendeen & A. Bruneau (eds.), Advances in Legume Systematics 9: 165-180.
Mennega, E.A., W.C.M. Tammens-de Rooij & M.J. Jansen-Jacobs (eds.). 1988. Checklist of woody plants of Guyana. Tropenbos Technical Series 2. Tropenbos Foundation, Ede.
Metcalfe, C.R. & L. Chalk. 1950. Anatomy of the Dicotyledons. Oxford University Press, London.
Pfeiffer, J. Ph. 1926. De houtsoorten van Suriname. Deel I. De Bussy, Amsterdam.
Polak, A.M. 1992. Major timber trees of Guyana. A field guide. Tropenbos Series 2. Tropenbos Foundation, Wageningen.
Record, S.R. & R.W. Hess. 1943. Timbers of the New World. Yale University Press, New Haven.
Reiders-Gouwentak, C.A. & J.F. Rijsdijk. 1968. Hout van Leguminosae uit Suriname. H. Veenman en Zonen N.V., Wageningen.
Stern, W.L. 1988. Index Xylariorum 3. IAWA Bull. n.s. 9(3).
Vink, A.T. 1965. Surinam Timbers. Surinam Forest Service, Paramaribo.

NUMERICAL LIST OF ACCEPTED TAXA

1. Abarema Pittier
   - 1-1a. A. barbouriana (Standl.) Barneby & J.W. Grimes var. barbouriana
   - 1-1b. A. barbouriana (Standl.) Barneby & J.W. Grimes var. arenaria (Ducke) Barneby & J.W. Grimes
   - 1-2. A. commutata Barneby & J.W. Grimes
   - 1-3. A. curvicarpa (H.S. Irwin) Barneby & J.W. Grimes var. curvicarpa
   - 1-4. A. ferruginea (Benth.) Pittier
   - 1-5. A. floribunda (Spruce ex Benth.) Barneby & J.W. Grimes
   - 1-6. A. gallorum Barneby & J.W. Grimes
   - 1-7a. A. jupunba (Willd.) Britton & Killip var. jupunba
   - 1-7b. A. jupunba (Willd.) Britton & Killip var. trapezifolia (Vahl) Barneby & J.W. Grimes
   - 1-8. A. laeta (Benth.) Barneby & J.W. Grimes
   - 1-9. A. longipedunculata (H.S. Irwin) Barneby & J.W. Grimes
   - 1-10. A. mataybifolia (Sandwith) Barneby & J.W. Grimes

2. Acacia Mill.
   - 2-1. A. alemquerensis Huber
   - 2-2. A. articulata Ducke
   - 2-3. A. farnesiana (L.) Willd.
   - 2-4. A. macracantha Willd.
   - 2-5. A. polyphylla DC.
   - 2-6a. A. tenuifolia (L.) Willd. var. tenuifolia
   - 2-6b. A. tenuifolia (L.) Willd. var. producta J.W. Grimes

3. Adenanthera L.
   - 3-1. A. pavonina L.

4. Albizia Durazz.
   - 4-1. A. barinensis Cárdenas
   - 4-2. A. glabripetala (H.S. Irwin) G.P. Lewis & P.E. Owen
   - 4-3. A. lebbeck (L.) Benth.
   - 4-4. A. niopoides (Spruce ex Benth.) Burkart
   - 4-5a. A. subdimidiata (Splitg.) Barneby & J.W. Grimes var. subdimidiata
   - 4-5b. A. subdimidiata (Splitg.) Barneby & J.W. Grimes var. minor Barneby & J.W. Grimes

5. Anadenanthera Speg.
   - 5-1. A. peregrina (L.) Speg.

6. Balizia Barneby & J.W. Grimes
   - 6-1. B. pedicellaris (DC.) Barneby & J.W. Grimes

7. Calliandra Benth.
   - 7-1. C. coriacea (Willd.) Benth.
   - 7-2. C. glomerulata H. Karst. var. glomerulata
   - 7-3. C. houstoniana (Mill.) Standl. var. calothyrsus (Meisn.) Barneby
   - 7-4. C. hymenaeodes (Pers.) Benth.
   - 7-5. C. rigida Benth.
   - 7-6. C. laxa (Willd.) Benth. var. stipulacea (Benth.) Barneby
   - 7-7. C. surinamensis Benth.
   - 7-8. C. tenuiflora Benth.

8. Cedrelinga Ducke
   - 8-1. C. cateniformis (Ducke) Ducke

9. Chloroleucon (Benth.) Britton & Rose
   - 9-1. C. acacioides (Ducke) Barneby & J.W. Grimes

10. Desmanthus Willd.
    - 10-1. D. pernambucanus (L.) Thell.
    - 10-2. D. virgatus (L.) Willd.

11. Entada Adans.
    - 11-1a. E. polystachya (L.) DC. var. polyphylla (Benth.) Barneby
    - 11-1b. E. polystachya (L.) DC. var. polystachya

12. Enterolobium Mart.
    - 12-1. E. cyclocarpum (Jacq.) Griseb.
    - 12-2. E. oldemannii Barneby & J.W. Grimes
    - 12-3. E. schomburgkii (Benth.) Benth.

13. Hydrochorea Barneby & J.W. Grimes
    - 13-1. H. corymbosa (Rich.) Barneby & J.W. Grimes
    - 13-2. H. gonggrijpii (Kleinhoonte) Barneby & J.W. Grimes

14. Inga Mill.
    - 14-1. I. acreana Harms
    - 14-2. I. acrocephala Steud.

14-3. I. alata Benoist
14-4. I. alba (Sw.) Willd.
14-5. I. albicoria Poncy
14-6. I. auristellae Harms
14-7. I. bourgonii (Aubl.) DC.
14-8. I. brachystachys Ducke
14-9. I. brevipes Benth.
14-10. I. calanthoides Amshoff
14-11. I. capitata Desv.
14-12. I. cayennensis Sagot ex Benth.
14-13. I. cordatoalata Ducke
14-14. I. disticha Benth.
14-15. I. edulis Mart.
14-16. I. fanchoniana Poncy
14-17. I. fastuosa (Jacq.) Willd.
14-18. I. flagelliformis (Vell.)Mart.
14-19. I. graciliflora Benth.
14-20. I. gracilifolia Ducke
14-21. I. grandiflora Ducke
14-22. I. heterophylla Willd.
14-23. I. huberi Ducke
14-24. I. ingoides (Rich.)Willd.
14-25. I. jenmanii Sandwith
14-26. I. lateriflora Miq.
14-27. I. laurina (Sw.) Willd.
14-28. I. leiocalycina Benth.
14-29. I. leptingoides Amshoff
14-30. I. lomatophylla (Benth.) Pittier
14-31. I. longiflora Spruce ex Benth.
14-32. I. longipedunculata Ducke
14-33. I. loubryana Poncy
14-34. I. macrophylla Humb. & Bonpl. ex Willd.
14-35. I. marginata Willd.
14-36. I. melinonis Sagot
14-37. I. mitaraka Poncy
14-38. I. nobilis Willd.
14-39. I. nouragensis Poncy
14-40. I. nubium Poncy
14-41. I. obidensis Ducke
14-42. I. paraensis Ducke
14-43. I. pezizifera Benth.
14-44. I. pilosula (Rich.) J.F. Macbr.
14-45. I. poeppigiana Benth.
14-46. I. retinocarpa Poncy

14-47. I. rhynchocalyx Sandwith
14-48. I. rubiginosa (Rich.) DC.
14-49. I. sarmentosa Glaziou ex Harms
14-50. I. sertulifera DC.
14-51. I. splendens Willd.
14-52. I. stipularis DC.
14-53. I. striata Benth.
14-54. I. suaveolens Ducke
14-55. I. thibaudiana DC.
14-56. I. umbellifera (Vahl) Steud. ex DC.
14-57. I. vera Willd. subsp. affinis (DC.) T.D. Penn.
14-58. I. virgultosa (Vahl) Desv.

15. Leucaena Benth.
15-1. L. leucocephala (Lam.) de Wit

16. Macrosamanea Britton & Rose ex Britton & Killip
16-1. M. kegelii (Meisn.) Kleinhoonte
16-2. M. pubiramea (Steud.) Barneby & J.W. Grimes var. pubiramea

17. Mimosa L.
17-1a. M. annularis Benth. var. odora Barneby
17-1b. M. annularis Benth. var. xinguënsis (Ducke) Barneby
17-2. M. bimucronata (DC.) Kuntze
17-3. M. camporum Benth.
17-4. M. casta L.
17-5. M. debilis Humb. & Bonpl. ex Willd.
17-6. M. diplotricha C. Wright ex Sauvalle
17-7. M. dormiens Humb. & Bonpl. ex Willd.
17-8a. M. guilandinae (DC.) Barneby var. guilandinae
17-8b. M. guilandinae (DC.) Barneby var. duckei (Huber) Barneby
17-9. M. invisa Mart. ex Colla var. spiciflora (H. Karst.) Barneby
17-10a. M. microcephala Humb. & Bonpl. ex Willd. var. cataractae (Ducke) Barneby
17-10b. M. microcephala Humb. & Bonpl. ex Willd. var. lumaria Barneby
17-11a. M. myriadenia (Benth.) Benth. var. myriadenia
17-11b. M. myriadenia (Benth.) Benth. var. dispersa Barneby
17-12. M. pellita Humb. & Bonpl. ex Willd.
17-13. M. polydactyla Humb. & Bonpl. ex Willd.
17-14a. M. pudica L. var. hispida Brenan

17-14b. M. pudica L. var. pastoris Barneby
17-14c. M. pudica L. var. tetrandra (Willd.) DC.
17-14d. M. pudica L. var. unijuga (Walp. & Duchass.) Griseb.
17-15. M. quadrivalvis L. var. leptocarpa (DC.) Barneby
17-16. M. rufescens Benth.
17-17. M. schomburgkii Benth.
17-18a. M. schrankioides Benth. var. sagotiana (Benth.) Barneby
17-18b. M. schrankioides Benth. var. schrankioides
17-19a. M. somnians Humb. & Bonpl. ex Willd. subsp. somnians var. somnians
17-19b. M. somnians Humb. & Bonpl. ex Willd. subsp. viscida (Willd.) Barneby var. viscida
17-20. M. surumuënsis Harms

18. Neptunia Lour.
18-1. N. natans (L.f.) Druce
18-2. N. plena (L.) Benth.

19. Parkia R. Br.
19-1. P. decussata Ducke
19-2. P. gigantocarpa Ducke
19-3. P. igneiflora Ducke
19-4. P. nitida Miq.
19-5. P. pendula (Willd.) Benth. ex Walp.
19-6. P. reticulata Ducke
19-7. P. ulei (Harms) Kuhlmann var. surinamensis Kleinhoonte
19-8. P. velutina Benoist

20. Pentaclethra Benth.
20-1. P. macroloba (Willd.) Kuntze

21. Piptadenia Benth.
21-1. P. floribunda Kleinhoonte
21-2. P. leucoxylon Barneby & J.W. Grimes

22. Pithecellobium Mart.
22-1. P. dulce (Roxb.) Benth.
22-2. P. roseum (Vahl) Barneby & J.W. Grimes
22-3. P. unguis-cati (L.) Benth.

23. Plathymenia Benth.
23-1. P. reticulata Benth.

24. Pseudopiptadenia Rauschert
    - 24-1. P. psilostachya (DC.) G.P. Lewis & M.P. Lima
    - 24-2. P. suaveolens (Miq.) J.W. Grimes

25. Samanea (Benth.) Merr.
    - 25-1. S. inopinata (Harms) Barneby & J.W. Grimes
    - 25-2. S. saman (Jacq.) Merr.

26. Stryphnodendron Mart.
    - 26-1. S. guianense (Aubl.) Benth.
    - 26-2. S. moricolor Barneby & J.W. Grimes
    - 26-3. S. polystachyum (Miq.) Kleinhoonte
    - 26-4. S. pulcherrimum (Willd.) Hochr.

27. Zapoteca H.M. Hern.
    - 27-1. Z. portoricensis (Jacq.) H.M. Hern. subsp. flavida (Urb.) H.M. Hern.

28. Zygia P. Browne
    - 28-1. Z. cataractae (Kunth) L. Rico
    - 28-2. Z. claviflora (Spruce ex Benth.) Barneby & J.W. Grimes
    - 28-3. Z. collina (Sandwith) Barneby & J.W. Grimes
    - 28-4. Z. eperuetorum (Sandwith) Barneby & J.W. Grimes
    - 28-5. Z. inaequalis (Humb. & Bonpl. ex Willd.) Pittier
    - 28-6. Z. inundata (Ducke) H.C. Lima ex Barneby & J.W. Grimes
    - 28-7. Z. juruana (Harms) L. Rico
    - 28-8a. Z. latifolia (L.) Fawc. & Rendle var. latifolia
    - 28-8b. Z. latifolia (L.) Fawc. & Rendle var. lasiopus (Benth.) Barneby & J.W. Grimes
    - 28-9. Z. palustris Barneby & J.W. Grimes
    - 28-10. Z. potaroënsis Barneby & J.W. Grimes
    - 28-11. Z. racemosa (Ducke) Barneby & J.W. Grimes
    - 28-12. Z. sabatieri Barneby & J.W. Grimes
    - 28-13. Z. tetragona Barneby & J.W. Grimes
    - 28-14. Z. unifoliolata (Benth.) Pittier

COLLECTIONS STUDIED
(Numbers in **bold** represent types)

GUYANA

Abraham, A.A., 206 (17-12).
Acevedo, P., 3293 (13-2); 3366 (14-55); 3450 (14-43).
Altson, R.A., 427 (16-2).
Andel, T.R. van, *et al.*, 776 (14-43); 831 (14-51); 978 (14-44); 980 (14-55); 1113 (14-25); 1132 (14-36); 1161 (14-28); 1162 (14-25); 1261 (14-28); 1322 (14-19); 1747 (14-55); 1761 (14-48).
Anderson, C.W., 88 (28-8b); 89 (11-1b); 335 (14-44).
Appun, C.F., 33 (20-1).
Boom, B., *et al.*, 1663 (1-1b); 7182 (17-12); 7200 (20-1); 7229 (17-11a); 7323 (1-7b); 8034 (18-1); 8041 (18-2); 9300 (14-35).
Boyan, J., see under FD.
Boyan, R., see under FD.
Clarke, D., *et al.*, 210 (14-48); 1124 (14-25); 1252 (14-53); 1815 (14-41); 2211 (14-51); 3029 (14-14); 3631 (19-8); 3981 (14-51); 4296 (14-43); 4304 (14-28); 4305 (14-11); 4499 (14-55); 4787 (14-14); 4905 (19-4); 4927 (14-52); 4991 (14-51); 6158 (14-55); 6201 (14-11); 6220 (14-28); 6505 (14-56); 6506 (14-38); 6604 (14-14); 6651 (14-56); 6796 (14-9); 6966, 7048 (19-7); 7100 (14-43); 7127 (14-8); 7509 (14-31); 7601 (14-56); 7636 (14-13); 7715 (14-31); 7725 (19-4); 7729 (14-50); 7747 (14-8); 7772 (14-13); 7826 (14-30); 7916 (14-54); 8011 (14-55); 8139 (14-6); 8496 (14-56); 8808 (14-54).
Cook, C.D.K., 117 (17-14c).
Cowan, R.S., *et al.*, **2016** (28-10); 2118 (17-11a); 2233 (14-50).
Cruz, J.S. de la, 1080 (14-26); 1081 (1-7b); 1097 (20-1); 1118, 1165, 1200 (16-2); 1239 (1-7b); 1421, 1575 (28-8b); 1576 (16-2); 1621 (6-1); 1802 (14-44); 1995 (20-1); 2036 (6-1); 2209, 2233 (14-26); 2245 (14-25); 2283 (28-8b); 2329 (20-1); 2356 (17-11a); 2358, 2390 (16-2); 2428, 2558, 2560 (14-26); 2593 (1-7b); 2623 (17-11a); 2681 (20-1); 2718 (1-7b); 2747 (16-2); 2763 (14-4); 2783 (16-2); 2832 (14-4); 2853, 3026 (16-2); 3179 (28-8b); 3246 (1-7b); 3252, 3421 (16-2); 3480 (25-2); 3488 (17-13); 3500 (1-7b); 3511 (17-13); 3653, 3934 (16-2); 3937 (20-1); 4100 (11-1b); 4228 (14-26); 4326 (16-2); 4388 (1-7b); 4440 (14-4); 4470 (28-8b).
Davis, D.H., 175 (16-2); 287 (18-2); 780 (22-2); 781 (2-5); 810 (4-5b); 1663 (17-20); 1678 (5-1).
Davis, T.A.W., see under FD.
Ek, R.C., *et al.*, 794 (14-55); 1147 (14-52).
Fanshawe, D.B., see under FD.
FD (Forest Dept. British Guiana), s.n. (Wilson-Browne 84) (17-9); s.n. (Wilson-Browne 108) (28-5); 244 (Davis) (14-7); 444 (Davis 488) (14-26); 483 (Wilson-Browne) (14-9); 766 (Hohenkerk 4) (14-27); 1249 (Sandwith 516) (14-15); 2038 (Lockie) (14-57); 2113 (Davis) (19-7); 2116 (19-4); 2188 (14-24); 2207 (14-55); 2292

SURINAME

FRENCH GUIANA

INDEX TO SYNONYMS, NAMES IN NOTES AND SOME TYPES

INDEX TO VERNACULAR NAMES

Alphabetic list of families of series A occurring in the Guianas

Defined as in Cronquist, 1981, and numbered in his sequence, with alternative names. Those published, with chronological fascicle number and year.

Family	No.	Fascicle
Abolbodaceae		
(see Xyridaceae	182)	15. 1994
Acanthaceae	156	23. 2006
(incl. Thunbergiaceae)		
(excl. Mendonciaceae	159)	
Achatocarpaceae	028	22. 2003
Agavaceae	202	
Aizoaceae	030	22. 2003
(excl. Molluginaceae	036)	22. 2003
Alismataceae	168	27. 2009
Amaranthaceae	033	22. 2003
Amaryllidaceae		
(see Liliaceae	199)	
Anacardiaceae	129	19. 1997
Anisophylleaceae	082	
Annonaceae	002	
Apiaceae	137	
Apocynaceae	140	
Aquifoliaceae	111	
Araceae	178	
Araliaceae	136	
Arecaceae	175	
Aristolochiaceae	010	20. 1998
Asclepiadaceae	141	
Asteraceae	166	
Avicenniaceae		
(see Verbenaceae	148)	4. 1988
Balanophoraceae	107	14. 1993
Basellaceae	035	22. 2003
Bataceae	070	
Begoniaceae	065	
Berberidaceae	016	
Bignoniaceae	158	
Bixaceae	059	
(incl. Cochlospermaceae)		
Bombacaceae	051	
Bonnetiaceae		
(see Theaceae	043)	
Boraginaceae	147	
Brassicaceae	068	
Bromeliaceae	189	p.p. 3. 1987
Burmanniaceae	206	6. 1989
Burseraceae	128	
Butomaceae		
(see Limnocharitaceae	167)	27. 2009
Buxaceae	115a	
Byttneriaceae		
(see Sterculiaceae	050)	
Cabombaceae	013	
Cactaceae	031	18. 1997
Caesalpiniaceae	088	p.p. 7. 1989
Callitrichaceae	150	
Campanulaceae	162	
(incl. Lobeliaceae)		
Cannaceae	195	1. 1985
Canellaceae	004	
Capparaceae	067	
Caprifoliaceae	164	
Caricaceae	063	
Caryocaraceae	042	
Caryophyllaceae	037	22. 2003
Casuarinaceae	026	11. 1992
Cecropiaceae	022	11. 1992
Celastraceae	109	
Ceratophyllaceae	014	
Chenopodiaceae	032	22. 2003
Chloranthaceae	008	24. 2007
Chrysobalanaceae	085	2. 1986
Clethraceae	072	
Clusiaceae	047	
(incl. Hypericaceae)		
Cochlospermaceae		
(see Bixaceae	059)	
Combretaceae	100	27. 2009
Commelinaceae	180	
Compositae		
(= Asteraceae	166)	
Connaraceae	081	
Convolvulaceae	143	
(excl. Cuscutaceae	144)	
Costaceae	194	1. 1985
Crassulaceae	083	
Cruciferae		
(= Brassicaceae	068)	
Cucurbitaceae	064	
Cunoniaceae	081a	
Cuscutaceae	144	
Cycadaceae	208	9. 1991
Cyclanthaceae	176	
Cyperaceae	186	
Cyrillaceae	071	27. 2009
Dichapetalaceae	113	27. 2009
Dilleniaceae	040	
Dioscoreaceae	205	
Dipterocarpaceae	041a	17. 1995
Droseraceae	055	22. 2003
Ebenaceae	075	
Elaeocarpaceae	048	
Elatinaceae	046	
Eremolepidaceae	105a	25. 2007

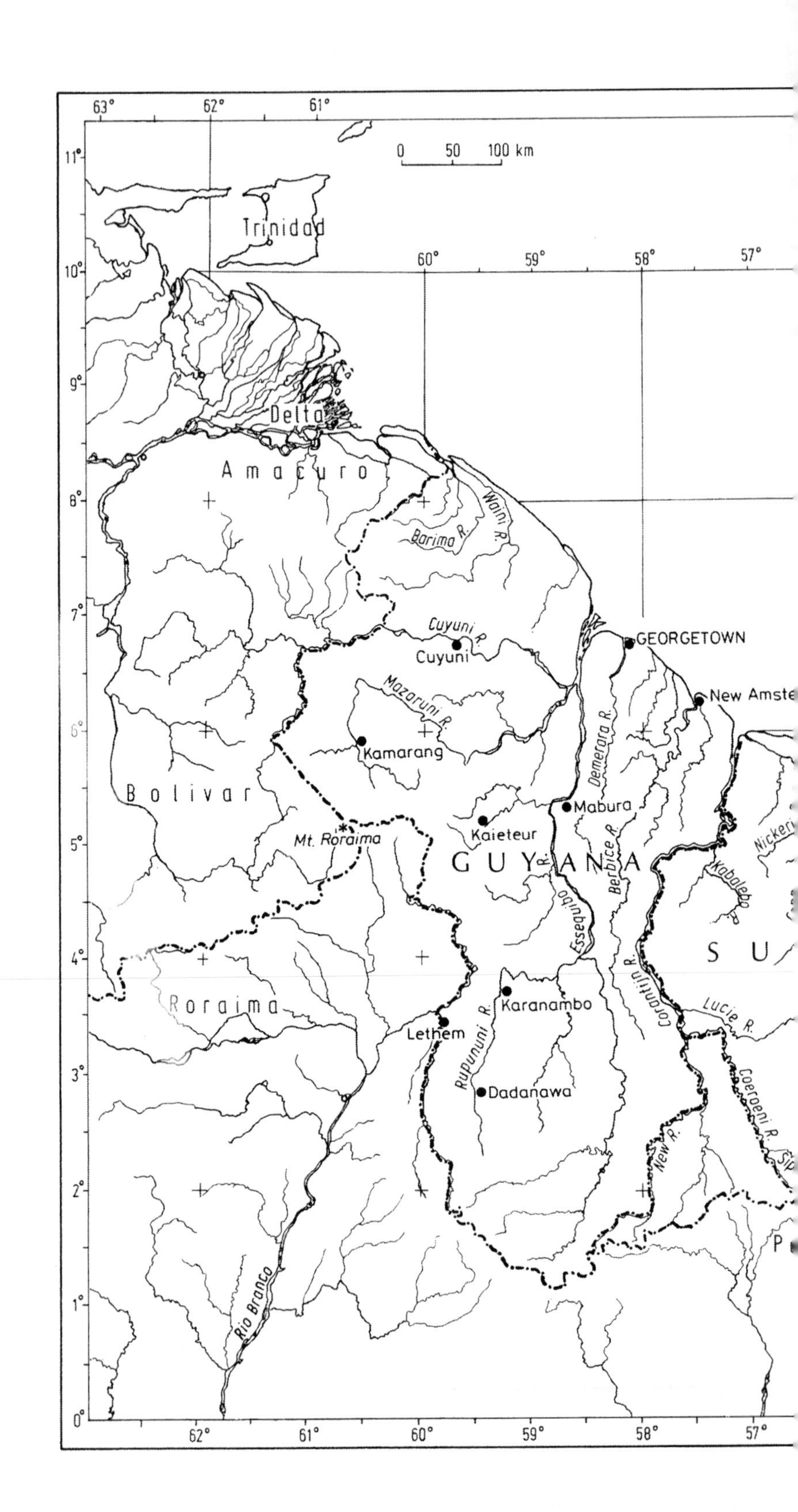

Trinidad
Delta
Amacuro
Bolivar
Roraima
GUYANA
Barima R.
Waini R.
Cuyuni R.
Cuyuni
Mazaruni R.
Kamarang
Kaieteur
Mt. Roraima
GEORGETOWN
New Amste
Mabura
Demerara R.
Berbice R.
Essequibo R.
Karanambo
Lethem
Rupununi R.
Dadanawa
Corantijn R.
New R.
Nickeri
Kabalebo R.
Lucie R.
Coeroeni R.
Rio Branco
0 50 100 km

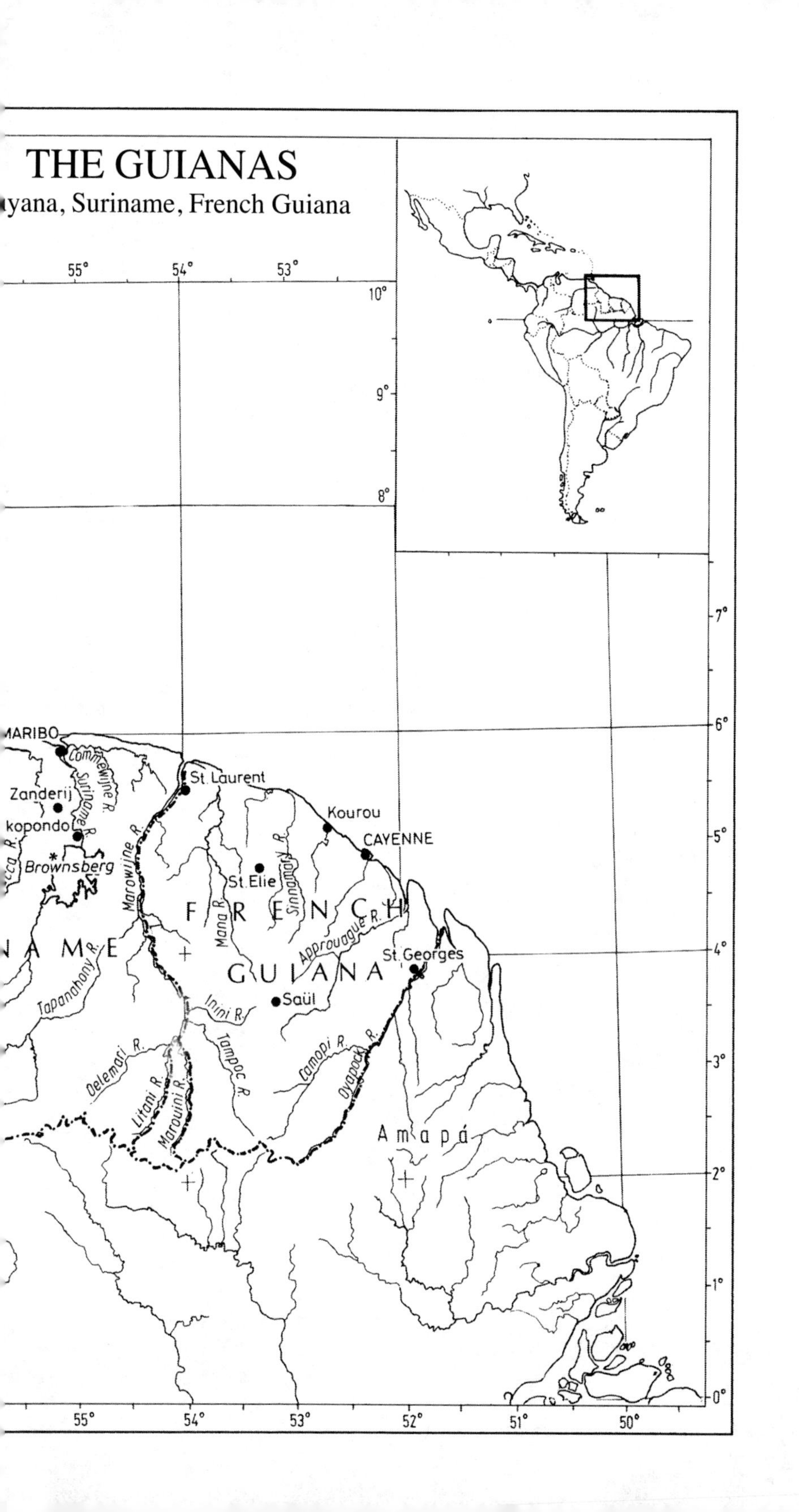
THE GUIANAS
ıyana, Suriname, French Guiana
55°
54°
53°
10°
9°
8°
7°
6°
5°
4°
3°
2°
1°
0°
52°
51°
50°
ΛARIBO
Commewijne R.
Suriname R.
Zanderij
kopondo
Brownsberg
Marowijne R.
St. Laurent
Kourou
CAYENNE
St. Elie
Sinnamary R.
Mana R.
F R E N C H
G U I A N A
Approuague R.
St. Georges
Saül
Inini R.
Tapanahony R.
N A M E
Oelemari R.
Litani R.
Marouini R.
Tampoc R.
Camopi R.
Oyapock R.
A m a p á